Elements of
Green's Functions and Propagation

Elements of Green's Functions and Propagation

Potentials, Diffusion, and Waves

G. BARTON

CLARENDON PRESS · OXFORD

Oxford University Press, Walton Street, Oxford OX2 6DP
Oxford New York
Athens Auckland Bangkok Bombay
Calcutta Cape Town Dar es Salaam Delhi
Florence Hong Kong Istanbul Karachi
Kuala Lumpur Madras Madrid Melbourne
Mexico City Nairobi Paris Singapore
Taipei Tokyo Toronto
and associated companies in
Berlin Ibadan

Published in the United States by
Oxford University Press Inc., New York

First published 1989
Reprinted 1991 (with corrections), 1995 (with corrections)

British Library Cataloguing in Publication Data
Barton, G. (Gabriel)
Elements of Green's functions and propagation.
1. Differential equations. Solution.
Green's functions
I. Title
515.3'5
ISBN 0 19 851998 2 (Pbk)

Library of Congress Cataloging in Publication Data
Barton Gabriel
Elements of Green's functions and propagation: potentials, diffusion, and waves / G. Barton.
p. cm. Bibliography: p. Includes index.
1. Green's functions. 2. Potential. Theoy of. 3. Heat equation.
4. Wave equation. 5. Poisson's equation. I. Title.
QC174.17.G68B37 1989 530.1'2—dc19 88–5224
ISBN 0 19 851998 2 (Pbk)

Printed and bound in Great Britain by
Biddles Ltd, Guildford and King's Lynn

This book belongs to my father

BARTON (BLAU) GYÖRGY

BUDAPEST 1894–1952

Apám szájából szép volt az igaz

Preface

This book stems from a short optional third-year course on 'Methods of theoretical physics' open to experimentalists and theorists alike. Granted that such a brief would force radical selection, it still seemed that the course would become too narrow if it tried to introduce substantially new mathematical techniques, or to cover dramatically new physics. The natural alternative is a more formal and thorough exploration of physics already familiar in an elementary context, and the complex of topics was chosen from just that point of view. The writer must, of course, make himself responsible for exploiting the opportunities that his particular choice affords.

In order to co-ordinate the material at the technical level, all that is needed beyond the common ground of mathematics-for-physics courses is an intelligent working knowledge of the Dirac delta-function. With this one can construct a small core of formalism centred on Green's functions, which then reveals a spectacular variety of physics; it also directs attention to some practically important aspects of the classic partial differential equations. Besides, such an approach should allow the student access to the truly basic method of the theorist: namely, the habit of purposeful dialogue with the formalism, paralleling the dialogue conducted by the experimenter in the laboratory.

Lastly, the exposition offers plenty of scope for illustrating the very various relationships that the physicist can profitably entertain with mathematics. The writer's own preferences can perhaps be discerned from the Reader's Guide (Chapter 0), and from some other comments dispersed through the book with proper apprehension.

As regards the selection, a somewhat longer book would have introduced integral equations, and a chapter on the Helmholtz equation applied to scattering. Further, it is arguable that electromagnetic rather than scalar radiation should be used to illustrate the wave equation. However, electromagnetic radiation is so well treated in so many books that it became impossible to resist the temptation of trying instead to illuminate it indirectly, by way of implicit contrast and analogy.

Disconcertingly, no text could be found at the right level for such a course, so that fairly extensive notes had to be distributed instead. The book is a revised and slightly expanded version of the notes. Sixteen lectures plus eight problems classes suffice to motivate and illustrate the main ideas; much of the detail, many worked examples, and most of the background material (including almost everything in the appendices) are best left to private study, which need not be immediate.

It is a pleasure to thank Nick Radcliffe for comments on the lectures when first delivered; Handel Davies for his advice on radiation and relativity (and for much other generous instruction over the years); and David Waxman for an unstintingly critical reading of the manuscript.

Brighton Gabriel Barton
December 1987

Contents

Appendices

I Introduction

Ainsi l'interprétation physique se montre un guide sûr là où l'analyste qui ne recourait pas à elle ne pouvait et ne peut que s'égarer.

HADAMARD (1964, page 156)

No amount of physical acumen suffices to justify a meaningless string of symbols.

VAN KAMPEN (1981, page 245)

0 Reader's guide

0.1 A note on what there is and why

- Physicists in their final year as undergraduates sometimes suspect that their mathematics courses, while fostering competence with important equations, may have underemphasized the pleasure and the practical advantages to be had from the analytic structures that in fact pervade much of our subject. This book is for such students. Without teaching them any additional mathematics other than simple properties of the delta-function, it aims to use some that they already have, in order to explore those parts of physics whose underlying concern is with the operators ∇^2, $\partial/\partial t$, and $\partial^2/\partial t^2$. Potentials, diffusion, and waves are the prime examples.

  In devising a unified (i.e. broadly efficient) approach to these phenomena, the theorist is challenged to combine his desire for clarity in their formal† treatment with proper respect for their diversity. Some knowledge of their basic physics and a feeling for its importance in its own right are therefore presupposed as part-motivation. The formalism, tempered in these familiar contexts, then proves equally profitable under the later demands of field theories well beyond the immediate scope of this book.

- *The prerequisites from physics* are introductory mechanics and electrostatics; a first acquaintance with heat conduction or diffusion; and some elementary experience with waves (on strings, in air, or electromagnetic). Quantum mechanics and relativity would furnish perspective, but are far from essential.

- *The prerequisites from mathematics* are a working knowledge at physics-undergraduate level of ordinary linear second-order differential equations; volume and surface integrals; vectors, and vector analysis up to Gauss's and Green's theorems and other identities involving grad and div; partial differential equations solved by separation of variables; Legendre polynomials and spherical harmonics; and the basic ideas about complete orthonormal sets of functions, as obtained from Fourier series and Fourier integrals.

To the writer's surprise, the main argument can readily dispense with the theory of functions of a complex variable. Cauchy's theorem and the

† Throughout, we use *formal* in its rightful acceptation: lack of form makes a discussion neither simple nor physical, but merely chaotic.

calculus of residues afford a shortcut here and there, but appear only in small-print sections and appendices, where they can be skipped without prejudice.

- The title evokes *propagation* as evolution in time, and also in the loose sense that, even with time-independent equations, data prescribed at one point contribute to the solution at other spatially remote points. Analysis soon reveals the basic elements from which an understanding of such systems is best put together: namely point sources described by *Dirac delta-functions,* and the *Green's functions* which describe their effects. The traditional undergraduate treatment tends to obscure much of the underlying simplicity by focusing on these elements late, or never, or worst of all apologetically. Having suffered learning and teaching from this tradition, the writer hopes that his book may help to discredit it even further.

  Accordingly, Chapter 1 introduces delta-functions in the requisite detail and in a manner intended to make them the natural tool of first resort through all that follows. Chapter 2 formulates many of the basic ideas needed later, but in the simpler setting of ordinary instead of partial differential equations. Chapter 3 merely previews the partial differential equations to come.

  The rest of the text then discusses the classic partial differential equations of physics, namely Poisson's equation (potential theory), the diffusion equation, and the wave equation (for linear non-dispersive waves). It ends with a chapter on the Helmholtz equation, which is linked to Poisson's by its mathematics, but to (monochromatic) waves by its physics. Diffusion sits where it does because it introduces time dependence with less initial effort than waves could.

- All these equations are tackled mainly (though not exclusively) by Green's-function techniques, which bring out their similarities, and then, by contrast, highlight their significant differences all the more effectively. Green's functions are equally rewarding when a given equation is examined successively in three dimensions, in two dimensions, and in one. Moreover, they encourage a spontaneously efficient common approach to propagation from prescribed initial conditions, volume-distributed sources, and surface-distributed sources or boundary data. This is the method we favour throughout.†

  The design is deliberately cumulative: ideas and methods are

† So much so that one title contemplated originally read 'Green's functions: the theory of sources, boundaries, and propagation'.

introduced and discussed when first required, and are not normally restated at length when they are applied to other equations later. Obviously, one need not absorb every detail at a first reading; but it is equally obvious that the cumulative effect must condition the advice that the next section offers on skipping.

- We shall use mathematics as the theoretical physicist generally does, not as a free-standing discipline, but as a means for handling relations reasonably, efficiently, and elegantly if possible. In this one is guided by the familiar formalism of differentiation, integration ('proper' and 'improper'), and the summation of infinite series, implemented as appropriate limits when necessary; but the order of these operations will be interchanged whenever convenient, without first enquiring in each case whether the interchange can be warranted rigorously and in advance. The results are eventually examined for consistency and for physical sense; and in doubtful cases one must at that stage be prepared to reassess one's mathematics. The less sophisticated are thus put on their guard where it matters, without having their time and vigilance gratuitously frittered away; indeed they may be motivated to pursue the mathematics further and with some discrimination. The more sophisticated may discover from such counterexamples the real point of some niceties they already know, and learn how to exploit these with decent restraint.†

  In short, for physicists mathematical rigour is a last resort, which we shall need very rarely; but physical insight and mathematical reasonableness are inextricably and equally essential. Revenge for slights to either is always prompt and automatic.

0.2 A note on how to read it

- The book holds more (perhaps by a third) than one would wish to absorb at a first reading. One possible approach is suggested by the following comments, which should, however, be considered in the light of the previous remarks about accumulation.

- *Small print* raises somewhat delicate mathematical points that can and probably should be skipped at first, except maybe by readers

† It is only fair to warn fledgeling mathematicians that they will feel that our definitions underdefine, that the pre-conditions for the theorems are hazy, and that the proofs do not (under such conditions of course cannot) prove. This is intentional: the definitions are designed to promote recognition within the natural habitats of physics; the theorems assert truths that are illuminating and fruitful under those conditions; and the proofs are arguments meant to elicit informed and intelligent though possibly cautious assent. Full logical rigour could be summoned, but its routine airing in public would distract too much attention from the essentials.

with a special affinity. Small-print sections are needed later only by other small-print sections.

- *The appendices* serve various functions. Some (e.g. H) provide detail required in problems, but are readable at best only as worked examples of techniques established (at least in principle) in the text. Others (e.g. I and L) deal with cases where especially careful mathematics is necessary in order to illustrate physics that is interesting also in its own right. Another kind of appendix (e.g. D, K, and O) supplies physical (rather than mathematical) rigour, where it is needed for a proper appreciation of just what the mathematical conclusions actually assert. The appendices can be postponed until required to answer some specific want.

  The text itself contains several readily recognizable and quite lengthy *case histories* (including some favourites of the writer, e.g. Sections 8.5, 11.3, and 13.5–6). These are meant to catch the reader's interest from the start, so that sheer curiosity about the outcome motivates her to follow the calculation through. The case histories will then double as worked examples; but if the reader's fancy is not engaged, then they too can be skipped at first reading.

- *The summaries* will prove uninformative until the appropriate sections have been digested. They appear partly for emphasis and partly as previews, so that a few centrally important relations are highlighted as such immediately on their first appearance. Until one has appreciated the significance of all the entries in a summary, one is unlikely to profit from the next chapter or from the next part.

- *The exercises* embedded in the text generally call for detailed or for alternative reasoning that might well have been spelled out in a book of unlimited length. They are intended to test whether an argument or a method has been understood in its immediate local settting, or in relation to ideas already familiar. The sensible reader will not proceed past these exercises without doing them, or at least spending some minutes visualizing how he could do them if he would. (By contrast, a handful of exercises call for the elucidation of rather searching questions, sometimes dressed as paradoxes. These may of course be postponed until they trouble the reader's conscience.)

  In physics it goes without saying that the only test as well as the reward of insight is the solution of problems, followed perhaps by the formulation of further problems. Accordingly, disregard of *the problems* automatically reassigns one to the sidelines. Some problems call only for the routine application of standard methods, while others present more of a challenge. All are accessible once the pertinent sections and appendices have been thought through.

1 The Dirac delta-function

Summary

Definition: If $f(x)$ is any well-behaved function, then

$$\int_{x_1}^{x_2} \mathrm{d}x\, f(x)\, \delta(x) = \begin{cases} f(0) & \text{if } x_1 < 0 < x_2 \\ 0 & \text{if the origin is outside the integration range} \end{cases}$$

Some properties: (a is any real constant)

- $\delta(x) = 0$ if $x \neq 0$; $\quad f(x)\,\delta(x) = f(0)\,\delta(x)$,
- $\int_{-\infty}^{\infty} \mathrm{d}x\, f(x)\,\delta(x-a) = f(a)$,
- $\delta(ax) = \dfrac{1}{|a|}\,\delta(x)$;
  in particular $\delta(-x) = \delta(x)$,
- $\delta(f(x)) = \sum_n \dfrac{\delta(x - x_n)}{|f'(x_n)|}$, where the sum runs over the real roots of $f(x) = 0$, and $f'(x) \equiv \mathrm{d}f/\mathrm{d}x$.
- In particular,

$$\delta(x^2 - a^2) = \frac{1}{2\,|a|}\,\{\delta(x-a) + \delta(x+a)\}.$$

Integral of δ: $\delta(x) = \dfrac{\mathrm{d}}{\mathrm{d}x}\,\theta(x)$, where

$\theta(x) \equiv 1 \quad \text{if } x > 0, \qquad \theta(x) \equiv 0 \quad \text{if } x < 0.$

Derivative of δ: Define $\delta'(x)$ by $\delta'(x) = 0$ if $x \neq 0$, and

$\int_{-\infty}^{\infty} \mathrm{d}x\, f(x)\,\delta'(x) = -f'(0)$. (Note: $\delta'(-x) = -\delta'(x)$.)

Representations:

- $$\delta(x) = \lim_{\varepsilon \to 0} \begin{cases} \dfrac{1}{\varepsilon}\,\theta\!\left(\dfrac{\varepsilon}{2} - |x|\right): & \text{square step} \\ \dfrac{1}{\varepsilon\sqrt{\pi}}\exp(-x^2/\varepsilon^2): & \text{Gaussian} \\ \dfrac{\varepsilon/\pi}{x^2 + \varepsilon^2}: & \text{Lorentzian} \\ \dfrac{\sin(x/\varepsilon)}{\pi x}: & \text{Dirichlet} \end{cases}$$

$$\delta(x) = \lim_{K\to\infty} \frac{1}{2\pi} \int_{-K}^{K} \mathrm{d}k \exp(ikx) = \frac{1}{2\pi} \int_{-\infty}^{\infty} \mathrm{d}k \exp(ikx): \qquad \text{Fourier}$$

Periodic representation: $\sum_{m=-\infty}^{\infty} \delta(x - 2m\pi) = \frac{1}{2\pi} \sum_{n=-\infty}^{\infty} \exp(inx)$

Representation through the closure property: Let $\{\phi_n(x)\}$ be any complete set of orthonormal functions over $x_1 < x < x_2$, i.e.

$$\int_{x_1}^{x_2} \mathrm{d}x\, \phi_n^*(x)\phi_{n'}(x) = \delta_{nn'} \quad \text{(orthonormality)}$$

and, for any function $f(x)$, $f(x) = \sum_n c_n \phi_n(x)$ (with appropriately chosen c_n): (completeness). Then

$$\sum_n \phi_n^*(x')\phi_n(x) = \delta(x' - x).$$

- Range $0 < x < L$, plus BCs $\phi(0) = 0 = \phi(L)$:

  $$\phi_n(x) = \sqrt{\frac{2}{L}} \sin\left(\frac{n\pi x}{L}\right) \Rightarrow \frac{2}{L} \sum_{n=1}^{\infty} \sin\left(\frac{n\pi x'}{L}\right) \sin\left(\frac{n\pi x}{L}\right) = \delta(x - x').$$

- With BCs $\phi'(0) = 0 = \phi'(L)$:

  $$\phi_0(x) = \sqrt{\frac{1}{L}}, \qquad \phi_{n\geq 1}(x) = \sqrt{\frac{2}{L}} \cos\left(\frac{n\pi x}{L}\right),$$

  $$\Rightarrow \frac{2}{L} \left\{ \frac{1}{2} + \sum_{n=1}^{\infty} \cos\left(\frac{n\pi x'}{L}\right) \cos\left(\frac{n\pi x}{L}\right) \right\} = \delta(x - x').$$

- With periodic BCs: $\phi_n(0) = \phi_n(L)$ and $\phi_n'(0) = \phi_n'(L)$ etc:

  $$\phi_n(x) = \sqrt{\frac{1}{L}} \exp(2\pi inx/L)$$

  $$\Rightarrow \frac{1}{L} \sum_{n=-\infty}^{\infty} \exp(2\pi in(x - x')) = \delta(x - x').$$

δ-function in 3D: Define $\int \mathrm{d}^3r\, f(\boldsymbol{r})\delta(\boldsymbol{r}) = f(0)$, etc. E.g. in Cartesians, $\delta(\boldsymbol{r}) = \delta(x)\,\delta(y)\,\delta(z)$,

$$\delta(\boldsymbol{r}) = \frac{1}{(2\pi)^3} \int \mathrm{d}^3k \exp(i\boldsymbol{k} \cdot \boldsymbol{r}).$$

Spherical polar coordinates: $(\theta, \phi) \equiv \Omega$, e.g. $f(\boldsymbol{r}) = f(r, \Omega)$;

$$\int \mathrm{d}^3r \cdots = \int_0^{\infty} \mathrm{d}r\, r^2 \int_0^{2\pi} \mathrm{d}\phi \int_0^{\pi} \mathrm{d}\theta \sin\theta \cdots;$$

define

$$\int_0^{2\pi} d\phi \int_0^{\pi} d\theta \sin\theta = \int_0^{2\pi} d\phi \int_{-1}^{1} d\cos\theta \equiv \int d\Omega \cdots$$

Then

$$\delta(\boldsymbol{r}-\boldsymbol{r}') = \frac{1}{r^2}\,\delta(r-r')\,\delta(\Omega-\Omega'), \qquad (*)$$

where

$$\delta(\Omega-\Omega') = \delta(\phi-\phi')\,\delta(\cos\theta-\cos\theta')$$
$$= \sum_{l=0}^{\infty}\sum_{m=-l}^{l} Y^*_{lm}(\Omega')Y_{lm}(\Omega).$$

$$\delta(\Omega-\Omega') = \frac{1}{4\pi}\sum_{l=0}^{\infty}(2l+1)P_l(\hat{\boldsymbol{r}}\cdot\hat{\boldsymbol{r}}').$$

(Recall $\int d\Omega\, Y^*_{lm}(\Omega)Y^*_{l'm'}(\Omega) = \delta_{ll'}\,\delta_{mm'}$,

$$Y_{lm}(\Omega) \equiv Y_{lm}(\hat{\boldsymbol{r}}) = \left[\frac{(2l+1)}{4\pi}\frac{(l-m)!}{(l+m)!}\right]^{\frac{1}{2}} P_l^{|m|}(\cos\theta)\exp(im\,\phi),$$

$$P_l(\hat{\boldsymbol{r}}\cdot\hat{\boldsymbol{r}}') = \frac{4\pi}{2l+1}\sum_{m=-l}^{l} Y^*_{lm}(\hat{\boldsymbol{r}}')Y_{lm}(\hat{\boldsymbol{r}}).)$$

Strong definition of radial $\delta(r)$:

$$\int_0^{\infty} dr\, f(r)\,\delta(r) = f(0)$$

whence in the degenerate case of (*):

$$\delta(\boldsymbol{r}) = \frac{\delta(r)}{r^2}\frac{1}{4\pi}.$$

- Identities: $\delta(r^2-a^2) = \dfrac{1}{2\,|a|}\,\delta(r-|a|)$,

  $\dfrac{1}{r}\,\delta(r) = -\delta'(r)$.

Plane polar coordinates (2D):

$$\int_0^{\infty} dr\, r \int_0^{2\pi} d\phi\, \delta(\boldsymbol{r}-\boldsymbol{r}') = 1$$

$$\delta(\boldsymbol{r}-\boldsymbol{r}') = \frac{1}{r}\,\delta(r-r')\,\delta(\phi-\phi')$$

$$\delta(\mathbf{r}-\mathbf{r}')=\frac{1}{r}\delta(r-r')\frac{1}{2\pi}\sum_{m=-\infty}^{\infty}\exp(im(\phi-\phi'));$$

under the strong definition of radial $\delta(r)$:

$$\delta(\mathbf{r})=\frac{1}{r}\delta(r)\frac{1}{2\pi}.$$

Section 1.4 and paragraphs in small print can and probably should be reserved for a second reading. The generalizations to three dimension spelled out in Section 1.4 are plausible and indeed almost obvious at first sight, even if they become less so on reflection: the reader may appreciate these technicalities more readily by returning to them only as they come into use, from Chapter 4 onwards.

1.1 Definition and properties

1.1.1 Introduction

In physics, it is often convenient to envisage some finite effect achieved in an arbitrarily short interval of time or space: for instance, impulsive forces in mechanics (always an idealization), or the charge density due to a point charge like an electron (possibly quite realistic). The requisite mathematics is embodied in the rules for the Dirac delta-function. The description of point sources by such functions is fundamental to all that follows: we shall see that it affords the most efficient and intuitively the most appealing way to understand how fields arise and how they propagate. There is also an immediate bonus through the light that delta-functions shed on ordinary Fourier integrals and series, as sketched in Appendix C.

Intuitively, the function $\delta(x)$ is defined to be zero when $x\neq 0$, and infinite at $x=0$ in such a way that the area under it is unity.† To express this concisely, we adopt the convention that η_1, η_2 appearing in integration limits are always strictly positive ($\eta_1>0$, $\eta_2>0$, with neither η_1 nor η_2 allowed to be zero). Then, for any such $\eta_{1,2}$,

$$\delta(x)=0 \quad \text{if} \quad x\neq 0; \qquad \int_{-\eta_1}^{\eta_2} \mathrm{d}x\,\delta(x)=1. \tag{1.1.1}$$

Obviously this entails

$$\int_{-\eta_1}^{-\eta_2} \mathrm{d}x\,\delta(x)=0=\int_{\eta_1}^{\eta_2} \mathrm{d}x\,\delta(x), \tag{1.1.2}$$

while the integral in (1.1.1) could equally well be written as

† See the cautionary comments in Section 1.5.

$\int_{-\infty}^{\infty} dx\, \delta(x) = 1$. We also adopt the normal rule for reversing the limits on a definite integral, by demanding

$$\int_{\eta_2}^{-\eta_1} dx\, \delta(x) = -\int_{-\eta_1}^{\eta_2} dx\, \delta(x) = -1. \qquad (1.1.3)$$

The defining relations (1.1.1–3) could be replaced by the following and equivalent relation. Let $f(x)$ be any well-behaved function, continuous (hence finite), and differentiable as often as later arguments may require. (Restrictions on its behaviour at infinity will be imposed in Section 1.2.) We call such an $f(x)$ a test function. Then

$$\int_{-\eta_1}^{\eta_2} dx\, \delta(x) f(x) = f(0) \int_{-\eta_1}^{\eta_2} dx\, \delta(x) = f(0), \qquad (1.1.4)$$

while $\int_{x_1}^{x_2} dx\, \delta(x) f(x)$ vanishes if the integration region does not include the origin. The first equality in (1.1.4) holds because $\delta(x \neq 0) = 0$ entails that there is no contribution to the integral from anywhere where $f(x) \neq f(0)$; consequently $f(x)$ may be replaced by $f(0)$, and $f(0)$, being a constant, can then be moved outside the integral. The last equality follows from (1.1.1).

Exercise: Show that, conversely, (1.1.4) implies (1.1.1).

One physical example is the force $F(t)$ delivering a finite impulse (momentum change) I over an arbitrarily short interval of time: $I = \int_{-\eta_1}^{\eta_2} dt\, F(t)$, in the limit $\eta_1 \to 0+$, $\eta_2 \to 0+$, $F \to \infty$, but with I fixed. (The notation $0+$ is explained in Appendix A.) Then $F(t) = I\,\delta(t)$. Another example is the one-dimensional charge density $\rho(x)$ (charge per unit length measured along the x-axis) due to a point charge Q situated at the origin: $Q = \int_{-\infty}^{\infty} dx\, \rho(x)$, but $\rho(x \neq 0) = 0$, whence $\rho(x) = Q\,\delta(x)$.

1.1.2 Weak and strong definitions of $\delta(x)$

Because $\delta(x)$ at this point is a new mathematical object of an unfamiliar kind (see the comments in Section 1.1.4), all its properties and all the rules for handling it must be justified carefully and explicitly from its definition. Observe that (1.1.1–4) leave undefined the values of the integrals when one of the limits is precisely zero. These relations and their consequences are therefore equally compatible with many different assignments of the integrals $\int_L$ and $\int_R$ defined by

$$\int_L \equiv \int_{-\eta_1}^{0} dx\, \delta(x), \qquad \int_R \equiv \int_0^{\eta_2} dx\, \delta(x), \qquad (1.1.5)$$

subject only to $\int_L + \int_R = 1$. We call eqns (1.1.1–4) the *weak definition* of $\delta(x)$. By contrast, any definition that does assign values to $\int_L$ and $\int_R$ separately we call a *strong definition*.

The weak definition is the one commonly found in textbooks, though without being explicitly qualified as 'weak'. It suffices for all applications where the integration limits are never zero. However, expressions shown to be equal under the weak definition ('weakly equal') can differ once a strong definition is adopted in addition (they may not be 'strongly equal'). Although very occasionally we shall stress the difference by writing weak equalities as $\overset{(w)}{=}$, this will not be done where the difference is irrelevant or clear from the context. Unfortunately, it is the universal custom to use the same symbol $\delta(x)$ for *any* function satisfying the weak definition, which leaves no generally recognized way to distinguish between two functions a and b such that $a(x) \overset{(w)}{=} \delta(x)$ and $b(x) \overset{(w)}{=} \delta(x)$, while nevertheless $a(x) \neq b(x)$ because a and b differ by assigning different values to $\int_L$ and $\int_R$ separately. This must be remembered particularly when considering the relation between $\delta(-x)$ and $\delta(x)$.

Luckily, in practice there is less scope for confusion than one might have feared. The point is that, provided no integration stops at zero, strong definition is unnecessary, and all distinctions introduced only by strong definitions can simply be ignored: in other words one need not distinguish $\overset{(w)}{=}$ from $=$. On the other hand, when integration limits are zero, then circumstances always dictate the appropriate strong definition, after which relations that are weakly true but strongly false are never encountered. This happens for instance in Section 1.4.2.

It may be worth repeating the elementary logical point that the weak definition demands less, and is therefore satisfied more easily (is satisfied by more functions) than the strong definition. Thus there are more relations that are true under the weak than under the strong definition: this is why there are indeed equations that are 'weakly true but strongly false'.

1.1.3 Properties of $\delta(x)$

This subsection plays safe by identifying explicitly all those relations that are only weakly true, i.e. relations that are guaranteed only if no integration limit is ever zero, but that may fail under some strong definitions. Relations not specifically identified as weak apply irrespective of any strong definition that might be adopted subsequently.

Where the proof is not given, it should be supplied as an exercise, either from first principles, or by reduction to a rule already established.

(i) *Physical dimensions.* The physical dimensions of a quantity are denoted by $[\cdots]$, and specified by powers of mass, length, and time (M, L, T). The dimensions of a dimensionless quantity, i.e. of a pure number, are written as [1]. Since (1.1.1) prescribes $[\delta(x)][\mathrm{d}x]=[1]$, we have $[\delta(x)]=[x]^{-1}$. Thus the dimensions of $\delta(x)$, if x is a single independent variable, are the inverse of those of the argument x. (Section 1.4.1 shows that this is no longer so for $\delta(\boldsymbol{r})$ if $\boldsymbol{r}$ is a vector in a space with more than one dimension.)

(ii) $f(x)\,\delta(x)=f(0)\,\delta(x);$ (1.1.6a)

in particular,

$$x\,\delta(x)=0. \tag{1.1.6b}$$

(iii)

$$\int_{x_1}^{x_2} \mathrm{d}x\, f(x)\,\delta(x-y)=\begin{cases} f(y) & \text{if } x_1<y<x_2; \\ 0 & \text{if } y \text{ is not included between } x_1 \text{ and } x_2. \end{cases} \tag{1.1.7}$$

Proof: To bring the definition of the δ-function to bear, we must change the integration variable to the argument of $\delta(\cdots)$, whatever that is. In this case, let $x-y=\xi$; then the integral becomes $\int_{x_1-y}^{x_2-y}\mathrm{d}\xi\, f(\xi+y)\,\delta(\xi)$, and the result follows from (1.1.4). ■
(The symbol ■ is explained in Appendix A.)

By the same token, $f(x)\,\delta(x-y)=f(y)\,\delta(x-y)$.

(iv) With a any real positive constant,

$$\delta(ax)=\frac{1}{a}\,\delta(x), \qquad (a>0). \tag{1.1.8}$$

Proof: We must show that $\delta(ax)$ has all the properties that the definition of $\delta(x)$ entails for $\delta(x)/a$. If $x\neq 0$, then $ax\neq 0$, so $\delta(ax)=0$. It remains only to show that both sides of (1.1.8) give the same result when integrated over any range $-\eta_1$ to η_2, however narrow, that includes the origin. The RHS gives $1/a$. The LHS gives $J\equiv\int_{-\eta_1}^{\eta_2}\mathrm{d}x\,\delta(ax)$. As in (iii) above, we must change the integration variable to $\xi=ax$, $x=\xi/a$; then

$$J=\int_{-a\eta_1}^{a\eta_2}\delta(\xi)\,\mathrm{d}\xi/a=\int_{-\tilde{\eta}_1}^{\tilde{\eta}_2}\mathrm{d}\xi\,\delta(\xi)/a=1/a. \qquad ■$$

The last step follows because $\tilde{\eta}_{1,2}\equiv a\eta_{1,2}$ are again positive constants, whence the defining relation (1.1.4) applies directly.

(v) The analogue of (1.1.8) with negative a applies only weakly:

$$\delta(-ax)\overset{(\mathrm{w})}{=}\frac{1}{a}\,\delta(x), \qquad (a>0). \tag{1.1.9}$$

Evidently (1.1.8–9) are both covered by

$$\delta(ax) \overset{(w)}{=} \frac{1}{|a|}\,\delta(x), \qquad (a \neq 0). \tag{1.1.10}$$

Proof of (1.1.9): If $x \neq 0$, then $-ax \neq 0$, hence $\delta(-ax) = 0$. It remains only to show that both sides give the same result when integrated from $-\eta_1$ to η_2. The RHS gives $1/a$. On the LHS we now change the integration variable to $\xi = -ax$, $x = -\xi/a$, and obtain

$$\begin{aligned} J &\equiv \int_{-\eta_1}^{\eta_2} \mathrm{d}x\,\delta(-ax) = \int_{a\eta_1}^{-a\eta_2} (-\mathrm{d}\xi/a)\,\delta(\xi) \\ &= \int_{-a\eta_2}^{a\eta_1} \mathrm{d}\xi\,\delta(\xi)/a = \int_{-\bar{\eta}_2}^{\bar{\eta}_1} \mathrm{d}\xi\,\delta(\xi)/a = 1/a. \end{aligned}$$

In the second step, the sign reversal on interchanging the integration limits follows from (1.1.3). ■

(vi) *Reflection properties.* $\delta(x)$ is a (weakly) even function:

$$\delta(-x) \overset{(w)}{=} \delta(x), \qquad \delta(x-y) \overset{(w)}{=} \delta(y-x). \tag{1.1.11a,b}$$

It is clear that (1.1.9–11) must indeed be written as weak relations, because as strong relations they could easily be falsified, say by defining $\int_{\mathrm{L}} = 0$ and $\int_{\mathrm{R}} = 1$ in (1.1.5), whence $\int_{-\infty}^{0} \mathrm{d}x\,\delta(x) = 0$, $\int_0^\infty \mathrm{d}x\,\delta(x) = 1$. But if $\delta(x)$ were a strongly even function, then these two integrals would have to be equal. Only with the special choice $\int_{\mathrm{L}} = \frac{1}{2} = \int_{\mathrm{R}}$ is $\delta(x)$ strongly even. Moreover, (1.1.11a) makes sense, and can be relevant, only if the variable x can indeed assume negative values. It ceases to apply for instance when x is the radial variable in polar coordinates, or if $x = y^2$ with y real (cf. Sections 1.4.2 and 1.4.3.)

(vii) If $g(x)$ has real roots x_n (i.e. if $g(x_n) = 0$), then

$$\delta(g(x)) \overset{(w)}{=} \sum_n \frac{\delta(x-x_n)}{|g'(x_n)|}, \qquad g'(x) \equiv \frac{\mathrm{d}g}{\mathrm{d}x}. \tag{1.1.12}$$

The sum runs over all real roots. (This relation is weak if some of the $g'(x_n)$ are negative.)

Proof: $\delta(g(x)) = 0$ unless $g(x) = 0$, i.e. unless x equals one of the x_n. Hence we need consider $\delta(g(x))$ only in the immediate vicinity of the points x_n, taking each in turn. As $x \to x_n$, we may approximate

$$g(x \to x_n) \sim g(x_n) + (x - x_n)g'(x_n) + \cdots = (x - x_n)g'(x_n) + \cdots,$$
$$\delta(g(x \to x_n)) = \delta((x-x_n)g'(x_n)) \overset{(w)}{=} \delta(x-x_n)/|g'(x_n)|.$$

(The notation $g(x \to x_n) \sim \cdots$ is explained in Appendix A.) The last step relies on (1.1.10). ■

(viii)

$$\delta(x^2 - a^2) = \frac{1}{2\,|a|}\{\delta(x-a) + \delta(x+a)\}. \tag{1.1.13}$$

This is a special case of (1.1.12). Without further definition (1.1.13) fails if $a = 0$, as (1.1.12) fails if $g(x)$ has coincident roots (where $g'(x)$ vanishes as well as $g(x)$). Such cases are dealt with by strong definitions, as in Section 1.4.2, Theorem (iii).

(ix) *The integral of* $\delta(x)$. We define the *step-function* $\theta(x)$ (sometimes called the Heaviside function, and written $H(x)$):

$$\theta(x) = \begin{cases} 0 & \text{if} \quad x < 0, \\ 1 & \text{if} \quad x > 0, \end{cases} \tag{1.1.14}$$

and the sign function $\varepsilon(x)$ (often written sgn (x)):

$$\varepsilon(x) = \begin{cases} -1 & \text{if} \quad x < 0, \\ +1 & \text{if} \quad x > 0. \end{cases} \tag{1.1.15}$$

Thus $\theta(x) + \theta(-x) = 1$, $\theta(x) - \theta(-x) = \varepsilon(x)$, and $|x| = x\varepsilon(x)$. These are weak definitions, leaving open the values of θ and ε at $x = 0$. If required, the appropriate strong definitions assigning these values are best made jointly with a strong definition of $\delta(x)$. (Often some mathematical formalism, e.g. a Fourier representation, might suggest the left-and-right averaged values, i.e. $\theta(0) = \frac{1}{2}$, $\varepsilon(0) = 0$. But there are no generally compelling reasons to adopt them.) It is easy to see that

$$\theta(x) = \int_{-\infty}^{x} \mathrm{d}x'\, \delta(x'), \tag{1.1.16a}$$

$$\varepsilon(x) = -1 + 2\int_{-\infty}^{x} \mathrm{d}x'\, \delta(x'), \tag{1.1.16b}$$

whence, by differentiation,

$$\delta(x) = \frac{\mathrm{d}}{\mathrm{d}x}\theta(x) = \frac{1}{2}\frac{\mathrm{d}}{\mathrm{d}x}\varepsilon(x). \tag{1.1.17}$$

Figure 1.1 sketches the functions $\theta(\pm x)$ and $\varepsilon(x)$, and their relation to $\delta(x)$. As an example, note $\mathrm{d}\,|x|/\mathrm{d}x = \mathrm{d}(x\varepsilon(x))/\mathrm{d}x = \varepsilon(x) + 2x\,\delta(x) = \varepsilon(x)$, which uses in turn the rule for differentiating a product, (1.1.17), and (1.1.6b).

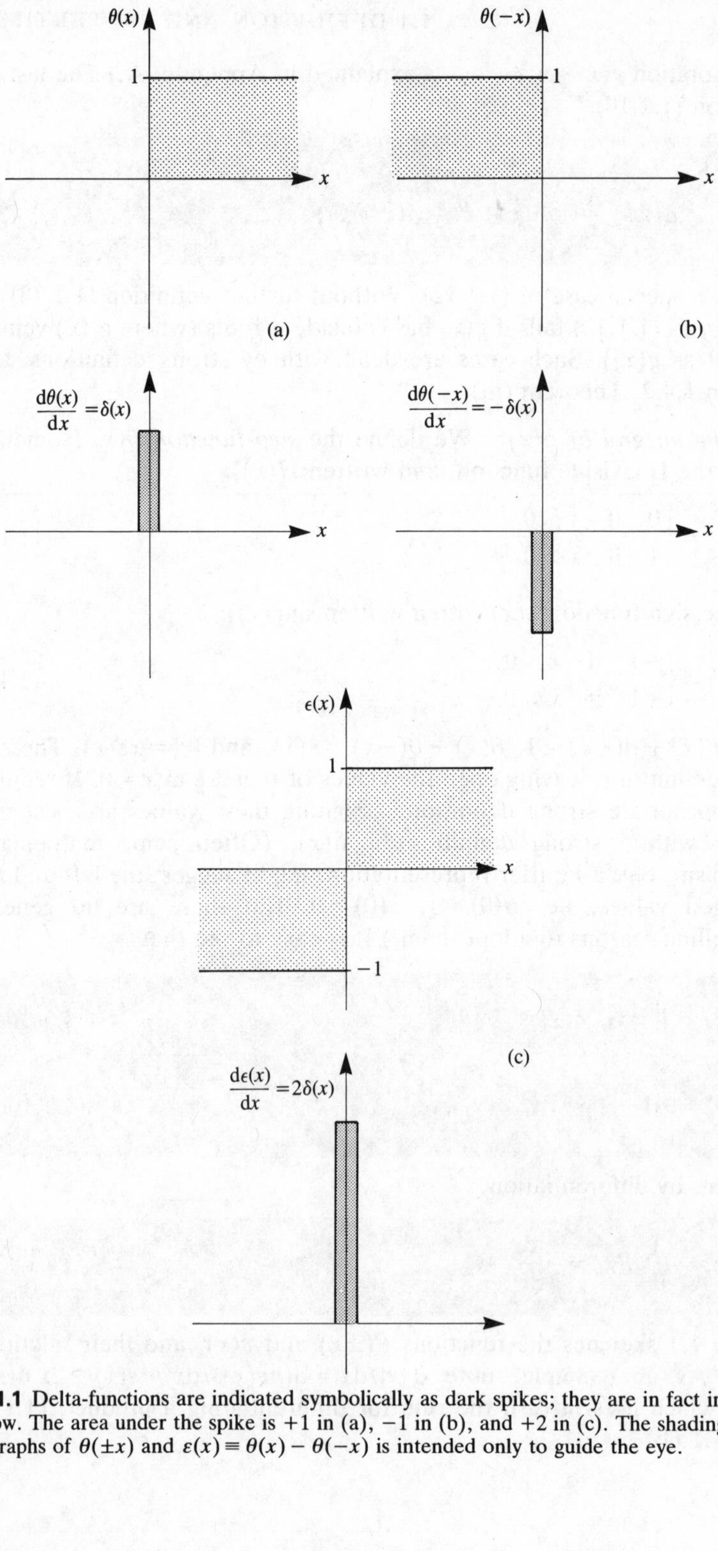

Fig. 1.1 Delta-functions are indicated symbolically as dark spikes; they are in fact infinitely narrow. The area under the spike is +1 in (a), −1 in (b), and +2 in (c). The shading under the graphs of $\theta(\pm x)$ and $\varepsilon(x) \equiv \theta(x) - \theta(-x)$ is intended only to guide the eye.

(x) *The derivative of* $\delta(x)$. Unlike the integral of $\delta(x)$, its derivative $\delta'(x)$ is not directly determined by the definition of $\delta(x)$ itself. It would be meaningless to ask what this derivative *is*: instead, one *defines* it so that $\delta'(x) = 0$ when $x \neq 0$, and so that the ordinary rules for integration by parts continue to apply to the product of $\delta'(x)$ with a differentiable test function $f(x)$. Thus we adopt the (weak) definition

$$\int_{-\eta_1}^{\eta_2} \mathrm{d}x\, f(x)\, \delta'(x) \equiv f(x)\, \delta(x) \Big|_{-\eta_1}^{\eta_2} - \int_{-\eta_1}^{\eta_2} \mathrm{d}x\, f'(x)\, \delta(x),$$

$$\int_{-\eta_1}^{\eta_2} \mathrm{d}x\, f(x)\, \delta'(x) = -f'(0). \tag{1.1.18}$$

In the middle expression, the integrated term vanishes because $\delta(x)$ vanishes at the endpoints $-\eta_1 \neq 0$ and $\eta_2 \neq 0$.

Exercises: (i) Find the analogues of (1.1.18) for integration ranges excluding the origin. (ii) Show that $x\, \delta'(x) = -\delta(x)$. A strong version of this appears in (1.4.14) below.

(xi) Because, according to (1.1.11), $\delta(x)$ under its weak definition is even, $\delta'(x)$ is an odd function.

Proof: Act on $\delta(x-y) \overset{(\mathrm{w})}{=} \delta(y-x)$ with $\partial/\partial x$. On the left one obtains $(\partial/\partial x)\, \delta(x-y) = \delta'(x-y)$; on the right, using the chain rule of differentiation, one obtains $(\partial/\partial x)\, \delta(y-x) = \delta'(y-x)\, \partial(y-x)/\partial x = -\delta'(y-x)$. (As always, the prime denotes the derivative of a function with respect to its argument; in this case, with respect to $(x-y)$ on the left, and with respect to $(y-x)$ on the right.) Equating the two sides and setting $y = 0$ we find

$$\delta'(x-y) \overset{(\mathrm{w})}{=} -\delta'(y-x) \Rightarrow \delta'(x) = -\delta'(-x). \tag{1.1.19}$$ ■

Section 1.4.5 describes an application.

(xii) Extending (1.1.18), we define the nth derivative $\delta^{(n)}(x)$ to vanish at $x \neq 0$, and so that, if the test function $f(x)$ and its first n derivatives are continuous, then

$$\int_{-\eta_1}^{\eta_2} \mathrm{d}x\, f(x)\, \delta^{(n)}(x) = (-1)^n f^{(n)}(0). \tag{1.1.20}$$

Exercise: Resolve the paradox presented by the Taylor expansion $\delta(x+a) = \sum_{n=0}^{\infty} a^n\, \delta^{(n)}(x)/n!$, where $\delta(x+a)$ is zero except at $x = -a$, while every term of the sum on the right is zero except at $x = 0$.

(xiii) Analogously to (1.1.10) for $\delta(ax)$ one has

$$\delta'(ax) \overset{(w)}{=} \varepsilon(a)\,\delta'(x)/a^2, \tag{1.1.21}$$

where $\delta'(ax)$ stands for $\mathrm{d}\delta(ax)/\mathrm{d}(ax)$.

Proof: For $a>0$ and $a<0$ in turn, we show that the two sides give the same result when multiplied by a test function $f(x)$ and integrated from $-\eta_1$ to $+\eta_2$. In virtue of (1.1.18), the RHS gives $-\varepsilon(a)f'(0)/a^2$. On the left we change the integration variable to $\xi = ax$, so that $x = \xi/a$. When $a>0$, the LHS gives

$$\int_{-\eta_1}^{\eta_2} \mathrm{d}x\, f(x)\,\delta'(ax) = \int_{-a\eta_1}^{a\eta_2} (\mathrm{d}\xi/a) f(\xi/a)\,\delta'(\xi)$$

$$= -\frac{1}{a}\frac{\mathrm{d}f(\xi/a)}{\mathrm{d}\xi}\bigg|_{\xi=0} = -\frac{1}{a^2}f'(0) = -\frac{1}{a^2}\,\varepsilon(a>0)f'(0).$$

When $a<0$, the LHS of (1.1.21) gives

$$\int_{|a|\eta_1}^{-|a|\eta_2} (\mathrm{d}\xi/a) f(\xi/a)\,\delta'(\xi) = -\int_{-|a|\eta_2}^{|a|\eta_1} (\mathrm{d}\xi/a) f(\xi/a)\,\delta'(\xi)$$

$$= \frac{1}{a}\frac{\mathrm{d}f(\xi/a)}{\mathrm{d}\xi}\bigg|_{\xi=0} = \frac{1}{a^2}f'(0) = -\frac{1}{a^2}\,\varepsilon(a<0)f'(0). \qquad \blacksquare$$

Exercise: Express $\delta^{(n)}(ax)$ in terms of $\delta^{(n)}(x)$.

(xiv) *The hierarchy of singularities.* There is a natural hierarchy amongst the singular functions we have just introduced. In this hierarchy $\ldots, |x|$, ($\varepsilon(x)$ or $\theta(x)$), $\delta(x)$, $\delta'(x), \ldots,$ each function is proportional to the derivative of the one before; thus $\mathrm{d}|x|/\mathrm{d}x = \varepsilon(x)$, $\mathrm{d}\varepsilon(x)/\mathrm{d}x = 2\delta(x)$, $\mathrm{d}\delta(x)/\mathrm{d}x = \delta'(x)$. In an obvious and familiar sense, differentiation makes these singular functions less smooth (more singular); conversely, integration makes them smoother. For instance, intuitively speaking, the sign-function $\varepsilon(x)$, which has merely a discontinuity at the origin, is less singular than $\delta(x)$, which has an infinite spike there. In this intuitive sense, if $G(x)$ say is a linear combination of such functions, then in any linear combination of $\int \mathrm{d}x\, G$, G, G', G'', $\ldots,$ the highest derivative always provides the most singular term.

Independently of its intuitive aspects, the arrangement of these functions according to the order of derivatives proves basic to the differential equation for Green's functions in Section 2.1.3. There, the important fact is that singularities weaker than $\delta(x)$ can make no contribution, in the limit, to the integral $\int_{-\eta_1}^{\eta_2} \mathrm{d}x \cdots$; for instance, contrast $\lim_{\eta_1,\eta_2\to 0} \int_{-\eta_1}^{\eta_2} \mathrm{d}x\,\delta(x) = 1$ with $\lim_{\eta_1,\eta_2\to 0} \int_{-\eta_1}^{\eta_2} \mathrm{d}x\,\varepsilon(x) = 0$.

1.1.4 Comments

Ultimately, the delta-function and its derivatives make sense only if multiplied by a sufficiently well-behaved test function, and then integrated over some finitely wide range of x; and it is only in this context that, ultimately, they are ever required in practice. Equations involving them without such integrations are just a convenient half-way stage; nevertheless such equations are taken seriously, because of the enormous flexibility they afford in calculations, and in exhibiting the structure of many mathematical relationships. Theories of delta-functions that are rigorous in their details but still quite accessible are given by Lighthill (1958) and by Schwartz (1966). In particular, such a theory is needed to guarantee in advance that sensible results emerge from the many interchanges of orders of operations (e.g. of limits and integrals) that we shall permit ourselves without further apology throughout this book (see also the comments at the end of Appendix C).

In fact, despite its name, $\delta(x)$ is not really a function at all: it is more properly described as a 'distribution' or a 'generalized function'. Often it is helpful to think of it as a 'functional': while an ordinary function f maps *numbers* x onto numbers $f(x)$, a functional maps (ordinary) *functions* f onto numbers. Thus, $\delta(x)$ regarded as a functional maps any ordinary (test) function f onto the number $f(0)$. That $\delta(x)$ cannot be an ordinary function is evident: while it might be regarded as mapping any number $x \neq 0$ onto 0, there is no number onto which $\delta(x)$ could be regarded as mapping the crucial number 0.

No meaning is attached, in general, to the product of two generalized functions whose singularities coincide. For instance, $\delta(x)\,\delta(x)$ and $\delta(x)\,\delta'(x)$ make no sense; by contrast, $\int_{-\infty}^{\infty} dx\, \delta(x-a)\,\delta(x-b) = \delta(a-b)$ does.

Finally, it can prove useful to relate the delta-function to convolutions. A convolution C is a special kind of operator, mapping functions onto other functions of the same variable, say g onto G, written symbolically as $Cg = G$. The convolution C is represented by a function $c(x-y)$, and one defines

$$G(x) = \int_{-\infty}^{\infty} dy\, c(x-y)g(y) = \int_{-\infty}^{\infty} dy\, c(y)g(x-y). \tag{1.1.22}$$

For general c, G is of course a function different from g. But in the special case where $c(x-y) = \delta(x-y)$, eqn (1.1.3) entails that

$$G(x) = \int_{-\infty}^{\infty} dy\, \delta(x-y)g(y) = g(x). \tag{1.1.23}$$

Thus, $\delta(x-y)$ regarded as a convolution represents the identity (unit) operator 1, which maps any test function onto the same function. With the symbolism used above, this is expressed as delta $\cdot\, g = 1 \cdot g = g$, for all g. (The Fourier transforms of convolutions are discussed in Appendix C.)

1.2 Representations of $\delta(x)$ and of $\delta'(x)$

Although, as just explained, $\delta(x)$ is not a function in the ordinary sense, it can be represented, in many different ways, as the limit of (a sequence of) ordinary functions, most often in the form

$$\delta(x) = \lim_{\varepsilon \to 0} F(x, \varepsilon), \tag{1.2.1a}$$

where

$$\int_{-\infty}^{\infty} dx\, F(x, \varepsilon) = 1 \tag{1.2.1b}$$

for any value of ε. (Do not confuse the parameter ε with the sign function.) Any representation is valid provided that, in the limit, it satisfies the integral relations defining $\delta(x)$. (In principle, (1.2.1b) could be relaxed to $\lim_{\varepsilon \to 0} \int_{-\infty}^{\infty} dx\, F(x, \varepsilon) = 1$.)

Representations are useful because, as we shall see, they turn up naturally in applications, and because they allow many calculations involving $\delta(x)$ to be performed explicitly in terms of familiar functions. Conversely, in some physical problems representative functions $F(x, \varepsilon)$ turn up in their own right with small but finite ε, and $\delta(x)$ can prove a convenient approximation to F. (This happens for instance with the Golden Rule for time-dependent perturbation theory in quantum mechanics, and with diffraction patterns in the geometric-optics limit of wave propagation.)

Any representation automatically implies some strong definition of $\delta(x)$, since it automatically prescribes a definite value for the integrals $\int_L$ and $\int_R$ in (1.1.5). This is irrelevant as long as $\delta(x)$ is to be used only weakly; but if $\delta(x)$ is to be used strongly, then one must restrict oneself to representations F that lead to the requisite strong definition.

Finally, from now on we require test functions $f(x)$ to vanish fast enough at infinity to ensure the convergence of any integral in which they occur. This does not rule out test functions increasing strongly with $|x|$ over any given finite region, provided that in any particular problem this region can be bounded in advance (i.e. it may be arbitrarily wide, but must not widen indefinitely as $\varepsilon \to 0$). Different problems require test functions decreasing, ultimately, at different rates, which generally are easy to specify in each particular problem, but which one need not prescribe in advance once and for all. (As a rule, we shall allow these asymptotic restrictions to remain tacit: for instance, even if x^2 were to appear as a test function, it should be understood that beyond some fixed limits it might have to change into another function that does vanish at infinity.) The power and convenience of the methods using delta-functions stem largely from the fact that physical systems and physically relevant regions of space and time are finite, so that they can tolerate such mathematical restrictions at arbitrarily large $|x|$, and yet remain insensitive to the precise way in which the restrictions are implemented.

In the examples below, the $F(x, \varepsilon)$ should be sketched as functions of x as an exercise. One can then observe how $F(x, \varepsilon)$ becomes higher and narrower as the limit is approached. By differentiating each F, one obtains the corresponding representation of $\delta'(x)$.

Exercise: Write down and sketch these.

Proofs of (1.2.1a) will be given only where they are not obvious. One method of proof† proceeds through two standard steps, (a) noting that $\lim_{\varepsilon\to 0} F(x, \varepsilon) = 0$ when $x \neq 0$, and (b) showing that $\lim_{\varepsilon\to 0} \int_{-\eta_1}^{\eta_2} dx\, F(x, \varepsilon) = 1$ for arbitrarily small η_1 and η_2. An alternative method relies on the ultimate asymptotic vanishing of the test functions $f(x)$.

(i) *Square step:*

$$F(x, \varepsilon) = \frac{1}{2\varepsilon}\,\theta(\varepsilon - |x|). \tag{1.2.2}$$

(ii) *Gaussian:*

$$F(x, \varepsilon) = \frac{1}{\varepsilon\sqrt{\pi}} \exp\left(-x^2/\varepsilon^2\right). \tag{1.2.3}$$

Proof: (a) As $\varepsilon \to 0$ with fixed $x \neq 0$, one has $F(x, \varepsilon) \to 0$, because the exponential (with its exponent $\to -\infty$) vanishes much faster than the prefactor $1/\varepsilon$ diverges. (b) Let $x = \varepsilon\xi$; then

$$\int_{-\eta_1}^{\eta_2} dx\, F(x, \varepsilon)$$

$$= \int_{-\eta_1/\varepsilon}^{\eta_2/\varepsilon} d\xi\, \varepsilon \frac{1}{\varepsilon\sqrt{\pi}} \exp\left(-\xi^2\right) \xrightarrow[(\varepsilon\to 0)]{} \frac{1}{\sqrt{\pi}} \int_{-\infty}^{\infty} d\xi\, \exp\left(-\xi^2\right) = 1. \quad \blacksquare$$

(iii) *Lorentzian (resonance shape):*

$$F(x, \varepsilon) = \frac{\varepsilon/\pi}{x^2 + \varepsilon^2}. \tag{1.2.4}$$

Proof: Step (a) is obvious. In (b), again let $x = \varepsilon\xi$; then

$$\int_{-\eta_1}^{\eta_2} dx\, F(x, \varepsilon) = \int_{-\eta_1/\varepsilon}^{\eta_2/\varepsilon} d\xi\, \varepsilon \frac{\varepsilon/\pi}{\varepsilon^2(\xi^2 + 1)} \to \frac{1}{\pi} \int_{-\infty}^{\infty} \frac{d\xi}{\xi^2 + 1} = 1. \quad \blacksquare$$

The other method proceeds by verifying that (1.2.1a) satisfies the

† See the cautionary comments in Section 1.5.

alternative definition (1.1.4) of $\delta(x)$, conveniently reformulated as $\int_{-\infty}^{\infty} \mathrm{d}x\, \delta(x) f(x) = f(0)$. We need merely write

$$\lim_{\varepsilon \to 0} \int_{-\infty}^{\infty} \mathrm{d}x \frac{\varepsilon/\pi}{x^2+\varepsilon^2} f(x) = \lim \int_{-\infty}^{\infty} \mathrm{d}\xi\, \varepsilon \frac{\varepsilon/\pi}{\varepsilon^2(\xi^2+1)} f(\varepsilon\xi)$$

$$= \frac{1}{\pi} \int_{-\infty}^{\infty} \mathrm{d}\xi \frac{1}{\xi^2+1} f(0) = f(0) \frac{1}{\pi} \int_{-\infty}^{\infty} \mathrm{d}\xi \frac{1}{\xi^2+1} = f(0). \tag{1.2.5}$$

■

The need for an asymptotic restriction on $f(x)$ is clear: for instance, with $f(x) = x^2$ the leftmost integral would simply fail to make sense.

Exercise: Use this method to prove (1.2.2–4).

(iv) *Dirichlet (diffraction peak):*

$$F(x, \varepsilon) = \frac{\sin(x/\varepsilon)}{\pi x}. \tag{1.2.6}$$

In contrast to examples (i–iii), as $\varepsilon \to 0$ with fixed $x \neq 0$, this F does not vanish monotonically. Instead, it oscillates with increasing frequency as a function of x over any range not including the origin. But as regards the integrals defining $\delta(x)$, such infinitely rapid oscillations of the integrand are equally effective, since they cause positive and negative contributions to the integral to cancel each other.

Proof: This relies on the special case $a = 1$ of Dirichlet's result (derived in Appendix B):

$$\int_{-\infty}^{\infty} \mathrm{d}x \frac{\sin(ax)}{x} = \pi\varepsilon(a) = \pm\pi \quad \text{if} \quad a \gtrless 0. \tag{1.2.7}$$

Adopting the second method, we need merely write

$$\int_{-\infty}^{\infty} \mathrm{d}x \frac{\sin(x/\varepsilon)}{\pi x} f(x) = \int_{-\infty}^{\infty} \mathrm{d}\xi\, \varepsilon \frac{\sin(\xi)}{\pi\xi\varepsilon} f(\varepsilon\xi)$$

$$= \frac{1}{\pi} \int_{-\infty}^{\infty} \mathrm{d}\xi \frac{\sin\xi}{\xi} f(0) = f(0) \frac{1}{\pi} \int_{-\infty}^{\infty} \mathrm{d}\xi \frac{\sin\xi}{\xi} = f(0).$$

■

The representations (i)–(iv) are illustrated in Fig. 1.2.

(v) *Fourier representation.* Using a standard integral, the Lorentz representative F in (1.2.4) can be rewritten

$$F(x, \varepsilon) = \frac{\varepsilon/\pi}{x^2+\varepsilon^2} = \frac{1}{2\pi} \int_{-\infty}^{\infty} \mathrm{d}k \exp(-\varepsilon|k|) \cos(kx), \tag{1.2.8}$$

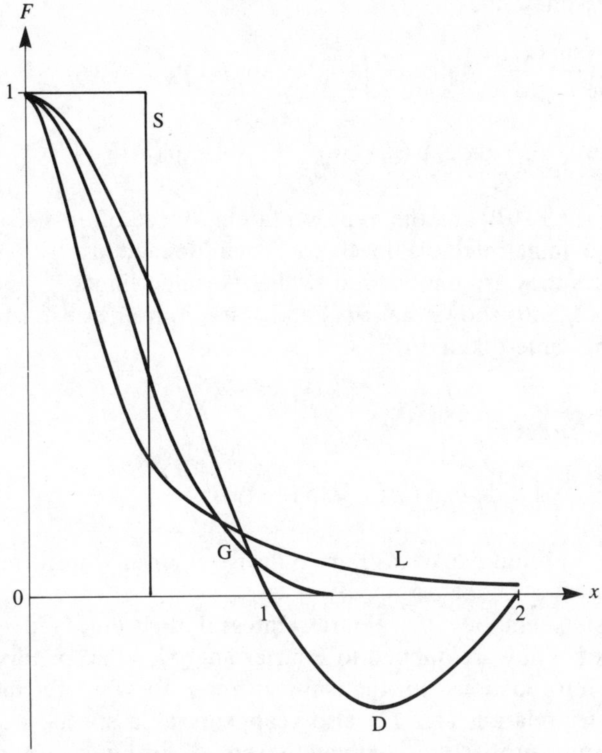

Fig. 1.2 Graphs of some representative functions $F(x, \varepsilon)$, labelled S for the square step (1.2.2), G for the Gaussian (1.2.3), L for the Lorentzian (1.2.4), and D for the Dirichlet function (1.2.6). These particular representations are symmetric ($F(x, \varepsilon) = F(-x, \varepsilon)$), and are shown only for $x \geqslant 0$. For S we have taken $\varepsilon = \varepsilon_S = \frac{1}{2}$, whence $F_S = \theta(\frac{1}{2} - |x|)$; the values of ε in the other functions have been adjusted to make $F(0, \varepsilon) = 1$. Thus $F_G = \exp(-\pi x^2)$, $F_L = 1/(1 + \pi^2 x^2)$, and $F_D = \sin(\pi x)/\pi x$.

$$\lim_{\varepsilon \to 0} F(x, \varepsilon) = \delta(x) = \frac{1}{2\pi} \int_{-\infty}^{\infty} dk \cos(kx), \tag{1.2.9}$$

$$\delta(x) = \frac{1}{2\pi} \int_{-\infty}^{\infty} dk \exp(ikx). \tag{1.2.10}$$

Exercise: Derive (1.2.9–10) in a similar way from the Gaussian representation.

The same result follows from the Dirichlet representation: with

$K = 1/\varepsilon \to \infty$, one finds

$$\delta[x] = \lim_{\varepsilon \to 0} \frac{\sin(x/\varepsilon)}{\pi x} = \lim_{\varepsilon \to 0} \frac{1}{2\pi} \int_{-1/\varepsilon}^{1/\varepsilon} dk \exp(ikx)$$

$$= \lim_{K \to \infty} \frac{1}{2\pi} \int_{-K}^{K} dk \exp(ikx) = \frac{1}{2\pi} \int_{-\infty}^{\infty} dk \exp(ikx). \tag{1.2.11}$$

Equations (1.2.9–10) are the representations most often used. Strictly speaking the final integrals in them are ill-defined as they stand; in doubtful cases they are understood as the indicated limits.

Equation (1.2.10) shows that $\delta(x)$ and 1 are Fourier transforms of each other. By the same token

$$\delta(x-y) = \frac{1}{2\pi} \int_{-\infty}^{\infty} dk \exp(ik(x-y))$$

$$= \frac{1}{2\pi} \int_{-\infty}^{\infty} dk \exp(ikx) \cdot \exp(-iky), \tag{1.2.12}$$

whence $\delta(x-y)$ and $\exp(-iky)$ are likewise Fourier transforms of each other.

These results embody the Fourier integral theorem, and extend to Fourier series. They are applied to Fourier analysis in Appendix C, since many first introductions to the subject tend to obscure the crucial structure. The relation (1.2.12) also reappears as a special case of the general closure property of complete sets of orthonormal functions, discussed in Section 1.3.

(vi) *Asymmetric representations.* In Examples (i–v), the functions $F(x, \varepsilon)$ are symmetric: $F(x, \varepsilon) = F(-x, \varepsilon)$. Accordingly they imply the strong definition $\int_L = \int_R = \frac{1}{2}$, and make $\delta(x)$ strongly even. But it is equally easy to construct asymmetric $F(x, \varepsilon)$: for instance, the Gaussian (1.2.3) could be replaced by the lop-sided function

$$F(x, \varepsilon) = 2\theta(x) \frac{1}{\varepsilon\sqrt{\pi}} \exp(-x^2/\varepsilon^2), \tag{1.2.13}$$

whose limit as $\varepsilon \to 0$ still satisfies the weak defining equations of $\delta(x)$, but which entails $\int_L = 0$, $\int_R = 1$.

Exercise: Prove this, and construct similar lop-sided examples from (1.2.1, 2, 4).

More instructive is the following representation which is peculiar to situations where x cannot be negative, and where, accordingly, one requires from the outset the strong definition

$$\delta(x) = 0 \quad \text{if} \quad x > 0, \qquad \int_0^{\eta_1} dx\, \delta(x) = 1. \tag{1.2.14}$$

We can then adopt

$$F(x, \varepsilon) = \varepsilon x^{\varepsilon-1}. \tag{1.2.15}$$

(The writer owes this example to Dr David Waxman.) Since $x^{\varepsilon-1}$ is complex for negative x, this, unlike (1.2.13), clearly makes no sense to the left of the origin: in order words $\int_R = 1$ while $\int_L$ is undefined, and the question whether $\delta(x)$ equals $\delta(-x)$ cannot even arise. Another inherently one-sided representation occurs in eqn (1.3.15) below.

(vii) *Representations of* $\delta'(x)$. While every representation $\lim_{\varepsilon\to 0} F(x, \varepsilon)$ of $\delta(x)$ automatically supplies a representation $\lim_{\varepsilon\to 0} \partial F(x, \varepsilon)/\partial x$ of $\delta'(x)$, the best way to visualize $\delta'(x)$ is also the most obvious, namely as

$$\delta'(x) = \lim_{a\to 0} \frac{1}{a} \{\delta(x + \tfrac{1}{2}a) - \delta(x - \tfrac{1}{2}a)\}. \tag{1.2.16}$$

If as in Section 1.1.1 we regard $\delta(x \mp \tfrac{1}{2}a)$ as the charge densities (per unit length) due to unit point charges at $x = \pm\tfrac{1}{2}a$ respectively, then we recognize $\delta'(x)$ as the *negative* of the charge density due to a *unit dipole* placed at the origin. Such a dipole can be constructed from a point charge q at $x = \tfrac{1}{2}a$ plus a point charge $-q$ at $x = -\tfrac{1}{2}a$, with fixed $qa = 1$, i.e. with $q = 1/a$, in the limit $a \to 0$. Expressed in symbols, this reads

$$\rho = \text{(charge/unit length, due to unit dipole at the origin)}$$

$$= \lim_{a\to 0} \left\{\frac{1}{a}\,\delta(x - \tfrac{1}{2}a) + \left(-\frac{1}{a}\right)\delta(x - (-\tfrac{1}{2}a))\right\} = -\delta'(x), \tag{1.2.17}$$

as claimed. The three-dimensional version is explored further in Section 1.4.5 below.

The representation (1.2.16) automatically entails a strong definition, which makes δ' strongly (as well as weakly) odd. It would be equally easy, and equally compatible with $\delta'(x) = -\rho_{\text{dipole}}$, to choose a representation under which δ' is not strongly odd: for instance the one-sided representation

$$\delta'(x) = \lim_{a\to 0} \frac{1}{a} \{\delta(x + 2a) - \delta(x + a)\}.$$

1.3 The closure property of complete orthonormal sets

A set of functions $\{\phi_n(x)\}$ defined over some interval, and subject to stated boundary conditions at its ends, is *complete* if any (suitably restricted) function $f(x)$ can be expressed as a linear combination of them, i.e. in the form

$$f(x) = \sum_n c_n \phi_n(x), \tag{1.3.1}$$

provided the coefficients c_n are chosen appropriately.

The set is *orthonormal* if, integrating over the range of x,

$$\int \mathrm{d}x\, \phi_n^*(x)\phi_{n'}(x) = \delta_{nn'}. \tag{1.3.2}$$

The Kronecker symbol $\delta_{nn'}$ is the discrete version of the delta-function:

$$\delta_{nn'} \equiv 0 \quad \text{if} \quad n \neq n'; \qquad \delta_{nn'} \equiv 1 \quad \text{if} \quad n = n'. \tag{1.3.3}$$

Many complete orthonormal sets (c.o.n.s.) are real, in which case the complex conjugation sign in (1.3.2) is irrelevant: e.g. in examples (i, ii) below. In other cases it may be possible (though not always convenient) to construct a new c.o.n.s. $\{\tilde{\phi}_\nu\}$, with real $\tilde{\phi}_\nu(x)$, every $\tilde{\phi}_\nu$ being a linear combination of just a finite number of the original ϕ_n. The reality properties of c.o.n.s. are discussed further in Appendix E.

The coefficients c_n in (1.3.1) are identified by appeal to (1.3.2). Multiply both sides of (1.3.1) by $\phi_p^*(x)$ and integrate: on the right, take $\int \mathrm{d}x$ under $\sum_n$:

$$\int \mathrm{d}x\, \phi_p^*(x) f(x) = \sum_n c_n \int \mathrm{d}x\, \phi_p^* \phi_n = \sum_n c_n\, \delta_{pn} = c_p. \tag{1.3.4}$$

This determines c_p in terms of $f(x)$.

C.o.n.s. are familiar from Fourier series, and in quantum mechanics as the sets of eignfunctions of Hermitean operators (observables). (Note that in formulae quoted from quantum mechanics we generally set $\hbar = 1$.) For instance, the energy eigenfunctions of the linear harmonic oscillator, i.e. of the Hamiltonian

$$H = \left\{ -\frac{1}{2m}\frac{\partial^2}{\partial x^2} + \frac{1}{2} m\omega^2 x^2 \right\},$$

are complete over the interval $-\infty < x < \infty$, subject to the boundary conditions $\phi(\pm\infty) = 0$; the radial energy eigenfunctions of the three-dimensional oscillator with angular momentum l, i.e. of

$$H = \left\{ -\frac{1}{2m}\left[\frac{\partial^2}{\partial r^2} + \frac{2}{r}\frac{\partial}{\partial r} - \frac{l(l+1)}{r^2}\right] + \frac{1}{2} m\omega^2 r^2 \right\},$$

are complete over $0 \leq r < \infty$, subject to $\phi(r \to 0) \sim r^l$ and $\phi(\infty) = 0$.

In the cases we shall meet, there are infinitely many ϕ_n; then the series (1.3.1) formed with the coefficients (1.3.4) converges uniformly to $f(x)$ if $f(x)$ is smooth enough and if at the ends of the interval it satisfies the same boundary conditions as the ϕ_n. (Uniform convergence means that a given accuracy ε can be achieved for all x by summing $N(\varepsilon)$ terms, where $N(\varepsilon)$ can be determined independently of x.) By contrast, if f is singular (e.g. discontinuous) at some point, or if it violates the boundary conditions, then in most cases the series still converges except at the singularity or the endpoint, but not uniformly: to achieve a pre-assigned accuracy ε, the requisite number $N(\varepsilon, x)$ of terms increases indefinitely as x moves nearer to these points. More important, and sufficient for physical applications, is the fact that even for such f the series still converges in the mean, i.e. the integrated squared deviation $\int \mathrm{d}x\, |f(x) - \sum^N c_n\phi_n(x)|^2$ tends to zero as $N \to \infty$.

Recall that the orthogonality of eigenfunctions belonging to different eigenvalues follows readily from the Hermitecity of H. By contrast, it can be quite cumbersome to prove that a set is complete if it is treated purely on its merits; but general and more manageable proofs emergy from the Sturm–Liouville theory of differential equations.

What is important to us is that *any* c.o.n.s. furnishes a representation of the delta-function, by virtue of the *closure property*

$$\sum_n \phi_n^*(x')\phi_n(x) = \delta(x - x'). \tag{1.3.5}$$

Proof:

$$f(x) = \sum_n c_n\phi_n(x) = \sum_n \left\{\int \mathrm{d}x'\, \phi_n^*(x')f(x')\right\}\phi_n(x)$$

$$f(x) = \int \mathrm{d}x' \left\{\sum_n \phi_n^*(x')\phi_n(x)\right\} f(x'). \tag{1.3.6}$$

But $f(x)$ is arbitrary (within some admissible class of function), i.e. it can be chosen independently over different regions; therefore its value $f(x)$ at one point x cannot be expressed in terms of its values over other regions x', contrary to what (1.3.6) appears to imply at first sight. The contradiction is removed only if the contents of the curly brackets satisfy (1.3.5). (Alternatively, we need merely compare (1.3.6) with the identity $f(x) = \int \mathrm{d}x'\, \delta(x - x')f(x')$.) ■

If the members of a complete orthonormal set are labelled by a continuously variable index k instead of a discrete index n, then the orthonormality condition (1.3.2) is replaced by

$$\int \mathrm{d}x\, \phi_k^*(x)\phi_{k'}(x) = \delta(k - k'), \tag{1.3.7a}$$

and the closure property (1.3.5) by

$$\int \mathrm{d}k\, \phi_k^*(x)\phi_k(x') = \delta(x - x'). \tag{1.3.7b}$$

The most important such set consists of the normed eigenfunctions

$$\phi_k(x) = (2\pi)^{-\frac{1}{2}} \exp(ikx) \tag{1.3.8a}$$

of the operator $-i\,\partial/\partial x$ (the momentum operator in quantum mechanics), whose eigenvalue k can assume any value from $-\infty$ to $+\infty$. These ϕ_k, and their N-dimensional generalizations $\phi_{\boldsymbol{k}}(\boldsymbol{r}) = (2\pi)^{-N/2} \exp(i\boldsymbol{k}\cdot\boldsymbol{r})$, are automatically eigenfunctions also of $-\nabla^2 = (-i\nabla)^2$; as such they will be reviewed in Section 4.5.2, and in constant use afterwards. With (1.3.8a), and with the corresponding replacement

$$\sum_n \cdots \to \int d^N k \cdots, \tag{1.3.8b}$$

eqn (1.3.7b) reduces to the Fourier representation (1.2.12) of the delta-function (or rather to its N-dimensional version, eqn (1.4.3) below). Similarly, the expansion (1.3.1) becomes the Fourier integral representation of $f(x)$ (apart from a different assignment of factors $1/2\pi$ in Appendix C).

There are also cases where some members of the c.o.n.s. carry a discrete label, while others carry a continuous label; then, on the left of (1.3.5), one has a sum over the discrete and an integral over the continuous labels. This happens for instance with the eigenfunctions of the attractive square-well Hamiltonian

$$H = \left\{ -\frac{1}{2m}\frac{\partial^2}{\partial x^2} - V_0\theta(a - |x|) \right\},$$

which always has at least one discrete negative eigenvalue (a bound state), in addition to a continuous (and doubly degenerate) spectrum extending from 0 to $+\infty$. In the following, any such sum, integral, or sum-plus-integral will be written simply as $\sum_n$.

Some familiar c.o.n.s. furnish particularly useful representations of the delta-function. Examples (i–iii) arise from trigonometric functions, i.e. from eigenfunctions of the operator $-d^2/dx^2$ over an interval $0 \leq x \leq L$. Different boundary conditions furnish different representations.

(i) *Dirichlet conditions:*

$$\phi(0) = 0 = \phi(L).$$

$$\phi_n(x) = (2/L)^{\frac{1}{2}} \sin(n\pi x/L), \qquad n = 1, 2, 3, \ldots$$

$$\sum_n \phi_n^*(x')\phi_n(x) = \frac{2}{L}\sum_n \sin\left(\frac{n\pi x'}{L}\right)\sin\left(\frac{n\pi x}{L}\right)$$

$$= \delta(x - x'), \qquad (0 \leq x, x' \leq L). \tag{1.3.9}$$

(ii) *Neumann conditions:*

$\phi'(0) = 0 = \phi'(L)$.

$$\phi_n = (1/L)^{\frac{1}{2}} \quad \text{if} \quad n = 0,$$

$$\phi_n = (2/L)^{\frac{1}{2}} \cos(n\pi x/L), \qquad n = 1, 2, 3, \ldots$$

$$\begin{aligned}\sum_n \phi_n^*(x')\phi_n(x) &= \frac{1}{L}\left\{1 + 2\sum_{n=1}^{\infty} \cos\left(\frac{n\pi x'}{L}\right)\cos\left(\frac{n\pi x}{L}\right)\right\} \\ &= \frac{1}{L}\sum_{n=-\infty}^{\infty} \cos\left(\frac{n\pi x'}{L}\right)\cos\left(\frac{n\pi x}{L}\right) \\ &= \delta(x - x'), \qquad (0 \leqslant x, x' \leqslant L).\end{aligned} \tag{1.3.10}$$

The last form of (1.3.10) may look more elegant, but it can be deceptive, because $\cos(n\pi x/L)$ and $\cos(-n\pi x/L)$ are really the same function (i.e. they are not linearly independent).

(iii) *Periodic boundary conditions.* To enforce periodicity (on solutions of our second-order differential equation $-\phi'' = \text{const.}\ \phi$) one must impose two conditions, namely $\phi(0) = \phi(L)$ and $\phi'(0) = \phi'(L)$. Then

$$\phi_n(x) = L^{-\frac{1}{2}} \exp[i2\pi nx/L], \qquad n = 0, \pm 1, \pm 2, \ldots,$$

$$\begin{aligned}\sum_n \phi_n^*(x')\phi_n(x) &= L^{-1}\sum_{n=-\infty}^{\infty} \exp[i2\pi n(x - x')/L] \\ &= L^{-1}\left\{1 + 2\sum_{n=0}^{\infty} \cos[2\pi n(x - x')/L]\right\} \\ &= \delta(x - x'), \qquad (0 \leqslant x, x' \leqslant L).\end{aligned} \tag{1.3.11}$$

Here, in contrast to examples (i) and (ii), ϕ_n and ϕ_{-n} are linearly independent (unless $n = 0$), and both are included not by choice but from necessity.

Setting $x' = 0$ one finds

$$L^{-1}\sum_{n=-\infty}^{\infty} \exp(i2\pi nx/L) = \delta(x), \qquad 0 \leqslant x \leqslant L. \tag{1.3.12}$$

(A more direct proof appears in Appendix C.)

Because the series (1.3.11–12) are periodic in x with period L, the restriction $0 \leqslant x \leqslant L$ can be relaxed simply by repeating the delta-functions on the right likewise with period L. For instance, (1.3.12) yields

$$L^{-1}\sum_{n=-\infty}^{\infty} \exp(i2\pi nx/L) = \sum_{m=-\infty}^{\infty} \delta(x - mL). \tag{1.3.13}$$

If the restriction $0 \leqslant x \leqslant L$ is now reimposed, one can drop all the terms with $m \neq 0$ on the right, because none of their arguments can then vanish.

Exercises: (i) Determine the representation of $\delta(x-x')$ over $0 \leqslant x, x' \leqslant L$ furnished by the mixed boundary conditions $\phi(0) = 0 = \phi'(L)$. (ii) Derive (1.3.13) by using the Poisson summation formula (C.12). (iii) Use Poisson's summation formula to relax the restrictions $0 \leqslant x, x' \leqslant L$ on (1.3.9–10).

(iv) *Legendre polynomials.* The Legendre polynomials $P_l(\mu)$ are eigenfunctions of the operator $L \equiv \{(1-\mu^2)\,d^2/d\mu^2 - 2\mu\, d/d\mu\}$, over the range $-1 \leqslant \mu \leqslant 1$, subject to the boundary conditions that they be non-singular at the endpoints (in spite of the fact that the operator L itself is singular there). Then $LP_l = l(l+1)P_l$, $l = 0, 1, 2, \ldots$. More is said about the P_l in Appendix A. The orthonormal functions, and the corresponding representation, are

$$\phi_l(\mu) = [\tfrac{1}{2}(2l+1)]^{\frac{1}{2}} P_l(\mu), \tag{1.3.14}$$

$$\sum_{l=0}^{\infty} \tfrac{1}{2}(2l+1) P_l(\mu') P_l(\mu) = \delta(\mu - \mu'), \qquad -1 \leqslant \mu, \mu' \leqslant 1.$$

Setting $\mu' = 1$ and using $P_l(1) = 1$ gives

$$\sum_{l=0}^{\infty} \tfrac{1}{2}(2l+1) P_l(\mu) = \delta(\mu - 1), \qquad -1 \leqslant \mu \leqslant 1. \tag{1.3.15}$$

Since $\mu \leqslant 1$, this is another one-sided representation, asserting that $\int_{\mu_1}^{1} d\mu\, f(\mu)\, \delta(\mu - 1) = f(1)$.

(v) In all such examples, the sum (1.3.5) up to N terms, call it $S_N(x, x')$, is of course an ordinary rather than a generalized function. What is asserted by (1.3.5) is that $\delta(x-x') = \lim_{N\to\infty} S_N(x, x')$. The limit $N \to \infty$ plays the same role here as did the limit $\varepsilon \to 0$ in the representations of Section 1.2.

1.4 The delta-function in three dimensions and in two

1.4.1 Three dimensions: Cartesian and polar coordinates

In principle the extension from one to several dimensions is obvious. In three dimensions (3D; similarly 2D and nD), integrations over volume are written, interchangeably, as $\int dV$ or $\int d^3r$. Then the weak definition of $\delta(\boldsymbol{r})$ reads

$$\delta(\boldsymbol{r}) = 0 \qquad \text{if } \boldsymbol{r} \neq 0,$$
$$\int_V d^3r\, \delta(\boldsymbol{r}) = 1 \qquad \text{if } V \text{ includes the origin.} \tag{1.4.1}$$

Strong definition will not be needed until Section 1.4.2.

The dimensions are $[\delta(\boldsymbol{r})] = [\text{volume}]^{-1}$, ($[\mathrm{L}]^{-n}$ in nD). Most of the results from Section 1.1 generalize straightforwardly.

In *Cartesian coordinates*, $\boldsymbol{r} \equiv (x, y, z)$, one has

$$\delta(\boldsymbol{r}) = \delta(x)\,\delta(y)\,\delta(z), \tag{1.4.2}$$

and the Fourier representation for instance becomes

$$\delta(\boldsymbol{r}) = (2\pi)^{-3} \int \mathrm{d}^3k \exp(i\boldsymbol{k}\cdot\boldsymbol{r}), \tag{1.4.3}$$

where $\boldsymbol{k}\cdot\boldsymbol{r} = k_1x + k_2y + k_3z$ is the ordinary scalar product.

Exercise: Verify (1.4.3), and try to find 3D analogues of the other representations in Section 1.2. Notice that this is particularly easy for the Gaussian representation.

In *polar coordinates* there are subtleties. First we need some notation and conventions. As usual, $\boldsymbol{r} \equiv (r, \theta, \phi)$ is related to Cartesians by $x = r\sin\theta\cos\phi$, $y = r\sin\theta\sin\phi$, $z = r\cos\theta$. The relation is made unique by the restrictions $0 \leqslant r$, $0 \leqslant \theta \leqslant \pi$ (whence $-1 \leqslant \cos(\theta) \leqslant 1$), and $0 \leqslant \phi < 2\pi$. Then the volume element becomes

$$\begin{aligned} \mathrm{d}V &= \mathrm{d}r\, r^2 \sin\theta\, \mathrm{d}\theta\, \mathrm{d}\phi = \mathrm{d}r\, r^2\, \mathrm{d}\cos\theta\, \mathrm{d}\phi \\ &\equiv \mathrm{d}r\, r^2\, \mathrm{d}\Omega, \end{aligned} \tag{1.4.4}$$

where, in the second and third forms, θ and $\cos(\theta)$, respectively, is regarded as the independent variable. We write

$$\Omega \equiv (\theta, \phi), \qquad \int \mathrm{d}\Omega \equiv \iint \mathrm{d}\cos\theta\, \mathrm{d}\phi, \tag{1.4.5}$$

so that Ω specifies the direction of the unit vector $\hat{\boldsymbol{r}}$ (sometimes we shall write Ω as $\Omega(\hat{\boldsymbol{r}})$ or $\Omega_{\hat{r}}$), and $\int \mathrm{d}\Omega$ denotes integration over solid angle: $\int \mathrm{d}\Omega\, 1 = 4\pi$. By convention, $\Omega = 0$ means $\theta = 0$, $\phi = 0$, i.e. $\hat{\boldsymbol{r}} = \hat{\boldsymbol{z}}$. Denoting by χ the angle between $\boldsymbol{r}$ and $\boldsymbol{r}'$, one has

$$\begin{aligned} \cos\chi &= \hat{\boldsymbol{r}}\cdot\hat{\boldsymbol{r}}' = \{\cos\theta\cos\theta' + \sin\theta\sin\theta'\cos(\phi - \phi')\} \\ &= \{\cos\theta\cos\theta' + \sin\theta\sin\theta'[\cos\theta\cos\phi' + \sin\phi\sin\theta']\} \end{aligned} \tag{1.4.6}$$

It is best to start by considering $\delta(\boldsymbol{r} - \boldsymbol{r}')$ with $\boldsymbol{r}' \neq 0$, rather than directly with $\delta(\boldsymbol{r})$, which we call the degenerate case. The (weak) definition requires that

$$\delta(\boldsymbol{r} - \boldsymbol{r}') = 0 \qquad \text{if } \boldsymbol{r} \neq \boldsymbol{r}', \tag{1.4.7a}$$

$$\begin{aligned} &\int_0^\infty \mathrm{d}r\, r^2 \int_0^{2\pi} \mathrm{d}\phi \int_{-1}^{1} \mathrm{d}\cos\theta\, \delta(\boldsymbol{r} - \boldsymbol{r}') \\ &\quad \equiv \int_0^\infty \mathrm{d}r\, r^2 \int d\Omega\, \delta(\boldsymbol{r} - \boldsymbol{r}') = 1. \end{aligned} \tag{1.4.7b}$$

Accordingly

$$\delta(\boldsymbol{r}-\boldsymbol{r}') = \frac{1}{r^2}\,\delta(r-r')\,\delta(\phi-\phi')\,\delta(\cos\theta-\cos\theta')$$
$$= \frac{1}{r^2}\,\delta(r-r')\,\delta(\Omega-\Omega'). \qquad (1.4.8\text{a, b})$$

This defines $\delta(\Omega-\Omega')$ by comparison. The proof is simply by substitution into (1.4.7). The factor $1/r^2$ could equally well be written as $1/r'^2$ or as $1/rr'$. One must be very careful to distinguish the 3D $\boldsymbol{r}$, $\boldsymbol{r}'$ and $\delta(\boldsymbol{r}-\boldsymbol{r}')$ from the radial (hence effectively one-dimensional) r, r' and $\delta(r-r')$.

The most fruitful representation of $\delta(\Omega-\Omega')$ stems from the closure property of the spherical harmonics $Y_{lm}(\Omega)$, which constitute a complete orthonormal set over the surface of the unit sphere (i.e. for functions of Ω alone). They are complete because they are the simultaneous eigenfunctions of $\boldsymbol{L}^2$ and L_z, where $\boldsymbol{L} = -i\boldsymbol{r}\wedge\boldsymbol{\nabla}$ is the orbital angular momentum operator in quantum mechanics; the boundary conditions are that the Y_{lm} be non-singular. More information about them appears in Appendix A; meanwhile, $\boldsymbol{L}^2 Y_{lm} = l(l+1)Y_{lm}$, $L_z Y_{lm} = mY_{lm}$.

The spherical harmonics satisfy the remarkable addition theorem

$$\sum_{m=-l}^{l} Y^*_{lm}(\Omega')Y_{lm}(\Omega) = \frac{2l+1}{4\pi}\,P_l(\hat{\boldsymbol{r}}\cdot\hat{\boldsymbol{r}}'). \qquad (1.4.9)$$

Identifying $n \equiv (l, m)$ in the closure relation (1.3.5), we find

$$\delta(\Omega-\Omega') = \sum_{l=0}^{\infty}\sum_{m=-l}^{l} Y^*_{lm}(\Omega')Y_{lm}(\Omega) \qquad (1.4.10)$$
$$= \frac{1}{4\pi}\sum_{l=0}^{\infty}(2l+1)P_l(\hat{\boldsymbol{r}}\cdot\hat{\boldsymbol{r}}') = \frac{1}{2\pi}\,\delta(1-\hat{\boldsymbol{r}}\cdot\hat{\boldsymbol{r}}'). \qquad (1.4.11\text{a, b})$$

The second step relies on (1.4.9), and the last on (1.3.15). Any of these forms can be substituted for $\delta(\Omega-\Omega')$ on the right of (1.4.8b).

1.4.2 Polar coordinates: the strong definition of $\delta(\boldsymbol{r})$

The degenerate case $\boldsymbol{r}'=0$ of $\delta(\boldsymbol{r}-\boldsymbol{r}')$ with polar coordinates occurs constantly, and is a prime example requiring a strong definition of the delta-function. Two points need attention as $\boldsymbol{r}'\to 0$ in equations like (1.4.8–11). First, we encounter the radial delta-function $\delta(r)$ at the endpoint of its range $0 \le r$. Obviously we need a strong definition, and we adopt the one-sided definition

$$\delta(r) = 0 \quad \text{if} \quad r \neq 0, \qquad \int_0^{\eta_1} \mathrm{d}r\, \delta(r) = 1. \tag{1.4.12}$$

Thus all of the peak under $\delta(r)$ is assigned to the physical region $0 \leqslant r$. This definition applies whenever such integrals are encountered from now on. (Some books choose, instead, the definition $\int_0^{\eta_1} \mathrm{d}r\, \delta(r) = \frac{1}{2}$. Their formulae are obtained from ours by the replacement $\delta(r) \to 2\,\delta(r)$.)

Second, the null vector $\boldsymbol{r}' = 0$ has no unique direction, which leaves Ω' undefined, and $\delta(\Omega - \Omega')$ apparently ambiguous. This difficulty is only apparent, because the ambiguity does not affect the value of the integral $\int \mathrm{d}^3r\, f(\boldsymbol{r})\, \delta(\boldsymbol{r})$ with any test function $f(\boldsymbol{r})$ that is well-defined at $\boldsymbol{r} = 0$, in the sense that $\lim_{r\to 0} f(r, \Omega) = f(\mathbf{0})$ is independent of the direction Ω along which $\boldsymbol{r}$ approaches the origin. (For instance, $(xy + b^2)/(x^2 + y^2 + z^2 + a^2)$ is well-defined, while $xy/(x^2 + y^2 + z^2)$ and z/r are not.) The definition of $\delta(\boldsymbol{r})$ does not cater for ill-defined test functions, and we do not consider such cases. For well-defined test functions, it is evidently enough (and compatible with (1.4.1)) to write

$$\delta(\boldsymbol{r}) = \delta(r)/4\pi r^2. \tag{1.4.13}$$

Starting from the general expression for $\delta(\boldsymbol{r} - \boldsymbol{r}')$ in (1.4.8), eqn (1.4.13) amounts to replacing the factor $\delta(\Omega - \Omega') = \sum_{l,m} Y^*_{lm}(\Omega')Y_{lm}(\Omega)$ by just the single term $l = 0 = m$, i.e. by $|Y_{00}|^2 = 1/4\pi$; *but this is appropriate only in the degenerate case* $\boldsymbol{r}' = 0$, *and only for use with test functions that are well-defined at the origin.* (It is easily checked that $\int \mathrm{d}r\, r^2\, \mathrm{d}\Omega\, \delta(r)/4\pi r^2 = 1$, as required.) One must remember that in general the sums in (1.4.10–11) representing $\delta(\Omega - \Omega')$ are by no means equivalent to the isotropic ($l = 0$) term alone. For instance, the entire sum must be retained in problems formulated on the surface of the unit sphere, where the test functions are well-defined functions of Ω alone.

The following consequences of the definition (1.4.12) prove widely useful.

(i) **Theorem:** Under integrations with respect to volume, i.e. under $\int_0 \mathrm{d}r\, r^2 \cdots$,

$$\delta(r)/r = -\delta'(r). \tag{1.4.14}$$

Proof: Multiply by a test function of the form $f(r) = g(r)/r$, with $g(r)$ well-behaved at $r = 0$ (whence f itself may, but need not, behave as badly as $1/r$ as $r \to 0$), and integrate over volume. We must show that both sides given the same result. The LHS gives

$$\int_0 \mathrm{d}r\, r^2 \frac{\delta(r)}{r}\frac{g(r)}{r} = \int_0 \mathrm{d}r\, \delta(r) g(r) = g(0).$$

The RHS of (1.4.14) gives

$$\begin{aligned}\int_0 \mathrm{d}r\, r^2(-\delta'(r))g(r)/r &= -\int_0 \mathrm{d}r\, \delta'(r)[rg(r)] \\ &= \int_0 \mathrm{d}r\, \delta(r) \frac{\mathrm{d}}{\mathrm{d}r}[rg(r)] \\ &= \int_0 \mathrm{d}r\, \delta(r)[rg'(r) + g(r)] \\ &= [rg'(r) + g(r)]|_{r=0} = g(0).\end{aligned}$$ ■

The second step relies on the definition (1.1.18) of δ', and the last step on (1.1.6b). (The weak one-dimensional version of this result was set as an exercise following (1.1.18) above.)

(ii) **Theorem:** If r is the radial variable in spherical, cylindrical, or plane polar coordinates (i.e. if $r \geqslant 0$), then

$$\delta(r^2 - a^2) = \frac{1}{2|a|}\, \delta(r - |a|). \tag{1.4.15}$$

Proof: By virtue of (1.1.13)

$$\begin{aligned}\delta(r^2 - a^2) &= \frac{1}{2|a|}\{\delta(r-a) + \delta(r+a)\} \\ &= \frac{1}{2|a|}\{\delta(r-|a|) + \delta(r+|a|)\}.\end{aligned}$$

But, subject to $r \geqslant 0$, the argument of $\delta(r + |a|)$ cannot vanish, and this component may be dropped. ■

(iii) When $a = 0$, the result (1.4.15) must be handled cautiously. (This also suggests a procedure appropriate to some cases of coincident roots in (1.1.12).)

Theorem:

$$\delta(r^2) = \delta(r)/2r. \tag{1.4.16}$$

Proof: As in Theorem (i) we multiply by $g(r)/r$ and integrate over volume. The RHS gives

$$\int_0 \mathrm{d}r\, r^2 \frac{\delta(r)}{2r}\frac{g(r)}{r} = \frac{1}{2}\int_0 \mathrm{d}r\, \delta(r) g(r) = \tfrac{1}{2} g(0).$$

On the left of (1.4.16) we change the integration variable to r^2:

$$\int_0 \mathrm{d}r\, r^2\, \delta(r^2)\frac{g(r)}{r} = \int_0 \tfrac{1}{2}\, \mathrm{d}(r^2)\, \delta(r^2) g(r) = \tfrac{1}{2} g(0).$$ ■

1.4.3 Polar coordinates in 2D

In 2D, $\boldsymbol{r} = (x, y) = (r, \phi)$, where $x = r\cos(\phi)$, $y = r\sin(\phi)$, and $0 \leqslant r$, $0 \leqslant \phi < 2\pi$. Then the appropriate 'volume' (actually surface) element is $\mathrm{d}V = \mathrm{d}r\, r\, \mathrm{d}\phi$. It is left as an exercise to verify the following 2D analogues of our 3D results:

$$[\delta(\boldsymbol{r})] = [L]^{-2};$$

$$\delta(\boldsymbol{r} - \boldsymbol{r}') = \frac{1}{r}\delta(r - r')\,\delta(\phi - \phi')$$

$$= \frac{1}{r}\delta(r - r')\frac{1}{2\pi}\sum_{m=-\infty}^{\infty} \exp\left[im(\phi - \phi')\right]; \tag{1.4.17}$$

and, for test functions well-defined at the origin,

$$\delta(\boldsymbol{r}) = \delta(r)/2\pi r. \tag{1.4.18}$$

1.4.4 The behaviour of $\delta(\boldsymbol{r})$ under general coordinate transformations

The change from Cartesian to polar coordinates, say in 2D, is a special case of the general 2D coordinate transformation from (x, y) to (ξ_1, ξ_2), where the ξ_i are functions of (x, y). We are interested only in transformations that are one-to-one (have a unique inverse), except perhaps at the endpoints of the ranges of ξ_1, ξ_2. (Thus ϕ has no unique value when $r = 0$.) The condition for a transformation to be of this kind is that the Jacobian J should not vanish: $J \neq 0$, where

$$J \equiv \left|\frac{\partial(x, y)}{\partial(\xi_1, \xi_2)}\right| \equiv \begin{vmatrix} \partial x/\partial\xi_1, & \partial x/\partial\xi_2 \\ \partial y/\partial\xi_1, & \partial y/\partial\xi_2 \end{vmatrix}. \tag{1.4.19}$$

Recall that the volume element expressed in terms of the new coordinates is $\mathrm{d}V = J(\xi_1, \xi_2)\, \mathrm{d}\xi_1\, \mathrm{d}\xi_2$.

Suppose now that $(x, y) = (0, 0)$ corresponds to $(\xi_1, \xi_2) = (a_1, a_2)$. Then we require $\delta(\xi_1, \xi_2)$ to vanish unless $(\xi_1, \xi_2) = (a_1, a_2)$, and to satisfy

$$1 = \int \mathrm{d}V\, \delta(\boldsymbol{r}) = \iint \mathrm{d}\xi_1,\, \mathrm{d}\xi_2 J(\xi_1, \xi_2)\, \delta(\boldsymbol{r}).$$

Accordingly

$$\delta(\boldsymbol{r}) = \delta(\xi_1 - a_1)\, \delta(\xi_2 - a_2)/J(\xi_1, \xi_2). \tag{1.4.20}$$

When the ranges of the ξ_i are restricted, one must adopt an appropriate strong definition of $\delta(\xi_i - a_i)$.

Exercise: Deduce the results in Sections 1.4.2, 3 from (1.4.20).

1.4.5 Properties and applications of $\nabla\,\delta(\boldsymbol{r}-\boldsymbol{r}')$

The 3D generalization of $\delta'(x)$ is $\nabla\,\delta(\boldsymbol{r})$, where $\nabla \equiv (\hat{\boldsymbol{x}}\,\partial/\partial x + \hat{\boldsymbol{y}}\,\partial/\partial y + \hat{\boldsymbol{z}}\,\partial/\partial z)$. (Similarly, we shall write $\nabla' \equiv (\hat{\boldsymbol{x}}\,\partial/\partial x' + \hat{\boldsymbol{y}}\,\partial/\partial y' + \hat{\boldsymbol{z}}\,\partial/\partial z')$.) The degenerate case $\boldsymbol{r}'=0$ is somewhat awkward to visualize directly, but presents no difficulties once $\nabla\,\delta(\boldsymbol{r}-\boldsymbol{r}')$ has been understood. We illustrate some of the manipulations that are called for in practice. In this section, only the weak definition of the delta-function is needed.

Just as in the 1D equations (1.1.11) and (1.1.19), $\delta(\boldsymbol{r}-\boldsymbol{r}')$ is an even function, while $\nabla\,\delta(\boldsymbol{r}-\boldsymbol{r}')$ is odd. Accordingly (watch the primes!)

$$\nabla\,\delta(\boldsymbol{r}-\boldsymbol{r}') = \nabla\,\delta(\boldsymbol{r}'-\boldsymbol{r}) = -\nabla'\,\delta(\boldsymbol{r}-\boldsymbol{r}') = -\nabla'\,\delta(\boldsymbol{r}'-\boldsymbol{r}). \tag{1.4.21}$$

With a test function $f(\boldsymbol{r})$, the leftmost expression leads to

$$\int \mathrm{d}^3r\, f(\boldsymbol{r})\nabla\,\delta(\boldsymbol{r}-\boldsymbol{r}') = -\int \mathrm{d}^3r\,(\nabla f(\boldsymbol{r}))\,\delta(\boldsymbol{r}-\boldsymbol{r}')$$
$$= -(\nabla f)\big|_{\boldsymbol{r}=\boldsymbol{r}'} = -\nabla' f(\boldsymbol{r}'). \tag{1.4.22}$$

The same result emerges, but in a different way, from the third expression in (1.4.21): this form suggests that ∇' be moved immediately outside the integral (which is with respect to d^3r, not d^3r'):

$$\int \mathrm{d}^3r\, f(\boldsymbol{r})(-\nabla'\,\delta(\boldsymbol{r}-\boldsymbol{r}')) = -\nabla' \int \mathrm{d}^3r\, f(\boldsymbol{r})\,\delta(\boldsymbol{r}-\boldsymbol{r}')$$
$$= -\nabla' f(\boldsymbol{r}'). \tag{1.4.23}$$

To gain insight into the function $\nabla\,\delta$, we think of it as we did of $\delta'(x)$ at the end of Section 1.2. Consider the charge density due to two point charges of magnitudes $\pm p/a$, situated at $\boldsymbol{r}' = \pm\frac{1}{2}\boldsymbol{a}$ respectively:

$$\rho = p\frac{1}{a}\{\delta(\boldsymbol{r}-(\boldsymbol{r}'+\tfrac{1}{2}\boldsymbol{a})) - \delta(\boldsymbol{r}-(\boldsymbol{r}'-\tfrac{1}{2}\boldsymbol{a}))\}. \tag{1.4.24}$$

Now set $\boldsymbol{a} = a\boldsymbol{n}$, where $\boldsymbol{n}$ is a unit vector, and take the limit $a \to 0$ with fixed p and fixed $\boldsymbol{n}$. This yields the charge density due to a point dipole $\boldsymbol{p} = (p/a)\boldsymbol{a} = (p/a)(a\boldsymbol{n}) = p\boldsymbol{n}$, situated at $\boldsymbol{r}'$:

$$\rho = (\text{charge density at } \boldsymbol{r}, \text{ due to a point dipole } \boldsymbol{p} \text{ at } \boldsymbol{r}')$$
$$= -\boldsymbol{p}\cdot\nabla\,\delta(\boldsymbol{r}-\boldsymbol{r}') = \boldsymbol{p}\cdot\nabla'\,\delta(\boldsymbol{r}-\boldsymbol{r}'). \tag{1.4.25}$$

The first expression is the obvious 3D analogue of (1.2.17); the second follows from (1.4.21).

Exercise: Derive (1.4.25) by taking the limit $\boldsymbol{a} \to 0$ of the Taylor expansion of (1.4.24) in powers of $\boldsymbol{a}$.

Typically, $\rho(\boldsymbol{r})$ is used to calculate the potential $\psi(\boldsymbol{s})$ at the point $\boldsymbol{s}$, due to the point dipole $\boldsymbol{p}$ at $\boldsymbol{r}'$. From Coulomb's law (anticipating Chapter 4) one has $4\pi\varepsilon_0\psi(\boldsymbol{s}) = \int d^3r\rho(\boldsymbol{r})/|\boldsymbol{r}-\boldsymbol{s}|$; thus $1/|\boldsymbol{r}-\boldsymbol{s}|$ plays the role of $f(\boldsymbol{r})$ in (1.4.22–23). Accordingly, using say the rightmost form of (1.4.25),

$$4\pi\varepsilon_0\psi(\boldsymbol{s}) = \int d^3r\,(\boldsymbol{p}\cdot\nabla'\,\delta(\boldsymbol{r}-\boldsymbol{r}'))/|\boldsymbol{r}-\boldsymbol{s}|$$

$$= (\boldsymbol{p}\cdot\nabla')\int d^3r\,\delta(\boldsymbol{r}-\boldsymbol{r}')/|\boldsymbol{r}-\boldsymbol{s}| = (\boldsymbol{p}\cdot\nabla')1/|\boldsymbol{r}'-\boldsymbol{s}|$$

$$= -\boldsymbol{p}\cdot(\boldsymbol{r}'-\boldsymbol{s})/|\boldsymbol{r}'-\boldsymbol{s}|^3 = \boldsymbol{p}\cdot(\boldsymbol{s}-\boldsymbol{r}')/|\boldsymbol{r}'-\boldsymbol{s}|^3. \qquad (1.4.26)$$

This is one of the well-known forms of the dipole potential. (Dipoles and higher multipoles are discussed further in Sections 2.6, 4.4.4, and 4.4.5.)

1.5 A caution regarding the definition of $\delta(x)$

The classic definition of the delta function, going back to Dirac, is eqn (1.1.1). Nevertheless it is incomplete:† one need merely observe that both these requirements are satisfied equally well by any linear combination $\{\delta(x) + c_1\delta'(x) + c_2\delta''(x) + \cdots\}$ of $\delta(x)$ itself and of its derivatives, with arbitrary coefficients for the latter. Loosely speaking, the point is that the derivatives of $\delta(x)$ do vanish for $x \neq 0$, and do have zero area, but cannot in general be ignored just on this account. To exclude such admixtures, one must supplement (1.1.1) with the additional constraints

$$\int_{-\eta_1}^{\eta_2} dx\, x^n\delta(x) = 0, \qquad n = 1, 2, 3, \ldots \qquad (1.5.1)$$

In view of (1.1.18), the constraint with $n = 1$ eliminates admixtures proportional to $\delta'(x)$, and so on. The alternative definition (1.1.4) is in no need of such refinement; therefore it becomes preferable in strict logic, though perhaps less informative on the intuitive level in a first encounter.

† The writer is grateful to Professor H. Genz for this observation.

Problems

1.1 Evaluate

(i) $\int_{-1}^{2} \mathrm{d}x\, \delta(x) \cos(2x)$; (ii) $\int_{-1}^{2} \mathrm{d}x\, \delta(2x) \cos(x)$;

(iii) $\int_{-\infty}^{\infty} \mathrm{d}x\, \delta'(x) \exp(ix)$; (iv) $\int_{0}^{\infty} \mathrm{d}x\, \delta'(2\tfrac{1}{2}x - 1) \tan^{-1} x$.

1.2 Evaluate

$$\int_0^1 \mathrm{d}x \int_0^1 \mathrm{d}y\, \delta(x-y)$$

(i) by repeated integration; or (ii) by changing variables to $\xi = (x+y)/\sqrt{2}$, $\eta = (x-y)/\sqrt{2}$.

1.3 Evaluate the integral

$$\int_0^\infty \mathrm{d}x\, \delta(\cos x) \exp(-x),$$

and verify that it equals $1/2 \sinh(\pi/2)$.

1.4 If $0 \leqslant x,\ y \leqslant 1$, determine the coefficients $c_n(y)$ in the expansion

$$\delta(x-y) = \sum_n c_n P_n(x),$$

where the $P_n(x)$ are the Legendre polynomials. You may quote

$$\int_{-1}^{1} \mathrm{d}x\, P_n(x) P_m(x) = 2\, \delta_{nm}/(2n+1).$$

1.5 Differentiate $|x|$, $\sin|x|$, $\cos|x|$.

Hint: Recall that $|x| = x\varepsilon(x)$.

1.6 (i) Determine $\psi(x)$ for all x, given that $\psi(-\infty) = 0$, and

$$\frac{\mathrm{d}\psi}{\mathrm{d}x} = \exp(-q\,|x|) + \delta(x).$$

(ii) By integrating the equation directly from $-\infty$ to $+\infty$, determine $\psi(+\infty)$, and check that it agrees with the value given by part (i).

1.7 Determine $\psi(x)$ for all x in terms of $\psi(-\infty)$, given that

$$\frac{d\psi}{dx} - \exp(-q\,|x|)\psi = \delta(x).$$

If $\psi(-\infty) = \exp(-3/q)$, verify that $\psi(+\infty) = 2\cosh(1/q)$.

1.8 Determine the reflection and transmission amplitudes $R(k)$ and $T(k)$ of the delta-function potential by finding that solution of

$$-\frac{\hbar^2}{2m}\frac{d^2\psi}{d^2x} + \frac{\hbar^2}{2m}\lambda\,\delta(x)\psi = \frac{\hbar^2k^2}{2m}\psi$$

which behaves like $\exp(ikx) + R\exp(-ikx)$ when $x \to -\infty$, and like $T\exp(ikx)$ when $x \to +\infty$. Verify that $|T|^2 + |R|^2 = 1$, and that $R(0) = -1$, $T(\infty) = 1$.

1.9 Evaluate

$$\lim_{K_1\to\infty}\lim_{K_2\to\infty}\int_{-\infty}^{\infty} dx \int_{-\infty}^{\infty} dy \exp(-(x^2+y^2)^{\frac{1}{2}})$$
$$\times \int_{-K_1}^{K_1} dk_1 \exp(ik_1(x-a)) \int_{-K_2}^{K_2} \exp(ik_2(y-b))$$

Hint: Start by implementing the two limits.

1.10 Evaluate
(i) $\int d^3r \exp(-\alpha r^2)\,\delta(r)$;
(ii) $\int d^3r \exp(-\alpha r^2)\,\delta(\mathbf{r})$;
(iii) $\int d^3r \exp(-\alpha r^2)\,\delta'(r)/r^n$, $\quad n = 0, 1, 2.$

Hint: In (iii), with $n = 0, 1$, it may help to use $\delta'(r) = -\delta(r)/r$. The results read 0, 4π respectively. The case with $n = 2$ requires careful thought.

1.11 Evaluate

$$\int d\Omega\, z^2\,\delta(\mathbf{r} - \mathbf{s}),$$

where $\mathbf{s} = (s_1, s_2, s_3)$ is a fixed vector, and the integral is taken over the solid angle of $\mathbf{r} \equiv (x, y, z) \equiv (r, \Omega)$. Express your result in terms of r and of s_1, s_2, s_3.

1.12 Evaluate

$$\int d^n r \frac{1}{(r^2+a^2)} \nabla\,\delta(\mathbf{r} - \mathbf{b}).$$

1.13 Verify that the equation $xf(x) = g(x)$ for $f(x)$ is satisfied by $f(x) = g(x)/x + A\,\delta(x)$, where A is an arbitrary constant.

1.14 Construct a representation of $\delta(x)$ by using the function $\sin^2(x/\varepsilon)/x^2$.

2 Ordinary differential equations

Summary

Second-order linear equations:

- Homogeneous:

$$-(\mathrm{d}^2/\mathrm{d}x^2 + q(x)\,\mathrm{d}/\mathrm{d}x + r(x))\phi \equiv L\phi = 0; \qquad (*)$$

- Inhomogeneous:

$$L\psi = f(x). \qquad (**)$$

Wronskian: ϕ_1 and ϕ_2 solve (*). $W(\phi_1, \phi_2) \equiv \phi_1\phi_2' - \phi_1'\phi_2 \equiv W(x)$.

- If ϕ_1 and ϕ_2 are linearly independent, $W = \text{const.}\exp(-\int \mathrm{d}xq) \neq 0$. [Special case $q = 0 \Rightarrow W(x) = \text{constant}$.]
- If ϕ_1 and ϕ_2 are linearly dependent, $W = 0$.

General solution of (**): $\psi = \phi_c + \psi_p$. The complementary function ϕ_c solves (*) and contains two adjustable parameters. The particular integral ψ_p is *any* solution of (**).

Green's function: $LG(x\,|\,x') = \delta(x - x')$; then

$$\psi_p(x) = \int \mathrm{d}x'\, G(x\,|\,x')f(x'). \qquad (***)$$

Boundary conditions on G are chosen to suit the problem.

Initial-value problem: Equation (**), plus $\psi(x_1) = a$, $\psi'(x_1) = u$. Let ϕ_1 solve (*) and satisfy $\phi_1(x_1) = a$, $\phi_1'(x_1) = u$. Let ϕ_2 be *any* solution of (*) linearly independent of ϕ_1. Then $\phi_c = \phi_1$,

$$G(x\,|\,x') = \theta(x - x')\{\phi_2(x')\phi_1(x) - \phi_1(x')\phi_2(x)\}/W(x'),$$

and ψ_p is given by (***).

Boundary-value problem: Consider only homogeneous BCs,

- either DBCs: $\psi(x_1) = 0 = \psi(x_2)$,
- or NBCs: $\psi'(x_1) = 0 = \psi'(x_2)$.

Dirichlet BVP: (see Table 2.1).

- GENERAL CASE: (*) with these BCs has no solution. Let ϕ_1 solve (*) with $\phi_1(x) = 0$, and let ϕ_2 solve (*) with $\phi_2(x_2) = 0$. Then ϕ_1 and ϕ_2 are linearly independent. The unique solution of (**) is

ψ_p given by (***), with

$$G(x \mid x') = -\{\theta(x' - x)\phi_2(x')\phi_1(x) + \theta(x - x')\phi_1(x')\phi_2(x)\}/W(x').$$

- SPECIAL CASE: (*) with these BCs does have a solution: ϕ_1 and ϕ_2 are linearly dependent: $W = 0$.

– *General special case:* (**) has no solution.

– *Special special case:* (**) does have a solution (subject to a special condition on $f(x)$); but this solution is not unique, because any multiple of ϕ_1 may be added to it.

2.1 Introduction

This chapter introduces, in a simple context, several ideas that will be in constant use later on. Especially important are the distinction between homogeneous and inhomogeneous problems; the concept of the Green's function; and the expansions constructed in Section 2.4, which recur essentially unchanged wherever the operator $-\nabla^2$ plays a role.

2.1.1 Some basic ideas and definitions

We shall be concerned only with linear second-order equations, both homogeneous and inhomogeneous.

Second order means no derivatives higher than the second.

Linearity is the essential restriction: $\psi(x)$, $\psi'(x)$, $\psi''(x)$ may appear, but only linearly, i.e. each only to the power 0 or 1, and no products of them: e.g. ψ^2, $\psi^{\frac{1}{2}}$, $\psi\psi''$ are all barred. (By contrast, generalization to higher order is straighforward at least in principle, while most first-order problems are very simple.)

Thus, without further loss of generality, we define the differential operator L and consider the equation

$$L \equiv -\left\{\frac{d^2}{dx^2} + q(x)\frac{d}{dx} + r(x)\right\}, \tag{2.1.1}$$

$$L\psi(x) = f(x), \tag{2.1.2}$$

where q, r, f are given functions of x. (The connection with standard Sturm–Liouville theory and notation is straightforward, but is not pursued here.) If necessary, the independent variable with respect to which L acts is indicated by a prime or a suffix: thus $L' = -\{d^2/dx'^2 + q(x')\,d/dx' + r(x')\}$. The minus sign in the definition of L is merely a convention inspired by Poisson's equation, and harmonizes the notation with Chapters 4–13.

Unless otherwise stated, $q(x)$ and $r(x)$ are taken as well-behaved; any singularities of these functions will be dealt with *ad hoc*.

Recall that the most general solution of (2.1.2) is

$$\psi = (\text{complementary function}) + (\text{particular integral})$$
$$\equiv \text{CF} + \text{PI} \equiv \phi_c + \psi_p, \tag{2.1.3}$$

where the PI can be *any* function that obeys (2.1.2), while the CF obeys the corresponding homogeneous equation

$$L\phi_c = 0. \tag{2.1.4}$$

An equation (more generally: a problem, as explained below) is *homogeneous* if any solution remains a solution when multiplied by an arbitrary constant. Homogeneity for linear equations implies the *superposition principle,* which means that not only numerical multiples of solutions, but also arbitrary linear combinations of solutions, are themselves solutions. The term $f(x)$ manifestly destroys homogeneity and the superposition principle: equations with non-zero f are *inhomogeneous.*

The complementary function contains two adjustable parameters (integration constants: two because the equation is of second order), which must be fixed by fitting two initial or two boundary conditions, ICs or BCs respectively. (Frequent abbreviations are listed in Appendix A.) Without such conditions the physics is not fully specified and the mathematical solution is not unique.

Initial-value problems (IVPs) require ψ either for $x \geqslant x_1$ or for $x \leqslant x_1$, and prescribe both ψ and ψ' at one and the same point x_1;

$$\psi(x_1) = a, \qquad \psi'(x_1) = u. \tag{2.1.5}$$

Boundary-value problems (BVPs) require ψ for $x_1 \leqslant x \leqslant x_2$ (it can happen that $x_1 \to -\infty$ or $x_2 \to +\infty$ or both); they usually prescribe *either* $\psi(x_1) = a$ and $\psi(x_2) = b$ (Dirichlet BCs, abbreviated DBCs, defining a Dirichlet problem), *or* $\psi'(x_1) = u$ and $\psi'(x_2) = v$ (Neumann BCs, abbreviated NBCs, defining a Neumann problem). Other variants occur, but less often: one could prescribe a linear combination $(\alpha\psi + \beta\psi')$ at the endpoints (Churchill BCs), or prescribe ψ at one endpoint and ψ' at the other (mixed BCs). These variants will not be pursued here.

Both mathematically and physically, IVPs and BVPs are very different. All IVPs have solutions; some BVPs have none.

A *problem* (as opposed to just an equation) is *homogeneous* only if $f = 0$ in (2.1.2) *and also* the prescribed initial or boundary values are zero. Otherwise the problem is *inhomogeneous* (i.e. it is so if any or all of f and the initial or boundary values are non-zero).

Homogeneous problems always have the so-called trivial solution

$\psi = 0$, which we shall always ignore: the statement that a homogeneous problem has no solution always means that it has none beside the trivial one. (Indeed homogeneous problems are *recognizable* as those that have the trivial solution.) The superposition principle applies only to the solutions of homogeneous problems.

Four preliminary examples follow: their variety should be noted, physical and mathematical.

(i) IVP: $q(x) = \gamma \equiv \eta/m$ and $r(x) = \omega^2$ are positive constants; x is time; $\psi(x)$ is the position of a particle having mass m; writing $-f(x) = F(x)/m$, one has

$$\psi'' + \gamma\psi' + \omega^2\psi = F(x)/m. \tag{2.1.6}$$

This is Newton's second law for a harmonic oscillator subject to a frictional force $-\eta\psi'$, with undamped (circular) frequency ω, and experiencing an applied force $F(x)$ which may be an arbitrary function of time. To determine the motion completely, the initial position and velocity must be given. There is always a solution, and it is unique.

(ii) BVP: $q = 0 = r$; x is position, and $\psi(x)$ the scalar potential,† due to a static source distribution $f(x) = \rho(x)$ that depends only on x:

$$\mathrm{d}^2\psi/\mathrm{d}x^2 = -\rho(x). \tag{2.1.7}$$

This is Poisson's equation in 1D. To specify the potential fully, we need BCs or ICs. One common problem has two conducting planes at $x = x_1$ and $x = x_2$, maintained at fixed potentials (possibly zero), whence $\psi(x_1)$ and $\psi(x_2)$ are given.

Exercise: Suppose (2.1.7) is complemented by the NBCs $\psi'(x_1) = u$, $\psi'(x_2) = v$. Show that the problem can have no solution unless $u - v = \int_{x_1}^{x_2} \mathrm{d}x\, \rho(x)$.

(iii) BVP: $q = 0$, $r = 2m[E - V(x)]/\hbar^2$; x is position, and ψ the x-dependent factor of an energy eigenfunction $\Psi(x, t) = \exp(-iEt/\hbar)\psi(x)$ in quantum mechanics, for a particle having mass m, energy E, and experiencing a potential $V(x)$:

$$\mathrm{d}^2\psi/\mathrm{d}x^2 + (2m/\hbar^2)[E - V(x)]\psi = 0. \tag{2.1.8}$$

If this equation is supplemented by the boundary conditions $\psi(\pm\infty) = 0$,

† Electrostatics is covered by our notation if we denote the charge density by ρ, and identify $\psi(\boldsymbol{r})$ with $\varepsilon_0\Phi(\boldsymbol{r})$, where $\Phi(\boldsymbol{r})$ is the electrostatic potential expressed in SI units. Then the electric field is $\boldsymbol{E} = -\mathrm{grad}\,\Phi = -\mathrm{grad}\,\psi/\varepsilon_0$; $\nabla^2\psi = -\rho$ implies the correct electrostatic form of Poisson's equation, namely $\nabla^2\Phi = -\rho/\varepsilon_0$; and the energy density is $u = \frac{1}{2}\varepsilon_0\boldsymbol{E}^2 = \frac{1}{2}\varepsilon_0(\nabla\Phi)^2 = (1/2\varepsilon_0)(\nabla\psi)^2$. However, we shall generally refer to ψ simply as the potential, to $-\nabla\psi$ as the field, and to $\frac{1}{2}(\nabla\psi)$ as the energy density.

it becomes the Schroedinger energy-eigenvalue problem for bound states; this has solutions only sometimes, i.e. only for special values of the parameter E. Note that this problem is homogeneous.

(iv) BVP: Equation (2.1.8), with $V(x)$ vanishing fast enough at infinity (at least as fast as $1/|x|^{1+\varepsilon}$, $\varepsilon > 0$) and with any given $E = \hbar^2 k^2/2m > 0$. Then the equation entails the asymptotic behaviour

$$\begin{aligned} \psi(x \to -\infty) &\sim A \exp(ikx) + B \exp(-ikx), \\ \psi(x \to +\infty) &\sim C \exp(ikx) + D \exp(-ikx). \end{aligned} \tag{2.1.9}$$

The equation itself eventually delivers two relations between the four constants A, B, C, D; thus, two more constraints must be prescribed by BCs reflecting the physics. To represent particles incident only from the left, one demands $D = 0$, and, purely for convenience, chooses $A = 1$. Then $B/A = B = R$ and $C/A = C = T$ are the reflection and transmission amplitudes, and $|R|^2$ and $|T|^2$ the ratios of the reflected and transmitted to the incident intensity. Though the equation is homogeneous, the problem is inhomogeneous, because of the prescribed coefficient $A = 1$ of the incident wave. The solution is unique.

Finally we need the notion of *linearly independent* solutions. Suppose $\phi_1(x)$ and $\phi_2(x)$ both solve the homogeneous equation (2.1.4): $L\phi_1 = 0 = L\phi_2$. (In Sections 2.1–3 only, we reserve the letters $\phi(\psi)$ for various solutions of the homogeneous (inhomogeneous) equation: thus $L\phi = 0$, $L\psi = f$.) We call them linearly independent (of each other) if $c_1\phi_1(x) + c_2\phi_2(x) = 0$ for all x, with c_1 and c_2 constants, implies $c_1 = 0 = c_2$.

Otherwise, ϕ_1 and ϕ_2 are *linearly dependent*; this means that each is a constant multiple of the other, neither c_1 nor c_2 being zero.

Theorem: The homogeneous equation has two and only two linearly independent solutions. (For a proof, see any good book on differential equations.)

Given any two solutions ϕ_1, ϕ_2, their *Wronskian* is defined by

$$W\{\phi_1, \phi_2\} \equiv \phi_1\phi_2' - \phi_1'\phi_2. \tag{2.1.10}$$

Theorem: If ϕ_1 and ϕ_2 are linearly dependent, then W is identically zero, i.e. $W = 0$ for all x.

Exercise: Prove this.

Theorem: If ϕ_1 and ϕ_2 are linearly independent, then $W\{\phi_1, \phi_2\} \neq 0$ for *any* x. Thus, evaluating W (for any x) is a test for linear independence.

Proof: The trick is to consider $\mathrm{d}W/\mathrm{d}x$:

$$\frac{\mathrm{d}W}{\mathrm{d}x} = \phi_1'\phi_2' + \phi_1\phi_2'' - \phi_1''\phi_2 - \phi_1'\phi_2' = \phi_1\phi_2'' - \phi_1''\phi_2. \tag{2.1.11}$$

Substitute for ϕ_2'' and ϕ_1'' from (2.1.4), with L from (2.1.1):

$$\frac{\mathrm{d}W}{\mathrm{d}x} = \phi_1(-q\phi_2' - r\phi_2) - (-q\phi_1' - r\phi_1)\phi_2$$
$$= -q(\phi_1\phi_2' - \phi_1'\phi_2) = -q(x)W(x). \tag{2.1.12}$$

Since ϕ_1 and ϕ_2 are, by assumption, linearly independent, W need not be identically zero, and we require the non-trivial solution:

$$\int \mathrm{d}W/W = \log W = -\int \mathrm{d}x\, q(x), \qquad W(x) = \exp\left\{-\int \mathrm{d}x q(x)\right\}. \tag{2.1.13}$$

This is *Abel's formula*. The indefinite integral in (2.1.13) is understood as $\int_a^x \mathrm{d}x' q(x')$, where a is an arbitrary constant. Since, by assumption, $q(x)$ is non-singular, $\int \mathrm{d}xq(x)$ is never infinite; hence the RHS cannot vanish for any x, and $W \neq 0$. ■

Often one meets the simple case $q = 0$ (for all x). Then, by (2.1.13),

$$W(x) = \text{constant} \qquad (\text{if } q = 0). \tag{2.1.14}$$

This applies for instance in examples (ii–iv) above, but not in example (i) unless damping is negligible so that $\gamma = 0$.

2.1.2 Green's functions and the matching conditions

We concentrate on methods for solving (2.1.2) with given $q(x)$, $r(x)$, and given BCs or ICs, but with arbitrary $f(x)$. In a sense we wish to answer questions about the effects of $f(x)$ before they are even asked, by setting up machinery that can process any $f(x)$ prescribed subsequently.

The basic idea is to consider a standardized problem related to (2.1.1), namely

$$LG(x \mid x') = -\{\partial^2/\partial x^2 + q(x)\,\partial/\partial x + r(x)\}G(x \mid x')$$
$$= \delta(x - x'), \tag{2.1.15}$$

where $G(x \mid x')$ is a *Green's function,* still to be subjected to ICs or BCs. Its dimensions are $[G(x \mid x')] = [x]^2[\delta(x)] = [x]$; in nD, this becomes $[G] = [x]^{-n+2}$. G delivers a particular integral: provided that $x_1 < x$, $x' < x_2$, one has

$$\psi_p(x) = \int_{x_1}^{x_2} \mathrm{d}x'\, G(x \mid x')f(x'). \tag{2.1.16}$$

Proof:

$$L\psi_p(x) = L\int_{x_1}^{x_2} dx'\, G(x\,|\,x')f(x') = \int_{x_1}^{x_2} dx'\,\{LG(x\,|\,x')\}f(x')$$

$$= \int_{x_1}^{x_2} dx'\,\delta(x-x')f(x') = f(x). \qquad \blacksquare$$

The one essentially novel requirement for solving (2.1.15) for G is to ensure, in the immediate vicinity of $x = x'$, that the expression on the left matches the delta-function on the right; Sections 2.2 and 2.3 provide illustrations. The basic idea stems from the hierarchy of singularities (Section 1.1.3, point (xiv)), and in particular from the observation that the highest derivative of a generalized function is the most singular. This suggests that, in (2.1.15), the delta-function must be matched by the second derivative $\partial^2 G/\partial x^2$. If so, then the first derivative $\partial G/\partial x$ will, like $\varepsilon(x-x')$ or $\theta(x-x')$, have a discontinuity but no infinity, and G itself will be continuous at $x = x'$.

It is easily shown by *reductio ad absurdum* that, indeed, G cannot be discontinuous (i.e., in terms of the hierarchy of singularities, that the most singular component of G is proportional to $|x-x'|$.

Proof: Suppose, on the contrary, that G is discontinuous, i.e. that it has a singular component proportional to $\varepsilon(x-x')$. Then, near $x = x'$, we can write $G = F(x) + \Gamma(x, x')$, where $F(x)$ and its first and second derivatives are continuous, while $\Gamma(x, x') \equiv \frac{1}{2}\varepsilon(x-x')\,\Delta$, with Δ the discontinuity of G (i.e. $\Delta = G(x'+\,|\,x') - G(x'-\,|\,x')$). In the present argument about singularities, F can be ignored. Differentiation yields $\partial\Gamma/\partial x = \delta(x-x')\,\Delta$, and $\partial^2\Gamma/\partial x^2 = \delta'(x-x')\,\Delta$, so that $\delta'(x-x')\,\Delta$ emerges as the strongest singularity on the left of (2.1.15). But there is no term to match this on the right; therefore $\Delta = 0$, and G cannot have been discontinuous after all. $\blacksquare$

To apply these ideas, integrate both sides of (2.1.15) with respect to x from $x' - \eta_1$ to $x' + \eta_2$, and then let $\eta_1 \to 0$, $\eta_2 \to 0$. For short, this is called integration from $x'-$ to $x'+$, and written $\int_{x'-}^{x'+} dx \cdots$. In the limit there is no contribution to the integral from the second and third terms on the left, because these integrands remain finite while the width of the integration region shrinks to zero. Only the first term survives, and one finds the *jump condition*

$$\int_{x'-}^{x'+} dx\left\{\frac{\partial^2}{\partial x^2} + q\frac{\partial}{\partial x} + r\right\}G(x\,|\,x') = \int_{x'-}^{x'+} dx\,\frac{\partial^2 G}{\partial x^2}$$

$$= \left.\frac{\partial G(x\,|\,x')}{\partial x}\right|_{x=x'+} - \left.\frac{\partial G(x\,|\,x')}{\partial x}\right|_{x=x'-}$$

$$= -\int_{x'-}^{x'+} dx\,\delta(x-x') = -1,$$

or in other words

$$\left.\frac{\partial G(x\,|\,x')}{\partial x}\right|_{x=x'+} - \left.\frac{\partial G(x\,|\,x')}{\partial x}\right|_{x=x'-} = -1. \tag{2.1.17}$$

Beside this we have the *continuity condition*

$$G(x'+\,|\,x') - G(x'-\,|\,x') = 0. \tag{2.1.18}$$

The matching process always consists of enforcing the continuity and the jump conditions. Equations (2.1.17, 18) follow from (2.1.15) alone: they are general truths, and apply equally to the Green's functions for IVPs (Section 2.2) and for BVPs (Section 2.3).

2.2 The Green's function for initial-value problems

2.2.1 Definition of the Green's function

Choose $x = 0$ at the initial point, and write the ICs as

$$\psi(0) = a, \qquad \psi'(0) = u. \tag{2.2.1}$$

The problem is to find a solution of $L\psi(x) = f(x)$ subject to (2.2.1), valid for all $x \geqslant 0$, for arbitrary $f(x)$.

The strategy is to satisfy (2.2.1) with an appropriate complementary function ϕ_c ($L\phi_c = 0$), and to define a Green's function so that it delivers a particular integral that does not again upset the ICs. Thus we require

$$\psi = \phi_c + \psi_p, \tag{2.2.2}$$

$$\phi_c(0) = a, \qquad \phi_c'(0) = u, \tag{2.2.3}$$

$$\psi_p(0) = 0, \qquad \psi_p'(0) = 0. \tag{2.2.4}$$

In the special case of homogeneous ICs, i.e. $a = 0 = u$, one can set $\phi_c = 0$, i.e. one needs no complementary function at all.

We write the particular integral as

$$\psi_p(x) = \int_0^\infty dx'\; G(x\,|\,x')f(x'), \tag{2.2.5}$$

where G obeys (2.1.15) (like all Green's functions); in order to guarantee (2.2.4) irrespective of f, we demand in addition

$$G(0\,|\,x') = 0, \qquad \frac{\partial}{\partial x}G(x\,|\,x')\big|_{x=0} \equiv G_x(0\,|\,x') = 0. \tag{2.2.6}$$

Clearly, for the present we are considering $G(x\,|\,x')$ as a function of x, with x' merely a parameter.

Notice that the ICs on G are always homogeneous, irrespective of the ICs on ψ and on ϕ_c.

2.2.2 Construction of the Green's function

$G(x \mid x')$ is defined by (2.1.15) and (2.2.6). To construct it, observe that it obeys $LG = 0$ for all x except $x = x'$. For $x < x'$ and $x > x'$, G can therefore be expressed as a linear combination of any pair ϕ_1, ϕ_2 of linearly independent solutions of $L\phi = 0$, though generally a different combination in the two regions. The pair ϕ_1, ϕ_2 can be chosen freely, without any reference to ϕ_c (though of course ϕ_c is necessarily expressible as a linear combination of them).

We deal separately with the two regions $x \lessgtr x'$, and then match the solutions across $x = x'$.

In the region $0 \leq x < x'$,

$$G(x \mid x') = 0, \qquad (x < x'). \tag{2.2.7}$$

Proof: $G = 0$ satisfies the equation $LG = 0$ (trivially), and it satisfies both the ICs at $x = 0$. But equation plus ICs define the solution uniquely. ■

In the region $x' < x$, write

$$G(x \mid x') = A\phi_1(x) + B\phi_2(x), \qquad (x' < x). \tag{2.2.8}$$

The solutions (2.2.7, 8) are matched across $x = x'$ as described in Section 2.1.2. Loosely speaking, this ensures that G obeys the differential equation at $x = x'$ (as well as for $x \gtrless x'$).

In our present IVP, (2.2.7) shows that $G(x \mid x')$ vanishes for all $x < x'$; so consequently does $\partial G(x \mid x')/\partial x$. In particular, this is still true at $x = x'-$; therefore the continuity and jump conditions (2.1.18, 17) reduce to

$$G(x'+ \mid x') = 0, \qquad \left.\frac{\partial G(x \mid x')}{\partial x}\right|_{x=x'+} = -1. \tag{2.2.9a, b}$$

It remains only to determine A and B by satisfying (2.2.9a, b). They give, respectively, $A\phi_1(x') + B\phi_2(x') = 0$ and $A\phi_1'(x') + B\phi_2'(x') = -1$, or in other words

$$\begin{bmatrix} \phi_1(x'), & \phi_2(x') \\ \phi_1'(x'), & \phi_2'(x') \end{bmatrix} \begin{bmatrix} A \\ B \end{bmatrix} = \begin{bmatrix} 0 \\ -1 \end{bmatrix}. \tag{2.2.10}$$

These are two simultaneous, linear, inhomogeneous algebraic equations for the two unknowns A and B. Thus there exists a unique solution provided only that $\det[\;] = W\{\phi_1(x'), \phi_2(x')\} \equiv W(x') \neq 0$, which is so, simply because ϕ_1, ϕ_2 are linearly independent (cf. the last theorem in

Section 2.1.1). The solution is

$$A = \phi_2(x')/W(x'), \qquad B = -\phi_1(x')/W(x'). \tag{2.2.11}$$

We amalgamate (2.2.7) and (2.2.8) into $G(x \mid x') = \theta(x - x') \times \{A\phi_1(x) + B\phi_2(x)\}$, substitute for A and B from (2.2.11), and write the result explicitly:

$$\begin{aligned} G(x \mid x') &= \theta(x - x')\{\phi_2(x')\phi_1(x) - \phi_1(x')\phi_2(x)\}/W \\ &= \theta(x - x')\frac{\{\phi_2(x')\phi_1(x) - \phi_1(x')\phi_2(x)\}}{\{\phi_1(x')\phi_2'(x') - \phi_2(x')\phi_1'(x')\}}. \end{aligned} \tag{2.2.12}$$

(Be careful to distinguish primes (on x') from derivatives (ϕ').)

Thus the solution to our IVP reads

$$\psi(x) = \phi_c(x) + \int_0^\infty dx'\, \theta(x - x')G(x \mid x')f(x'), \tag{2.2.13}$$

$$\psi(x) = \phi_c(x) + \int_0^x dx'\, G(x \mid x')f(x'); \tag{2.2.14}$$

the second form follows because the step-function in the first form simply restricts the integration variable to $x' < x$.

Exercise: Verify by explicit substitution that (2.2.14) satisfies both the differential equation and the ICs. When differentiating the integral one must remember that d/dx acts both on the integration limit and on the factor G of the integrand.

2.2.3 Comments

It is important to realize that nothing can get in the way of the explicit construction culminating in (2.2.13, 14): it always works, i.e. we have solved the most general linear non-singular IVP. (That this is possible in principle is not so surprising, since $\psi(0)$ and $\psi'(0)$ suffice to integrate the equation rightwards from $x = 0$, as one learns in numerical analysis.) 'Solution' here means 'reduction to quadratures', i.e. to the evaluation of definite integrals: $G(x \mid x')$ can be calculated once and for all, numerically if necessary; then, once $f(x)$ is given, we need merely evaluate the integral $\int_0^x dx'\, G(x \mid x')f(x')$, numerically if necessary; but we do not at this stage need to solve equations any more. Likewise, ϕ_c can be determined, irrespective of f, once the initial values are given.

To appreciate the physical implications, consider ψ as the position of a unit-mass damped oscillator, as in Section 2.1.1, example (i). For simplicity suppose it at rest at the origin, i.e. in stable equilibrium, at time $x = 0$, so that, in (2.2.1), $a = 0 = u$. Then the complementary function ϕ_c is zero. Now let a unit impulse be delivered at time x'; this means $F(x) = -f(x) = \delta(x - x')$. Then $\psi = 0$ for $x < x'$, followed by motion for $x > x'$ according to $\psi(x) = -G(x \mid x')$.

For arbitrary f, we can regard $F(x')\,dx' = -f(x')\,dx'$ as the impulse delivered in the time interval dx' around x'. The resulting motion $\int_0^x dx' f(x') G(x \mid x')$ is just the linear superposition of the motions resulting from all these successive impulses up to the present time x. This is so because, and only because, our equation is linear.

The property that $G(x \mid x') = 0$ for $x < x'$ embodies *causality*: the effect (non-zero ψ and ψ') cannot precede the *cause* (the impulse at time x').

Functions like G are also called 'response functions' or 'influence functions'.

The key results (2.2.12, 14) are illustrated by the worked example of the undamped harmonic oscillator, Section 2.5.1. In addition, Appendix D describes what is perhaps the most spectacular application of this technique in physics, the Einstein–Langevin theory of Brownian motion.

2.3 The Green's function for boundary-value problems

For BVPs, unlike IVPs, a Green's function may not exist, either because the problem has no solution, or because the solution is not unique. The situation becomes particularly clear in the special but important case where the differential operator L is Hermitean, as discussed in Section 2.4.

For simplicity, in the present section we confine ourselves to homogeneous DBCs

$$\psi(x_1) = 0 = \psi(x_2). \tag{2.3.1}$$

(NBCs can lead to slight complications, of the kind encountered with Poisson's equation in Chapter 6; these will cause no difficulties of principle once Section 2.4 has been understood. The generalization to inhomogeneous BCs (where the prescribed values of $\psi(x_1)$ or $\psi(x_2)$ are not both zero) can be supplied by the methods used later to derive the magic rules in Chapters 5 and 6.)

The problem is to solve, in the range $x_1 \leqslant x \leqslant x_2$,

$$L\psi(x) = f(x) \tag{2.3.2}$$

subject to (2.3.1), for arbitrary $f(x)$. It will appear presently that the existence and nature of the solution depend critically on whether the associated homogeneous problem

$$L\phi = 0 \tag{2.3.3}$$

subject to the same (homogeneous) BCs has a solution. We shall distinguish between the *general case,* where (2.3.3) has no solution, and the *special case,* where it has.

The strategy is to look for ψ in the form

$$\psi(x) = \int_{x_1}^{x_2} \mathrm{d}x' \, G(x \mid x') f(x') \tag{2.3.4}$$

where

$$LG(x \mid x') = \delta(x - x'), \quad G(x_1 \mid x') = 0, \quad G(x_2 \mid x') = 0. \tag{2.3.5a, b, c}$$

This G obeys the same differential equation as did the (different) G in the IVP, but now it satisfies the BCs (2.3.5b, c) inspired by (2.3.1).

One should watch chiefly for the differences between the following procedure and that for IVPs in Section 2.2. Until further notice, we continue to regard G as a function of x, with x' just a parameter.

As before, we construct G from the solutions of the homogeneous equation (2.3.3), which is satisfied by G for $x < x'$ and $x > x'$. Let ϕ_1 be a solution that obeys the left-hand BC; similarly, let ϕ_2 be a solution that obeys the right-hand BC:

$$L\phi_{1,2} = 0, \qquad \phi_1(x_1) = 0, \qquad \phi_2(x_2) = 0. \tag{2.3.6a, b, c}$$

We do not yet know whether ϕ_1 and ϕ_2 are linearly independent.

Comparing (2.3.6) with (2.3.5) we see that G is of the form

$$G(x \mid x') = \theta(x' - x) C\phi_1(x) + \theta(x - x') D\phi_2(x), \tag{2.3.7}$$

since this satisfies both the BCs, whatever the values of the constants C and D (which can of course depend on x', though not on x).

C and D are determined by matching across $x = x'$, exactly as for the IVP in Section 2.2. The continuity and jump conditions (2.1.18, 17) are identically the same as before; they yield, respectively, $D\phi_2(x') - C\phi_1(x') = 0$, $D\phi_2'(x') - C\phi_1'(x') = -1$. Thus

$$\begin{bmatrix} -\phi_1(x'), & \phi_2(x') \\ -\phi_1'(x'), & \phi_2'(x') \end{bmatrix} \begin{bmatrix} C \\ D \end{bmatrix} = \begin{bmatrix} 0 \\ -1 \end{bmatrix}. \tag{2.3.8}$$

Naturally one hopes that (2.3.8) has a unique solution. The condition for this is

$$\begin{aligned} \det[\;] &= \{-\phi_1(x')\phi_2'(x') + \phi_2(x')\phi_1'(x')\} \\ &= -W\{\phi_1(x'), \phi_2(x')\} \equiv -W(x') \neq 0. \end{aligned} \tag{2.3.9}$$

Abel's formula (2.1.13) ensures that this condition is satisfied either for all x' or for none. If it is satisfied, then the solution reads $C = -\phi_2(x')/W(x')$, $D = -\phi_1(x')/W(x')$, whence

$$G(x \mid x') = -\frac{\{\theta(x' - x)\phi_2(x')\phi_1(x) + \theta(x - x')\phi_1(x')\phi_2(x)\}}{\{\phi_1(x')\phi_2'(x') - \phi_1'(x')\phi_2(x')\}}. \tag{2.3.10}$$

Exercise: Check explicitly that (2.3.10) indeed satisfies all the defining conditions (2.3.5).

Evidently we must distinguish the general case $W \neq 0$ from the special case $W = 0$. That this nomenclature is apt is best seen if $q = 0 \Rightarrow W =$ constant; then W depends only on the parameters of the problem contained in r. Thus $W = 0$ imposes a constraint on any such parameters, often amounting to a resonance condition: even if the constraint can be satisfied by certain special parameter values, any small change in these values restores the general case $W \neq 0$.

The general case and the special case are one manifestation of the '*Fredholm alternative*', which pervades the theory of all linear operators (algebraic and integral as well as differential): see also Section 2.4 and especially Table 2.1.

In the general case, $W \neq 0$, our problem is solved, by (2.3.4, 10). In full, the solution reads

$$\psi(x) = -\int_{x_1}^{x_2} \mathrm{d}x'\, f(x')\{\theta(x'-x)\phi_2(x')\phi_1(x) + \theta(x-x')\phi_1(x')\phi_2(x)\}/W(x'), \tag{2.3.11}$$

$$\psi(x) = -\left\{\phi_1(x)\int_{x}^{x_2} \mathrm{d}x'\, f(x')\phi_2(x')/W(x') + \phi_2(x)\int_{x_1}^{x} \mathrm{d}x'\, f(x')\phi_1(x')/W(x')\right\}, \tag{2.3.12}$$

where

$$W(x') = \{\phi_1(x')\phi_2'(x') - \phi_1'(x')\varphi_2(x')\}.$$

In the special case, $W = 0$ expresses the fact that ϕ_1 and ϕ_2 defined by (2.3.6) are actually the same function: in other words, the corresponding homogeneous problem ($L\phi = 0$, $\phi(x_1) = 0 = \phi(x_2)$) then has a solution (satisfying both BCs, and not only one). In this special case, G does not exist, whence our attempt at a solution in the form (2.3.4) fails. Again there are two possibilities:

(i) *either* the *general special case,* where the inhomogeneous problem has no solution; in fact it has none unless $f(x)$ satisfies some constraints;

(ii) *or* the *special special case,* where $f(x)$ does satisfy these constraints. Then our inhomogeneous problem has solutions, but they are not unique: in fact there are infinitely many. Obviously so: for if we have one solution to the inhomogeneous problem, then we can add to it any numerical multiple of the solution to the corresponding homogeneous problem, and the result still satisfies all the requirements (i.e. both the equation and the BCs). This is illustrated in Sections 2.4 and 2.5.

We shall not spell out the conditions on f for the special special case to obtain, except (in Section 2.4) for the important subclass of problems where the operator L is Hermitean.

To illustrate the results of this section, the BVP Green's function for the undamped oscillator is determined in Section 2.5.2.

Meanwhile, Poisson's equations in 1D with homogeneous DBCs serves to exemplify the general case. The operator in this example is $L = -d^2/dx^2$, as in (2.1.7). For simplicity we take $x_1 = 0$, $x_2 = a$. The homogeneous solutions $\phi_{1,2}$ defined by (2.3.6) can be identified by inspection: $\phi_1 = x$, $\phi_2 = (a - x)$. Then $W \equiv \phi_1\phi_2' - \phi_1'\phi_2 = x(-1) - 1(a - x) = -a$. Consequently G, eqn (2.3.7), becomes

$$\begin{aligned} G(x \mid x') &= -\{\theta(x' - x)(a - x')x/a + \theta(x - x')x'(a - x)/a\} \\ &= -x_<(a - x_>)/a, \end{aligned} \qquad (2.3.13)$$

where we have introduced the convenient notation

$$x_< \equiv \min(x, x'), \qquad x_> \equiv \max(x, x'), \qquad (2.3.14)$$

i.e. $x_<$ ($x_>$) is the lesser (greater) of x and x'. This notation will recur frequently.

Exercise: Sketch $G(x \mid x')$ as a function of x, for a representative selection of values of x' between 0 and a.

The end-result (2.3.12) now reads

$$\psi(x) = (1/a)\left\{x \int_x^a dx' f(x')(a - x') + (a - x) \int_0^x dx' f(x')x'\right\}. \qquad (2.3.15)$$

Exercises: (i) Work through this problem explicitly from first principles, i.e. starting from (2.3.4–7), and without quoting any of the relations (2.3.8–12). (ii) Do the same but with NBCs.

2.4 Construction of the boundary-value-problem Green's function from the eigenfunctions of L when L is Hermitean

So far the operator L from (2.1.1) has featured only in the two equations $L\psi = f$ and $L\psi = 0$. At this point we extend the enquiry to the eigenvalue equation

$$L\phi^{(n)}(x) = \lambda_n\phi^{(n)}(x), \qquad (2.4.1)$$

supplemented by homogeneous BCs at x_1 and x_2. (From here on, the letter ϕ is no longer reserved for solutions of the homogeneous equation

$L\phi = 0$. In later sections (2.4.1) will be written $L\phi_n = \lambda_n \phi_n$, but for the moment we write $\phi^{(n)}$ in order to forestall confusion with the functions ϕ_1, ϕ_2 which appeared in Sections 2.1–3 as solutions of $L\phi = 0$.) The BCs could equally well be Dirichlet, Neumann, or any other, but they must be homogeneous. They are taken as such throughout this section. The eigenvalue problem thus defined has solutions only for certain special values of the constant λ_n. When $x_2 - x_1$ is finite, the λ_n are discrete. (For $L = -\nabla^2$ under Dirichlet BCs, this is proved in Problem 5.12.)

This tack is profitable only if L is a Hermitean operator, which means that for *any* pair of functions $a(x)$, $b(x)$ that obey the BCs,

$$\int_{x_1}^{x_2} \mathrm{d}x\, a^*(x)[Lb(x)] = \int_{x_1}^{x_2} \mathrm{d}x\, [La(x)]^* b(x). \tag{2.4.2}$$

In practice L tends to be Hermitean if it governs a system with neither diffusion nor energy dissipation. A Hermitean operator has a set of eigenfunctions $\{\phi^{(n)}\}$ that is complete, and may be taken as orthonormal:

$$\int_{x_1}^{x_2} \mathrm{d}x\, \phi^{(n)*}(x)\phi^{(n')}(x) = \delta_{nn'}. \tag{2.4.3}$$

Appendix E discusses the reality properties of the set $\{\phi^{(n)}\}$ and the allied possibilities of degeneracy (which means that more than one linearly independent eigenfunction corresponds to the same eigenvalue). These properties (but only these) are strongly conditioned by the dimensionality of the space, i.e. by the number of independent variables. In 1D, i.e. for ordinary (not partial) differential equations, one finds that (i) there is no degeneracy; (ii) if L is real (as well as Hermitean), then the set $\{\phi^{(n)}\}$ is necessarily real (in contrast to nD, where it may but need not be chosen as real).

It is important, and will be easy to see, that the relations constructed in this section apply to any Hermitean L, regardless not only of the detailed form of L, but also of the dimensionality. With $\boldsymbol{r}$ a vector in nD space, we need merely replace $\phi^{(n)}(x)$ by $\phi^{(n)}(\boldsymbol{r})$, $\int_{x_1}^{x_2} \mathrm{d}x \cdots$ by $\int_V \mathrm{d}^n r$, and remember in case of degeneracy that sums $\sum_n$ run over all linearly independent eigenfunctions. The key results will be used again and again in other contexts, without repeating the derivations.

The basic idea is to exploit the fact that the set $\{\phi^{(n)}\}$ is complete, by expressing $G(x \mid x')$ regarded as a function of x in the form

$$G(x \mid x') = \sum_n c_n \phi^{(n)}(x), \tag{2.4.4}$$

where the coefficients c_n depend on x' but not on x. This expansion guarantees that G obeys the correct BCs (2.3.5b, c), since the $\phi^{(n)}$ do. It remains to satisfy the differential equation (2.3.5a), namely $LG(x \mid x') = \delta(x - x')$. On the left we substitute (2.4.4) and take L under the sum; on the right, we use the closure property $\delta(x - x') = \sum_m \phi^{(m)*}(x')\phi^{(m)}(x)$:

$$LG(x \mid x') = L\sum_m c_m\phi^{(m)}(x) = \sum_m c_m L\phi^{(m)}(x)$$
$$= \sum_m c_m\lambda_m\phi^{(m)}(x) = \delta(x - x') = \sum_m \phi^{(m)*}(x')\phi^{(m)}(x),$$

$$\sum_m c_m\lambda_m\phi^{(m)}(x) = \sum_m \phi^{(m)*}(x')\phi^{(m)}(x). \tag{2.4.5}$$

We can equate the coefficients of $\phi^{(n)}$ on the two sides (multiply by $\phi^{(n)*}(x)$, integrate over x, and appeal to (2.4.3)); one finds $c_n = \phi^{(n)*}(x')/\lambda_n$, whence

$$G(x \mid x') = \sum_n \frac{1}{\lambda_n}\phi^{(n)*}(x')\phi^{(n)}(x). \tag{2.4.6}$$

This is a crucial result and will recur frequently. Very often it affords the only convenient grip on G, especially when there are no closed-form expressions like those illustrated in Section 2.5.

If L is real, then all the $\phi^{(n)}$ are or can be chosen to be real, the complex-conjugation sign in (2.4.6) is irrelevant, and G is symmetric in x and x':

$$G(x \mid x') = G(x' \mid x) \qquad (\text{if } L^* = L). \tag{2.4.7}$$

Evidently (2.4.6) makes no sense if one of the eigenvalues is zero. If so, we shall often denote the zero eigenvalue by λ_0. (In many such cases the non-zero eigenvalues are all positive, so that one can and does write $\lambda_0 = 0 < \lambda_1 \leq \lambda_2 \leq \lambda_3 \ldots$. However, the illustration in Section 2.5 is an exception to this rule.) But a situation with $\lambda_0 = 0$ is precisely the special case described in Section 2.3, since $L\phi^{(0)} = \lambda_0\phi^{(0)} = 0$ identifies $\phi^{(0)}$ precisely as a solution of the homogeneous equation satisfying *both* the BCs.

Thus, when $\lambda_0 = 0$, no Green's function exists. But in the present case of Hermitean L, one can nevertheless deal with this special case quite painlessly. First we construct the function

$$H(x \mid x') \equiv {\sum_n}' \frac{1}{\lambda_n}\phi^{(n)*}(x')\phi^{(n)}(x), \tag{2.4.8}$$

where the primed sum is defined to omit the terms with zero eigenvalues; in the most common cases this means simply $\sum' \equiv \sum_{(n\neq 0)}$. Observe that H

obeys the differential equation

$$
\begin{aligned}
LH(x \mid x') &= \sum_{n\neq 0} \frac{1}{\lambda_n} \phi^{(n)*}(x') L\phi^{(n)}(x) \\
&= \sum_{n\neq 0} \frac{1}{\lambda_n} \phi^{(n)*}(x') \lambda_n \phi_n(x) \\
&= \sum_{n\neq 0} \phi^{(n)*}(x') \phi^{(n)}(x) \\
&= \sum_{n} \phi^{(n)*}(x') \phi_n(x) - \phi^{(0)*}(x')\phi^{(0)}(x),
\end{aligned}
$$

$$
LH(x \mid x') = \delta(x-x') - \phi^{(0)*}(x')\phi^{(0)}(x). \tag{2.4.9}
$$

Thus H is not a Green's function, because the RHS of (2.4.9) is not simply $\delta(x-x')$. *We call H the pseudo Green's function.*

Theorem: In the special case with a zero eigenvalue (where $L\phi^{(0)}=0$), the inhomogeneous BVP $L\psi=f$ has a solution if and only if

$$
\int_{x_1}^{x_2} \mathrm{d}x\, \phi^{(0)*}(x) f(x) = 0. \tag{2.4.10}
$$

If solutions do exist, they are of the form

$$
\psi(x) = C\phi^{(0)}(x) + \int_{x_1}^{x_2} \mathrm{d}x'\, H(x \mid x') f(x'), \tag{2.4.11}
$$

where C is an arbitrary constant.

Proof: To verify the sufficiency of the condition (the 'if'), act on both sides of (2.4.11) with L, and show that indeed $L\psi=f$:

$$
\begin{aligned}
L\psi &= \int_{x_1}^{x_2} \mathrm{d}x'\, \{LH(x \mid x')\} f(x') \\
&= \int_{x_1}^{x_2} \mathrm{d}x'\, \{\delta(x-x') - \phi^{(0)*}(x')\phi^{(0)}(x)\} f(x') \\
&= f(x).
\end{aligned}
$$

The first step relies on $L\phi^{(0)}=0$, the second on (2.4.9), and the last on (2.4.10).

The necessity of the condition (the 'only if') is verified by *reductio ad absurdum*. Suppose a solution exists while (2.4.10) fails. Since $\{\phi^{(n)}\}$ is complete, write $\psi = \sum_n b_n \phi^{(n)}$, and substitute this on the left of $L\psi = f$. One finds $\sum_n b_n L\phi^{(n)} = \sum_n b_n \lambda_n \phi^{(n)} = \sum_{n\neq 0} b_n \lambda_n \phi^{(n)} = f$. Multiply both sides by $\phi^{(0)*}$, and integrate. On the left this gives LHS $=$ $\sum_{n\neq 0} b_n \lambda_n \int_{x_1}^{x_2} \mathrm{d}x\, \phi^{(0)*}\phi^{(n)} = \sum_{n\neq 0} b_n \lambda_n \delta_{n0} = 0$. On the right it gives

RHS $= \int_{x_1}^{x_2} dx\, \phi^{(0)*} f \neq 0$, by assumption. Since LHS $\neq$ RHS, no solution can have existed in the first place. ■

The various possibilities for BVPs with Hermitean L, i.e. the Fredholm alternatives, are summarized in Table 2.1.

It is clear that, regarding any inhomogeneous BVP $L\psi = f$, one of the first questions to ask is whether L has zero as one of its eigenvalues.

Exercises: (i) Show that for Poisson's equation in 1D, zero is an eigenvalue with homogeneous NBCs but not with homogeneous DBCs. (ii) For the same problem, verify by explicit Fourier analysis that the series (2.4.6) indeed sums to the Green's function already known in closed form from (2.3.13).

Our formalism suggests that G, if it exists, is just the inverse of the operator L. It is easy to state this in precise form. For simplicity we take L as real. While L is a differential operator, G is an integral operator, which acts on any function $b(x)$ satisfying the BCs to turn it into another such function $a(x)$ according to $a(x) \equiv \int_{x_1}^{x_2} dx'\, G(x \mid x') b(x')$. We write this symbolically as $a = Gb$. We assert

$$GL = 1 = LG, \quad \text{i.e.} \quad G = L^{-1}, \tag{2.4.12a, b, c}$$

Table 2.1 The Fredholm alternative for the boundary-value problem $L\psi(x) = f(x)$, with Hermitean L, and homogeneous boundary conditions at x_1, x_2. Associated eigenvalue equation: $L\phi^{(n)} = \lambda_n \phi^{(n)}(x)$

Homogeneous equation: $L\psi = 0$	Constraints on the inhomogeneous term $f(x)$?	Inhomogeneous equation: $L\psi = f$
GENERAL CASE (no λ_n is zero): no solution exists	None	A solution exists, and is unique: $\psi(x) = \int_{x_1}^{x_2} dx'\, G(x \mid x') f(x')$, $G(x \mid x') = \sum_n \phi^{(n)*}(x') \phi^{(n)}(x)/\lambda_n$
SPECIAL CASE ($\lambda_0 = 0$): Solutions exist, namely $C\phi^{(0)}(x)$, C arbitrary	GENERAL SPECIAL CASE: $\int_{x_1}^{x_2} dx\, \phi^{(0)*}(x) f(x) \neq 0$	No solution exists
	SPECIAL SPECIAL CASE: $\int_{x_1}^{x_2} dx\, \phi^{(0)*}(x) f(x) = 0$	Solutions exist but they are not unique: $\psi(x) = C\phi^{(0)}(x) + \int_{x_1}^{x_2} dx'\, H(x \mid x') f(x')$, C arbitrary. $H(x \mid x') = \sum_{n\,(\lambda_n \neq 0)} \phi^{(n)*}(x') \times \phi^{(n)}(x)/\lambda_n$

where (2.4.12a, b), respectively, are shorthand for

$$\int_{x_1}^{x_2} \mathrm{d}x'\, G(x \mid x')L'b(x') = b(x) = L\int_{x_1}^{x_2} \mathrm{d}x'\, G(x \mid x')b(x'). \tag{2.4.13a, b}$$

While (2.4.13b) follows immediately from $LG(x \mid x') = \delta(x - x')$ as above, (2.4.13a) is verified by first integrating by parts to transfer the derivatives $\mathrm{d}/\mathrm{d}x'$ in L' so that they act on G instead of b. But, by virtue of (2.4.7), $L'G(x \mid x') = L'G(x' \mid x) = \delta(x' - x)$, whence (2.4.13a) follows.

The fact that G and L are inverse to each other is made clear most elegantly by translating eqns (2.4.1, 3, 5, 6) into the Dirac notation for abstract operators customary in quantum mechanics. The translation is accomplished by identifying $\phi^{(n)}(x)$ as $\langle x \mid n\rangle$, $\phi^{(n)*}(x)$ as $\langle n \mid x\rangle$, and $\int_{x_1}^{x_2} \mathrm{d}x\, a^*(x)b(x)$ as $\langle a \mid b\rangle$. Then $L\,|n\rangle = \lambda_n\,|n\rangle$, and $\langle n \mid n'\rangle = \delta_{nn'}$. Since the set $\{|n\rangle\}$ is complete, the unit operator (by virtue of the closure relation), and the operator L itself, may be written

$$1 = \sum_n |n\rangle\langle n|, \qquad L = \sum_n \lambda_n\,|n\rangle\langle n|. \tag{2.4.14a, b}$$

Similarly, (2.4.6) becomes

$$G = \sum_n \frac{1}{\lambda_n}|n\rangle\langle n|, \tag{2.4.15}$$

and the relationships (2.4.12) follow automatically. For instance,

$$\begin{aligned} GL &= \sum_n (1/\lambda_n)\,|n\rangle\langle n| \sum_m \lambda_m\,|m\rangle\langle m| \\ &= \sum_{n,m} (\lambda_m/\lambda_n)\,|n\rangle\langle n \mid m\rangle\langle m| = \sum (\lambda_m/\lambda_n)\,|n\rangle\,\delta_{nm}\,\langle m| \\ &= \sum_n |n\rangle\langle n| = 1, \end{aligned}$$

as expected.

A zero eigenvalue obviously stalls this process: operators with a zero eigenvalue have no inverses, essentially because division by zero is meaningless. Instead of G one is led naturally to consider

$$H = \sum{}' \frac{1}{\lambda_n}|n\rangle\langle n|, \tag{2.4.16}$$

which acts as the inverse of L in the restricted subspace spanned by those $|n\rangle$ that correspond to $\lambda_n \neq 0$. If the inhomogeneous term $|f\rangle$ in the equation $L\,|\psi\rangle = |f\rangle$ is wholly in this subspace, i.e. if it satisfies (2.4.10), $\langle 0 \mid f\rangle = 0$, then $|\psi\rangle = H\,|f\rangle$ solves the equation, but so does $C\,|0\rangle + H\,|f\rangle$ as asserted by (2.4.11).

Exercise: Write down HL and LH in Dirac notation, and compare them with the unit operator.

2.5 Examples: the forced undamped harmonic oscillator

The results of Sections 2.2–4 are illustrated by the forced undamped harmonic oscillator, governed by (2.1.6) with $\gamma = 0$:

$$-L\psi = (\mathrm{d}^2/\mathrm{d}x^2 + \omega^2)\psi(x) = -f(x) \equiv F(x)/m. \tag{2.5.1}$$

Recall that ω is the natural frequency, x is time, and $F(x)$ the applied force.

2.5.1 The initial-value problem

Here one imposes the ICs (2.2.1): $\psi(0) = a$, $\psi'(0) = u$. According to (2.2.2), $\psi = \phi_c + \psi_p$, and as in (2.2.3) we choose the complementary function ϕ_c to satisfy the ICs. The most general solution of $L\phi_c = 0$ is

$$\phi_c(x) = A \sin(\omega x + \delta), \tag{2.5.2}$$

containing the two adjustable parameters A and δ. These are determined by the ICs as

$$A = (a^2 + u^2/\omega^2)^{\frac{1}{2}}, \qquad \delta = \tan^{-1}(\omega a/u). \tag{2.5.3}$$

For the particular integral ψ_p we need the Green's function (2.2.12). Any pair of linearly independent solutions ϕ_1, ϕ_2 of $L\phi = 0$ will serve, and we choose

$$\phi_1 = \sin(\omega x), \qquad \phi_2 = \cos(\omega x). \tag{2.5.4}$$

Exercise: Go through the argument that follows with $\phi_1 = \sin(\omega x + \delta_1)$ and $\phi_2 = \sin(\omega x + \delta_2)$, and show that the end-results are the same for any choice of δ_1 and δ_2 provided only $\delta_1 \neq \delta_2$. In other words, a specially enlightened choice of ϕ_1 and ϕ_2 may prove convenient but is not essential.

ϕ_1 and ϕ_2 are indeed linearly independent, since (cf. eqns (2.1, 10))

$$W = \{\phi_1\phi_2' - \phi_1'\phi_2\} = -\omega\{\sin^2(\omega x) + \cos^2(\omega x\} = -\omega \neq 0. \tag{2.5.5}$$

Then G is given by (2.2.12):

$$\begin{aligned} G(x \mid x') &= \theta(x - x')\{\phi_2(x')\phi_1(x) - \phi_1(x')\phi_2(x)\}/W \\ &= \theta(x - x')\{\cos(\omega x')\sin(\omega x) - \sin(\omega x')\cos(\omega x)\}/(-\omega) \\ &= -\theta(x - x')\sin[\omega(x - x')]/\omega. \end{aligned} \tag{2.5.6}$$

Assembling ϕ_c and ψ_p according to (2.2.2) we obtain the explicit end-result

$$\begin{aligned} \psi(x) &= (a^2 + u^2/\omega^2)^{\frac{1}{2}} \sin[\omega x + \tan^{-1}(\omega a/u)] \\ &\quad + \int_0^x \mathrm{d}x' \frac{\sin[\omega(x - x')]}{\omega} \frac{F(x')}{m}. \end{aligned} \tag{2.5.7}$$

Exercises: (i) Verify by explicit substitution that (2.5.7) satisfies the differential equation and the ICs. (ii) Solve the IVP for the damped oscillator, governed by (2.1.6) with $\gamma \neq 0$. (iii) Determine the solution for a free particle by taking the limit $\omega \to 0$.

2.5.2 The boundary-value problem

As in Section 2.3 we consider only DBCs, and set $x_1 = 0$, $x_2 = T$:

$$\psi(0) = 0, \qquad \psi(T) = 0. \tag{2.5.8c, b}$$

For instance, a (small-amplitude) pendulum is to be vertical at times 0 and T. In the process of solving (2.5.1, 8) we shall have to determine, in terms of f, the requisite initial *velocity* $\psi'(0)$ that will guarantee the final BC $\psi(T) = 0$.

The two solutions ϕ_1 and ϕ_2 of $L\phi = 0$, both of the form (2.5.2), but satisfying the initial and the final BC respectively, are

$$\phi_1(x) = \sin[\omega x], \qquad \phi_2(x) = \sin[\omega(T - x)]. \tag{2.5.9a, b}$$

Hence their Wronskian W, and the Green's function (2.3.10) when it exists, are given by

$$\begin{aligned} W &\equiv \{\phi_1\phi_2' - \phi_1'\phi_2\} \\ &= -\omega\{\sin[\omega x]\cos[\omega(T-x)] + \cos[\omega x]\sin[\omega(T-x)]\} \\ &= -\omega\sin[\omega x + \omega(T-x)] = -\omega\sin[\omega T], \end{aligned} \tag{2.5.10}$$

$$\begin{aligned} G(x \mid x') &= \frac{1}{\omega\sin[\omega T]}\{\theta(x'-x)\sin[\omega x]\sin[\omega(T-x)] \\ &\quad + \theta(x-x')\sin[\omega(T-x)]\sin[\omega x']\} \\ &= \frac{\sin[\omega x_<]\sin[\omega(T-x_>)]}{\omega\sin[\omega T]}. \end{aligned} \tag{2.5.11}$$

The special case when G does not exist arises if

$$\begin{aligned} &W = 0 \Rightarrow \sin[\omega T] = 0 \Rightarrow \omega T = n\pi, \\ &T = n\pi/\omega = n\tau/2, \qquad n = 1, 2, 3, \ldots, \end{aligned} \tag{2.5.12}$$

where $\tau = 2\pi/\omega$ is the natural period. Thus the special case obtains if the pendulum would return to its initial position at time T in the absence of an applied force. The general case obtains for all other values of ωT.

In the general case, the solution (2.3.12) reads

$$\begin{aligned} \psi(x) = -\frac{1}{\omega\sin[\omega T]}&\{\sin[\omega x]\int_x^T dx' f(x')\sin[\omega(T-x')] \\ &+ \sin[\omega(T-x)]\int_0^x dx' f(x')\sin[\omega x']\}. \end{aligned} \tag{2.5.13}$$

The requisite initial velocity is

$$\psi'(0) = -\frac{1}{\sin[\omega T]}\int_0^T dx'\, f(x')\sin[\omega(T-x')]. \tag{2.5.14}$$

Exercise: Verify this from (2.5.13).

2.5.3 Expansion in eigenfunctions of L

Since L is Hermitean, the BVP is amenable to the methods of Section 2.4. Subject to the DBCs (2.5.8), the orthonormal eigenfunctions and the eigenvalues of L are

$$\phi^{(n)}(x) = (2/T)^{\frac{1}{2}}\sin(n\pi x/T), \tag{2.5.15}$$

$$\lambda_n = [(n\pi/T)^2 - \omega^2], \tag{2.5.16}$$

with $n = 1, 2, 3, \ldots$.

Accordingly, in the general case, (2.4.6) reads

$$G(x \mid x') = \frac{2}{T}\sum_{n=1}^{\infty}\frac{\sin(n\pi x'/T)\sin(n\pi x/T)}{[(n\pi/T)^2 - \omega^2]}. \tag{2.5.17}$$

Explicit (if somewhat tedious) Fourier analysis shows that this expression is indeed equal to (2.5.11).

Exercise: Verify this.

In applications, the important question is, which form is more convenient? The answer may depend on the circumstances, and on one's programming experience and tastes.

In the special case, the natural notation of (2.5.15, 16) now labels the offending zero eigenvalue not as in Section 2.4 by zero, but by the integer $N = \omega T/\pi$. According to (2.4.10), there is no solution unless $\int_0^T dx \sin(N\pi x/T) f(x) = 0$. If $f(x)$ does satisfy this constraint, then the solutions are given by (2.4.11, 8). With C arbitrary and ω expressed as $N\pi/T$, this becomes

$$\psi(x) = C\sin(N\pi x/T) + \int_0^T dx'\, f(x')\frac{2}{T}\sum_{(n\neq N)}^{\infty}\frac{\sin(n\pi x'/T)\sin(n\pi x/T)}{(n^2 - N^2)(\pi^2/T^2)}. \tag{2.5.18}$$

Finally it is instructive to write down an explicit solution of the differential equation in the general special case, where the constraint on f fails, and to observe just how and why this failure makes the BVP insoluble. To this end, choose $\omega T = \pi$ (i.e. $N = 1$) and $-f(x) = \sin(\omega x)$, and consider the most general solution of $\psi'' + \omega^2\psi = \sin \omega x$ subject merely to the left-hand BC $\psi(0) = 0$. The solution reads

$$\psi(x) = B \sin(\omega x) - (1/2\omega)x \cos(\omega x), \tag{2.5.19}$$

where $B = u/\omega + 1/2\omega^2$ reflects the value of the initial velocity u (which is not prescribed). The point is simply that the right-hand BC $\psi(T) = 0$ cannot now be satisfied by any choice of B, because $\psi(T) = \psi(\pi/\omega) = \{B \sin \pi - (\pi/2\omega^2) \cos \pi\} = \pi/2\omega \neq 0$, quite independently of B.

2.6 Poisson's equation with higher multipole layers

The expressions (2.2.14) and (2.3.4) solve the IVP and the BVP, respectively, even if the inhomogeneous term f contains derivatives of the delta-function. However, it is more illuminating to consider such inhomogeneities *ab initio* in their own right; the new features become clear from the 1D Poisson equation with the singularities at the origin. In this section we are concerned only with the constraints that such singularities impose on the solution in their immediate vicinity, and need not distinguish between IVPs and BVPs. Choosing signs by hindsight, we set the problem

$$\mathrm{d}^2\psi/\mathrm{d}x^2 = -\alpha_0\,\delta(x) + \alpha_1\,\delta'(x) - \alpha_2\,\delta''(x) + \ldots. \tag{2.6.1}$$

The physical meaning of the RHS, e.g. in electrostatics, is immediate. The term $\alpha_0\,\delta(x)$ represents an infinitely extended distribution of charge in the yz-plane, carrying charge α_0 per unit area. Similarly, $-\alpha_1\,\delta'(x)$ represents a sheet of dipoles carrying dipole moment α_1 per unit area (α_1 is positive when the dipole points to the right: cf. (1.4.25)); $\alpha_2\,\delta''(x)$ represents a sheet of quadrupoles, i.e. two dipole layers of strengths $\pm\alpha_1$ separated by a distance a, in the limit $a \to 0$, $\alpha_1 \to \infty$, with $\alpha_1 a = \alpha_2$ fixed; and so on.

For $x < 0$ and $x > 0$, the RHS of (2.6.1) vanishes, whence

$$\psi(x<0) = A_\mathrm{L} + B_\mathrm{L}x, \qquad \psi(x>0) = A_\mathrm{R} + B_\mathrm{R}x. \tag{2.6.2}$$

It will turn out that eqn (2.6.1) entails two relations between the four constants $A_{\mathrm{L,R}}$, $B_{\mathrm{L,R}}$; our chief aim is to determine these relations. The physics of any specific problem will impose two further conditions: in IVPs, both conditions on the same side of the origin, and in BVPs one on either side. The snag is that in fact eqn (2.6.1) entails more than two conditions, which cannot all be satisfied merely by adjusting the constants featured in (2.6.2).

The standard case, which we have already studied, has $\alpha_0 = 1$, $\alpha_{n\geq 1} = 0$, and leads to the appropriate Green's function. Our two relations are then supplied by continuity (2.1.18), entailing $A_\mathrm{L} = A_\mathrm{R}$, and by the jump condition (2.1.17) multiplied by α_0, entailing $B_\mathrm{R} - B_\mathrm{L} = -\alpha_0$. But in general the argument whereby these relations were deduced fails for (2.6.1), since they were found by matching ψ'' to $-\alpha_0\,\delta(x)$, whereas the hierarchy of singularities (Section 2.1.2) demands that ψ'' be matched to the highest derivative of $\delta(x)$ present on the RHS. To avoid a dead end, one must abandon (2.6.2) and look for a solution in the form

$$\begin{aligned}\psi(x) = \theta(-x)(A_\mathrm{L} + B_\mathrm{L}x) + \theta(x)(A_\mathrm{R} + B_\mathrm{R}x)\\ -\beta_2\,\delta(x) + \beta_3\,\delta'(x) - \ldots.\end{aligned} \tag{2.6.3}$$

Differentiating twice, we obtain

$$\begin{aligned}\psi'(x) &= -\delta(x)(A_L + B_L x) + B_L\theta(-x) + \delta(x)(A_R + B_R x) + B_R\theta(x)\\ &\quad - \beta_2\,\delta'(x) + \beta_3\,\delta''(x) + \cdots\\ &= \delta(x)(A_R - A_L) + B_L\theta(-x) + B_R\theta(x) - \beta_2\,\delta'(x) + \beta_3\,\delta''(x) + \ldots,\\ \psi''(x) &= \delta(x)(B_R - B_L) + \delta'(x)(A_R - A_L) - \beta_2\,\delta''(x) + \beta_3\,\delta'''(x) + \ldots,\end{aligned} \tag{2.6.4}$$

where $x\,\delta(x) = 0$ has been used. Equating (2.6.4) to the RHS of (2.6.1), we find

$$B_R - B_L = -\alpha_0, \qquad A_R - A_L = \alpha_1, \tag{2.6.5a, b}$$

$$\alpha_n = \beta_n, \qquad (n \geq 2). \tag{2.6.6}$$

The implications become clearer in terms of the notation

$$A \equiv A_R + A_L, \qquad \Delta A \equiv A_R - A_L = \alpha_1, \tag{2.6.7a}$$

$$B \equiv B_R + B_L, \qquad \Delta B \equiv B_R - B_L = -\alpha_0; \tag{2.6.7b}$$

$$A_R = \tfrac{1}{2}(A + \Delta A) = \tfrac{1}{2}(A + \alpha_1), \tag{2.6.8a}$$

$$A_L = \tfrac{1}{2}(A - \Delta A) = \tfrac{1}{2}(A - \alpha_1); \tag{2.6.8b}$$

$$B_R = \tfrac{1}{2}(B + \Delta B) = \tfrac{1}{2}(B - \alpha_0), \tag{2.6.8c}$$

$$B_L = \tfrac{1}{2}(B - \Delta B) = \tfrac{1}{2}(B + \alpha_0). \tag{2.6.8d}$$

Then the solution (2.6.3) can be written as

$$\begin{aligned}\psi(x) &= \tfrac{1}{2}\theta(-x)[(A - \alpha_1) + (B + \alpha_0)x] + \tfrac{1}{2}\theta(x)[(A + \alpha_1) + (B - \alpha_0)x]\\ &\quad - \alpha_2\,\delta(x) + \alpha_3\,\delta'(x) + \ldots,\end{aligned} \tag{2.6.9}$$

$$\psi(x) = \tfrac{1}{2}(A + Bx) + \tfrac{1}{2}\varepsilon(x)\alpha_1 - \tfrac{1}{2}\varepsilon(x)x\alpha_0 - \alpha_2\,\delta(x) + \alpha_3\,\delta'(x) + \ldots. \tag{2.6.10}$$

Most remarkably, ΔA and ΔB remain unaffected by the higher delta-function derivatives on the right of (2.6.1). Therefore, in the end, the conventional electrostatic rules remain in force: a charge layer α_0 by itself leaves the potential ψ continuous, but causes a jump α_0 in the field $-\psi'$ (as dictated by Gauss's law); conversely, a dipole layer α_1 causes a jump α_1 in the potential, but leaves the field continuous. (See also Figs 2.1a, b.) Though these are the same conclusions that one would have reached simply by ignoring the higher derivatives on the right of (2.6.1), we stress again that mathematically such an argument would have been inconsistent.

Exercise: Derive (2.6.5, 6) from the standard Green's-function solutions, first of the IVP, and then of the BVP. (The initial point and the boundary points must not, of course, coincide with the origin.)

Evidently, observations on the potential or the field made at $x < 0$ and $x > 0$ are completely unaffected by the presence of quadrupolar or higher multipolar layers. By the same token, only charge and dipole layers are needed for

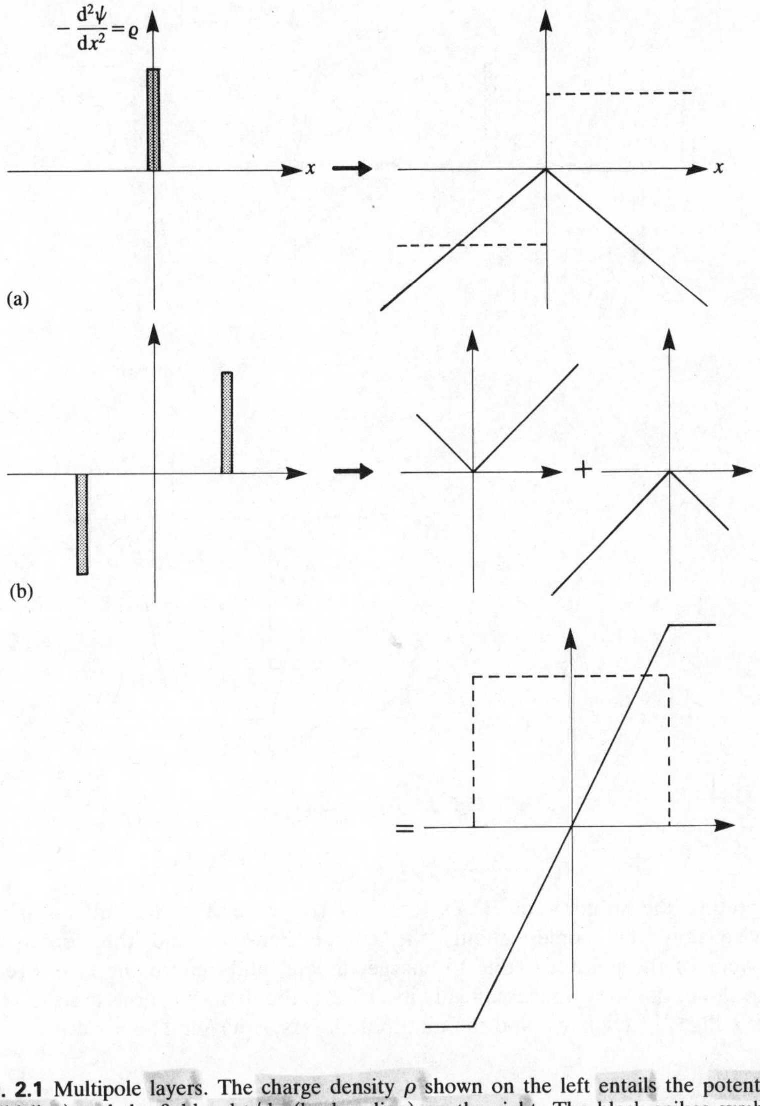

Fig. 2.1 Multipole layers. The charge density ρ shown on the left entails the potential ψ (solid line) and the field $-d\psi/dx$ (broken line) on the right. The black spikes symbolize delta functions. (a) Charge layer. (b) Dipole layer, represented as two finitely-separated finite-strength charge layers of opposite signs (see Section 1.2, point (vii)). The potentials due to the two delta-peaks are shown separately on the top line; their sum, i.e. the total potential ψ, and $-d\psi/dx$, are shown on the bottom line. The potential extends beyond the dipole layer, but the field vanishes everywhere outside it. (c) Quadrupole layer, represented as two equal-strength but oppositely-directed dipole layers. The potentials due to the two dipole layers are shown separately on the top line; their sum ψ, and $-d\psi/dx$, are shown on the bottom line. Potential and field both vanish everywhere outside the quadrupole layer. The same is true automatically for an octupole layer (two equal-strength but opposite-sign quadrupole layers), and so on for all higher-order multipole layers, conformably with the analytic proof given in the text.

(c)

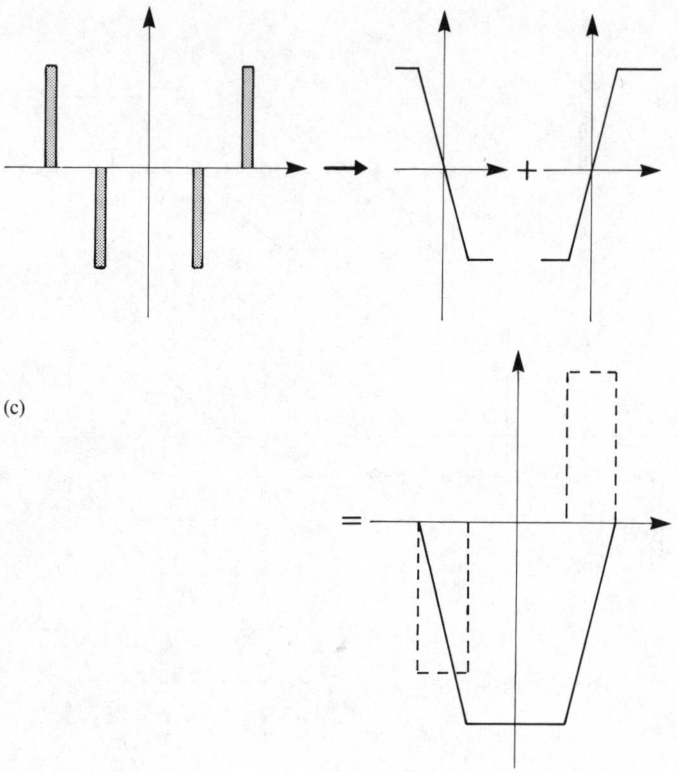

Fig. 2.1 *(continued)*

interpreting the solution of BVPs later on (cf. especially Section 4.4.5). To visualize how this comes about, the charge densities and the consequent behaviour of the potentials due to charge, dipole, and quadrupole layers (each acting alone) are shown schematically in Fig. 2.1; the delta-functions are shown as (finitely high) heavy lines, and the multipole layers as having finite widths.

Problems

2.1 (i) If $y_1(x)$ solves the equation

$$y'' + q(x)y' + r(x)y = 0,$$

show that a second linearly independent solution $y_2(x)$ is given by

$$y_2(x) = y_1(x) \int_a^x \mathrm{d}x' \, W(x')/y_1^2(x'),$$

where

$$W(x) = \exp\left\{-\int_b^x \mathrm{d}x' \, q(x')\right\},$$

and a, b are arbitrary constants.

Hint: Use Abel's formula for the Wronskian W, and the fact (which you should verify) that $W/y_1^2 = \mathrm{d}/\mathrm{d}x\,(y_2/y_1)$.

(ii) One solution of the equation

$$y'' - \left(\frac{1}{x} + 1\right)y' + \frac{1}{x}y = 0$$

is $y_1 = \exp(x)$. By using the results of part (i), find a second linearly independent solution y_2; verify by direct substitution that your y_2 is indeed a solution.

2.2 *The no-degeneracy theorem for 1D Schroedinger bound states.* Bound states in a potential $V(x)$ $(V(\pm\infty) = 0)$ have wavefunctions obeying

$$-\frac{\hbar^2}{2m}\psi'' + V\psi = E\psi,$$

where $E < 0$, and $\psi(\pm\infty) = 0$. Show that there cannot be two linearly independent solutions for the same value of E.

Hint: Show that the Wronskian is independent of x, and evaluate it at $x = \infty$.

2.3 (i) Find the Green's function $G(x \mid x')$ defined by

$$\frac{\mathrm{d}}{\mathrm{d}x}G(x \mid x') + \lambda G(x \mid x') = \delta(x - x'),$$

$$G = 0 \quad \text{for} \quad x < x'.$$

Hint: Solve the equation for $x > x'$, and then determine and apply the appropriate jump condition across $x = x'$.

(ii) By using G, construct the solution of the equation

$$\frac{dy}{dx} + \lambda y = \mu, \qquad y(0) = 0,$$

where μ is a given constant.
(iii) Verify your result by solving the equation for y by the standard elementary method.

2.4 *Brownian motion of an overdamped oscillator.* In a very viscous medium, the inertial force my'' can sometimes be neglected relative to the resistance. For a particle in an oscillator potential one then has the following (approximate) equation of motion:

$$\eta y'(t) + m\omega^2 y(t) = F(t),$$

where $F(t)$ is the fluctuating Langevin force satisfying

$$\langle F(t) \rangle = 0, \qquad \langle F(t)F(t') \rangle = 2\eta kT\,\delta(t - t').$$

(i) Determine the Green's function defined by

$$(\eta\, d/dt + m\omega^2)G(t \mid t') = \delta(t - t'),$$

$$G = 0 \quad \text{for} \quad t < t'.$$

(ii) For an ensemble of particles all starting from the origin at time zero, use G to calculate their mean square displacement as a function of time. Examine $\langle y^2 \rangle$ as $t \to \infty$, in the light of the equipartition of energy.

2.5 Consider the differential equation

$$\left(\frac{d^2}{dx^2} - a^2\right)\psi(x) = f(x),$$

subject to the boundary conditions $\psi(\infty) = 0 = \psi(-\infty)$. $f(x)$ is given, and $a^2 > 0$.
(i) Determine the appropriate Green's function by solving

$$\left(\frac{\partial^2}{\partial x^2} - a^2\right)G(x \mid x') = \delta(x - x')$$

subject to $G(\infty \mid x') = 0 = G(-\infty \mid x')$.
Verify that your solution has the symmetry $G(x \mid x') = G(x' \mid x)$.
(ii) If $f(x) = 1$ when $|x| < 1$, and zero everywhere else, use G to determine the form of the solution appropriate to the region $x > 1$.

2.6 The Green's function for Poisson's equation in 1D is defined by

$$-\frac{\partial^2}{\partial x^2} G(x \mid x') = \delta(x - x'),$$

$$(0 \leqslant x, x' \leqslant a)$$

plus the boundary conditions $G(0 \mid x') = 0 = G(a \mid x')$. Equation (2.3.13) gives G in closed form.
(i) Verify by substitution that this expression satisfies the differential equation.
(ii) Write down the eigenfunction expansion of G, by the method of Section 2.4.
(iii) Verify that the expansion in part (ii) is indeed the Fourier series of the closed expression (2.3.13).
(iv) Check that the results of parts (i) and (ii) agree with the limit $\omega \to 0$ of the harmonic-oscillator Green's function from Section 2.5.2.

2.7 A simple-harmonic oscillator starts from rest at the origin ($x(0) = 0 = \dot{x}(0)$), and is driven exactly on resonance; its equation of motion for $t \geqslant 0$ is

$$\ddot{x}(t) + \omega^2 x(t) = \alpha \sin(\omega t).$$

(i) Construct the appropriate Green's function by the method of Section 2.3, and use it to determine $x(t)$.
(ii) Solve the equation of motion directly, and verify that your two solutions agree.

2.8 An 'inverted oscillator' starts from rest at the origin ($x(0) = 0 = \dot{x}(0)$), and is driven harmonically; its equation of motion is

$$\ddot{x}(t) - \omega^2 x(t) = \alpha \sin(\Omega t).$$

(i) Construct the appropriate Green's function, by the method of Section 2.3, and use it to determine $x(t)$.
(ii) Solve the equation of motion directly, and verify that your two solutions agree.

2.9 As Problem 2.7, but for the underdamped oscillator with arbitrary driving frequency, whose equation of motion is

$$\ddot{x} + \gamma\dot{x} + \omega^2 x = \alpha \sin(\Omega t).$$

3 Partial differential equations: a preview

Summary

Mathematically, this summary is neither complete nor exact, but it covers almost all cases you are likely to meet in practice.

Definitions:

- ψ is required in a region V of $\boldsymbol{r}$. The closed surface S surrounds V, and is stationary. S may consist of several disjoint pieces. Parts or all of S may be at infinity. $\mathrm{d}\boldsymbol{S}$ points along the outward normal (away from V).
- $\psi_S(\boldsymbol{r})$ (or $\psi_S(\boldsymbol{r}, t)$) denotes ψ at a point on S; $\partial_n \psi_S$ denotes, at a point on S, the component of $\nabla\psi$ along $\mathrm{d}\boldsymbol{S}$ (i.e. along the *out*ward normal). [Do not confuse this suffix S with the partial derivative symbol.] See Appendix G on notation.
- ρ is a prescribed source density. t_0 is the starting time.
- An *equation* is homogeneous if $\rho = 0$. A *problem* is homogeneous only if $\rho = 0$ and all prescribed boundary values (see below) are zero. Otherwise the problem is inhomogeneous.
- DBCs prescribe ψ_S on all of S.
- NBCs prescribe $\partial_n \psi_S$ on all of S.
  [Other possibilities exist but are seldom met.]

Poisson's equation (elliptic): $-\nabla^2\psi = \rho$. Called the Laplace equation if $\rho = 0$.

- BCS: D or N, but not both, because in general they would be mutually inconsistent. NBCs leave ψ undetermined to within an additive constant. The homogeneous problem has only the solution $\psi = 0$ (D), or $\psi = \text{constant}$ (N).

Diffusion equation (parabolic): $(\partial/\partial t - D\,\nabla^2)\psi(\boldsymbol{r}, t) = \rho(\boldsymbol{r}, t)$.

- BCs: as for Poisson's equation, i.e. either $\psi_S(\boldsymbol{r}, t)$ or $\partial_n \psi_S(\boldsymbol{r}, t)$ *for all* $t > t_0$.
- Initial values required: $\psi(\boldsymbol{r}, t_0)$ throughout V. Then ψ can be found for all $t > t_0$, but *not* in general for all $t < t_0$. (This equation is not invariant under time-reversal.)
  (In the 'hyperspace' spanned by $\boldsymbol{r}$ and t jointly, the initial conditions are on the same footing as the BCs. Thus, (initial +

boundary) ≡ 'hyperboundary' conditions are required over an *open* hypersurface. Hyper BCs specified over a closed hypersurface would in general be inconsistent (i.e. we may not additionally specify $\psi(\boldsymbol{r}, t_1)$ throughout V at some $t_1 > t_0$).)

Wave equation (hyperbolic): $\left(\frac{1}{c^2}\frac{\partial^2}{\partial t^2} - \nabla^2\right)\psi(\boldsymbol{r}, t) = \rho(\boldsymbol{r}, t)$.

- BCs: *for finite V*, as for diffusion. (Then only discrete frequencies are allowed.) For *infinite V* (parts or all of S at ∞), outgoing-wave conditions at infinity.
- Initial conditions: both $\psi(\boldsymbol{r}, t_0)$ and $\psi_t(\boldsymbol{r}, t_0)$ throughout V. Then ψ can be found for all t, both $t > t_0$ and $t < t_0$. (This equation is invariant under time-reversal.)

Helmholtz equation (elliptic): $-(\nabla^2 + k^2)\psi(\boldsymbol{r}) = \rho(\boldsymbol{r})$. Very like Poisson's equation in most respects, except that even the homogeneous problem has non-trivial solutions for special values of k^2. In infinite regions: 'outgoing-wave' conditions at infinity.

The chief object of this chapter is to introduce the partial differential equations occupying the rest of the book; to articulate some rather general questions about them that arise almost immediately; and to outline some notation and some points of view that prove useful in addressing them. Accordingly, the chapter should be read mainly as a preview, particularly of the sheer variety of the physics and the mathematics to come: the reader should look to later chapters for a fuller account of the physics, and for backing to many mathematical facts that at this preliminary stage will be merely asserted.

3.1 Notation

A point in a space of 3, 2, or 1 dimension is denoted by the vector $\boldsymbol{r}$. The region of $\boldsymbol{r}$ where the unknown function $\psi(\boldsymbol{r})$ or $\psi(\boldsymbol{r}, t)$ is required is written V; the closed surface surrounding V is written S. S may consist of several disconnected pieces: e.g. as in Fig. 3.1 of two concentric spheres, with V the region between them. The vector element of surface $\mathrm{d}\boldsymbol{S}$ points along the outward normal: e.g. away from the centre on the larger and towards the centre on the smaller sphere. Parts or all of S may be at infinity: e.g. the larger sphere may have infinite radius.

Integrals with respect to volume are written, interchangeably, as $\int_V \mathrm{d}V$ or $\int_V \mathrm{d}^n r$ (the second form in nD, where $n = 3$, 2, or 1). Integrals over the surface are written as $\int \mathrm{d}S$. If $S = S_1 + S_2$ as in Fig. 3.1, then

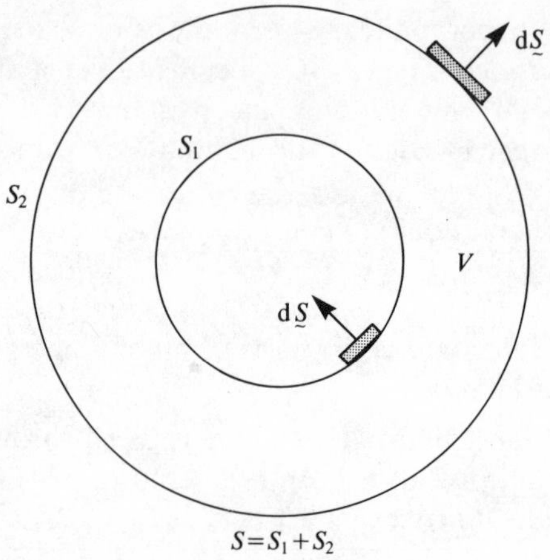

Fig. 3.1

$\int_S \mathrm{d}S = \int_{S_1} \mathrm{d}S + \int_{S_2} \mathrm{d}S$. The component of grad ψ along the (outward) normal (unit vector $\boldsymbol{n}$) is written as $\partial\psi/\partial n$ or $\partial_n\psi$, and similarly for other vectors. For instance, Gauss's theorem reads

$$\int_V \mathrm{d}V\, \nabla^2\psi = \int_S \mathrm{d}S\, \partial_n\psi = \int_S \mathrm{d}\boldsymbol{S} \cdot \boldsymbol{\nabla}\psi.$$

Finally, we shall sometimes use $\psi_S(\boldsymbol{r})$ (or $\psi_S(\boldsymbol{r}, t)$) and $\partial_n\psi_S(\boldsymbol{r})$ (or $\partial_n\psi_S(\boldsymbol{r}, t)$) for a function ψ and its outward-normal derivative at points $\boldsymbol{r}$ on the boundary S of V. (The awkwardness of this notation reflects an inherent dilemma discussed further in Appendix G.) The suffix S must not be confused with the usual partial-derivative suffix as in $\psi_x \equiv \partial\psi/\partial x$, etc.

Note that in a 3D problem, S is a surface as commonly understood (a 2D manifold). In a 2D problem, S is a closed plane curve (a 1D manifold). In a 1D problem, S consists of just the two endpoints of the interval V, and $\int_S \mathrm{d}S$ degenerates into a sum of the integrand at these two points, e.g. A and B as shown in Fig. 3.2. Then the direction of $\mathrm{d}\boldsymbol{S}$ is positive at B and negative at A; e.g.

$$\int_S \mathrm{d}\boldsymbol{S} \cdot \boldsymbol{\nabla}\psi = \frac{\partial\psi}{\partial x}(B) - \frac{\partial\psi}{\partial x}(A).$$

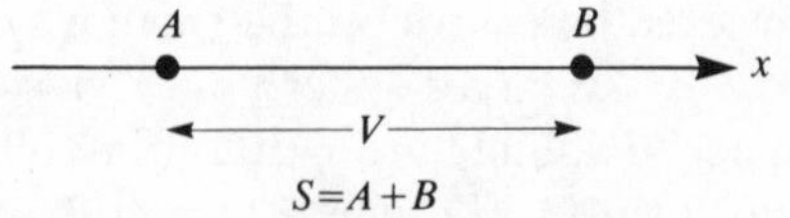

Fig. 3.2

The functions we seek will always be required to be single-valued. This must be borne in mind if V is not singly connected, e.g. if V is a torus (doughnut) or (the circumference of) a circle.

3.2 A preview of some standard equations

All the following equations are linear and of second order. We continue to observe the convention that the differential operator L on the left contains the Laplace operator ∇^2 prefaced by a minus sign.

(i) *Poisson's equation* reads

$$-\nabla^2\psi(\mathbf{r}) = \rho(\mathbf{r}). \tag{3.2.1}$$

The special homogeneous case with $\rho = 0$ is called the Laplace equation. The 'source density' ρ is prescribed; ψ is to be determined. In electrostatics (3.2.1) expresses Gauss's law (see Section 2.1, example (ii)); other manifestations of Poisson's equation are described in Section 4.1.

(ii) *The diffusion equation*, also called the heat equation, reads

$$\{\partial/\partial t - D\nabla^2\}\psi(\mathbf{r}, t) = \rho(\mathbf{r}, t). \tag{3.2.2}$$

Here ψ is the concentration, say of a solute: D is the diffusion constant entering Fick's law for the diffusion current density $\mathbf{j} = -D \operatorname{grad} \psi$; and ρ is the source-strength density (i.e. an amount $\rho(\mathbf{r}, t)\,\delta V\,\delta t$ of solute is injected in the interval δt into the volume element δV around $\mathbf{r}$). Then (3.2.2) expresses the conservation of the solute. Other manifestations of the equation are described in Section 8.1.

When nothing changes with time, the diffusion equation reduces to Poisson's equation.

(iii) *Schroedinger's equation* for a non-relativistic particle reads

$$\left\{-i\hbar\frac{\partial}{\partial t} + \left(-\frac{\hbar^2}{2m}\nabla^2 + V(\mathbf{r})\right)\right\}\psi(\mathbf{r}, t) = 0. \tag{3.2.3}$$

This equation is always homogeneous. When $V(\mathbf{r})$ vanishes everywhere, the equation is called 'free'. The free Schroedinger equation becomes the diffusion equation under the replacements

$$t \to -it, \qquad \hbar/2m \to D. \tag{3.2.4}$$

Unlike all the other equations named here, (3.2.3) and the Schroedinger wave function ψ are essentially complex, on account of the i on the left.

(iv) *The wave equation.* For scalar waves, e.g. for sound waves, where ψ is the velocity potential ($\mathbf{v} = -\operatorname{grad} \psi$), this reads

$$\left\{\frac{1}{c^2}\frac{\partial^2}{\partial t^2} - \nabla^2\right\}\psi(\mathbf{r}, t) = \rho(\mathbf{r}, t). \tag{3.2.5}$$

Here, c is the phase velocity of plane waves; the equation applies only if c is independent of $\boldsymbol{r}$, t, and of frequency (or equivalently of wavenumber). On the right, ρ is the source-strength density (i.e. a volume $\rho(\boldsymbol{r}, t)\,\delta V\,\delta t$ of fluid is injected in the interval δt into the (geometric) volume element δV. By virtue of Maxwell's equations, the same equation applies (with $\rho = 0$) if ψ is any *Cartesian* component of the electromagnetic fields $\boldsymbol{E}$ or $\boldsymbol{B}$ *in vacuo*. Other manifestations of the wave equation are described in Section 10.1.

When there is no time-dependence, the wave equation like the diffusion equation reduces to Poisson's equation. This is trivial. What is not trivial is that (3.2.5) reduces to Poisson's equation also in the so-called non-relativistic limit $c \to \infty$. In that limit the differential operator on the left of (3.2.5) ceases to refer to time at all. Hence, mathematically, t ceases to play the role of an independent variable, and assumes that of a mere parameter that may be needed to specify ρ. In particular, t no longer affects the manner in which the solution $\psi(\boldsymbol{r}, t)$ depends on the data $\rho(\boldsymbol{r}, t)$.

(v) *The Helmholtz equation.* Particularly for the wave equation one is often interested in the special case of simple-harmonic time variation, with $\psi(\boldsymbol{r}, t) = \exp(-i\omega t)\phi(\boldsymbol{r})$. Inhomogeneous problems have such solutions if $\rho(\boldsymbol{r}, t) = \exp(-i\omega t)\rho_\omega(\boldsymbol{r})$, and if any inhomogeneous boundary conditions are likewise simple harmonic. The physically requisite solution of a classical problem is usually the real or the imaginary part of such a complex ψ; but in quantum mechanics the solution of the Schroedinger equation is truly complex. As regards harmonic solutions, the wave equation reduces to the Helmholtz equation

$$-\{\nabla^2 + k^2\}\phi(\boldsymbol{r}) = \rho_\omega(\boldsymbol{r}) \tag{3.2.6}$$

with $k^2 = \omega^2/c^2$; the free Schroedinger equation also reduces to this, but with $k^2 = 2mE/\hbar^2$, and always with $\rho_\omega = 0$.

Obviously Poisson's equation is a special case, with $k^2 = 0$. Chapters 5 and 6 will show that this case is indeed very special.

Recall that the *normal modes* of a system are precisely the simple harmonic motions $\exp(-i\omega_n t)\phi_n(\boldsymbol{r})$ it can execute in the absence of external forcing either through an inhomogeneous source term in the equation or through inhomogeneous boundary conditions. Thus they are given by the solutions of the homogeneous Helmholtz equation subject to homogeneous boundary conditions. In this context the equation is best rearranged as follows, identifying $k^2 = \lambda_n$:

$$-\nabla^2\phi_n(\boldsymbol{r}) = \lambda_n\phi_n(\boldsymbol{r}). \tag{3.2.7}$$

This eigenvalue problem is fundamental to much that follows, and will be discussed in Section 4.5 and in Chapters 5 and 6. We anticipate four important facts. (i) For systems confined to finite volumes V, such simple-harmonic motion is possible only at certain discrete frequencies, determined by the eigenvalues λ_n in (3.2.7). That these eigenvalues are discrete is proved in Problem 5.12. (ii) The λ_n cannot be negative; (iii) under Dirichlet boundary conditions no λ_n can be zero; (iv) under Neumann boundary conditions, the lowest λ_n always is zero. The results (ii–iv) are derived in Section 4.5.

3.3 3D, 2D, and 1D

The real world is 3D. Nevertheless 2D (and 1D) problems arise in at least three significantly different ways.

(i) We may have a 3D system where, often for reasons of symmetry, none of the data and none of the constraints vary with one (or two) of the three Cartesian coordinates. For instance, we may have to solve the wave equation with a source density independent of the z-coordinate; we might have $\rho(\boldsymbol{r}, t) = \rho(x, y, t)$, with ρ non-zero only within some infinitely long right cylinder parallel to the z-axis, and varying with time exactly in phase at all values of z. Then it is obvious from the outset that the crux of the matter is to find solutions ϕ that are likewise independent of z:

$$\left\{\frac{1}{c^2}\frac{\partial^2}{\partial t^2} - \frac{\partial^2}{\partial x^2} - \frac{\partial^2}{\partial y^2}\right\}\phi(x, y, t) = \rho(x, y, t). \tag{3.3.1}$$

Such problems always stem from some idealization: for instance, it is unlikely that a source cylinder is infinitely long in fact, or that the phase of ρ is exactly the same for all z however large.

Often it is from choice rather than absolute necessity that one confines oneself to z-independent solutions like ϕ. For instance, if the homogeneous 3D equation $\Box^2\chi = 0$ has solutions $\chi(x, y, z, t)$ that do depend on z, then $(\phi + \chi)$ satisfies the inhomogeneous 3D equation just as ϕ does: $\Box^2(\phi + \chi) = \rho$. However, the essence of the problem is the discovery of a particular integral ψ satisfying $\Box^2\psi = \rho$, rather than of complementary functions χ; and the particular integral is evidently simplest when it is independent of z.

(ii) We may have a system like a stretched membrane (a drumhead) driven by a vibrator. Of course the 2D wave equation (3.3.1) which, naturally, one starts from, again stems from an idealization, because the thickness of the membrane is neglected, and so therefore is the possibility of waves travelling across rather than along the membrane. Nevertheless, physically the situation is very different from (i), because now there exists no line of sources extending to infinity in some not-explicitly-mentioned direction.

Thus, though the mathematics of the two cases is the same, the physics is very different. Later on, we shall have to bear this difference in mind when we try to give a physical interpretation of differences of behaviour in different dimensions.

(iii) As regards Poisson's equation under Neumann boundary conditions, a long narrow 2D strip becomes effectively 1D when the width of the strip drops below some critical value, depending on the smallest distance measurable lengthwise. The same happens as regards the diffusion equation, except that the critical width then depends also on the smallest measurable time interval. Similar reductions can operate from 3D to 2D. The detailed explanation requires some technicalities, and is deferred to Sections 6.5 and 9.4.5.

In an obvious sense this effect is complementary to (i): here it is the small dimension of the system that disappears from the equations, whereas in (i) it was a dimension along which the system extends to infinity.

One can also envisage fictitious worlds where space has any number of dimensions, not only one, two, or three, but also four or more. It turns out that from many points of view the most significant distinction is between spaces of even and odd dimensionality: for instance, 3D and 1D are more closely akin to each other than either is to 2D (at least this is true analytically even if not topologically).

3.4 Time and space variables

Mathematically, it is of course tempting at first sight to try and treat all the independent variables, t as well as $\boldsymbol{r}$, on essentially the same footing. For instance, under the substitutions

$$x_4 \equiv ict, \qquad x_{1,2,3} \equiv x, y, z \tag{3.4.1}$$

the wave operator transforms as follows:

$$\left\{\frac{1}{c^2}\frac{\partial^2}{\partial t^2} - \nabla^2\right\} = -\left\{\frac{\partial^2}{\partial x_1^2} + \frac{\partial^2}{\partial x_2^2} + \frac{\partial^2}{\partial x_3^2} + \frac{\partial^2}{\partial x_4^2}\right\}. \tag{3.4.2}$$

Then one might be tempted think of the solution of Poisson's equation in 2D, and of the wave equation in 1D, as of essentially analogous problems, namely to determine, from some boundary conditions and from the source terms, the requisite function over some region, e.g. the regions R shown in Fig. 3.3. When time is one of the independent variables, we call such a region a 'hypervolume', and its boundary a 'hypersurface'. In Fig. 3.3b, the hypersurface is closed, and consists of (i) the spatial region $0 \leqslant x \leqslant x_1$ at time 0; (ii) the same spatial region at time t_1; (iii) the point $x = 0$ between the times $t = 0$ and $t = t_1$; and (iv) the

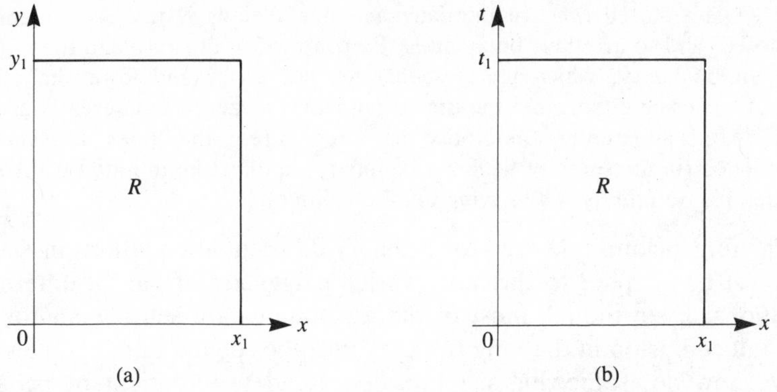

Fig. 3.3

point $x = x_1$ between the same two times. The symbol V is reserved for regions in ordinary 3D, 2D, or 1D space; R denotes a region where ψ is required, irrespective of whether this is a region in hyperspace (as for the diffusion or the wave equation), or in ordinary space (as for Poisson's equation).

Exercise: Describe the hyperboundary in the 3D problem analogous to Fig. 3.3b, where the independent variables are x, y, z, t.

Conditions that ψ must satisfy on a hypersurface will be called 'hyperboundary conditions'. Thus, if R is a region of hyperspace, say as in Fig. 3.3b, then hyperboundary conditions might include both ordinary boundary conditions (along the left-hand and right-hand edges of R), and also initial conditions (along the bottom edge). In principle they could include final conditions as well (along the top edge).

However, though the terminology of hyperspace can be convenient, most often such formal analogies between $\boldsymbol{r}$ and t are deceptive: the approach they suggest proves inefficient in problems with stationary boundaries , which are the only type we shall consider. In particular, we shall see that (ordinary) boundary conditions on the one hand and initial conditions on the other hand enter the problems in essentially different ways, even though the equation itself may be made to look symmetric, as in (3.4.2).

In calling surfaces or hypersurfaces and volumes or hypervolumes open or closed (as above), the words are used in their ordinary sense, for instance to distinguish, say in 2D, a curve with two distinct endpoints from a closed loop. This is different from the usage in set theory, where a closed set is one that includes all its limit points (e.g. $0 < x < 1$ is an open set (or interval), while

$0 \leqslant x \leqslant 1$ is a closed set). In our language, the regions R in Figs 3.1 and 3.3 are closed, and so are their boundaries. By contrast, in the xy-plane the x-axis is an open boundary, which merely separates the upper and lower halfplanes, without enclosing either; and the upper halfplane, regarded indifferently as $y > 0$ or as $y \geqslant 0$, is an open region. Notice that a region (e.g. the upper halfplane) can be bounded (in the sense of having a boundary), and yet be infinite (in the sense of extending to infinity, and having infinite volume).

The nomenclature devised for ordinary differential equations in Section 2.1 must be adapted to the more varied exigencies of partial differential equations. Even though most of the adaptations are self-explanatory, to forestall confusion in the long run they must be spelled out at some stage, as we now do, somewhat pedantically. Readers impatient of pedantry might prefer to skip the rest of this section now, and to return here only from the end of the chapter.

We saw in Chapter 2 that, with ordinary differential equations, BCs and ICs are mutually exclusive alternatives: depending on the physical problem, one or the other type of condition is needed in order to select a unique solution-function $\psi(x)$ of the one independent variable x. By contrast, to select a unique solution of a time-dependent partial differential equation in hyperspace, we shall need hyperboundary conditions of both types, i.e. both ordinary boundary conditions and initial conditions. These were illustrated in the second paragraph of this section, with reference to Fig. 3.3b.

Generalizing the nomenclature of Section 2.1, we say that *Dirichlet* (D) conditions prescribe the value of ψ on a hyperboundary; *Neumann* (N) conditions prescribe the (hyper)normal derivative of ψ, i.e. $\partial_n \psi$ on ordinary boundaries S (the vertical edges of R in Fig. 3.3b), and $\partial\psi/\partial t$ on constant-time hyperboundaries (like the bottom edge of R). By convention, $\partial_n \psi$ is positive if it points outwards from V. (We do not adopt any such convention for time-derivatives, because, except for a few asides, we shall consider only initial but not final conditions.) Churchill and mixed hyperboundary conditions can also occur; this should be borne in mind, though, for brevity, we shall not acknowledge them explicitly.

We need also to define *Cauchy* conditions: Cauchy BCs prescribe both $\psi(\boldsymbol{r}, t)$ and $\partial_n \psi(\boldsymbol{r}, t)$ for $\boldsymbol{r}$ on S but for all t, while Cauchy ICs prescribe both $\psi(\boldsymbol{r}, t)$ and $\partial_t \psi(\boldsymbol{r}, t_1)$ at given $t = t_1$, but for all $\boldsymbol{r}$.

With the equations not involving time, we continue to call a *problem* homogeneous only if both the equation and the BCs are so: i.e. there must be no source term in the equation, and all the quantities prescribed on S must have the value zero. As with ordinary differential equations, in finite volumes such homogeneous problems have solutions only for special parameter values, e.g. for the eigenvalues λ_n of the operator $-\nabla^2$ as governed by eqn (3.2.7).

By contrast, with time-dependent equations we call a problem homogeneous if the equation and the BCs are so, but irrespective of the ICs. This definition of homogeneity with respect to hyperspace is *not* the formal analogue of the definition given in the preceding paragraph with respect to ordinary space. But it turns out to be the only one worth formulating: in the time-dependent problems we meet in practice, we shall see that if all the initial values were zero as well as all the boundary values and all the source terms, then there would be no non-trivial solution at all.

A final disclaimer is appropriate about the operators, call them L, on the left of our time-dependent equations (3.2.2–5). For us, their role as operators in eigenvalue problems in hyperspace is of next to no importance: for instance, we shall not use expansions in terms of their eigenfunctions, so that very little of any direct relevance turns on the question whether such an L acting on functions of both t and $\mathbf{r}$ is Hermitean under some suitable choice of hyperboundary conditions. What will prove crucial, in time-dependent as well as in time-independent problems, is that $-\nabla^2$ is a Hermitean operator in a region V of ordinary space, when subject to ordinary homogeneous BCs on S. As regards the dependence of solutions on $\mathbf{r}$, this makes available the full armoury of complete sets $\{\phi_n(\mathbf{r})\}$ already introduced in Section 2.4. If one chooses an approach along such lines, then it is only in their account of the time-dependence that partial differential equations display their most characteristic differences from ordinary differential equations.

3.5 Hyperboundary conditions, and the classification of partial differential equations

3.5.1 'Well-posed' problems

Try an analogy with ordinary differential equations (ODEs). There, we want $\psi(x)$ over a region of the x-axis, i.e. over a *line*, which is bounded by two *points*. To determine a solution uniquely, we need either the value of the function at these two points, or else $\psi(x)$ and $\psi'(x)$ at one of them. In either case, the general solution of just the equation contains *two* adjustable (initially arbitrary) *constants*.

For partial differential equations (PDEs), think for simplicity of only two independent variables, as in Fig. 3.3: two space variables for Poisson's or the Helmholtz equation, or one space and one time variable for the diffusion or the wave equation. Now we want ψ in a two-dimensional region, i.e. over an *area* in the xy- or the xt-plane, which is bounded by a one-dimensional *curve*. Thus we might guess that in order to specify a solution uniquely, we need to be given the function and its first partial derivatives along this curve. In fact, since the function along the curve automatically determines the partial derivative in the tangential direction, this amounts to specifying the function, plus its partial derivative in the direction normal to the curve. *If* this conjecture were correct, then the general solution to just the equation would include *two* adjustable *functions* of one independent variable each.

Unfortunately, things are not so simple. A general theory does exist starting with $\boldsymbol{r}$, t on an equal footing, but it ramifies quite rapidly, and in any case it ends up with $\boldsymbol{r}$ and t on a very unequal footing. This is the 'method of characteristics', which will not be used here. We can dispense with it because we consider only stationary boundaries, and because all our equations are not only linear, but have constant coefficients.

Instead, we shall simply state for each type of equation what hyperboundary conditions it needs. Luckily these are suggested by physical intuition, if applied with care.

In fact, for solutions to be useful in our kind of physics, one needs hyperboundary conditions such that (i) solution exist; (ii) they are unique; (iii) they are stable.

(i) *Existence* implies that the conditions do *not overdetermine* the problem, i.e. that they are not incompatible.

Traditionally, physicists have not concerned themselves with existence proofs, and nor shall we. This does have an admitted disadvantage even from a strictly practical point of view. Lacking a prior existence proof, but given a would-be solution that has been subjected to the equation and to the hyperboundary conditions in turn (in either order), we should still, in principle, verify by substitution that it satisfies both requirements, since these might have been incompatible. All the solutions we shall find do (or would) pass an appropriate existence test; however, for several of them this is not obvious by mere inspection, and existence proofs would save some trouble (see Sections 7.1 and 7.4.1).

On the other hand, before it can be asserted that a solution does or does not exist, obviously it is necessary to define the class of functions to which an acceptable solution should belong; and mathematicians have an unfortunate history of defining such classes prematurely, and so narrowly that they exclude solutions perfectly acceptable to the physicist. Hence the above proviso that existence tests must be 'appropriate'. We shall illustrate this in Section 7.2.

(ii) *Uniqueness* implies that the conditions do *not underdetermine* the solution. This is important, because it assures us that any solution, no matter how obtained, is the correct solution. Uniqueness proofs will be given, in later chapters.

(iii) *Stability* means that arbitrarily small changes in the data (sources, ICs, or BCs) do not, for *any* physically relevant values of $\boldsymbol{r}$ and t, lead to finite (eventually: large) changes in the solution (see Section 7.4.2).

Up to about a generation ago, most workers seem to have thought it obvious that PDEs with unstable solutions could play no part in physics, largely on the grounds that input data are always of limited accuracy, but must none the less suffice for prediction. By contrast, more recently there has been much interest in simple systems of ODEs governing physical behaviour that appears random ('chaotic'), precisely because it is sensitive to undetectably small changes in the initial conditions. Admittedly such equations are always non-linear; but the mere fact that chaotic processes can be studied with profit implies that unstable problems are no longer self-evidently irrelevant to physics. Nevertheless, in this book we consider only stable problems.

Following Hadamard, problems that satisfy the requirements (i), (ii), (iii) listed above are called *well-posed*. We shall see that it is easy to sin against all three.

3.5.2 Elliptic, parabolic, and hyperbolic equations

To identify what problems are well-posed, we need just the beginnings of the general classification of second-order (but not as yet necessarily linear) PDEs. For simplicity we continue to work with only two independent variables ξ and η, where ξ is x, while η can be either the second Cartesian variable $r_2 = y$, or else time t. Write the equation, without further loss of generality, as

$$\left\{A\frac{\partial^2}{\partial\xi^2} + B\frac{\partial^2}{\partial\xi\,\partial\eta} + C\frac{\partial^2}{\partial\eta^2}\right\}\psi(\xi, \eta)$$
$$= \left(\text{function of } \xi, \eta, \psi, \frac{\partial\psi}{\partial\xi}, \frac{\partial\psi}{\partial\eta}\right). \tag{3.5.1}$$

Then the equation is called

$$\left.\begin{array}{lll} \text{elliptic} & \text{if} & B^2 - 4AC < 0; \\ \text{parabolic} & \text{if} & B^2 - 4AC = 0; \\ \text{hyperbolic} & \text{if} & B^2 - 4AC > 0. \end{array}\right\} \tag{3.5.2}$$

(These names presumably allude to what emerges if, on the left of (3.5.1), one replaces $\partial/\partial\xi \to \xi$, etc. If A, B, C are constants, then $(A\xi^2 + B\xi\eta + C\eta^2) = \text{constant}$ is the equation for the eponymous conic section.) Thus Poisson's and the Helmholtz equation are elliptic; the diffusion equation is parabolic; and the wave equation is hyperbolic.

Like the rest of the general theory, this classification comes fully into its own only if the equations are non-linear, or if A, B, C themselves are functions of ξ, η, which is not the case in any of our problems.

We proceed to the hyperboundary conditions appropriate under the general requirements (i), (ii), (iii) for a problem to be well-posed, but only for regions bounded either by constant-time hypersurfaces, or by fixed (stationary-in-time) geometric surfaces (or jointly by both). The assertions will be partially substantiated by uniqueness proofs, and by some dicusssions of existence and stability, when these equations are dealt with in later chapters.

3.5.3 Poisson's equation (elliptic)

The problem is well-posed if it asks for $\psi(\boldsymbol{r})$ in some arbitrary region V, as illustrated in the 2D case in Fig. 3.4, subject to either DBCs *or* NBCs on the *closed* surface S bounding V. Part or all of S may be at infinity. Here we recall the remarks following the proof of (2.4.11), and those at the end of Section 3.2 above, and state that with DBCs a (unique and stable) solution always exists because zero is never an eigenvalue of $-\nabla^2$; with NBCs, zero is always an eigenvalue, and a solution exists only if the prescribed boundary values satisfy a consistency requirement, as discussed in Section 6.1.

Cauchy BCs on a closed surface S would overdetermine the problem. On an open surface, DBCs or NBCs would be insufficient, and Cauchy conditions would be unstable.

Accordingly, in a problem well-posed in n dimensions, the most general solution of just the equation contains only one adjustable function of $(n-1)$ variables, and not two such functions, in spite of the fact that the equation is second order. Thus, for Poisson's equation, the conjecture voiced in Section 3.5.1 is not realized. Section 7.4 discusses its failure in some detail; meanwhile the conclusion is familiar from electrostatics, where the solution is uniquely determined once we

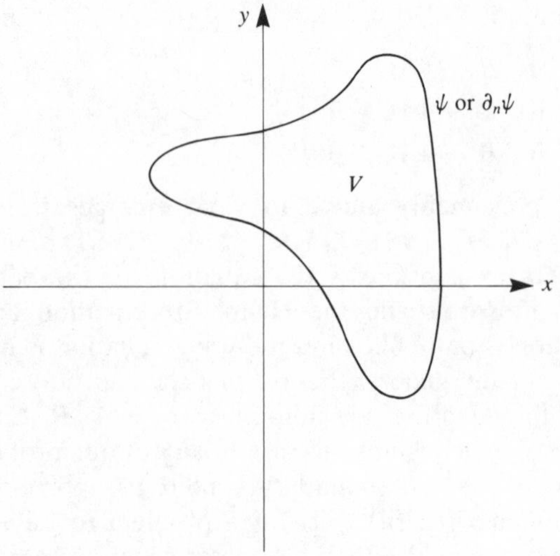

Fig. 3.4

prescribe either the potential ψ_S or the normal component of the field $-\partial_n\psi_S$.

If S recedes to infinity, then in physically relevant cases one requires, in 3D, that $\psi_S = 0$. We shall see that in 2D one might have to settle for the weaker condition that the field, $-\text{grad}\ \psi_S$, vanish; and in 1D, for the still weaker condition that grad ψ_S remain bounded.

3.5.4 The diffusion equation (parabolic)

The problem is well-posed if it asks for $\psi(x, t)$ in the region R shown in Fig. 3.5a, subject for all t between t_1 and t_2 either to DBCs or to NBCs at $x = x_1$ and $x = x_2$, prescribing, along the vertical edges of R, either $\psi(x_{1,2}, t)$ or $\psi_x(x_{1,2}, t)$ but not both; and subject also to Dirichlet ICs prescribing $\psi(x, t_1)$ along the lower horizontal edge, i.e. for x between x_1 and x_2. Under these conditions a unique and stable solution always exists. No further conditions at all may be imposed at the final time $t = t_2$, i.e. along the broken line marking the top boundary of R, for they would overdetermine the problem. Thus, for parabolic equations hyperboundary conditions are required along an *open* hypersurface, i.e. only along the solid-drawn boundaries of R.

Given $\psi(x, t_1)$ and thereby $\nabla^2\psi(x, t_1)$, the equation (3.2.2) itself determines $\psi_t(x, t_1)$. Hence there can be no question of ICs prescribing ψ_t independently of ψ; in other words, Cauchy ICs would be trivially inconsistent in general. Hindsight will show this to reflect the fact that the equation is only of first order in $\partial/\partial t$. The feature that only one function is prescribed along the geometric boundary (i.e. along the vertical sides in Fig. 3.5) is common to the diffusion and to Poisson's equations.

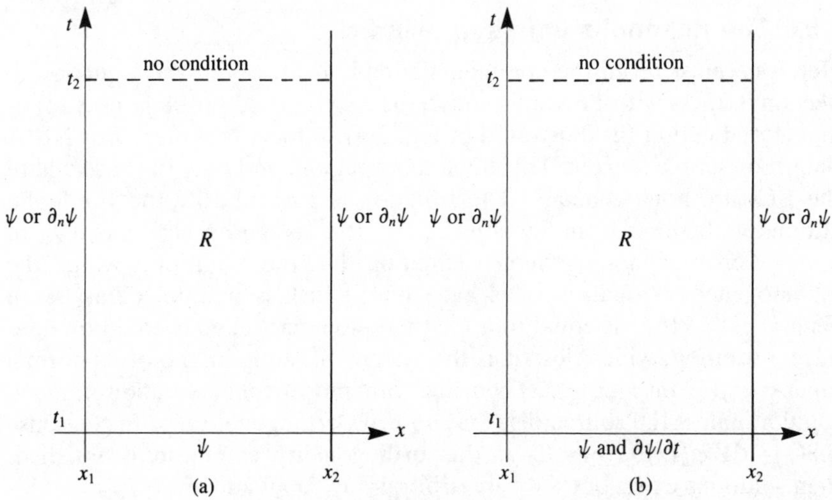

Fig. 3.5

We see that the conjecture from ODEs again proves misleading, because, for the diffusion as for Poisson's equation, only one adjustable function of one variable enters the general solution, and not two.

3.5.5 The wave equation (hyperbolic)

As with the diffusion equation, $\psi(x, t)$ is wanted in the region R of Fig. 3.5b, and subject to the same BCs along the vertical sides; but along the lower horizontal one must prescribe Cauchy ICs, i.e. both $\psi(x, t_1)$ and $\psi_t(x, t_1)$. By hindsight, this reflects the fact that the equation is of second order in $\partial/\partial t$. Under these conditions a unique stable solution always exists. No conditions may be prescribed along the top horizontal, for they would again overdetermine the problem. Mathematically, because the wave equation is reversible, one could prescribe ψ and ψ_t at t_2 instead of t_1; but physically this option is rarely interesting, because the future is rarely known. There are no such options with the diffusion equation, which is irreversible. This is discussed in Sections 9.5 and 10.6.

For the wave equation, the conjecture from ODEs is almost correct, since the ICs (albeit not the BCs) do require, in nD, two adjustable functions of $(n-1)$ variables each. But the analogy with ODEs is not perfect, because other data are needed in addition, namely ψ or $\partial_n\psi$ on S.

If $x_1 \to -\infty$, or $x_2 \to +\infty$, or both, the physics of the problem must supply the appropriate BCs at infinity. For instance, we might have ψ localized in some finite region at $t = t_1$, and impose outgoing-wave-only BCs at $x = \pm\infty$; or we might have incoming waves just from the left, and impose the BC that there be only right-moving waves at $+\infty$.

3.5.6 The Helmholtz equation (elliptic)

Here one must recall the comments already made about this equation in Section 3.2. As with Poisson's equation, a well-posed problem asks for ψ in a closed region (as illustrated in Fig. 3.4), subject to either D or NBCs on S. For simplicity consider only a finite region, and only the case where the BCs are homogeneous. The situation is governed by the Fredholm alternative familiar from Sections 2.3, 4. If k^2 is not an eigenvalue λ_n of $-\nabla^2$, then the homogeneous equation has no solution, while the inhomogeneous equation does have one, which is unique. (This is the general case.) If k^2 is equal to a λ_n, then the homogeneous equation does have a solution, which describes the system vibrating in one of its normal modes $\exp(-i\omega_n t)\phi_n(\boldsymbol{r})$. Then the inhomogeneous equation has no solution unless the source distribution ρ_ω is orthogonal to ϕ_n in the sense that $\int_V dV\, \phi_n^*(\boldsymbol{r})\rho_\omega(\boldsymbol{r}) = 0$. If this orthogonality condition is satisfied, then solutions exist, but they are arbitrary up to an addend $C\phi_n(\boldsymbol{r})$. (This is the special special case). If the orthogonality condition fails, then no solution exists. (This is the general special case.) The reason is that the amplitude of such a resonantly driven system increases with time, so that the motion is not simple harmonic, contrary to the assumption that leads to the Helmholtz equation in the first place. (A very similar example occurred at the end of Section 2.5; see also Section 10.5.2 below.) There is no paradox, because the underlying wave equation describes such resonant responses quite routinely (see Section 10.5.2). Notice that even the best-posed Helmholtz equation leads to a BVP in ordinary space, which (in a finite region) may have no solution; by contrast, the well-posed wave equation, which with respect to time leads to an initial-value problem, does always have a solution.

It is evident that as regards elliptic equations in finite regions, the interesting qualitative differences between 1D and nD (i.e. between ODEs and PDEs) show up already *à propos* of Poisson's equation. Consequently, when we return to the Helmholtz equation in Chapter 13, we shall be concerned mainly with the phenomena it governs in infinite regions.

II Potentials

Summary

Poisson's equation: $-\nabla^2\psi(\boldsymbol{r}) = \rho(\boldsymbol{r})$.

BCs: either D (i.e. ψ_S given), or N (i.e. $\partial_n\psi_S$ given), always on a *closed* surface S; or mixed. But Cauchy conditions (both ψ_S and $\partial_n\psi_S$ given) are inconsistent in general.

Green's function: $-\nabla^2 G(\boldsymbol{r} \mid \boldsymbol{r}') = \delta(\boldsymbol{r} - \boldsymbol{r}')$; plus BCs:
for D problems, $G_D(\boldsymbol{r} \text{ on } S \mid \boldsymbol{r}') = 0$;
for N problems, $\partial_n G_N(\boldsymbol{r} \text{ on } S \mid \boldsymbol{r}') = -1/A$,
where $A \equiv$ (area of S).

Magic rules:

- *For* D *problems*:

$$\psi(\boldsymbol{r}) = \int_V \mathrm{d}V'\, G_D(\boldsymbol{r}' \mid \boldsymbol{r})\rho(\boldsymbol{r}') - \int_S \mathrm{d}S'\, \psi_S(\boldsymbol{r}')\, \partial'_n G_D(\boldsymbol{r}' \mid \boldsymbol{r})$$

  ($\boldsymbol{r}$ in V; in $\int_S \cdots$, $\boldsymbol{r}'$ on S).
  With $-\nabla^2\phi_p = \lambda_p\phi_p$, $\phi_p = 0$ on S,

$$G_D(\boldsymbol{r} \mid \boldsymbol{r}') = \Sigma_p\, \phi_p^*(\boldsymbol{r}')\phi_p(\boldsymbol{r})/\lambda_p.$$

- *For* N *problems*:

$$\psi(\boldsymbol{r}) = \int_V \mathrm{d}V'\, G_N(\boldsymbol{r}' \mid \boldsymbol{r})\rho(\boldsymbol{r}') + \int_S \mathrm{d}S'\, (\partial'_n\psi_S(\boldsymbol{r}'))G_N(\boldsymbol{r}' \mid \boldsymbol{r}) + \langle\psi\rangle_S,$$

  where

$$\langle\psi\rangle_S \equiv \frac{1}{A}\int_S \mathrm{d}S\, \psi_S$$

  is the surface-average of ψ over S.

Pseudo Green's function for N problems: With $-\nabla^2\phi_p = \lambda_p\phi_p$, $\partial_n\phi_p = 0$ on S, define $H(\boldsymbol{r} \mid \boldsymbol{r}') = \Sigma'_p\, \phi_p^*(\boldsymbol{r}')\phi_p(\boldsymbol{r})/\lambda_p$, excluding ϕ_p's having $\lambda_p = 0$. Then

$$\psi(\boldsymbol{r}) = \int_V \mathrm{d}V'\, H(\boldsymbol{r}' \mid \boldsymbol{r})\rho(\boldsymbol{r}') + \int_S \mathrm{d}S'\, (\partial'_n\psi_S(\boldsymbol{r}'))H(\boldsymbol{r}' \mid \boldsymbol{r}) + \langle\psi\rangle_V,$$

where

$$\langle\psi\rangle_V \equiv \frac{1}{V}\int_V \mathrm{d}V\, \psi$$

is the volume-average of ψ over V.

Green's functions in unbounded space (S at infinity): $\boldsymbol{R} \equiv \boldsymbol{r} - \boldsymbol{r}'$

$$G_0^{(3)}(\boldsymbol{r} \mid \boldsymbol{r}') = \frac{1}{4\pi R}; \qquad G_0^{(2)}(\boldsymbol{r} \mid \boldsymbol{r}') = -\frac{1}{2\pi} \log R;$$

$$G_0^{(1)}(x \mid x') = -\tfrac{1}{2}|X| \equiv -\tfrac{1}{2}|x - x'|.$$

Kirchhoff representation (not a solution!):

$$\left\{\int_V \mathrm{d}V'\, G_0(\boldsymbol{r}' \mid \boldsymbol{r})\rho(\boldsymbol{r}') - \int_S \mathrm{d}S' \left[\psi(\boldsymbol{r}')\, \partial_n' G_0(\boldsymbol{r}' \mid \boldsymbol{r}) - (\partial_n' \psi(\boldsymbol{r}')) G_0(\boldsymbol{r}' \mid \boldsymbol{r})\right]\right\}$$

$$= \begin{cases} \psi(\boldsymbol{r}) & \text{if } \boldsymbol{r} \text{ is in } V, \\ 0 & \text{if } \boldsymbol{r} \text{ is not in } V. \end{cases}$$

4 Poisson's equation: I. Introduction

Since Poisson's is the first partial differential equation we meet, its treatment will be comparatively slow and explicit, serving to spell out ideas that later on will be exploited more briskly. Readers wishing to proceed as quickly as possible to the other equations, in Chapter 8 and beyond, need only Chapter 4, Chapter 5 (without Section 5.3.4), Section 7.5, and Appendix G.

4.1 Poisson's equation in physics

Poisson's equation reads

$$-\nabla^2\psi(\boldsymbol{r}) = \rho(\boldsymbol{r}). \tag{4.1.1}$$

The special homogeneous case without sources is called the Laplace equation:

$$\nabla^2\psi(\boldsymbol{r}) = 0; \tag{4.1.2}$$

it can apply only in limited regions.

In electrostatics, ρ is the charge density, the electric field is $E = -\text{grad}\ \psi$, and (4.1.1) expresses Gauss's law $\text{div}\ E = \rho$. The energy density in the field is $U = \frac{1}{2}(\text{grad}\ \psi)^2$. See the footnote in Section 2.1, regarding units.

For the steady irrotational flow of an incompressible fluid, $\boldsymbol{v}(\boldsymbol{r}) = -\text{grad}\ \psi$ is the fluid velocity at $\boldsymbol{r}$, so that a volume $\boldsymbol{v} \cdot \delta\boldsymbol{A}$ of fluid crosses the vector element of area $\delta\boldsymbol{A}$ per unit time; $\rho\ \delta V$ is the volume of fluid injected into δV per unit time; and (4.1.1) expresses the conservation of the fluid (by volume). With σ the mass density, the kinetic energy per unit volume is $\frac{1}{2}\sigma\ (\boldsymbol{\nabla}\psi)^2$.

In elasticity, $\psi(x)$ can denote the transverse displacement, say in the y-direction, of a stretched string under non-zero tension T (even when undisplaced). We confine ourselves to the approximation where the slope of the string is small enough to admit $\mathrm{d}\psi/\mathrm{d}x \equiv \tan\theta \approx \theta$. Then, to first order, the tension does not vary with the displacement, and produces a net transverse force

$$T\{\psi'(x+\delta x) - \psi'(x)\} \approx T\psi''(x)\ \delta x$$

on an element of length δx. If the externally applied transverse force per unit length is written as $F(x)$, static equilibrium demands $F(x)\,\delta x + T(\mathrm{d}^2\psi/\mathrm{d}x^2)\,\delta x = 0$, leading to (4.1.1) with $\rho(x) = F(x)/T$. The increase in length of the element δx is

$$\delta l = \{(\delta x)^2 + [\psi(x+\delta x) - \psi(x)]^2\}^{\frac{1}{2}} \approx \tfrac{1}{2}\,\delta x(\mathrm{d}\psi/\mathrm{d}x)^2,$$

whence the increase, due to the displacement, in the energy stored per unit length is $u = T\,\delta l/\delta x = \frac{1}{2}T(\mathrm{d}\psi/\mathrm{d}x)^2$.

Similarly, in 2D, the transverse displacement of an initially flat stretched membrane is governed by (4.1.1), with $T\rho(x, y)$ the externally applied transverse force per unit area. The underlying approximation is $|\nabla\psi| \ll 1$, and $u = \frac{1}{2}T(\nabla\psi)^2$. (In 2D, the tension T is defined as follows. Draw a line element of length δs on the (undisplaced) membrane: $T\,\delta s$ is the force (a pull) exerted in the plane of the membrane, along the normal to δs, by the material on either side of the line element on the material on the other side. Thus $[T] = [\text{force}]/[L]$.) See e.g. Coulson and Jeffrey (1977) for a derivation of the wave equation for the stretched string and membrane; Poisson's equation follows in the special case of equilibrium (time-independent ψ).

As discussed in Chapter 3, in order to determine ψ uniquely over the region V, one needs BCs on the closed surface S of V. DBCs prescribing ψ_S are the most common in electrostatics: e.g. $\psi_S = 0$ on an earthed conductor. NBCs are the more common in fluid mechanics: e.g. $v_n = -\partial_n\psi_S$ on an impenetrable wall. In elasticity, DBCs prescribe the displacement at the ends or along the rim of the system, while NBCs prescribe the externally applied transverse force there. On parts of S extending to infinity, the prescribed values are almost always zero in 3D, but we shall see that in 2D and 1D this is not so. Section 7.4 shows the penalties for attempts to impose wrong types of BCs.

An awkwardness in the notation for prescribed boundary values is discussed in Appendix G.

4.2 Uniqueness

Theorem: The solution of a problem under DBCs, if it exists, is unique. The solution of a problem under NBCs, if it exists, is unique up to an additive constant.

Proof: Let eqn (4.1.1) be satisfied throughout V by two functions ψ_1 and ψ_2; and let both ψ_1 and ψ_2 satisfy the BCs, so that $\psi_{1S} = \psi_{2S}$ for DBCs, or $\partial_n\psi_{1S} = \partial_n\psi_{2S}$ for NBCs. We consider the difference

$$\phi(\mathbf{r}) = \psi_1(\mathbf{r}) - \psi_2(\mathbf{r}); \tag{4.2.1}$$

it obeys the Laplace equation, since

$$\nabla^2\phi = \nabla^2\psi_1 - \nabla^2\psi_2 = -\rho + \rho = 0, \tag{4.2.2}$$

and satisfies *homogeneous* BCs, since $\phi_S = \psi_{1S} - \psi_{2S} = 0$, or $\partial_n\phi_S = \partial_n\psi_{1S} - \partial_n\psi_{2S} = 0$.

The basic idea is to consider the integral

$$J \equiv \frac{1}{2}\int_V dV(\text{grad}\,\phi)^2, \tag{4.2.3}$$

which, up to a proportionality constant, is the energy corresponding to ϕ. (This is not the same as the difference between the energies corresponding to ψ_1 and ψ_2.) A standard vector identity followed by appeal to (4.2.2) gives

$$\text{div}\,(\phi\,\text{grad}\,\phi) = (\text{grad}\,\phi)^2 + \phi\,\nabla^2\phi = (\text{grad}\,\phi)^2.$$

Substituting this into (4.2.3) and using Gauss's theorem we obtain

$$J = \frac{1}{2}\int_V dV\,\text{div}\,(\phi\,\text{grad}\,\phi) = \frac{1}{2}\int_S dS\,\phi_S\,\partial_n\phi_S = 0: \tag{4.2.4}$$

the surface integral vanishes because either ϕ_S or $\partial_n\phi_S$ vanishes. But the integrand in (4.2.3) is non-negative; such an integral can vanish only if the integrand vanishes everywhere (because it is impossible for individually positive and negative contributions to cancel, there being no negative contributions). Thus grad $\phi = 0$ everywhere, whence ϕ is a constant. Under NBCs this is all we can show. Under DBCs, ϕ vanishes on the boundary S; hence its constant value is actually zero. ■

Section 3.5.1 has already commented on the role of uniqueness in 'well-posed' problems. Some other implications will emerge in Section 7.5 when the Kirchhoff representation is contrasted with the magic rules. Comments on existence and stability are deferred to Sections 7.1.2–4 and Appendix I.

Strictly speaking, our proof shows only that grad ϕ cannot be non-zero over any part of V having finite n-dimensional volume; for instance, the proof does not rule out the possibility that grad ϕ might be non-zero say at a number of discrete points in V, since this need not prevent the integral in (4.2.3) from vanishing. Indeed, Section 7.3 shows that the value of ϕ itself may be reassigned arbitrarily at discrete points of V, without thereby affecting the solution at any other point.

The uniqueness theorem is deliberately non-committal as to whether solutions exist. In fact, Section 4.5.1 shows that the Dirichlet BVP does always have a solution, since it always realizes the 'general case' of the Fredholm alternative set out in Section 2.4; and that the Neumann BVP always realizes the 'special case', having a solution (i.e. realizing the 'special special case') only if the data ρ and $\partial_n \psi_S$ obey the self-consistency condition imposed by Gauss's law, namely

$$\int_V \mathrm{d}V\,\rho = -\int_S \mathrm{d}S\,\partial_n \psi_S. \tag{4.2.5}$$

Regarding existence, some caution at this stage is indicated, because Appendix I shows that, in the variational method, arguments akin to the uniqueness proof can easily be misapplied to suggest that a solution exists even when in fact, as with Cauchy BCs, it does not.

4.3 The Laplace equation: harmonic functions

4.3.1 Examples of harmonic functions

Functions that satisfy the Laplace equation (4.1.2) in a region V are said to be *harmonic* in V.

We list some especially important harmonic functions. For the moment we are not concerned with any boundary conditions that might be imposed eventually; we shall merely point out that some of these functions are harmonic only in regions that exclude the origin, while it will be obvious that some cannot occur if the region of interest extends to infinity and if the solution (or perhaps its gradient) is required to vanish there, or to remain bounded.

In 1D, the Laplace equation reduces to the ordinary differential equation $\psi'' = 0$, whose most general solution is

$$\psi(x) = cx + d. \tag{4.3.1a}$$

In connected regions excluding the origin this is equivalent to

$$\psi(x) = a\,|x| + b. \tag{4.3.1b}$$

In 2D, we note the exponential solutions in terms of Cartesian coordinates

$$\psi_k(\mathbf{r}) = \exp(ikx)\{b^{(+)}(k)\exp(ky) + b^{(-)}(k)\exp(-ky)\}, \tag{4.3.2}$$

or their real and imaginary parts separately. Since the $\exp(ikx)$ form a complete set of functions of x, any 2D solution of the Laplace equation can be expressed as a linear combination of such functions, i.e. in the form $\int_{-\infty}^{\infty} \mathrm{d}k\,\psi_k(\mathbf{r})$, with appropriately chosen $b^{(\pm)}(k)$.

In terms of plane polar coordinates, we note

$$\psi_0(\mathbf{r}) = a_0 \log r + b_0, \tag{4.3.3a}$$

$$\psi_m(\mathbf{r}) = \exp(im\phi)\{a_m r^{-|m|} + b_m r^{|m|}\}, \qquad m = \pm 1, \pm 2, \ldots. \tag{4.3.3b}$$

The terms in $\log r$ and $r^{-|m|}$ (i.e. those assigned coefficients a) are not harmonic in regions V that include the origin, since $-\nabla^2$ acting on these functions gives not zero but $\delta(\boldsymbol{r})$ or one of its derivatives. In other words, these functions are solutions not of the Laplace equation, but of Poisson's equation with a charge density describing a point charge or a point multipole at the origin. In yet other words they are, as we shall see later, Green's functions or derivatives of Green's functions. (The same applies to $a\,|x|$ in (4.3.1b).) For instance, as solutions of the Laplace equation, both the a and the b terms are admissible in the region between two concentric circles; only the b terms are admissible inside the inner circle; and, generally, only the a terms are admissible outside the outer circle.

Because the exp (im ϕ) constitute a complete set of functions of ϕ, any solution of the Laplace equation can be expressed as a linear combination of these functions, i.e. in the form $\sum_{-\infty}^{\infty} \psi_m(\boldsymbol{r})$, with appropriately chosen a_m and b_m.

In 3D, the obvious analogues of (4.3.2) are

$$\psi_{\boldsymbol{\kappa}}(\boldsymbol{r}) = \exp\left(i(\kappa_1 x + \kappa_2 y)\right)\{b^{(+)}(\boldsymbol{\kappa}) \exp(\kappa z) + b^{(-)}(\boldsymbol{\kappa}) \exp(-\kappa z)\},$$
$$\boldsymbol{\kappa} \equiv (\kappa_1, \kappa_2), \qquad \kappa \equiv (\kappa_1^2 + \kappa_2^2)^{\frac{1}{2}}. \tag{4.3.4}$$

The solutions (4.3.2, 4) are special cases of

$$\psi_{\boldsymbol{K}}(\boldsymbol{r}) = \exp(\boldsymbol{K} \cdot \boldsymbol{r}) \tag{4.3.5a}$$

featuring a complex vector $\boldsymbol{K} = \boldsymbol{K}_1 + i\boldsymbol{K}_2$ of zero length, i.e. such that

$$\boldsymbol{K}^2 = \boldsymbol{K}_1^2 - \boldsymbol{K}_2^2 + 2i\boldsymbol{K}_1 \cdot \boldsymbol{K}_2 = 0, \tag{4.3.5b}$$

$$\boldsymbol{K}_1^2 = \boldsymbol{K}_2^2, \qquad \boldsymbol{K}_1 \cdot \boldsymbol{K}_2 = 0. \tag{4.3.5c}$$

Clearly the real vectors $\boldsymbol{K}_1$ and $\boldsymbol{K}_2$ are of equal length and mutually orthogonal; but otherwise their orientation and their common magnitude are arbitrary.

In cylindrical polar coordinates there are two different standard sets of harmonic functions, depending on z through $\exp(\pm kz)$ or $\exp(\pm ikz)$ respectively, and on the radial coordinate through Bessel functions (see e.g. Jackson (1962), Section 3.7, or Morse and Feshbach (1953), pp. 1259–63).

In spherical polars, one has

$$\psi_{lm}(\boldsymbol{r}) = Y_{lm}(\Omega)\{a_{lm} r^{-(l+1)} + b_{lm} r^l\},$$
$$l = 0, 1, 2, \ldots, \qquad m = 0, \pm 1, \ldots, \pm l. \tag{4.3.6}$$

Recall that the spherical harmonics Y_{lm} are homogeneous polynomials of order l in the direction cosines x/r, y/r, and z/r of the vector $\boldsymbol{r}$; those for $l = 0, 1, 2$ are listed in Appendix A (see also Appendix E).

The remarks about (4.3.2, 3) made above apply equally to (4.3.4, 6).

4.3.2 General properties of harmonic functions

Harmonic functions have remarkable properties, exploited in many applications.

(i) **Theorem:** If $\psi(\boldsymbol{r})$ is harmonic in V, then so are $\partial\psi/\partial x$, $\partial\psi/\partial y$, and $\partial\psi/\partial z$.

Proof: From $\nabla^2 = \partial^2/\partial x^2 + \partial^2/\partial y^2 + \partial^2/\partial z^2$ it follows that

$$\nabla^2(\partial/\partial x) - (\partial/\partial x)\,\nabla^2 = 0; \tag{4.3.7a}$$

in other words ∇^2 and $\partial/\partial x$ commute. Similarly, ∇^2 commutes with $\partial/\partial y$ and $\partial/\partial z$. Hence

$$\nabla^2(\partial\psi/\partial x) = (\partial/\partial x)(\nabla^2\psi) = 0. \quad \blacksquare \tag{4.3.7b}$$

Corollary: If $\psi(\boldsymbol{r})$ is harmonic in V, then so are its partial derivatives with respect to x, y, z, to all orders.

Proof: From the theorem, by induction. ■

It is important to realize that these arguments do not apply to the partial derivatives of ψ with respect to curvilinear coordinates: e.g. $\partial\psi/\partial r$ is not harmonic, because $\partial/\partial r$ does not commute with the radial part $\nabla_r^2 \equiv r^{-2}\,\partial/\partial r(r^2\,\partial/\partial r)$ of ∇^2. For instance, (4.3.6) reminds one explicitly that $(\partial/\partial r)(r^2Y_{20}) = 2rY_{20}$ is not harmonic.

As an illustration useful for dipole potentials and fields, and for quadrupole potentials, we note that for any constant vectors $\boldsymbol{p}$ and $\boldsymbol{q}$ one has

$$(\boldsymbol{p}\cdot\nabla)\frac{1}{r} = -(\boldsymbol{p}\cdot\boldsymbol{r})\frac{1}{r^3}, \tag{4.3.8a}$$

$$(\boldsymbol{q}\cdot\nabla)(\boldsymbol{p}\cdot\nabla)\frac{1}{r} = -(\boldsymbol{p}\cdot\boldsymbol{q})\frac{1}{r^3} + 3(\boldsymbol{p}\cdot\boldsymbol{r})(\boldsymbol{q}\cdot\boldsymbol{r})\frac{1}{r^5}, \tag{4.3.8b}$$

where the coefficients of p_i, q_i, and of p_iq_j (with $i, j = x, y, z$) are harmonic except at the origin. The potential (4.3.8a) should be compared with (1.4.26).

(ii) A function which is harmonic in a region V containing the origin, but is otherwise arbitrary, can be expressed as a linear combination of the ψ_{lm}, eqn (4.3.6), with only the b_{lm} non-zero:

$$\psi(\boldsymbol{r}) = \sum_{lm} b_{lm}r^lY_{lm}(\Omega). \tag{4.3.9}$$

(Sums $\sum_{l=0}^{\infty}\sum_{m=-l}^{l}$ are written as $\sum_{lm}$ for short.) Consider, on the one hand, the value at the origin:

$$\psi(0) = b_{00}Y_{00} = b_{00}(4\pi)^{-\frac{1}{2}}; \tag{4.3.10}$$

and on the other hand, the surface average of ψ over a sphere of radius R centred on the origin. Writing this average temporarily as $\langle\psi\rangle$, and noting

$$\int d\Omega\, Y_{lm}(\Omega) = \delta_{l0}\,\delta_{m0}\,Y_{00}\int d\Omega$$

$$= \delta_{l0}\,\delta_{m0}\,Y_{00}\cdot 4\pi = \delta_{l0}\,\delta_{m0}(4\pi)^{\frac{1}{2}},$$

we have

$$\langle\psi\rangle = \frac{1}{4\pi R^2}\int dS\,\psi(R,\Omega) = \frac{1}{4\pi R^2}\int R^2\,d\Omega\,\psi(R,\Omega)$$

$$= \frac{1}{4\pi}\sum_{lm} b_{lm}R^l\int d\Omega\, Y_{lm} = b_{00}(4\pi)^{-\frac{1}{2}}. \tag{4.3.11}$$

Comparison with (4.3.10) shows that $\psi(0) = \langle\psi\rangle$. As long as the sphere is wholly within V, it is clearly irrelevant to this relationship whether or not the centre of the sphere happens to have been chosen as the origin. To display the result in general, we introduce a special symbol for such averages:

$$M_{r,R}\{\psi\} \equiv \begin{Bmatrix}\text{average value of } \psi \text{ over the surface of}\\ \text{a sphere of radius } R \text{ and centre at } \boldsymbol{r}\end{Bmatrix}. \tag{4.3.12}$$

Then our result can be formulated as follows: if $\psi(\boldsymbol{r})$ is harmonic in and on a sphere of radius R about $\boldsymbol{r}$, then

$$\psi(\boldsymbol{r}) = M_{r,R}\{\psi\}. \tag{4.3.13}$$

Another proof will emerge in (5.3.27).

Evidently, (4.3.13) implies the weaker result that $\psi(\boldsymbol{r})$ is equal to the volume average over such a sphere.

Exercise: Formulate and prove the corresponding results in 2D and in 1D.

(iii) **Theorem:** A function $\psi(\boldsymbol{r})$ harmonic in V can have neither maxima nor minima inside V. Consequently $\psi(\boldsymbol{r})$ attains its greatest and least values on the boundary S of V.

Proof: Around an arbitrary point $\boldsymbol{r}$ inside V, draw a sphere of radius R, such that the sphere too is wholly inside V. In eqn (4.3.13) the average M must lie between the greatest and the least values assumed by ψ on the sphere; by contemplating smaller and smaller values of R for fixed $\boldsymbol{r}$, we conclude that arbitrarily close to $\boldsymbol{r}$ there is a point where ψ is greater and another point where ψ is smaller than it is at $\boldsymbol{r}$. Therefore $\boldsymbol{r}$ can be neither a maximum nor a minimum of ψ. ■

A popular pretend-proof of this important no-maximum-or-minimum theorem runs as follows.

'A necessary condition for a maximum (minimum) of ψ is that ψ_{xx}, ψ_{yy}, ψ_{zz} all be strictly negative (positive). Consequently $\nabla^2\psi = \psi_{xx} + \psi_{yy} + \psi_{zz}$ cannot be zero. The same applies in 2D and 1D.'

Though suggestive of the correct result, this argument is false, because, contrary to what it asserts, it is perfectly possible for a function $\psi(\boldsymbol{r})$ to have an extremum at a point where $\nabla^2\psi = 0$. For instance, $\psi = r^4$ has a minimum at $\boldsymbol{r} = 0$, where

$$\nabla^2\psi = \frac{1}{r^2}\frac{\partial}{\partial r}\left(r^2\frac{\partial r^4}{\partial r}\right) = 20r^2$$

vanishes. A watertight version of the argument is given by Weinberger (1965), Section 12.

Harmonic functions are a natural generalization to nD of *monotonic* functions of a single independent variable. To see this, recall that a continuous function $\psi(x)$ is said to be monotonic over a region $x_1 \leqslant x \leqslant x_2$ if it has no maximum or minimum for $x_1 < x < x_2$; in that case $\psi(x)$ attains its greatest and least values at x_1 and x_2, i.e. at the boundaries of the region, which is precisely the property asserted in Theorem (iii). Indeed it would be more informative to describe solutions of the Laplace equation as monotonic instead of harmonic.

(iv) In 2D, but only in 2D, very powerful methods for harmonic functions emerge from the theory of functions of a complex variable. Since we shall make no use of them except once in Section 7.4.2, we sketch only the merest beginnings of this approach, referring for the rest to texts on electromagnetism or potential theory (e.g. Panofsky and Phillips 1955, Chapter 4).

Suppose the function $h(z)$ of the complex variable $z = x + iy$ is analytic in a region V of the complex plane. This means in particular that $h(z)$ is free of singularities in V, and has a well-defined derivative $\mathrm{d}h/\mathrm{d}z \equiv h'(z)$. Suppose further that h has the reflection property $h^*(z) = h(z^*) = h(x - iy)$, which entails that h is real when z is real, i.e. $h^*(x) = h(x)$. Then the real and imaginary parts of $h(z)$ are harmonic in V.

To see this, start from the chain rule of differentiation:

$$\frac{\partial h}{\partial x} = h'(z)\frac{\partial z}{\partial x} = h', \qquad \frac{\partial h}{\partial y} = h'(z)\frac{\partial z}{\partial y} = ih'; \tag{4.3.14}$$

$$\frac{\partial^2 h}{\partial x^2} = h''(z)\frac{\partial z}{\partial x} = h'', \qquad \frac{\partial^2 h}{\partial y^2} = ih''(z)\frac{\partial z}{\partial y} = -h''. \tag{4.3.15}$$

Thus

$$\nabla^2 h = \partial^2 h/\partial x^2 + \partial^2 h/\partial y^2 = 0. \tag{4.3.16}$$

The same reasoning applied to $h(x - iy) = h(z^*) = h^*(z)$ leads to

$$\nabla^2 h(x - iy) = \nabla^2 h^*(z) = 0. \tag{4.3.17}$$

Taking the sum and the difference, we find, as claimed, that

$$\begin{aligned} &\nabla^2(h + h^*)/2 = \nabla^2 \,\mathrm{Re}\, h(x + iy) = 0, \\ &\nabla^2(h - h^*)/2i = \nabla^2 \,\mathrm{Im}\, h(x + iy) = 0. \qquad \blacksquare \end{aligned} \tag{4.3.18}$$

For instance, consider $h(z) = z^m \equiv (|z| \exp(i\phi))^m$, where $|z| = (x^2 + y^2)^{\frac{1}{2}} = r$, and $\phi = \tan^{-1}(y/x)$. This identifies as harmonic the functions $\mathrm{Re}(z^m) = r^m \cos(m\phi)$ and $\mathrm{Im}\, z^m = r^m \sin(m\phi)$, which we recognize as the real and imaginary parts of $\psi_m(\mathbf{r})$ from (4.3.3).

Exercise: The series of harmonic functions $\sum_{n=0}^{\infty} z^n$ and $\sum_{n=1}^{\infty} z^n/n$ converge for $|z| < 1$; hence they are harmonic in the unit circle. Determine their real and imaginary parts, and verify by explicit differentiation that these indeed satisfy the Laplace equation.

4.4 The Green's function G_0 in unbounded space

4.4.1 Introduction

The discussion of ordinary differential equations in Chapter 2 suggests that in the solution of inhomogeneous problems a useful role is played by the Green's function, defined for Poisson's equation (4.1.1) by

$$-\nabla^2 G(\mathbf{r}\,|\,\mathbf{r}') = \delta(\mathbf{r} - \mathbf{r}'), \tag{4.4.1}$$

plus BCs appropriate to the problem. Equation (4.1.1) is satisfied by writing

$$\psi(\mathbf{r}) = \int \mathrm{d}V'\, G(\mathbf{r}\,|\,\mathbf{r}')\rho(\mathbf{r}') \equiv f(\mathbf{r}). \tag{4.4.2}$$

Here, $f(\mathbf{r})$ stands for the integral regarded as an explicit construct from the data ρ. Substitution into (4.1.1) confirms that

$$-\nabla^2 f = \int \mathrm{d}V'\, [-\nabla^2 G(\mathbf{r}\,|\,\mathbf{r}')]\rho(\mathbf{r}')$$

$$= \int \mathrm{d}V'\, \delta(\mathbf{r} - \mathbf{r}')\rho(\mathbf{r}') = \rho(\mathbf{r}). \qquad \blacksquare$$

The choice of the next topic presents a dilemma. From a strictly logical point of view it would be best to consider first the Green's functions in bounded regions, and only then to proceed to unbounded regions (bounding surface S at infinity); this is so especially in 2D and 1D, where the BCs at infinity have some peculiar features. However, point, line, and plane charges in the absence of physical boundaries are so familiar that we choose to deal first with Green's functions in unbounded regions. We call these the *free* Green's functions G_0, identified as such by the subscript. If necessary, the dimensionality is indicated by a superscript: thus $G_0^{(3)}$ in 3D. It is relatively easy to secure explicit and simple expressions for the G_0, which in turn help to guide intuition and to formulate physical interpretations for the Green's functions obtained later for bounded regions.

4.4.2 G_0 by direct solution of Poisson's equation

Define

$$\boldsymbol{R} \equiv \boldsymbol{r} - \boldsymbol{r}'. \tag{4.4.3}$$

When the source point is at the origin, i.e. in the special case $\boldsymbol{r}' = 0$, $\boldsymbol{R}$ is the same as the position vector $\boldsymbol{r}$ of the field point.

In 3D, with S at infinity, D and N BCs turn out equivalent: we simply require G_0 together with all its derivatives to vanish as $r \to \infty$ for fixed $\boldsymbol{r}'$. Then the solution of (4.4.1) is familiar:

$$G_0^{(3)}(\boldsymbol{r} \mid \boldsymbol{r}') = 1/4\pi R. \tag{4.4.4}$$

Together with (4.4.2) this yields $\psi(\boldsymbol{r}) = \int \mathrm{d}V' \, \rho(\boldsymbol{r}')/4\pi \, |\boldsymbol{r} - \boldsymbol{r}'|$, which is merely another way of asserting that the potential at $\boldsymbol{r}$ is the linear combination of the potentials $\mathrm{d}V' \, \rho(\boldsymbol{r}')/4\pi \, |\boldsymbol{r} - \boldsymbol{r}'|$ due to all the source elements $\mathrm{d}V' \, \rho(\boldsymbol{r}')$ at $\boldsymbol{r}'$.

For convenience, we move the origin to $\boldsymbol{r}'$; until further notice, $\boldsymbol{r}' = 0$ and $\boldsymbol{R} = \boldsymbol{r}$. Then (4.4.1, 4) reduce to the following statement:

Theorem:

$$-\nabla^2 G_0^{(3)}(\boldsymbol{r} \mid 0) = \delta(\boldsymbol{r}), \qquad G_0^{(3)}(r \to \infty) = 0 \tag{4.4.5a, b}$$

$$\Rightarrow \quad G_0^{(3)}(\boldsymbol{r} \mid 0) = 1/4\pi r. \tag{4.4.5c}$$

Proof: Adopt spherical polar coordinates, and look for a spherically symmetric, i.e. Ω-independent solution of (4.4.5a, b). Acting on functions of r alone, ∇^2 reduces to

$$\nabla_r^2 \equiv \frac{1}{r^2}\frac{\partial}{\partial r}\left(r^2 \frac{\partial}{\partial r}\right).$$

When $r \neq 0$, (4.4.5a) reduces to $\nabla_r^2 G_0 = 0$, and $1/r$ satisfies this because

$$-\nabla_r^2\left(\frac{1}{r}\right) = -\frac{1}{r^2}\frac{\partial}{\partial r}\left[r^2 \frac{\partial}{\partial r}\frac{1}{r}\right]$$

$$= -\frac{1}{r^2}\frac{\partial}{\partial r}\left[r^2\left(-\frac{1}{r^2}\right)\right]$$

$$= -\frac{1}{r^2}\frac{\partial}{\partial r}[-1] = 0.$$

To verify that (4.4.5c) satisfies (4.4.5a) also at $r=0$, we integrate both sides of (4.4.5a) with respect to volume over a sphere W centred on the origin, with an arbitrarily small radius η. On the right, the definition of $\delta(\boldsymbol{r})$ yields unity: $\int_W \mathrm{d}V\,\delta(\boldsymbol{r})=1$. On the left, one finds

$$\begin{aligned}-\int_W \mathrm{d}V\,\nabla^2 G_0 &= -\int \mathrm{d}V\,\operatorname{div}\operatorname{grad} G_0\\ &= -\int_{(r=\eta)} \mathrm{d}\boldsymbol{S}\cdot\boldsymbol{\nabla} G_0\\ &= -\int \mathrm{d}\boldsymbol{S}\cdot(-\hat{\boldsymbol{r}}/4\pi\eta^2)\\ &= \int \eta^2\,\mathrm{d}\Omega/4\pi\eta^2 = 1. \quad \blacksquare\end{aligned} \tag{4.4.6}$$

If $F(\boldsymbol{r})$ is any function behaving like C/r when $r\to 0$, then by the same argument $-\nabla^2 F$ contains a component $4\pi C\,\delta(\boldsymbol{r})$. For instance, at points $\boldsymbol{r}\neq 0$ the function $F(\boldsymbol{r})=\exp(-\mu r)/r$ satisfies the homogeneous Helmholtz equation $-(\nabla^2+\mu^2)F=0$. But, if V includes $\boldsymbol{r}=0$, then the true equation obeyed by F is $-(\nabla^2+\mu^2)F=4\pi\,\delta(\boldsymbol{r})$.

In 2D, we adopt (by hindsight) the BC that grad G remain bounded at infinity; then the familiar solution of (4.4.1) is

$$G_0^{(2)}(\boldsymbol{r}\,|\,\boldsymbol{r}')=\frac{1}{2\pi}\log\left(\frac{1}{R}\right)+A=\frac{1}{2\pi}\log\left(\frac{a}{R}\right), \tag{4.4.7}$$

where A, a are arbitrary constants. (Strictly speaking, the middle expression in (4.4.7) is not admissible, because R is a dimensional quantity, and cannot therefore feature alone in the argument of a logarithm.)

Exercise: Verify (4.4.7), using

$$\nabla_r^2=\frac{1}{r}\frac{\partial}{\partial r}\left(r\frac{\partial}{\partial r}\right).$$

Evidently $G_0^{(2)}$ itself diverges as $r\to\infty$; Section 4.4.3 below casts some light on why this should be so. *A fortiori*, $G_0^{(2)}$ cannot vanish at infinity, and no homogeneous DBCs could have been imposed there. However, $\boldsymbol{\nabla} G_0^{(2)}(\boldsymbol{r}\,|\,\boldsymbol{r}')=-\hat{\boldsymbol{R}}/2\pi R=-\boldsymbol{R}/2\pi R^2$, whence homogeneous NBCs at infinity are obeyed automatically.

In 1D, (4.4.1) reads

$$-\partial^2 G_0^{(1)}(x\,|\,x')/\partial x^2=\delta(x-x'). \tag{4.4.8}$$

Integrating twice, and admitting two integration constants $A(x')$ and $B(x')$ which can actually be functions of x', one finds

$$G_0^{(1)}(x \mid x') = -\tfrac{1}{2}\,|x - x'| + A(x')x + B(x'). \tag{4.4.9a}$$

Two boundary conditions remain to be chosen; by hindsight one adopts those that prove most widely useful. First, we make G_0 symmetric in x and x', as it is in 3D and 2D: this enforces $A(x') = C$, $B(x') = Cx'$, where C is a constant independent of both x and x'. Hence

$$G_0^{(1)}(x \mid x') = -\tfrac{1}{2}\,|x - x'| + C(x + x'). \tag{4.4.9b}$$

Second, we choose to make $G_0^{(1)}$ an even function of x at large distances (i.e. we demand $G(x \to \infty \mid x') = G(x \to -\infty \mid x;))$, which entails $C = 0$:

$$G_0^{(1)}(x \mid x') = -\tfrac{1}{2}\,|x - x'|. \tag{4.4.9c}$$

Then $G_0^{(1)}$ describes only the flux from the unit point source at x' (i.e. it excludes any flux whose sources are at infinity); thereby it involves x and x' only in the combination $X \equiv x - x'$. As in 2D, $G_0^{(1)}$ cannot vanish at infinity (i.e. both at $x = +\infty$ and at $x = -\infty$); and in 1D, neither can

$$\partial G_0^{(1)}(x \mid x')/\partial x = -\tfrac{1}{2}\varepsilon(x - x'). \tag{4.4.10}$$

Comparing the asymptotic behaviours of G_0 in 3D, 2D, and 1D, one could say, speaking very loosely, that the lower-dimensional spaces are so cramped that the flux from a unit source cannot diverge fast enough for G_0 or for grad G_0 to vanish even at infinity. An echo of this sounds through Section 8.5 on diffusion, and again through the Green's functions for the wave equation.

Under our BCs, $G_0(\mathbf{r} \mid \mathbf{r}')$ is always a function of $|\mathbf{R}| = |\mathbf{r} - \mathbf{r}'|$ alone. (In 1D it is particularly clear to what extent this results from choice.) Consequently, G_0 is, trivially, symmetric in the sense that

$$G_0(\mathbf{r} \mid \mathbf{r}') = G_0(\mathbf{r}' \mid \mathbf{r}). \tag{4.4.11}$$

The Green's functions have been obtained, above, by solving differential equations. A useful alternative technique obtains the 3D Green's function from its Fourier integral representation, as in Appendix F.2; it is worth studying, because for the diffusion and the wave equations it becomes by far the easiest and the most transparent technique to use.

Exercises: (i) Verify that the Green's functions (4.4.4, 7, 9) have the correct dimensions. (ii) Examine them in the light of the solutions of the Laplace equation listed in (4.3.1b, 3, 6).

4.4.3 Embedding: G_0 in 2D and 1D from G_0 in 3D

In Section 4.4.2 we determined $G_0^{(3)}$, $G_0^{(2)}$, and $G_0^{(1)}$ by solving the defining equation $-\nabla^2 G_0(\boldsymbol{r} \mid \boldsymbol{r}') = \delta(\boldsymbol{r} - \boldsymbol{r}')$ independently in 3D, 2D, and 1D. The discussion in Section 3.3 suggests an alternative approach. For instance, the 2D equation

$$-(\partial^2/\partial x^2 + \partial^2/\partial y^2)G_0^{(2)}(\boldsymbol{r} \mid \boldsymbol{r}') = \delta(x - x')\,\delta(y - y')$$

can be viewed as governing the potential due to an infinitely long line source, of unit strength per unit length, extending along the entire z-axis in 3D. By symmetry, this potential is independent of z. Thus the formally 2D problem is embedded in the true 3D space; we call this the *method of embedding*. (Following Hadamard, it is generally called the *method of descent,* since one is 'descending' from 3D to 2D.) It proves particularly fruitful later, for the wave equation.

Let us consider first the potential due to such a line source stretched from $z = -L$ to $z = L$, and calculate it at the point $P = (x, y, 0)$ in the median plane, i.e. in the xy-plane. This situation is shown in Fig. 4.1. $G_0^{(2)}$ will be the limit of $\psi(P)$ as $L \to \infty$. From (1.4.18) for $\delta(\boldsymbol{r})$ in plane polar coordinates, we see that the 3D source density is

$$\rho(\boldsymbol{r}') = [\delta(r')/2\pi r']\theta(L - |z'|);$$

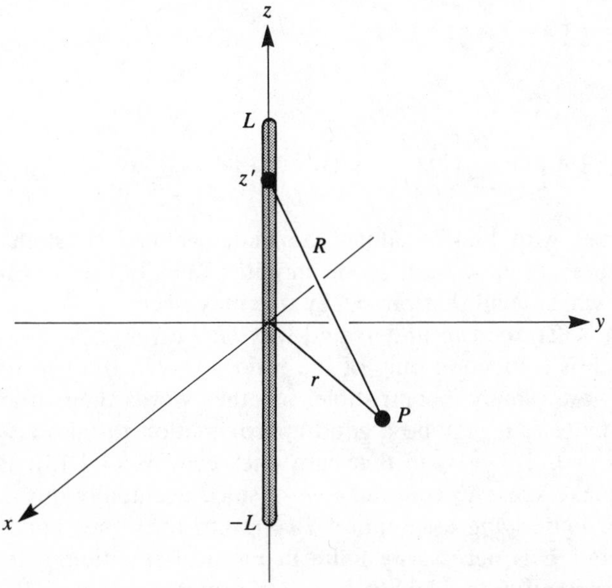

Fig. 4.1

then (4.4.2) with

$$G = G_0^{(3)} = 1/4\pi\,|\boldsymbol{r} - \boldsymbol{r}'| = 1/4\pi(r^2 + z'^2)^{\frac{1}{2}}$$

yields

$$\psi(P) = \int_{-\infty}^{\infty} \mathrm{d}z' \int_0^{\infty} \mathrm{d}r'\, r' \int_0^{2\pi} \mathrm{d}\phi' \frac{1}{4\pi(r^2 + z'^2)^{\frac{1}{2}}} \frac{\delta(r')}{2\pi r'} \theta(L - |z'|)$$

$$= \frac{1}{4\pi} \int_{-L}^{L} \mathrm{d}z' \frac{1}{(r^2 + z'^2)^{\frac{1}{2}}}$$

$$= \frac{1}{4\pi} \int_{-\sinh^{-1}(L/r)}^{\sinh^{-1}(L/r)} \frac{\mathrm{d}\alpha \cosh \alpha}{(1 + \sinh^2 \alpha)^{\frac{1}{2}}}$$

$$= \frac{1}{4\pi} 2 \sinh^{-1}(L/r)$$

$$= \frac{1}{2\pi} \log \left\{ \frac{L}{r} + \left(\frac{L^2}{r^2} + 1 \right)^{\frac{1}{2}} \right\},$$

$$\psi(P) = \frac{1}{2\pi} \log \left\{ \frac{L}{r} \left[1 + \left(1 + \frac{r^2}{L^2} \right)^{\frac{1}{2}} \right] \right\}. \tag{4.4.12}$$

Since $G_0^{(2)} = \lim_{L\to\infty} \psi(P)$, L/r is large,

$$\left\{ \cdots \right\} \sim \frac{L}{r} \left(1 + 1 + \frac{r^2}{2L^2} + \cdots \right) \sim 2L/r,$$

and

$$G_0^{(2)}(\boldsymbol{r} \mid 0) = \lim_{L\to\infty} \frac{1}{2\pi} \left\{ \log \frac{1}{r} + \log(2L) \right\}, \tag{4.4.13}$$

which agrees with (4.4.7), albeit the undetermined constants A and a written there are now seen to be infinite. Luckily this is irrelevant to grad $G_0^{(2)}$, which is all that physically one may need.

From (4.4.12) one can understand why $G_0^{(2)}$ diverges as $r \to \infty$. On the right, which is a function only of the ratio $\lambda \equiv r/L$, the two limits $r \to \infty$ and $L \to \infty$ are plainly incompatible; in other words their order matters. For fixed finite r, it may be a good approximation to take $L \gg r$, i.e. to consider $\lambda \to 0$, $1/\lambda \to \infty$; in that case one recovers (4.4.13). But then it does not make sense to consider $r \to \infty$, since eventually this is bound to violate the underlying assumption $L \gg r$. On the other hand, for fixed finite L (and L is necessarily finite in physical situations), it may be a good approximation to take $r \gg L$, i.e. to consider $\lambda \to \infty$, $1/\lambda \to 0$. But in that case, on the right of (4.4.12) one has $\{\cdots\} \sim \{1 + L/r + \cdots\}$,

whence

$$\psi(P) \underset{r\to\infty}{\simeq} \frac{1}{2\pi}\log(1+L/r) \sim \frac{(2L)}{4\pi r}, \tag{4.4.14}$$

which is simply the expected Coulomb potential due, at far-distant points, to our finite source of total strength $2L$.

By embedding one can similarly obtain $G_0^{(1)}$ from $G_0^{(3)}$. Now we consider the potential at a point $P=(0, 0, z)$ on the z-axis, due to a circular disk in the xy-plane, centred on the origin and of radius L, carrying unit source strength per unit area, as shown in Fig. 4.2. Then $G_0^{(1)} = \lim_{L\to\infty} \psi(P)$. We give the calculation without further comment:

$$\psi(P) = \int_0^L \mathrm{d}r'\, r' \int_0^{2\pi} \mathrm{d}\phi' \frac{1}{4\pi(z^2+r'^2)^{\frac{1}{2}}}$$

$$= \frac{2\pi}{4\pi}\int_0^{L^2} \tfrac{1}{2}\,\mathrm{d}r'^2 \frac{1}{(z^2+r'^2)^{\frac{1}{2}}},$$

$$\psi(P) = \tfrac{1}{2}\{(L^2+z^2)^{\frac{1}{2}} - |z|\}, \tag{4.4.15a}$$

$$G_0^{(1)} = \lim_{L\to\infty} \psi(P) = \tfrac{1}{2}\{L - |z|\} = \text{constant} - \tfrac{1}{2}|z|, \tag{4.4.15b}$$

as prescribed by (4.4.9) except for the name of the variable.

Exercise: Obtain $G_0^{(1)}$ from $G_0^{(2)}$ by embedding.

Recall that in Section 3.3 we actually described three different views of

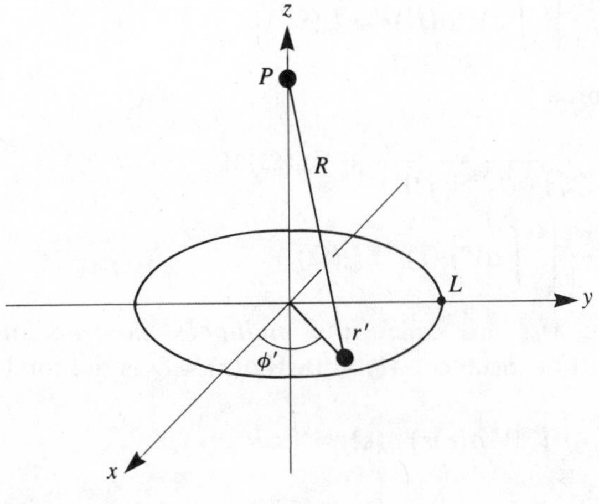

Fig. 4.2

reduced dimensionality. Embedding fits view (i); the independent calculations of Section 4.4.2 in 3D, 2D, and 1D fit view (ii); view (iii), of the unwanted dimensions frozen out by NBCs, will be illustrated in Sections 6.5 and 9.4.5.

4.4.4 Example: multipole potentials

The standard formula

$$\frac{1}{4\pi R} = \frac{1}{4\pi}\sum_{l=0}^{\infty} P_l(\hat{\boldsymbol{r}}\cdot\hat{\boldsymbol{r}}')\frac{r_<^l}{r_>^{l+1}}$$

$$= \sum_{lm}\frac{1}{2l+1}Y_{lm}(\Omega)Y_{lm}^*(\Omega')\frac{r_<^l}{r_>^{l+1}} \tag{4.4.16}$$

allows $\psi(\boldsymbol{r})$ in 3D to be written

$$\psi(\boldsymbol{r}) = \int \mathrm{d}V'\, G_0^{(3)}(\boldsymbol{r}\,|\,\boldsymbol{r}')\rho(\boldsymbol{r}') = \int \mathrm{d}V'\,\frac{\rho(\boldsymbol{r}')}{4\pi R}$$

$$= \sum_{lm}\frac{1}{2l+1}Y_{lm}(\Omega)\int \mathrm{d}V'\,\rho(\boldsymbol{r}')Y_{lm}^*(\Omega')\frac{r_<^l}{r_>^{l+1}}. \tag{4.4.17}$$

(As in (2.3.14), $r_<$ ($r_>$) is the lesser (greater) of r and r'.)

Suppose now that $\rho(\boldsymbol{r}')$ either vanishes beyond some finite distance, say a, or that it decreases beyond a faster than any inverse power of r': $\exp(-r'/a)$ is an example of such a *localized* source distribution. As $r\to\infty$ i.e. in practice for $r \gg a$, this entails $r_< = r'$, $r_> = r$. Accordingly, adopting convenient normalizations by hindsight, we write

$$\psi(\boldsymbol{r}\to\infty) \sim \sum_{lm}\frac{1}{[4\pi(2l+1)]^{\frac{1}{2}}}\frac{1}{r^{l+1}}Y_{lm}(\Omega)$$

$$\times\left\{\left[\frac{4\pi}{2l+1}\right]^{\frac{1}{2}}\int \mathrm{d}V'\rho(\boldsymbol{r}')r'^{l}Y_{lm}^*(\Omega')\right\}, \tag{4.4.18a}$$

or in other words

$$\psi(\boldsymbol{r}\to\infty) \sim \sum_{lm}\frac{1}{[4\pi(2l+1)]^{\frac{1}{2}}}\frac{1}{r^{l+1}}Y_{lm}(\Omega)M_{lm}, \tag{4.4.18b}$$

$$M_{lm} \equiv \left[\frac{4\pi}{2l+1}\right]^{\frac{1}{2}}\int \mathrm{d}V\,\rho(\boldsymbol{r})r^l Y_{lm}^*(\Omega). \tag{4.4.18c}$$

The constants M_{lm} are called the *multipole moments* of the source distribution ρ. For instance, $M_{00} = \int \mathrm{d}V\,\rho(\boldsymbol{r}) = Q$ is the total charge, and

$$M_{10} = (4\pi/3)^{\frac{1}{2}}\int \mathrm{d}V\,\rho(\boldsymbol{r})rY_{10}(\Omega) = \int \mathrm{d}V\,\rho(\boldsymbol{r})z$$

is the z-component of the total dipole moment $\boldsymbol{p} \equiv \int \mathrm{d}V\rho(\boldsymbol{r})\boldsymbol{r}$.

Exercise: Express p_x and p_y in terms of the M_{lm}.

In practice one would choose the origin at some point well inside the source distribution. Though the value of the potential at a given point does not of course depend on the choice of origin, the values of the individual multipole moments do in general depend on this choice. To this rule there are two important exceptions. First, $M_{00} = Q$ is obviously independent of the choice of coordinates; in particular it is invariant under translations $\boldsymbol{r} \to \boldsymbol{r} - \boldsymbol{a}$. Second, if (but only if) $Q = 0$, then the M_{lm}, i.e. the total dipole moment $\boldsymbol{p} = \int \mathrm{d}V\rho(\boldsymbol{r})\boldsymbol{r}$, is likewise translation-invariant. To see this, note that the transformation $\boldsymbol{r} \to \boldsymbol{r} - \boldsymbol{a}$ induces $\boldsymbol{p} \to \int \mathrm{d}V\rho(\boldsymbol{r})(\boldsymbol{r} - \boldsymbol{a}) = \boldsymbol{p} - \boldsymbol{a}Q$, which is indeed equal to $\boldsymbol{p}$ (i.e. invariant) if and only if $Q = 0$.

Of course, whatever the values of the constants M_{lm}, the expression, call it χ, on the right of (4.4.18b) as it stands is harmonic except at $\boldsymbol{r} = 0$, being a linear combination of standard harmonic functions from (4.3.6). One might ask what kind of source distribution $\rho(\boldsymbol{r})$ leads to the potential $\chi(\boldsymbol{r})$. Formally, this question is answered by Poisson's equation: $\rho = -(4\pi)^{-1}\nabla^2\chi$; but it is instructive to consider it in a little more detail.

From Poisson's equation we know that ρ is zero except at $\boldsymbol{r} = 0$. If ρ were simply proportional to $\delta(\boldsymbol{r})$, then one would have $M_{00} = Q$, while all the higher multipole moments would vanish because of the factors r^l in the integrand of M_{lm}. To represent non-vanishing point multipoles of order $l \geqslant 1$, ρ must contain lth-order derivatives of the delta-function, or the same singularities expressed in a different form. We give two examples, one in a Cartesian and one in a spherical basis (see Appendices A and E).

First, consider the source distribution

$$\rho = -p\hat{\boldsymbol{z}} \cdot \nabla\,\delta(\boldsymbol{r}) = -p\,\partial\delta(\boldsymbol{r})/\partial z,$$

describing a point dipole $\boldsymbol{p} = p\hat{\boldsymbol{z}}$ at the origin. (We know already, from Section 1.4.5, that this induces the potential $\psi = pz/r^3$.) On substitution into (4.4.18c), ρ yields

$$\begin{aligned} M_{10} &= \left(\frac{4\pi}{3}\right)^{\frac{1}{2}} \int \mathrm{d}^3r\left(-p\frac{\partial\,\delta(\boldsymbol{r})}{\partial z}\right) r Y_{10}(\Omega) \\ &= \int \mathrm{d}^3r\left(-p\frac{\partial\,\delta(\boldsymbol{r})}{\partial z}\right) z = p, \end{aligned}$$

while all the other multipole moments $(l, m) \neq (1, 0)$ vanish.

Second, we determine the source distribution responsible for any prescribed M_{lm}, using spherical polar coordinates throughout. In view of

(4.4.18c) one requires

$$M_{lm} = \left(\frac{4\pi}{2l+1}\right)^{\frac{1}{2}} \int_0^\infty \mathrm{d}r\, r^2 \int \mathrm{d}\Omega\, \rho(\boldsymbol{r}) r^l Y_{lm}^*(\Omega),$$

with $\rho(\boldsymbol{r})$ zero except at $r = 0$. This is achieved by

$$\rho = \rho_{lm} \equiv M_{lm} \cdot \left(\frac{2l+1}{4\pi}\right)^{\frac{1}{2}} \frac{\delta(r)}{r^2} \frac{1}{r^l} Y_{lm}(\Omega) \tag{4.4.19}$$

under the usual strong definition of $\delta(r)$ from Section 1.4.2. For instance, $\delta(r)/r^2 = 4\pi\,\delta(\boldsymbol{r})$ (eqn (1.4.13)) shows that $\rho_{00} = M_{00}\,\delta(\boldsymbol{r})$, as expected. The singular factors $r^{-l} Y_{lm}$, with $l \geqslant 1$, are effectively equivalent to l derivatives acting on $\delta(\boldsymbol{r})$. Equation (1.4.14) for instance is suggestive of this fact.

4.4.5 Example: source and dipole layers

We determine how $\psi(\boldsymbol{r})$ and its normal derivative vary as the field point $\boldsymbol{r}$ crosses a smooth surface Σ carrying a two-dimensional layer of sources, or of dipoles normal to Σ. (The surface is called Σ rather than S because it need not be closed.) The results can be foreseen from the 1D problem already solved in Section 2.6, to which the general case eventually reduces by virtue of the fact that a small enough portion of a smooth surface is effectively flat.

Consider first the case where Σ carries a source strength $\alpha_0(\boldsymbol{r}')\,\mathrm{d}S'$ on the surface element of area $\mathrm{d}S'$ at $\boldsymbol{r}'$. The resultant potential at $\boldsymbol{r}$ is

$$\psi(\boldsymbol{r}) = \int_\Sigma \mathrm{d}S'\, \alpha_0(\boldsymbol{r}')/4\pi R. \tag{4.4.20}$$

We are concerned with ψ only in the immediate vicinity of a point P on Σ. Move the origin to P (whence $\alpha_0(P) = \alpha_0(0)$), and take the tangent plane as the xy-plane, so that the z-axis points along the surface normal. Subdivide $\Sigma = \Sigma_c + \Sigma_f$, where Σ_c is a circle centred on P, with radius η small enough for Σ_c to be treated as flat; all the rest of Σ belongs to the 'far' region Σ_f. (Once chosen, η is taken as fixed; we do not now contemplate the limit of vanishing η.) Correspondingly, $\psi = \psi_c + \psi_f$. We consider only the case where $\alpha_0(\boldsymbol{r}')$ is continuous, i.e. $\lim_{\boldsymbol{r}'\to 0} \alpha_0(\boldsymbol{r}') = \alpha_0(P)$ does not depend on the direction along Σ from which P is approached.

With the field point close enough to P, i.e. with r small enough, ψ_f is clearly a continuous and arbitrarily-often differentiable function of $\boldsymbol{r}$, since the integral $\int_{\Sigma_f} \mathrm{d}S' \alpha_0(\boldsymbol{r}')/4\pi R$ involves no vanishing denominators. Hence we need consider only ψ_c. When η is small enough, α_0, being continuous by assumption, varies negligibly over Σ_c, and we can write

$$\psi_c(\boldsymbol{r}) = \frac{\alpha_0(P)}{4\pi} \int_{\Sigma_c} \mathrm{d}^2 r' \Big/ |\boldsymbol{r} - \boldsymbol{r}'|. \qquad (4.4.21)$$

As $\boldsymbol{r} \to 0$, the tangential derivatives of ψ_c vanish by symmetry (i.e. ψ_c ceases to vary with x and y). Hence we need consider only how ψ_c varies along the z-axis, i.e. we need $\psi_c(\boldsymbol{r})$ only for $\boldsymbol{r} = (0, 0, z)$, and obtain essentially the same explicit formula as in (4.4.15a):

$$\begin{aligned} \psi_c(0, 0, z) &= (\alpha_0(P)/4\pi) \int_0^{\eta} 2\pi \, \mathrm{d}r' \, r' \Big/ (r'^2 + z^2)^{\frac{1}{2}} \\ &= \tfrac{1}{2}\alpha_0(P)\{(\eta^2 + z^2)^{\frac{1}{2}} - |z|\}. \end{aligned} \qquad (4.4.22)$$

By inspection, ψ_c is continuous as the field point $\boldsymbol{r}$ crosses Σ, i.e. as z varies through zero. In other words,

$$\lim_{z \to 0+} \psi_c = \lim_{z \to 0-} \psi_c = \tfrac{1}{2}\eta\alpha_0(P).$$

By contrast, the normal derivative $\partial_n \psi_c \equiv \partial\psi_c/\partial z$ is discontinuous. From (4.4.22) one has

$$\partial\psi_c(0, 0, z)/\partial z = \tfrac{1}{2}\alpha_0(P)\{z/(\eta^2 + z^2)^{\frac{1}{2}} - \varepsilon(z)\}. \qquad (4.4.23)$$

As $z \to 0\pm$, only the sign function survives:

$$\lim_{z \to 0\pm} \frac{\partial \psi_c}{\partial z} = -\tfrac{1}{2}\alpha_0(P) \lim_{z \to 0\pm} \varepsilon(z) = \mp \tfrac{1}{2}\alpha_0(P), \qquad (4.4.24)$$

$$\left.\frac{\partial \psi_c}{\partial z}\right|_{0+} - \left.\frac{\partial \psi_c}{\partial z}\right|_{0-} = -\alpha_0(P). \qquad (4.4.25)$$

Accordingly, the jump in $\partial\psi/\partial z$ depends only on α_0; in particular, it is independent of the radius η chosen for the small circle Σ_c. The LHS of (4.4.25) is in fact the *sum* of the normal derivatives on the two sides, each taken in the direction away from the surface. In view of $\partial\psi/\partial z = -E_z$, this is just the elementary electrostatic result usually derived by appeal to a Gaussian pill-box straddling Σ.

Next, consider Σ carrying normally-oriented dipoles, with dipole moment $\mathrm{d}\boldsymbol{p} \equiv \alpha_1(\boldsymbol{r}')\,\mathrm{d}\boldsymbol{S}'$ on $\mathrm{d}\boldsymbol{S}'$. (There is no need to consider dipoles parallel to the surface, because they are equivalent to a source layer such as we have just considered, plus possibly a line of sources on the edges of Σ, which can be dealt with separately.) Combining all the elementary dipole potentials $(-\mathrm{d}\boldsymbol{p} \cdot \boldsymbol{\nabla})(1/4\pi R)$, we obtain

$$\psi(\boldsymbol{r}) = -\int_{\Sigma} \mathrm{d}\boldsymbol{S}' \, \alpha_1(\boldsymbol{r}') \cdot \boldsymbol{\nabla}(1/4\pi R). \qquad (4.4.26)$$

Again we need consider only $\psi_c(\boldsymbol{r})$ at $\boldsymbol{r} = (0, 0, z)$:

$$\psi_c(0, 0, z) = -\frac{\alpha_1(P)}{4\pi}\int_0^\eta 2\pi\, dr'\, r' \frac{\partial}{\partial z}\frac{1}{(r'^2 + z^2)^{\frac{1}{2}}}$$

$$= \tfrac{1}{2}\alpha_1(P)\int_0^\eta dr'\, r' \frac{z}{(r'^2 + z^2)^{\frac{3}{2}}}$$

$$= \tfrac{1}{2}\alpha_1(P) z \left\{\frac{1}{|z|} - \frac{1}{(\eta^2 + z^2)^{\frac{1}{2}}}\right\},$$

$$\psi_c(0, 0, z) = \tfrac{1}{2}\alpha_1(P)\{\varepsilon(z) - z/(\eta^2 + z^2)^{\frac{1}{2}}\}. \tag{4.4.27}$$

As $z \to 0\pm$, only the sign function survives:

$$\lim_{z\to 0\pm} \psi_c = \pm\tfrac{1}{2}\alpha_1(P), \tag{4.4.28}$$

$$\psi_c(0+) - \psi_c(0-) = \alpha_1(P). \tag{4.4.29}$$

Accordingly, $\psi(\boldsymbol{r})$ increases discontinuously by α_1 as the field point $\boldsymbol{r}$ crosses the surface in the direction of the local dipole moments $\alpha_1\, d\boldsymbol{S}$.

Exercise: Check the dimensions of (4.4.25, 29).

As regards the normal derivative, (4.4.27) entails

$$\frac{\partial \psi_c}{\partial z} = \tfrac{1}{2}\alpha_1(P)\left\{2\,\delta(z) - \frac{1}{(\eta^2 + z^2)^{\frac{1}{2}}} + \frac{z^2}{(\eta^2 + z^2)^{\frac{3}{2}}}\right\}. \tag{4.4.30}$$

The $\delta(z)$-proportional term is irrelevant at any point off the surface, and in particular to the limits $z \to 0\pm$. Accordingly,

$$\lim_{z\to 0+} \partial\psi_c/\partial z = \lim_{z\to 0-} \partial\psi_c/\partial z = -\alpha_1/2\eta,$$

whence $\partial\psi/\partial z$ is continuous across Σ.

The continuity and discontinuity conditions just found are of course valid independently of how the coordinate system is chosen. To summarize: across a surface Σ carrying sources $\alpha_0\, dS$ and dipoles $\alpha_1\, d\boldsymbol{S}$ on the element $d\boldsymbol{S}$, the potential jumps by α_1 when Σ is crossed in the direction $d\boldsymbol{S}$ (eqn (4.4.29)); the normal derivative jumps in such a way that its values in the direction away from Σ on the two sides add up to $-\alpha_0$ (eqn (4.4.25)). These conclusions should be compared with Section 2.6, which applies when Σ is an infinite plane, and α_0 and α_1 are constants independent of position; Fig. 2.1 in particular illustrates the provenance of the delta-function in (4.4.30).

4.5 The eigenvalue problem for $-\nabla^2$

4.5.1 Finite regions

We now revert to a finite region V bounded by a closed surface S. Section 3.2 has explained how the eigenvalue equation

$$-\nabla^2\phi_n(\boldsymbol{r}) = \lambda_n\phi_n(\boldsymbol{r}) \tag{4.5.1}$$

subject to homogeneous boundary conditions determines the normal modes of the wave equation. As in Section 2.4, the set of eigenfunctions $\{\phi_n\}$ serves as a basis for representing the Green's function for Poisson's equation; in fact, it proves equally valuable, if in a somewhat different fashion, for constructing the propagators and Green's functions for the diffusion and the wave equation. Anticipating such uses, we consider some general properties of the λ_n and the ϕ_n that are independent of the shape and size of V.

Theorem: $-\nabla^2$ is a Hermitean operator under homogeneous DBCs or NBCs.

Proof: We must establish (cf. Section 2.4) that for any pair of functions $a(\boldsymbol{r})$, $b(\boldsymbol{r})$ obeying the BCs, one has

$$\int_V \mathrm{d}V\, a^*(\nabla^2 b) = \int_V \mathrm{d}V\, (\nabla^2 a)^* b. \tag{4.5.2}$$

This is easily done by starting from Green's theorem, noting first that $(\nabla^2 a)^* = \nabla^2 a^*$ because ∇^2 is real:

$$\int_V \mathrm{d}V\, \{a^*(\nabla^2 b) - (\nabla^2 a^*)b\} = \int_S \mathrm{d}S\, \{a^*\partial_n b - (\partial_n a^*)b\} = 0. \tag{4.5.3}$$

The surface integral vanishes because, on S, either a^* and b vanish (under DBCs), or $\partial_n a^*$ and $\partial_n b$ vanish (under NBCs). ■

By an argument familiar perhaps from quantum mechanics, the Hermitean nature of $-\nabla^2$ implies that the λ_n are real, and that the ϕ_n can be taken as orthonormal: $\int_V \mathrm{d}V\, \phi_n^*\phi_{n'} = \delta_{nn'}$. In quantum mechanics one is familiar also with the Hermitean operator for momentum, $\boldsymbol{p} = -i\,\boldsymbol{\nabla}$, whence $\boldsymbol{p}^2 = -\nabla^2$; this suggests in turn that $-\nabla^2$ is a non-negative operator, implying in particular that its eigenvalues λ_n are non-negative. In actual fact, according to quantum mechanics $\boldsymbol{p}$ is not a physical observable in finite regions (since $-i\boldsymbol{\nabla}\psi$ need not obey the BCs merely because ψ does), so that this argument is not compelling. Nevertheless the conclusion is correct.

Theorem: The eigenvalues of $-\nabla^2$ are non-negative:

$$\lambda_n \geqslant 0. \tag{4.5.4}$$

Proof: Multiply both sides of (4.5.1) by ϕ_n^* and integrate over V. On the right this yields $\lambda_n \int_V dV\, |\phi_n|^2$. On the left it yields

$$-\int_V dV\, \phi_n^* \nabla^2 \phi_n = -\int_V dV\, \{\mathrm{div}\,(\phi_n^* \nabla \phi_n) - |\nabla \phi_n|^2\}$$

$$= -\int_S d\boldsymbol{S}\, \phi_n^* \cdot \nabla \phi_n + \int_V dV\, |\nabla \phi_n|^2$$

$$= \int_V dV\, |\nabla \phi_n|^2;$$

the surface integral vanishes by the same argument as in the preceding proof. Thus we find

$$\lambda_n = \int_V dV\, |\nabla \phi_n|^2 \Big/ \int_V dV\, |\phi_n|^2. \tag{4.5.5}$$

This formula makes sense because the denominator on the right cannot vanish (by assumption ϕ_n is not the trivial solution $\phi_n = 0$); since the numerator cannot be negative, neither can λ_n. ■

Theorem:

(i) Under DBCs, no eigenvalue can be zero.
(ii) Under NBCs, the lowest eigenvalue always is zero.

Proof: The argument is practically the same as the uniqueness proof in Section 4.2.

(i) Under DBCs, ϕ_n vanishes on S; but, being the non-trivial solution, it does not vanish everywhere, whence the numerator in (4.5.5) is strictly positive.
(ii) The NBCs $\partial_n \phi_S = 0$, like (4.5.1), feature only grad ϕ_n but not ϕ_n itself. Hence the normed function $\phi_0 = \text{constant} = V^{-\frac{1}{2}}$ satisfies both the equation and the BCs, and is an acceptable solution. Since grad (constant) = 0, the numerator in (4.5.5) vanishes. ■

Theorem: The eigenvalues and eigenfunctions of $-\nabla^2$ are discrete; in other words their number is (only) countably infinite, and they can be labelled by the integers n.

For the proof see Problems 5.12 and 5.13.

Let us for the moment write the eigenvalues under NBCs as μ_n, and reserve λ_n for the eigenvalues under DBCs. By virtue of the last theorem, each set of eigenvalues can be arranged in a non-decreasing sequence

$$0 < \lambda_1 \leq \lambda_2 \leq \lambda_3, \ldots, \tag{4.5.6a}$$

$$0 = \mu_0 \leq \mu_1 \leq \mu_2, \ldots. \tag{4.5.6b}$$

According to the Fredholm alternative as explained in Section 2.4 (setting $L=-\nabla^2$ there), $\lambda_n \neq 0$ asserts that under DBCs one is never in the special case; in other words a Green's function and a solution to the general inhomogeneous problem always exist. By contrast, under NBCs, the result $\mu_0=0$ asserts that one is always in the special case, so that a Green's function subject to homogeneous NBCs can never exist. The condition for the special special case to obtain, i.e. for an inhomogeneous Neumann problem to have a solution, is that the data should obey Gauss's law $\int_V dV\,\rho = -\int_S dS\,\partial_n\psi$. Chapter 6 discusses this and shows how solutions can be represented either through the pseudo Green's function $H(\mathbf{r}\,|\,\mathbf{r}')$ already introduced in Section 2.4, or else by defining a Neumann Green's function subject to suitably chosen inhomogeneous BCs. The additive constant left arbitrary in the Neumann solution by the uniqueness proof (Section 4.2) is simply the term $C\phi_0$ in eqn (2.4.11), since in the present case $\phi_0=$ constant ($=V^{-\frac{1}{2}}$ by normalization).

As regards the set of eigenfunctions $\{\phi_n\}$ we recall the closure property from Section 2.4, i.e. $\Sigma_n \phi_n^*(\mathbf{r}')\phi_n(\mathbf{r}) = \delta(\mathbf{r}-\mathbf{r}')$, and the reality properties from Appendix E.

The normal-mode frequencies of the wave equation are given by setting λ_n or μ_n equal to $k_n^2 = \omega_n^2/c^2$. Thus our results guarantee that all frequencies ω_n are real: in other words the homogeneous wave equation has no solutions growing or decaying exponentially in time.

There are interesting inequalities between the λ_n and the μ_n (Levine and Weinberger 1986). We quote some of these without proof:

$$\mu_n \leq \lambda_n, \qquad n=0,1,2,\ldots; \tag{4.5.7}$$

$$\mu_1 < \lambda_1; \tag{4.5.8}$$

and, in N dimensions, for finite regions with smooth convex boundaries,

$$\mu_{n+N} \leq \lambda_{n+1}, \qquad n=0,1,2,\ldots. \tag{4.5.9}$$

Evidently, (4.5.8) asserts that the lowest *non-zero* eigenvalue under NBCs is lower than the lowest (and automatically non-zero) eigenvalue under DBCs.

Exercise: Check whether the inequalities (4.5.7–9) apply to the first few eigenvalues in a cube (which is not a region with smooth boundaries), and in a sphere.

4.5.2 Unbounded space

Section 1.3 has already anticipated that in unbounded n-dimensional space the normed eigenfunctions and the eigenvalues k^2 of $-\nabla^2$ are given

by

$$-\nabla^2 \phi_{\boldsymbol{k}}(\boldsymbol{r}) = k^2 \phi_{\boldsymbol{k}}(\boldsymbol{r}), \tag{4.5.10}$$

$$\phi_{\boldsymbol{k}}(\boldsymbol{r}) = (2\pi)^{-n/2} \exp(i\boldsymbol{k} \cdot \boldsymbol{r}), \tag{4.5.11}$$

$$\begin{aligned} \int \mathrm{d}^n r\, \phi^*_{\boldsymbol{k}}(\boldsymbol{r}) \phi_{\boldsymbol{k}'}(\boldsymbol{r}) &= (2\pi)^{-n} \int \mathrm{d}^n r \exp[i(\boldsymbol{k} - \boldsymbol{k}') \cdot \boldsymbol{r}] \\ &= \delta(\boldsymbol{r} - \boldsymbol{r}'), \end{aligned} \tag{4.5.12}$$

where $\boldsymbol{k} \equiv (k_1, k_2, \ldots, k_n)$, with the k_i ranging continuously from $-\infty$ to $+\infty$. The sum over eigenfunctions that for finite volumes we generally write as Σ_p (p here instead of n to avoid confusion with the dimensionality) becomes

$$\Sigma_p \cdots \to \int \mathrm{d}^n k \cdots \equiv \int_{-\infty}^{\infty} \mathrm{d}k_1 \cdots \int_{-\infty}^{\infty} \mathrm{d}k_n \ldots . \tag{4.5.13}$$

For instance, the closure relation reads

$$\sum_p \phi^*_p(\boldsymbol{r}') \phi_p(\boldsymbol{r}) \to \int \frac{\mathrm{d}^n k}{(2\pi)^n} \exp(i\boldsymbol{k} \cdot (\boldsymbol{r} - \boldsymbol{r}')) = \delta(\boldsymbol{r} - \boldsymbol{r}'), \tag{4.5.14}$$

while the expansion of an arbitrary function F as a linear combination of the $\phi_{\boldsymbol{k}}$ is, in essence, just the Fourier integral representation of F.

All these relations should be familiar by now; we have collected them here chiefly for ease of reference when constructing unbounded-space Green's functions later on.

Problems

4.1 Determine the charge distribution $\rho(\boldsymbol{r})$ and the total charge $Q \equiv \int dV\, \rho(\boldsymbol{r})$ responsible for the potentials $\psi(\boldsymbol{r}) = \alpha \exp(-\lambda r)$ and $\psi(\boldsymbol{r}) = (\beta/r) \exp(-\lambda r)$,

(i) in 3D;
(ii) in 2D.
(iii) Determine Q directly from ψ, without using the explicit expressions for $\rho(\boldsymbol{r})$, and check that it agrees with the results from (i, ii).

4.2 (i) Determine $\psi(\boldsymbol{r})$ given that $\nabla^2\psi = 0$ for $z \neq 0$, $\psi(x, y, 0) = \sin(kx)$, and $\psi \to 0$ as $|z| \to \infty$.
(ii) What is the charge distribution responsible for this potential?

4.3 A point dipole $\boldsymbol{p} = \hat{\boldsymbol{z}}p$ is situated at the origin. Calculate the potential $\psi(r, \theta, \phi)$, and deduce that the electric field $\boldsymbol{E} \equiv -\boldsymbol{\nabla}\psi$ is given by

$$(E_r, E_\theta, E_\phi) = \frac{p}{4\pi G_0 r^3}(2\cos\theta, \sin\theta, 0).$$

Evaluate $\nabla^2 E_r$ and $\nabla^2 E_\theta$, and show that they fail to vanish even at $\boldsymbol{r} \neq 0$.

4.4 In the 2D region $a \leqslant r \leqslant 3a$, ψ obeys $\nabla^2\psi = 0$, while $\psi(a, \phi) = \cos(\phi)$, $\psi(3a, \phi) = \cos(3\phi)$.

(i) Without doing any calculations, explain briefly why one need not take seriously the suggestion that $\psi(2a, \phi) = \text{constant} \times \cos(2\phi)$.
(ii) Determine ψ throughout the region, by expressing ψ as a linear combination of the functions ψ_m in (4.3.3).

4.5 Consider the 3D source distributions $\rho(\boldsymbol{r}) = \lambda \exp(-r/a)$, and $\rho(\boldsymbol{r}) = \lambda(z/r) \exp(-r/a)$.

(i) For each, calculate the total charge $Q \equiv \int dV\, \rho(\boldsymbol{r})$, and, if $Q = 0$, also the dipole moment $\boldsymbol{p} = \int dV\, \rho(\boldsymbol{r})\boldsymbol{r}$.
(ii) Calculate $\psi(r, \Omega)$ (for all r), and verify that $\psi(r \to \infty, \Omega)$ behaves as expected from (i).

Hint: In (ii), use the polar expansion of the Green's function (see

Appendix H):

$$G_0^{(3)}(\boldsymbol{r} \mid \boldsymbol{r}') = \frac{1}{4\pi R}$$
$$= \sum_{l=0}^{\infty} \sum_{m=-l}^{l} \frac{1}{(2l+1)} \frac{r_<^l}{r_>^{l+1}} Y_{lm}^*(\Omega') Y_{lm}(\Omega).$$

4.6 For the 2D source distributions $\rho(\boldsymbol{r}) = \lambda(x/r) \exp(-r/a)$, and $\rho(\boldsymbol{r}) = \lambda(2x^2/r^2 - 1) \exp(-r/a)$, determine the potential $\psi(r, \phi)$ (for all r).

Hint: Use the polar expansion of the Green's function:

$$G_0^{(2)}(\boldsymbol{r} \mid \boldsymbol{r}') = \frac{1}{2\pi} \log \frac{a}{R} = \log\left(\frac{a}{r_>}\right) + \sum_{m=1}^{\infty} \frac{\cos[m(\phi - \phi')]}{m} \left(\frac{r_<}{r_>}\right)^m.$$

4.7 The hemisphere $r^2 = a^2$, $z \geqslant 0$ carries a uniform charge density σ per unit area.

(i) Determine $\psi(0, 0, z)$ and sketch it as a function of z.
(ii) Write down the two leading terms of $\psi(0, 0, z)$ as $z/a \to +\infty$, and again as $z/a \to -\infty$.
(iii) Where is the centre of charge?
(iv) Write down the two leading terms of $\psi(r, \Omega)$ as $r \to \infty$ in any direction; check that along the z-axis the results agree with (ii).
(v) Explain how in the equatorial plane ψ follows immediately from very familiar results, and write down $\psi(x, 0, 0)$ for all x.

Hint: For (iii), recall that a system with non-zero total charge has zero dipole moment with respect to its centre of charge.

4.8 The hemisphere $r^2 = a^2$, $z \geqslant 0$ carries a uniform dipole layer of strength $a\sigma$ pointing outward (the dipole moment residing on an element of area $\mathrm{d}S$ is $a\sigma\hat{\boldsymbol{r}}\,\mathrm{d}S = a^3\sigma\hat{\boldsymbol{r}}\,\mathrm{d}\Omega$).

(i) What is the total dipole moment $\boldsymbol{p}$ of the system?
(ii) Determine $\psi(0, 0, z)$ and sketch it as a function of z.
(iii) As $|z|/a \to \infty$, verify that $\psi(0, 0, z)$ approaches the potential due to a point dipole $\boldsymbol{p}$ situated at the origin.

Hint: The integration in (ii) is somewhat tedious (this of course serves to motivate (iii)). Check that at $z = 0$ your expression reduces to the easily-obtained result $\psi(0) = -2\pi\sigma a$.

(See also Problems 13.1, 2, 3 on eigenvalues of $-\nabla^2$.)

5 Poisson's equation: II. Dirichlet problems

This and the next two chapters may be easier to read in the light of Appendix G about our notation for boundary data.

5.1 The Green's function: definition and general properties

Our problem is to solve Poisson's equation

$$-\nabla^2\psi(\boldsymbol{r}) = \rho(\boldsymbol{r}) \tag{5.1.1}$$

in a region V, subject to prescribed values ψ_S on the closed surface S bounding V. From Sections 4.2 and 4.5.1 we know that, according to the Fredholm alternative, a unique solution exists, because there is no solution to the corresponding homogeneous problem (with ρ and ψ_S everywhere zero). Our tool is the Dirichlet Green's function defined by

$$-\nabla^2 G_D(\boldsymbol{r}\,|\,\boldsymbol{r}') = \delta(\boldsymbol{r}-\boldsymbol{r}') \tag{5.1.2a}$$

plus *homogeneous* DBCs

$$G_D(\boldsymbol{r}\,|\,\boldsymbol{r}') = 0 \text{ for } \boldsymbol{r} \text{ on } S, \text{ and } \boldsymbol{r}' \text{ anywhere in } V. \tag{5.1.2b}$$

The expression $\psi(\boldsymbol{r}) = \int_V dV' G_D/\boldsymbol{r}\,|\,\boldsymbol{r}')\rho/\boldsymbol{r}') \equiv f_D(\boldsymbol{r})$ obviously solves the problem when the prescribed ψ_S is zero everywhere on S. To see this, note that the integral vanishes for $\boldsymbol{r}$ on S because $G_D(\boldsymbol{r}\,|\,\boldsymbol{r}')$ does (by (5.1.2b)); and that it obeys (5.1.1) by the argument that was given for (4.4.2). What is by no means obvious at this stage is that G_D can help to accommodate arbitrary non-zero (inhomogeneous) boundary data as well. This will be shown in Section 5.2.

G_D naturally depends on the shape and size of V; but, since it obeys (5.1.2a), one can always choose to express it in the form

$$\begin{aligned} G_D(\boldsymbol{r}\,|\,\boldsymbol{r}') &= G_0(\boldsymbol{r}\,|\,\boldsymbol{r}') + \chi(\boldsymbol{r}\,|\,\boldsymbol{r}') \\ &= 1/4\pi R + \chi(\boldsymbol{r}\,|\,\boldsymbol{r}'), \end{aligned} \tag{5.1.3}$$

where

$$\nabla^2\chi(\boldsymbol{r}\,|\,\boldsymbol{r}') = 0 \tag{5.1.4a}$$

$$\chi(\boldsymbol{r}\,|\,\boldsymbol{r}') = -G_0(\boldsymbol{r}\,|\,\boldsymbol{r}') = -1/4\pi\,|\boldsymbol{r}-\boldsymbol{r}'| \quad \text{for} \quad \boldsymbol{r} \text{ on } S; \tag{5.1.4b}$$

(5.1.4a) follows because $-\nabla^2 G_0$ by itself balances the delta function on the right of (5.1.2a). (Recall $\boldsymbol{R} = \boldsymbol{r} - \boldsymbol{r}'$.) Thus, χ is a harmonic function such that the combination (5.1.3) satisfies the BC (5.1.2b).

Writers scared of delta-functions often define G_D by (5.1.3, 4.2b) instead of (5.1.2a, b). The physics of the subdivision (5.1.3) will be discussed later.

The harmonic nature of χ implies upper and lower bounds on G_D. The proofs exploit the properties of harmonic functions discussed in Section 4.3.2.

Theorem:

$$0 < G_D(\boldsymbol{r} \,|\, \boldsymbol{r}') < 1/4\pi R. \tag{5.1.5a, b}$$

Proof: (a) Consider the region W between S and a sphere S_2 of arbitrarily small radius η centred on $\boldsymbol{r}'$. (Thus $R = \eta$ on S_2.) In particular, consider $G_D(\boldsymbol{r} \,|\, \boldsymbol{r}')$ with $\boldsymbol{r}$ on S_2 in the limit $\eta \to 0$. Since $\chi(\boldsymbol{r} \,|\, \boldsymbol{r}')$ is harmonic it does not diverge anywhere in V; in particular χ does not diverge on S_2. Therefore G_D on S_2 is dominated by its other component: $G_D \to G_0 = 1/4\pi\eta \to \infty$. On the other hand, with $\boldsymbol{r}$ on S, the BC entails $G_D = 0$. Now G_D, being harmonic throughout W, assumes its greatest and least values on the boundary of W, which consists of $S + S_2$; thus its greatest value (on S_2) tends to $+\infty$, and its least value (on S) is 0. Therefore G_D cannot be negative. ■

An alternative argument runs as follows. If G_D were negative in some subregion of W, then at some point of this subregion it would have to assume its most negative value, i.e. it would have to have a minimum. But harmonic functions have no minima. ■

(b) χ is harmonic throughout V. On S, $\chi = -1/4\pi R < 0$; hence its greatest and least values are both negative, and χ is negative everywhere in V. Hence, by (5.1.3), $G_D < G_0$. ■

Theorem: Consider two closed surfaces S_{inner} and S_{outer}, such that S_{inner} is entirely surrounded by S_{outer}. Let $G_{D,\text{inner}}(\boldsymbol{r} \,|\, \boldsymbol{r}')$ and $G_{D,\text{outer}}(\boldsymbol{r} \,|\, \boldsymbol{r}')$ be the Green's functions for the regions bounded by S_{inner} and S_{outer} respectively, and let $\boldsymbol{r}$, $\boldsymbol{r}'$ both be within S_{inner}. Then

$$G_{D,\text{inner}}(\boldsymbol{r} \,|\, \boldsymbol{r}') < G_{D,\text{outer}}(\boldsymbol{r} \,|\, \boldsymbol{r}'). \tag{5.1.6}$$

Exercise: Prove this.

So far $g_D(\boldsymbol{r} \,|\, \boldsymbol{r}')$ has been considered as a function of $\boldsymbol{r}$, with $\boldsymbol{r}'$ playing the role of a parameter. But in fact G_D is a symmetric function of $\boldsymbol{r}$ and $\boldsymbol{r}'$, a property often called *reciprocity*.

Theorem:

$$G_D(\boldsymbol{r} \mid \boldsymbol{r}') = G_D(\boldsymbol{r}' \mid \boldsymbol{r}). \tag{5.1.7}$$

Proof: Apply Green's theorem to $G_D(\boldsymbol{r} \mid \boldsymbol{r}')$ and $G_D(\boldsymbol{r} \mid \boldsymbol{r}'')$ both regarded as functions of $\boldsymbol{r}$:

$$-\int_V dV\, \{G_D(\boldsymbol{r} \mid \boldsymbol{r}')\nabla^2 G_D(\boldsymbol{r} \mid \boldsymbol{r}'') - (\nabla^2 G_D(\boldsymbol{r} \mid \boldsymbol{r}'))G_D(\boldsymbol{r} \mid \boldsymbol{r}'')\}$$

$$= -\int_S dS\, \{G_D(\boldsymbol{r} \mid \boldsymbol{r}')\, \partial_n G_D(\boldsymbol{r} \mid \boldsymbol{r}'') - (\partial_n G_D(\boldsymbol{r} \mid \boldsymbol{r}'))G_D/\boldsymbol{r} \mid \boldsymbol{r}'')\}$$

$$= 0. \tag{5.1.8a}$$

The surface integral vanishes because the integrand of each term has either $G_D(\boldsymbol{r} \mid \boldsymbol{r}')$ or $G_D(\boldsymbol{r} \mid \boldsymbol{r}'')$ as a factor, and both these vanish for $\boldsymbol{r}$ on S. In the volume integrals we substitute, from (5.1.2a), $\nabla^2 G_D(\boldsymbol{r} \mid \boldsymbol{r}') = -\delta(\boldsymbol{r} - \boldsymbol{r}')$ and $\nabla^2 G_D(\boldsymbol{r} \mid \boldsymbol{r}'') = -\delta(\boldsymbol{r} - \boldsymbol{r}'')$. This yields

$$\int_V dV\, \{G_D(\boldsymbol{r} \mid \boldsymbol{r}')\, \delta(\boldsymbol{r} - \boldsymbol{r}'') - \delta(\boldsymbol{r} - \boldsymbol{r}')G_D(\boldsymbol{r} \mid \boldsymbol{r}'')$$

$$= G_D(\boldsymbol{r}'' \mid \boldsymbol{r}') - G_D(\boldsymbol{r}' \mid \boldsymbol{r}'') = 0, \tag{5.1.8b}$$

which is just (5.1.7) except that the variable called $\boldsymbol{r}$ in (5.1.7) is called $\boldsymbol{r}''$ in (5.1.8b). ■

The symmetry (5.1.7) is most remarkable. It shows that, however asymmetrically the points $\boldsymbol{r}$ and $\boldsymbol{r}'$ are situated relative to the boundary, the potential at $\boldsymbol{r}$ due to a point source placed at $\boldsymbol{r}'$ equals the potential at $\boldsymbol{r}'$ due to the same source placed at $\boldsymbol{r}$.

In particular, symmetry plus the BC (5.1.2b) imply

$$G_D(\boldsymbol{r} \mid \boldsymbol{r}') = 0 \quad \text{when} \quad \boldsymbol{r} \text{ or } \boldsymbol{r}' \text{ is on } S. \tag{5.1.9}$$

Since G_0 is automatically symmetric, being a function of R alone, the symmetry of G_D entails that of χ in the subdivision (5.1.3):

$$\chi(\boldsymbol{r} \mid \boldsymbol{r}') = \chi(\boldsymbol{r}' \mid \boldsymbol{r}). \tag{5.1.10}$$

We are now in a position to explore the physical significance of χ. In electrostatics (for units see the footnote in Section 2.1), $G_D(\boldsymbol{r} \mid \boldsymbol{r}')$ is the potential at $\boldsymbol{r}$ due to the presence of a unit point charge at $\boldsymbol{r}'$, both $\boldsymbol{r}$ and $\boldsymbol{r}'$ being inside the earthed conducting closed surface S. Then G_0 is the Coulomb potential exerted *directly* by the point charge; χ is the Coulomb potential exerted by the *induced* charges that must be situated on S in order to enforce the BCs (5.1.2b) in the presence of the point charge. Thus χ is also due to the point charge, but *indirectly*.

The subdivision (5.1.3) becomes essential instead of optional if one enquires into the force on a single point charge Q situated at $\boldsymbol{r}'$, and into its energy $U(\boldsymbol{r}')$. The total field at $\boldsymbol{r}$ is $\boldsymbol{E}(\boldsymbol{r}) = -Q\boldsymbol{\nabla}G_{\text{D}}(\boldsymbol{r} \mid \boldsymbol{r}')$; the field exerted by the induced charges is $\boldsymbol{E}_{\text{ind}}(\boldsymbol{r}) = -Q\boldsymbol{\nabla}\chi(\boldsymbol{r} \mid \boldsymbol{r}')$; since the point charge can exert no force directly on itself (at least not while it is at rest), the net force on it is

$$\boldsymbol{F}(\boldsymbol{r}') = Q\boldsymbol{E}_{\text{ind}}(\boldsymbol{r}') = -Q^2\boldsymbol{\nabla}\chi(\boldsymbol{r} \mid \boldsymbol{r}')\big|_{\boldsymbol{r}=\boldsymbol{r}'}, \tag{5.1.11}$$

where the differentiation with respect to $\boldsymbol{r}$ must be carried out first, at fixed $\boldsymbol{r}'$, *before* setting $\boldsymbol{r} = \boldsymbol{r}'$. However, the symmetry (5.1.10) allows (5.1.11) to be written in the simpler form

$$\boldsymbol{F}(\boldsymbol{r}') = -\boldsymbol{\nabla}'U(\boldsymbol{r}'), \qquad U(\boldsymbol{r}') \equiv \tfrac{1}{2}Q^2\chi(\boldsymbol{r}' \mid \boldsymbol{r}'). \tag{5.1.12a, b}$$

Proof: First we note that

$$\begin{aligned}\boldsymbol{\nabla}'\tfrac{1}{2}\chi(\boldsymbol{r}' \mid \boldsymbol{r}') &= \tfrac{1}{2}\{\boldsymbol{\nabla}\chi(\boldsymbol{r} \mid \boldsymbol{r}') + \boldsymbol{\nabla}'\chi(\boldsymbol{r} \mid \boldsymbol{r}')\}\big|_{\boldsymbol{r}=\boldsymbol{r}'} \\ &= \tfrac{1}{2}\{\boldsymbol{\nabla}\chi(\boldsymbol{r}' \mid \boldsymbol{r}) + \boldsymbol{\nabla}'\chi(\boldsymbol{r} \mid \boldsymbol{r}')\}\big|_{\boldsymbol{r}=\boldsymbol{r}'}.\end{aligned}$$

Since each gradient operator now acts on the second argument of χ, the two terms coincide at $\boldsymbol{r} = \boldsymbol{r}'$; this yields the desired equality $\boldsymbol{\nabla}'\tfrac{1}{2}\chi(\boldsymbol{r}' \mid \boldsymbol{r}') = \boldsymbol{\nabla}'\chi(\boldsymbol{r} \mid \boldsymbol{r}')\big|_{\boldsymbol{r}=\boldsymbol{r}'}$. The results (5.1.11, 12) will be illustrated below in Section 5.3.3 on the image potential. ■

Alternatively, (5.1.12b) for the potential energy follows from the standard expression for the total Coulomb energy of the system, written symbolically as $\frac{1}{2}\sum_i q_i\psi(\boldsymbol{r}_i)$, where the sum runs over all charges q_i. In our case $\sum_i$ includes an integral over the induced charges on S; however, since $\psi(\boldsymbol{r}) = QG_{\text{D}}(\boldsymbol{r} \mid \boldsymbol{r}')$ vanishes on S, the only non-zero contribution to the sum is the term $\frac{1}{2}Q \cdot QG_{\text{D}}(\boldsymbol{r}' \mid \boldsymbol{r}')$ arising from the point charge $q_i = Q$ itself, situated at $\boldsymbol{r} = \boldsymbol{r}'$. However, this expression includes the self-energy $\frac{1}{2}Q^2G_0(\boldsymbol{r}' \mid \boldsymbol{r}')$ of the point charge, which, albeit infinite, is independent of its position. Since such a position-independent energy cannot contribute to the force on the charge, we simply drop it, obtaining $\frac{1}{2}Q^2\{G_{\text{D}}(\boldsymbol{r}' \mid \boldsymbol{r}') - G_0(\boldsymbol{r}' \mid \boldsymbol{r}')\} = \frac{1}{2}Q^2\chi(\boldsymbol{r}' \mid \boldsymbol{r}')$ as written in (5.1.12b). ■

By contrast, when the charge density ρ is smooth, forces and energies can be calculated without complications from self-energies. The force on the charges in a volume element δV can then be written simply as $\boldsymbol{E}(\boldsymbol{r})\rho(\boldsymbol{r})\,\delta V = -(\boldsymbol{\nabla}\psi)\rho\,\delta V$, and ψ can be calculated by the magic rule from Section 5.2 below. In such continuous cases the force exerted by each (infinitesimal) charge element on itself cancels automatically, self-energies vanish, and there is no need to appeal to the subdivision (5.1.3).

Finally, G_{D} can be constructed as in Section 2.4 from the eigenfunctions of $-\nabla^2$ discussed in Section 4.5.

Theorem:

$$G_D(\mathbf{r} \mid \mathbf{r}') = \sum_n \frac{1}{\lambda_n} \phi_n^*(\mathbf{r}')\phi_n(\mathbf{r}). \qquad (5.1.13)$$

Proof: The expression on the right always makes sense, because, as shown in Section 4.5, no λ_n is zero. The sum obeys the BCs because the ϕ_n do. It obeys the defining differential equation (5.1.2a) because

$$-\nabla^2 \sum \lambda_n^{-1}\phi_n^*(\mathbf{r}')\phi_n(\mathbf{r}) = -\sum \lambda_n^{-1}\phi_n^*(\mathbf{r}')\nabla^2\phi_n(\mathbf{r})$$
$$= \sum \lambda_n^{-1}\phi_n^*(\mathbf{r}')\lambda_n\phi_n(\mathbf{r}) = \sum \phi_n^*(\mathbf{r}')\phi_n(\mathbf{r}) = \delta(\mathbf{r}-\mathbf{r}');$$

the last step follows from the closure property of the complete orthonormal set $\{\phi_n\}$. ∎

The representation (5.1.13) is often useful in actually evaluating G_D.

As shown in Appendix E, one may (though one need not) choose the set $\{\phi_n\}$ as real. Let us now do so. Then (5.1.13) reads $G_D(\mathbf{r} \mid \mathbf{r}') = \sum \lambda_n^{-1}\phi_n(\mathbf{r}')\phi_n(\mathbf{r}) = G_D(\mathbf{r}' \mid \mathbf{r})$, which, on mere inspection, affords an alternative proof of the symmetry (5.1.7).

Although this second proof appears to be more effortless than the first, it is so only if the convergence of the series (5.1.13) is taken on trust, and if one does not need to establish the precise sense in which the series represents the Green's function. (See the comments on convergence in Appendix C, and early in Section 3.) By contrast, the earlier proof relies only on the defining properties of G_D.

5.2 Solution of the general inhomogeneous problem: the magic rule

We are now in a position to solve the general inhomogeneous problem defined by Poisson's equation (5.1.1) plus prescribed ψ_S. One applies Green's theorem to G_D and to the (as yet unknown) solution ψ. By hindsight, interchange the previous roles of $\mathbf{r}$ and $\mathbf{r}'$ (relying on the symmetry (5.1.7)):

$$\int_V dV'\{\psi(\mathbf{r}')\nabla'^2 G_D(\mathbf{r}' \mid \mathbf{r}) - (\nabla'^2\psi(\mathbf{r}'))G_D(\mathbf{r}' \mid \mathbf{r})\}$$
$$= \int_S d\mathbf{S}' \cdot \{\psi(\mathbf{r}')\nabla' G_D(\mathbf{r}' \mid \mathbf{r}) - (\nabla'\psi(\mathbf{r}'))G_D(\mathbf{r}' \mid \mathbf{r})\}. \qquad (5.2.1)$$

The disposition of primed and unprimed variables should be watched carefully. ∇' differentiates with respect $\mathbf{r}'$; as always, $\int dV'$ and $\int dS'$ signify integrations with respect to $\mathbf{r}'$, i.e. $\int d^n\mathbf{r}'$ and $\int d^{(n-1)}\mathbf{r}'$ respectively.

On the left of (5.2.1), eqns (5.1.2a) and (5.1.1) remove the Laplace operators, yielding

$$\int_V dV' \{\psi(\mathbf{r}')[-\delta(\mathbf{r}'-\mathbf{r})]-[-\rho(\mathbf{r}')]G_D(\mathbf{r}' \mid \mathbf{r})\}$$

$$= -\psi(\mathbf{r}) + \int_V dV' \, \rho(\mathbf{r}')G_D(\mathbf{r}' \mid \mathbf{r}). \tag{5.2.2}$$

Equating to the RHS of (5.2.1) and rearranging, one finds

$$\psi(\mathbf{r}) = \int_V dV' \, \rho(\mathbf{r}')G_D(\mathbf{r}' \mid \mathbf{r}) - \int_S d\mathbf{S}' \cdot \{\psi(\mathbf{r}')\nabla' G_D(\mathbf{r}' \mid \mathbf{r})$$
$$- (\nabla'\psi(\mathbf{r}'))G_D(\mathbf{r}' \mid \mathbf{r})\}. \tag{5.2.3}$$

Up to this point we have used only the differential equation but not the BCs. Thus (5.2.3) applies subject to (5.1.1) and (5.1.2a) alone: for instance, it applies equally if G_D is replaced by G_0 (or, as in Chapter 6, by G_N). But now we appeal to the BC (5.1.9) on G_D, which shows that the second surface integral vanishes, because G_D vanishes with $\mathbf{r}'$ on S. Thus we obtain what we call the *magic rule*:

$$\psi(\mathbf{r}) = \int_V dV' \, \rho(\mathbf{r}')G_D(\mathbf{r}' \mid \mathbf{r}) - \int_S dS' \, \psi_S(\mathbf{r}') \, \partial_n' G_D(\mathbf{r}' \mid \mathbf{r})$$
$$\equiv f_D(\mathbf{r}) + g_D(\mathbf{r}). \tag{5.2.4}$$

Here, f_D and g_D stand for the volume and the (negative of the) surface integral, respectively, regarded as explicit constructs from the data ρ and ψ_S. (Compare f_D with f defined in (4.4.2): in Chapter 4 we worked with the analogous free-space volume integral $f_0(\mathbf{r}) \equiv \int dV' \, \rho(\mathbf{r}')G_0(\mathbf{r} \mid \mathbf{r}')$.) When $\mathbf{r}$ is inside V (not on S), one has

$$-\nabla^2 f_D = \rho, \qquad \nabla^2 g_D = 0. \tag{5.2.5a, b}$$

Exercise: Prove (5.2.5) by acting on f_D and g_D with $-\nabla^2$, and then taking the Laplace operator under the integral.

The magic rule solves the problem, because $G_D(\mathbf{r} \mid \mathbf{r}')$ is assumed known, being determined by the region V alone, irrespective of the data. Then, given the data ρ and ψ_S, we simply substitute them into f_D and g_D and evaluate these integrals, not needing to solve equations any more at this stage. In common parlance the solution has been reduced to quadratures.

(i) *Source points and field points.* Generally, we reserve the symbol $\mathbf{r}$ for the point where the solution $\psi(\mathbf{r})$ is wanted, called the field point. The symbol $\boldsymbol{r}'$ is generally reserved for points where data are specified, called source points. Equation (5.2.4) illustrates this distinction. The wish to observe it in the magic rule was the reason for the disposition of primed and unprimed variables in (5.2.1).

(ii) grad *and* grad′. Though $G_D(\boldsymbol{r}' \mid \boldsymbol{r}) = G_D(\boldsymbol{r} \mid \boldsymbol{r}')$ (e.g. in f_D), note that (e.g. in g_D) $\nabla' G_D(\boldsymbol{r}' \mid \boldsymbol{r}) \neq \nabla G_D(\boldsymbol{r}' \mid \boldsymbol{r})$. This is shown explicitly by the expansion (5.1.13), say with real $\{\phi_n\}$:

$$\nabla' G_D = \sum \lambda_n^{-1} \nabla' \phi_n(\boldsymbol{r}') \phi_n(\boldsymbol{r}) \neq \nabla G_D = \sum \lambda_n^{-1} \phi_n(\boldsymbol{r}') \nabla \phi_n(\boldsymbol{r}).$$

Exercise: Compare ∇G_0 with $\nabla' G_0$.

(iii) *Outward normal.* In G_D, $d\boldsymbol{S}'$ points outwards from V; it is easy, by carelessness, to get the sign of g_D wrong.

(iv) *The homogeneous problem.* If $\rho = 0$ everywhere in V, then f_D vanishes. If $\psi_S = 0$ everywhere on S, then g_D vanishes. If both ρ and ψ_S vanish everywhere, then we have the totally homogeneous Laplace–Dirichlet problem; the magic rule shows that then the (unique) solution is identically zero: $\psi = 0$ throughout V.

(v) *Self-consistency.* It is not in the least obvious on mere inspection that $\psi(\boldsymbol{r})$ as given by (5.2.4) approaches the initially prescribed boundary value ψ_S when the field point $\boldsymbol{r}$ approaches the surface (from inside V). Of course it must do so if a solution to our problem *exists*, because such a solution is then (uniquely) given by (5.2.4). But, without an existence proof, it is conceivable at this stage that the desired limit is not approached; in that case one would have to conclude that the problem has been overspecified and has no solution at all; or in other words that the magic rule is not self-consistent. In fact the rule is self-consistent, but the proof is postponed to Section 7.1; it will exploit results from the method of images from Section 5.3.3 below.

Anticipating such confirmation, we see that f_D solves the inhomogeneous (Poisson) equation under homogeneous BCs, while g_D solves the homogeneous (i.e. the Laplace) equation under inhomogeneous DBCs. The solution (5.2.4) of the general inhomogeneous Dirichlet problem is the sum of these two terms, by virtue of the fact that the equation is linear.

(vi) *Smoothing.* Suppose (for simplicity) that $\rho = 0 = f_D$. The prescribed function ψ_S is arbitrary, and can be discontinuous; e.g. ψ_S might be constant and non-zero on one part of S, and zero in immediately adjacent parts. But $\nabla' G_D(\boldsymbol{r}' \mid \boldsymbol{r})$ is a continuous function of $\boldsymbol{r}$ (except at $\boldsymbol{r} = \boldsymbol{r}'$); therefore $\psi(\boldsymbol{r}) = g_D(\boldsymbol{r})$ is a continuous function of $\boldsymbol{r}$ once $\boldsymbol{r}$ is off the surface by however small a distance. In this sense discontinuities of ψ on the boundary are smoothed out inside V. This is not obvious from the outset: for instance, as regards evolution in time, the analogous assertion is true for the (parabolic) diffusion equation, but false for the (hyperbolic) wave equation.

The behaviour of $\psi(\mathbf{r})$ inside V near a discontinuity of ψ_S is discussed in Section 7.2 below.

5.3 Applications

5.3.1 Introduction

It was stressed above that once the Green's function is known, the magic rule answers any question by quadratures, i.e. by the mere evaluation of integrals. Examples appear amongst the problems, and should be worked through carefully in order to gain indispensable experience with and appreciation of the magic rule. Many other examples appear in, for example, Jackson (1962), Chapters 1, 2, 3, and Sneddon (1957), Chapter 4, though not all are formulated there explicitly in terms of Green's functions: recall that any invitation to calculate the potential due to a point source calls for a Green's function, and that any BVP can be construed as calling for the application of the magic rule. Recall also that the determination of $\psi(\mathbf{r})$ is not always the end of the matter; often one needs grad ψ, which can be harder to find. Of course, though all such problems can be solved by Green's functions, this is not always the most convenient method, as illustrated in Section 5.4 below. The knack of choosing the best approach comes only with experience.

Here we concentrate primarily on determining G_D itself. All we shall say can be found in books on electromagnetism, and we merely rearrange and reexpress it systematically in the language of Green's functions.

Perhaps the most widespread method is the expansion of G_D in eigenfunctions of $-\nabla^2$, or an improved version, both discussed in Section 5.3.2 below. By contrast, in the special but important cases where the boundary is an infinite plane, or a sphere, the method of images delivers G in closed form rather than as an infinite series, an enormous advantage in any subsequent computation. Images are discussed in Sections 5.3.3 and 5.3.4.

All these methods, in fact all methods that are manageable analytically (wholly or in part), depend on the separability of the operator ∇^2 in a coordinate system adapted to the boundaries. We say that a coordinate system is so adapted if the boundaries are surfaces on which one of the coordinates is constant (e.g. the curved surface of a right circular cylinder has $r = \text{constant}$ in cylindrical polars, and its flat ends have $z = \text{constant}$, provided the axis of the cylinder has been chosen as the z-axis). The Laplace operator is said to be separable in coordinates (ξ_1, ξ_2, ξ_3) if one can write it as

$$\nabla^2\psi = \frac{1}{h_1h_2h_3}\left\{\frac{\partial}{\partial\xi_1}\left(\frac{h_2h_3}{h_1}\frac{\partial\psi}{\partial\xi_1}\right)+\frac{\partial}{\partial\xi_2}\left(\frac{h_3h_1}{h_2}\frac{\partial\psi}{\partial\xi_2}\right)+\frac{\partial}{\partial\xi_3}\left(\frac{h_1h_2}{h_3}\frac{\partial\psi}{\partial\xi_3}\right)\right\},$$
(5.3.1)

where all the hs may be functions of all the ξs. The distinguishing feature is that ∇^2 separates into three terms, in each of which there are differentiations only with respect to one of the three independent variables. Only if (5.3.1) applies can one find solutions of the Helmholtz equation in the separated form

$$\psi(\xi_1, \xi_2, \xi_3) = A(\xi_1)B(\xi_2)C(\xi_3). \tag{5.3.2}$$

Cartesian, spherical polar, and cylindrical polar coordinates are the familiar examples.

Remarkably, in 3D only a limited set of coordinate systems admit (5.3.1, 2). Morse and Feshbach (1953), Section 5.1 explain why this is and which they are. (See also Page (1955), Chapter 6.) The information is perhaps more often valuable in a negative sense: if the surfaces in a problem are not constant-coordinate surfaces of any separable system, then one can stop looking for solutions in the convenient form (5.3.2). Such problems yield only to numerical methods, which are beyond our scope here.

Finally, spectacularly powerful methods exist for solving 2D problems, by exploiting the theory of functions of a complex variable (cf. the remarks at the end of Section 4.3); they gain much of their power from the flexibility of conformal transformations. We do not here discuss or apply such special 2D methods (except once in Section 7.4.2 below): see any book on electrostatics or potential theory. Unfortunately, this approach does not generalize in any way to 3D.

5.3.2 The eigenfunction expansion and its variants in rectangular regions

We illustrate the expansion (5.1.13) in 2D by applying it to the rectangle $0 \leqslant x \leqslant a$, $0 \leqslant y \leqslant b$. Instead of a single label to identify the eigenfunctions of $-\nabla^2$, it is now convenient to introduce two labels n, m. Subject to DBCs, one has

$$\phi_{n,m}(x, y) = \sqrt{\left(\frac{2}{a}\right)}\sin\left(\frac{n\pi x}{a}\right)\sqrt{\left(\frac{2}{b}\right)}\sin\left(\frac{m\pi y}{b}\right), \tag{5.3.3.a}$$

$n, m = 1, 2, 3, \ldots$

$$\lambda_{n,m} = [(n\pi/a)^2 + (m\pi/b)^2]. \tag{5.3.3b}$$

Then (5.1.13) yields

$$G_D(x, y \mid x', y')$$

$$= \frac{2}{a}\frac{2}{b}\sum_{n=1}^{\infty}\sum_{m=1}^{\infty} \frac{\sin\left(\frac{n\pi x}{a}\right)\sin\left(\frac{n\pi x'}{a}\right)\sin\left(\frac{m\pi y}{b}\right)\sin\left(\frac{m\pi y'}{b}\right)}{\left[\left(\frac{n\pi}{a}\right)^2 + \left(\frac{m\pi}{b}\right)^2\right]}. \tag{5.3.4}$$

This can be awkward to use, because the double sum may take a long time to evaluate. In 3D, one would have a triple sum.

There is a related version of G_D in the form of a single sum (double sum in 3D): here one pays, by doing more analysis, for an expression requiring less numerical work. Roughly speaking, the idea is to accommodate one of the factors of $\delta(\boldsymbol{r} - \boldsymbol{r}') = \delta(x - x')\,\delta(y - y')$ on the right of (5.1.2a) through a series expansion, as above, but to accommodate the other by solving an ordinary differential equation for each term of this series. (In 3D one would secure a double instead of a triple sum.)

Though this description sounds rather abstruse, the simplest example makes the approach quite plain. We write

$$G_D(x, y \mid x', y') = \frac{2}{a}\sum_{n=1}^{\infty} \sin\left(\frac{n\pi x}{a}\right)\sin\left(\frac{n\pi x'}{a}\right) f_n(y \mid y'). \tag{5.3.5}$$

This is just a Fourier series for G_D regarded as a function of x; sines are chosen in view of the DBCs at the edges $x = 0$, $x = a$. The dependence on x' then follows from the known symmetry (5.1.7), which also leads us to anticipate that $f_n(y \mid y') = f_n(y' \mid y)$. Represent $\delta(x - x')$ by the corresponding sine series, given by (1.3.9) with $L = a$; substitution of this and of (5.3.5) into (5.1.2a) yields

$$-\nabla^2 G_D = \frac{2}{a}\sum \sin\left(\frac{n\pi x}{a}\right)\sin\left(\frac{n\pi x'}{a}\right)\left\{\left(\frac{n\pi}{a}\right)^2 f_n(y \mid y') - \frac{\partial^2}{\partial y^2} f_n(y \mid y')\right\}$$

$$= \frac{2}{a}\sum \sin\left(\frac{n\pi x}{a}\right)\sin\left(\frac{n\pi x'}{a}\right)\delta(y - y'). \tag{5.3.6}$$

Equating the coefficients of corresponding sines, we find

$$\left\{-\frac{\partial^2}{\partial y^2} + \left(\frac{n\pi}{a}\right)^2\right\} f_n(y \mid y') = \delta(y - y'), \tag{5.3.7a}$$

while the BCs (5.1.2b) entail

$$f_n(0 \mid y') = 0 = f_n(b \mid y'). \tag{5.3.7b}$$

This is nothing but an ordinary differential equation for a 1D Green's function, which we can solve by the technique familiar from Chapter 2.

The solution reads

$$f_n(y \mid y') = \frac{(a/n\pi)}{\sinh(n\pi b/a)}\left\{\theta(y'-y)\sinh\left(\frac{n\pi y}{a}\right)\sinh\left(\frac{n\pi(b-y')}{a}\right)\right.$$
$$\left. + \theta(y-y')\sinh\left(\frac{n\pi y'}{a}\right)\sinh\left(\frac{n\pi(b-y)}{a}\right)\right\}$$
$$= \frac{(a/n\pi)}{\sinh(n\pi b/a)}\sinh\left(\frac{n\pi y_<}{a}\right)\sinh\left(\frac{n\pi(b-y_>)}{a}\right). \tag{5.3.8}$$

Exercise: Derive (5.3.8) by solving (5.3.7a, b) from first principles.

Accordingly, the single-sum alternative to (5.3.4) reads

$$G_D(x, y \mid x', y') = \frac{2}{\pi}\sum_{n=1}^{\infty}\frac{1}{n\sinh(n\pi b/a)}\sin\left(\frac{n\pi x}{a}\right)\sin\left(\frac{n\pi x'}{a}\right)$$
$$\times\sinh\left(\frac{n\pi y_<}{a}\right)\sinh\left(\frac{n\pi(b-y_>)}{a}\right). \tag{5.3.9}$$

Evidently, every term of the single sum satisfies the Laplace equation except at $y = y'$; this contrasts with the double sum (5.3.4), whose individual terms are solutions of the Helmholtz equation $(\nabla^2 + k^2)\psi = 0$, with different k^2 for different terms.

Analogous methods for the sphere are used in Appendix H, Section H.4: namely series expansions for the dependence on angles, and a differential equation for the dependence on r.

5.3.3 The method of images for a halfspace

In 3D we require $G_D^{(3)}(\boldsymbol{r} \mid \boldsymbol{r}')$ in the halfspace $z \geqslant 0$, i.e. the solution of (5.1.2a) under the BC $G_D^{(3)}(x, y, 0 \mid x', y', z') = 0$. In this example the halfspace V is bounded by the xy-plane, plus say an infinite hemisphere on which (in 3D) G_D is likewise required to vanish.

The basic idea is to look for G_D in the form (5.1.3). It is obvious (once it has been pointed out) that if for $\chi(\boldsymbol{r} \mid \boldsymbol{r}')$ we write the potential, at $\boldsymbol{r}$, due to a unit *negative* point source situated at $\tilde{\boldsymbol{r}}' \equiv (x', y' - z')$ as shown in Fig. 5.1, then, by symmetry, G_D vanishes on the median plane between $\boldsymbol{r}'$ and $\tilde{\boldsymbol{r}}'$, i.e. precisely on the xy-plane, as required. Moreover, in the region V, i.e. for $z \geqslant 0$, χ is harmonic, since $-\nabla^2\chi(\boldsymbol{r} \mid \boldsymbol{r}') = -\delta(\boldsymbol{r} - \tilde{\boldsymbol{r}}')$, and the delta-function is non-zero only at $\tilde{\boldsymbol{r}}'$, which lies outside V. Accordingly, defining

$$\tilde{\boldsymbol{R}} \equiv \boldsymbol{r} - \tilde{\boldsymbol{r}}' \tag{5.3.10}$$

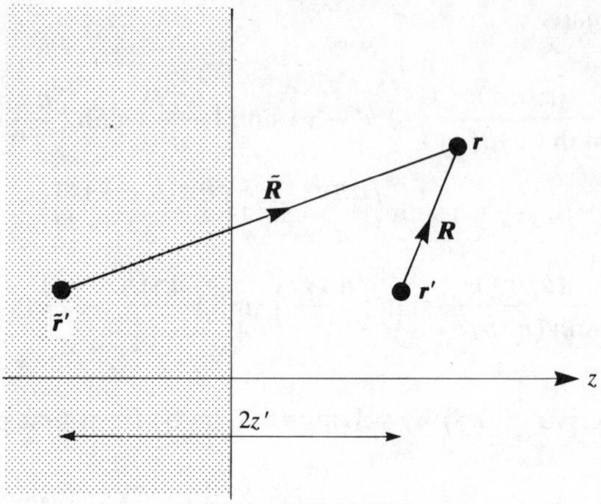

Fig. 5.1

by analogy to $\boldsymbol{R} = \boldsymbol{r} - \boldsymbol{r}'$, we have $\chi = -G_0(\boldsymbol{r} \mid \tilde{\boldsymbol{r}}')$, whence

$$G_{\mathrm{D}}^{(3)}(\boldsymbol{r} \mid \boldsymbol{r}') = G_0^{(3)}(\boldsymbol{r} \mid \boldsymbol{r}') - G_0^{(3)}(\boldsymbol{r} \mid \tilde{\boldsymbol{r}}') = \frac{1}{4\pi}\left(\frac{1}{R} - \frac{1}{\tilde{R}}\right)$$

$$= \frac{1}{4\pi}\left\{\frac{1}{|\boldsymbol{r} - \boldsymbol{r}'|} - \frac{1}{|\boldsymbol{r} - \tilde{\boldsymbol{r}}'|}\right\}$$

$$= \frac{1}{4\pi}\{[(x - x')^2 + (y - y')^2 + (z - z')^2]^{-\frac{1}{2}}$$

$$- [(x - x')^2 + (y - y')^2 + (z + z')^2]^{-\frac{1}{2}}\}. \tag{5.3.11}$$

One says that $G_{\mathrm{D}}^{(3)}$ in (5.3.11) has been constructed by the method of images, because, if the xy-plane were a mirror, then according to geometrical optics the image of the source point $\boldsymbol{r}'$ would be located at the position $\tilde{\boldsymbol{r}}'$ of the fictitious source responsible for χ.

Consider a single point charge Q in front of an earthed conducting plane. The force $\boldsymbol{F}$ on this charge is given by (5.1.12); with $\chi = -1/4\pi\tilde{R}$ as above, this yields

$$U(\boldsymbol{r}') = \tfrac{1}{2}Q^2\left(-\frac{1}{4\pi\tilde{R}}\right)\Bigg|_{\boldsymbol{r}=\boldsymbol{r}'} = -\frac{1}{2}\cdot\frac{Q^2}{4\pi\,|\boldsymbol{r}' - \tilde{\boldsymbol{r}}'|}$$

$$= -\frac{1}{2}\cdot\frac{Q^2}{4\pi(2z')} = -\frac{Q^2}{4\pi}\cdot\frac{1}{4z'}, \tag{5.3.12a}$$

$$\boldsymbol{F} = -\boldsymbol{\nabla}'U = -\hat{\boldsymbol{z}}\frac{Q^2}{4\pi}\cdot\frac{1}{4z'^2} = -\hat{\boldsymbol{z}}\frac{Q^2}{4\pi(2z')^2}. \tag{5.3.12b}$$

As one might have expected, this is just the inverse-square-law force that would be exerted on Q by a charge $-Q$ at the image position.

Exercise: Verify (5.3.12b) from (5.1.11) and (5.3.11), i.e. without appeal to (5.1.12).

In the 2D halfspace (halfplane) $x \geqslant 0$, the analogue of (5.3.11) is

$$G_D^{(2)}(\boldsymbol{r} \mid \boldsymbol{r}') = G_0^{(2)}(\boldsymbol{r} \mid \boldsymbol{r}') - G_0^{(2)}(\boldsymbol{r} \mid \tilde{\boldsymbol{r}}')$$
$$= \frac{1}{2\pi} \log\left(\frac{\tilde{R}}{R}\right) = \frac{1}{2\pi} \log \frac{[(x+x')^2 + (y-y')^2]^{\frac{1}{2}}}{[(x-x')^2 + (y-y')^2]^{\frac{1}{2}}}. \qquad (5.3.13)$$

The 1D analogue, in the halfspace (halfline) $x \geqslant 0$, is

$$G_D^{(1)}(x \mid x') = G_0^{(1)}(x \mid x') - G_0^{(1)}(x \mid -x')$$
$$= -\tfrac{1}{2}\,|x - x'| + \tfrac{1}{2}\,|x + x'| = \theta(x' - x)x + \theta(x - x')x' = x_<. \qquad (5.3.14)$$

Exercise: Verify (5.3.13) and (5.3.14).

Observe that, as $r \to \infty$, the image Green's functions vanish even in 2D and 1D, by contrast with the behaviour of G_0. This happens because the overall induced charge on the boundary is just -1, so that the total charge of the system is zero.

With the image Green's function (5.3.11), the surface integral $g_D^{(3)}$ in the magic rule (5.2.4) simplifies quite remarkably. Noting that $\partial_n' = -\partial/\partial z'$ and $\int dS' \cdots = \iint dx'\, dy' \ldots$, differentiating (5.3.11) with respect to z', setting $z' = 0$, and substituting into $g_D^{(3)}$, we obtain

$$g_D^{(3)}(\boldsymbol{r}) \equiv \int dS'\, \psi_S(x', y', 0) \left(\frac{\partial G_D^{(3)}(\boldsymbol{r} \mid \boldsymbol{r}')}{\partial z'}\right)\bigg|_{z'=0}; \qquad (5.3.15)$$

$$\frac{\partial G_D^{(3)}}{\partial z'} = \frac{1}{4\pi} \frac{\partial}{\partial z'} \left\{\frac{1}{[(x-x')^2 + (y-y')^2 + (z-z')^2]^{\frac{1}{2}}} - \frac{1}{[(x-x')^2 + (y-y')^2 + (z+z')^2]^{\frac{1}{2}}}\right\}$$
$$= \frac{1}{4\pi}\left\{\frac{(z-z')}{[(x-x')^2 + (y-y')^2 + (z-z')^2]^{\frac{3}{2}}} + \frac{(z+z')}{[(x-x')^2 + (y-y')^2 + (z+z')^2]^{\frac{3}{2}}}\right\},$$

$$\frac{\partial G_D^{(2)}(\boldsymbol{r} \mid \boldsymbol{r}')}{\partial z'}\bigg|_{z'=0} = \frac{1}{4\pi} \frac{2z}{[(x-x')^2 + (y-y')^2 + z^2]^{\frac{3}{2}}}$$
$$= \frac{1}{2\pi} \frac{z}{|\boldsymbol{r} - \boldsymbol{r}'|^3}, \qquad (5.3.16)$$

$$g_D^{(3)}(\boldsymbol{r}) = \int dS'\, \psi_S(x', y', 0) \frac{1}{2\pi} \frac{z}{|\boldsymbol{r} - \boldsymbol{r}'|^3}, \qquad (5.3.17)$$

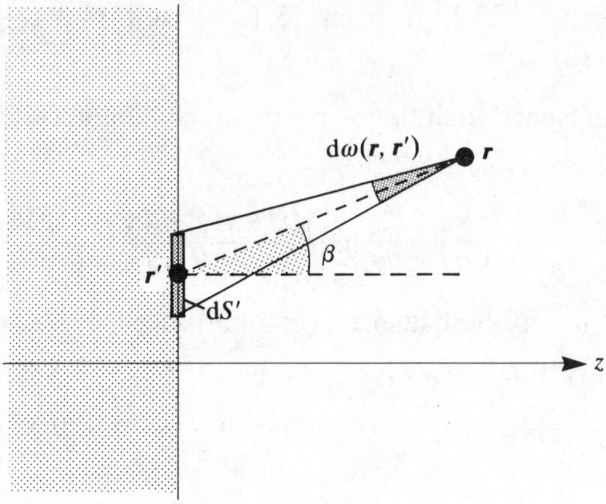

Fig. 5.2

where we recall that $\boldsymbol{r}' = (x', y', 0)$. Defining $\cos\beta = z/|\boldsymbol{r} - \boldsymbol{r}'|$, we see from Fig. 5.2 that $\mathrm{d}S' z/|\boldsymbol{r} - \boldsymbol{r}'|^3 = \mathrm{d}S' \cos\beta/|\boldsymbol{r} - \boldsymbol{r}'|^2$ is just the solid angle $\mathrm{d}\omega(\boldsymbol{r}, \boldsymbol{r}')$ subtended by the surface element $\mathrm{d}S'$ at the field point $\boldsymbol{r}$. (We write $\mathrm{d}\omega(\boldsymbol{r}, \boldsymbol{r}')$ with a comma instead of an upright bar as a reminder that it is defined only when $\boldsymbol{r}'$ is on S, albeit $\boldsymbol{r}$ can be anywhere in V.) Accordingly,

$$g_{\mathrm{D}}^{(3)}(\boldsymbol{r}) = (1/2\pi) \int_S \mathrm{d}\omega\,(\boldsymbol{r}, \boldsymbol{r}')\psi_S(\boldsymbol{r}'). \tag{5.3.18a}$$

In particular, if $\psi_S = \alpha$ is a constant over a part S_1 of S, then the contribution of S_1 is just $\alpha\omega_1/2\pi$, with ω_1 the total solid angle subtended at $\boldsymbol{r}$ by S_1. If $\psi_S = \alpha$ over all of S, then $\omega = 2\pi$, and we recover the position-independent result $g_{\mathrm{D}}^{(3)}(\boldsymbol{r}) = \alpha$ that might have been expected.

In 2D, it follows similarly from (5.3.13) that

$$g_{\mathrm{D}}^{(2)}(\boldsymbol{r}) = (1/\pi) \int \mathrm{d}\omega\,(\boldsymbol{r}, \boldsymbol{r}')\psi_S(\boldsymbol{r}'), \tag{5.3.18b}$$

where $\mathrm{d}\omega(\boldsymbol{r}, \boldsymbol{r}')$ is now the (ordinary) angle subtended at $\boldsymbol{r}$ by the boundary line-element $\mathrm{d}S'$ at $\boldsymbol{r}'$.

Exercise: Prove this.

Even in regions less simple than a halfspace, provided they are bounded by planes, it often proves useful to express G_{D} in terms of G_0, by introducing more than one image. A straightforward example with just three images occurs in Problem 5.9. More remarkable is the *multiple-image series* for the Green's function in say the 3D slab $0 \leqslant z \leqslant L$. Given the source point at $\boldsymbol{r}' = (x', y', z')$, one introduces opposite-sign images at $(x', y', 2NL - z')$, $(N = 0, \pm 1, \pm 2, \ldots)$ and same-sign images at $(x', y', 2NL + z')$, $(N = \pm 1, \pm 2, \ldots)$. Source and images together yield (see Problem 5.10)

$$G_D^{(3)}(\boldsymbol{r} \mid \boldsymbol{r}') = \sum_{N=-\infty}^{\infty} \{G_0^{(3)}(x, y, z \mid x', y', 2NL + z') - G_0^{(3)}(x, y, z \mid x', y', 2NL - z')\}. \quad (5.3.19)$$

Exercise: Sketch the positions of the first few images, indicating also their signs.

The 2D and 1D versions are given simply by replacing $G_0^{(3)}$ by $G_0^{(2,1)}$ respectively.

Such multiple-image series tend to be useful, i.e. rapidly-converging, precisely when the convergence of the normal-mode expansions like (5.3.4), or of their modified versions like (5.3.9), is slow. (This complementarity is even more striking for the diffusion equation, as we shall see in Section 9.4.) Illuminating discussions of (5.3.19) are given by Pumplin (1969) and Glasser (1970).

5.3.4 The method of images for spheres and circles

We start with the *exterior problem* for the sphere. This requires $G_D^{(3)}(\boldsymbol{r} \mid \boldsymbol{r}')$ in a 3D region V outside a sphere S of radius a and centre C, under the BCs that G_D vanish when $\boldsymbol{r}$ is on the sphere, and also at infinity. The basic idea is the same as in Section 5.3.3: for the harmonic function χ in (5.1.3) we try the potential due to a point source q situated at some point $\boldsymbol{r}'$ *inside* the sphere (i.e. again outside V). By symmetry, we expect the image point $\tilde{\boldsymbol{r}}'$ to lie on the line through C and the source point $\boldsymbol{r}'$, as shown in Fig. 5.3. Again we define $\boldsymbol{R} = \boldsymbol{r} - \boldsymbol{r}'$, $\tilde{\boldsymbol{R}} = \boldsymbol{r} - \tilde{\boldsymbol{r}}'$. For convenience, we move the origin to C; this makes $\boldsymbol{r}'$ and $\tilde{\boldsymbol{r}}'$ parallel, allowing us to write $\tilde{\boldsymbol{r}}' = \alpha \boldsymbol{r}'$. Then

$$G_D^{(3)}(\boldsymbol{r} \mid \boldsymbol{r}') = G_0^{(3)}(\boldsymbol{r} \mid \boldsymbol{r}') + \tilde{q} G_0^{(3)}(\boldsymbol{r} \mid \tilde{\boldsymbol{r}}') = \frac{1}{4\pi}\left(\frac{1}{R} + \frac{\tilde{q}}{\tilde{R}}\right) \quad (5.3.20\text{a, b})$$

$$= \frac{1}{4\pi}\left\{\frac{1}{[r^2 + r'^2 - 2\boldsymbol{r} \cdot \boldsymbol{r}']^{\frac{1}{2}}} + \frac{\tilde{q}}{[r^2 + \tilde{r}'^2 - 2\boldsymbol{r} \cdot \boldsymbol{r}']^{\frac{1}{2}}}\right\} \quad (5.3.20\text{c})$$

$$= \frac{1}{4\pi}\left\{\frac{1}{[r^2 + r'^2 - 2\boldsymbol{r} \cdot \boldsymbol{r}']^{\frac{1}{2}}} + \frac{\tilde{q}}{[r^2 + \alpha^2 r'^2 - 2\alpha \boldsymbol{r} \cdot \boldsymbol{r}']^{\frac{1}{2}}}\right\} \quad (5.3.20\text{d})$$

$$= \frac{1}{4\pi}\left\{\frac{1}{[r^2 + r'^2 - 2\boldsymbol{r} \cdot \boldsymbol{r}']^{\frac{1}{2}}} + \frac{\tilde{q}/\alpha^{\frac{1}{2}}}{[r^2/\alpha + \alpha r'^2 - 2\boldsymbol{r} \cdot \boldsymbol{r}']^{\frac{1}{2}}}\right\}. \quad (5.3.20\text{e})$$

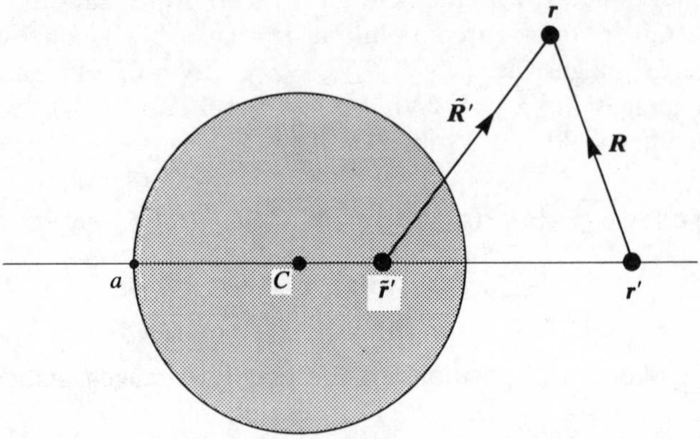

Fig. 5.3

While (5.3.20a, b, c) apply irrespective of the choice of origin, (5.3.20d, e) apply only with the origin at C so that indeed $\tilde{\boldsymbol{r}}' = \alpha \boldsymbol{r}'$.

With the origin at the centre of the sphere, the BC prescribes that $G_D^{(3)}$ as given by (5.3.20e) vanish when $r^2 = a^2$, irrespective of $\boldsymbol{r} \cdot \boldsymbol{r}'$ (i.e. irrespective of the angle between $\boldsymbol{r}$ and $\boldsymbol{r}'$). This can happen only if the two terms on the right cancel identically, which in turn imposes the two conditions

$$(a^2 + r'^2) = (a^2/\alpha + \alpha r'^2), \qquad \tilde{q}/\alpha^{\frac{1}{2}} = -1. \qquad (5.3.21\text{a, b})$$

For α, (5.3.21a) gives the quadratic equation

$$\{(r')^2\alpha^2 - (a^2 + r'^2)\alpha + a^2\} = 0,$$

whose roots are $\alpha = 1$ and $\alpha = a^2/r'^2$. The root $\alpha = 1$ would make G_D vanish identically for all $\boldsymbol{r}$ and $\boldsymbol{r}'$; hence we want the other root, which entails

$$\alpha = a^2/r'^2, \qquad \tilde{r}'r' = a^2, \qquad \tilde{q} = -a/r'. \qquad (5.3.22\text{a, b, c})$$

The relation $\tilde{r}'r' = a^2$ is familiar in geometry: the points $\tilde{\boldsymbol{r}}'$ and $\boldsymbol{r}'$ are *inverses* of each other with respect to the sphere. (For the sphere it would have been more accurate to speak of the method of inversion rather than of images.) Substituting from (5.3.22) into (5.3.20e) we obtain (still with the centre as origin)

$$G_D^{(3)}(\boldsymbol{r} \mid \boldsymbol{r}') = \frac{1}{4\pi}\left\{\frac{1}{[r^2 + r'^2 - 2\boldsymbol{r} \cdot \boldsymbol{r}']^{\frac{1}{2}}} - \frac{1}{[r^2r'^2/a^2 + a^2 - 2\boldsymbol{r} \cdot \boldsymbol{r}']^{\frac{1}{2}}}\right\}. \qquad (5.3.23)$$

This makes explicit both the symmetry in $\boldsymbol{r}$ and $\boldsymbol{r}'$, and the fact that $G_D^{(3)}$

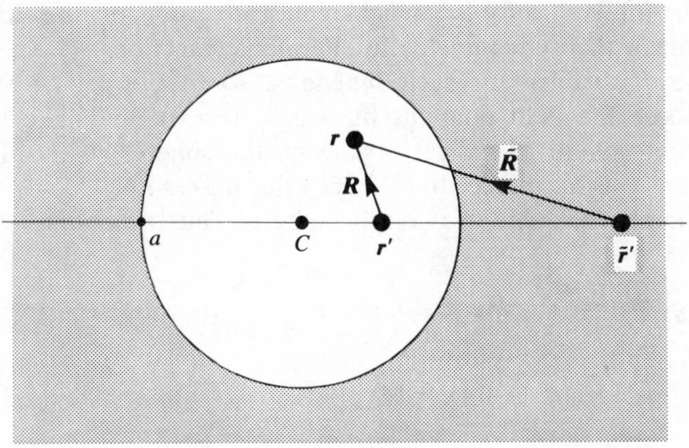

Fig. 5.4

vanishes when $r = a$. The second term is just $\tilde{q}/4\pi\tilde{R} = -a/4\pi r'\tilde{R}$, reminding one that it is not $\tilde{R}$ that is a symmetric function of $\boldsymbol{r}$ and $\boldsymbol{r}'$, but the combination

$$r'\tilde{R} = [r^2r'^2 + a^4 - 2a^2\boldsymbol{r}\cdot\boldsymbol{r}']^{\frac{1}{2}}. \tag{5.3.24}$$

All the equations (5.3.20–24) apply equally to the *interior problem* where $\boldsymbol{r}, \boldsymbol{r}'$ lie inside the sphere, and $\tilde{\boldsymbol{r}}'$ outside, as shown in Fig. 5.4.

Exercise: Verify this explicitly.

The surface integral $g_{\rm D}^{(3)}$ in the magic rule (5.2.4) requires the normal (i.e. radial) derivative

$$\left.\frac{\partial G_{\rm D}^{(3)}}{\partial r'}\right|_{r'=a} = \frac{r^2 - a^2}{4\pi a R^3} = \frac{r^2 - a^2}{4\pi a[r^2 + a^2 - 2ar\cos(\chi)]^{\frac{3}{2}}}, \tag{5.3.25}$$

where χ is the angle between $\boldsymbol{r}$ and $\boldsymbol{r}'$. Since, for the interior problem, $\partial_n' = +\partial/\partial r'$, this yields

$$g_{\rm D}^{(3)}(\boldsymbol{r}) = -\frac{(r^2 - a^2)}{4\pi a}\int_S {\rm d}S'\,\psi_S(\boldsymbol{r}')/R^3 \tag{5.3.26a}$$

$$= \frac{(a^2 - r^2)a}{4\pi}\int_S {\rm d}\Omega'\,\psi_S(a, \Omega')/R^3, \qquad (r \leqslant a), \tag{5.3.26b}$$

called Poisson's integral.

Exercise: Derive (5.3.25); it may help to start from $\partial G_{\rm D}/\partial r'|_{r'=a} = \hat{\boldsymbol{r}}'\cdot\boldsymbol{\nabla}'G_{\rm D}|_{r'=a}$.

Poisson's integral affords an independent proof of the important spherical-mean theorem (4.3.13) for harmonic functions. (Recall that ρ and thereby f_D vanish for such functions, so that $\psi = g_D$.) We need merely choose the field point as the origin ($\boldsymbol{r} = 0$), and take it as the centre of our sphere of radius a, drawing the sphere wholly inside the region where ψ is harmonic. In (5.3.26b) this makes $r = 0$, $R = a$, and ψ_S becomes simply the value of ψ on the sphere. Thus we obtain

$$\psi(0) = g_D^{(3)}(0) = \frac{a^3}{4\pi}\int d\Omega'\ \psi_S(a, \Omega')/a^3$$

$$= \int d\Omega'\ \psi_S(a, \Omega')/4\pi = M_{0,a}\{\psi\}. \qquad (5.3.27) \quad \blacksquare$$

Note that for the exterior problem one has $\partial_n' = -\partial/\partial r'$, whence $g_D^{(3)}$ is then given by the *negative* of (5.3.26).

The expansion of $G_D^{(3)}$ in spherical harmonics is considered in Appendix H, together with the analogous 2D expansion for the circle.

Finally, we quote the 2D Dirichlet Green's function for a circle of radius a centred on the origin. It features an image source again at the inverse point $\tilde{\boldsymbol{r}}'$ given by $\tilde{r}'r' = a$, so that Figs 5.3 and 5.4 are again appropriate. But now one finds

$$G_D^{(2)}(\boldsymbol{r} \mid \boldsymbol{r}') = \frac{1}{2\pi}\left\{\log\left(\frac{a}{R}\right) - \log\left(\frac{a}{\tilde{R}}\right) + \log\left(\frac{r'}{a}\right)\right\} \qquad (5.3.28a)$$

$$= \frac{1}{2\pi}\log\left(\frac{r'\tilde{R}}{aR}\right) = \frac{1}{2\pi}\log\frac{[r^2r'^2/a^2 + a^2 - 2\boldsymbol{r}\cdot\boldsymbol{r}']^{\frac{1}{2}}}{[r^2 + r'^2 - 2\boldsymbol{r}\cdot\boldsymbol{r}']^{\frac{1}{2}}}. \qquad (5.3.28b, c)$$

Exercise: Verify that $G_D^{(2)}$ vanishes when $r = a$, irrespective of $\boldsymbol{r}\cdot\boldsymbol{r}'$.

Since the coefficient of $\log(a/\tilde{R})$ is equal in magnitude to that of $\log(a/R)$, the image source is now equal in magnitude to the true source (but again of opposite sign); and we see that (as compared with 3D) there is an extra constant term (i.e. a term independent of $\boldsymbol{r}$), namely the third in (5.3.28a). As in 3D, the same expression applies for both the exterior and the interior problem. For the normal derivative analogous to (5.3.25) one now finds

$$\left.\frac{\partial G_D^{(2)}}{\partial r'}\right|_{r'=a} = \frac{(r^2 - a^2)}{2\pi a R^2} = \frac{(r^2 - a^2)}{2\pi a[r^2 + a^2 - 2ar\cos(\phi - \phi')]}. \qquad (5.3.29)$$

Accordingly, Poisson's integral for the interior of the circle (i.e. the 2D analogue of (5.3.26)) reads

$$g_D^{(2)} = \frac{(a^2 - r^2)}{2\pi a} \int_S dS' \, \psi_S(\mathbf{r}')/R^2 \tag{5.3.30a}$$

$$= \frac{(a^2 - r^2)}{2\pi} \int d\phi' \, \psi_S(a, \phi')/R^2, \qquad (r \leqslant a). \tag{5.3.30b}$$

Exercise: Derive (5.3.29) and (5.3.30).

To illustrate (5.3.26), consider the interior problem for a sphere free of sources, with $\psi_S = 0$ except on a small cap of area $a^2\,\Delta\Omega$ at the north pole, and with $\psi_S = \Psi$ on this cap. For simplicity, idealize further by letting $\Delta\Omega \to 0$ and $\Psi \to \infty$ in such a way that $\Psi\,\Delta\Omega = \alpha$ remains fixed, whence $\psi_S(\mathbf{r}') = \alpha\,\delta(\Omega')$. By appeal to (5.3.26b) this entails

$$\psi^{(3)}(\mathbf{r}) = g_D^{(3)}(\mathbf{r}) = \frac{(a^2 - r^2)a}{4\pi} \int d\Omega' \, \alpha\,\delta(\Omega')/R^3$$

$$= \frac{\alpha a(a^2 - r^2)}{4\pi R^3} = \frac{\alpha a(a^2 - r^2)}{4\pi[r^2 + a^2 - 2ar\cos\theta]^{\frac{3}{2}}}. \tag{5.3.31}$$

In the special case where $\mathbf{r}$ lies on the axis, one has $r = |z|$, $R^3 = (a - z)^3$, whence $\psi^{(3)} = \alpha a(a + z)/4\pi(a - z)^2$.

By (5.3.30) the solution to the analogous problem for a circle reads

$$\psi^{(2)}(\mathbf{r}) = \alpha(a^2 - r^2)/2\pi[r^2 + a^2 - 2ar\cos\phi], \tag{5.3.32}$$

which on the axis reduces to $\alpha(a + x)/2\pi(a - x)$.

Exercise: In the examples (5.3.31, 32), let χ be the angle between the axis and the vector $\mathbf{R}$ drawn from the pole to the field point. Adopt R and χ as independent variables, and discuss the behaviour of ψ as $R \to 0$ at fixed χ.

5.4 Boundary-value problems for the Laplace equation without Green's functions

To solve the homogeneous equation, i.e. the Laplace equation, subject to inhomogeneous BCs, one can dispense with Green's functions and bypass the magic rule. The magic-rule solution is given by the surface integral g_D defined in (5.2.4). The alternative is to expand ψ directly in an appropriately chosen complete set of harmonic functions, and to determine the expansion coefficients by fitting the data ψ_S. The harmonic functions given in Section 4.3 often prove convenient. We clarify the procedure with two examples.

First, consider the 3D halfspace $z \geqslant 0$; define $\boldsymbol{r}_\parallel = (x, y)$, and prescribe $\psi_S(x, y, 0) \equiv u(x, y)$. If u falls to zero fast enough as $r_\parallel \to \infty$, then one can and does require that ψ vanish as $r \to \infty$. Look for ψ as a linear combination of the harmonic functions (4.3.4). Then

$$\psi(\boldsymbol{r}) = \int \mathrm{d}^2\kappa\, b(\boldsymbol{\kappa}) \exp(i\boldsymbol{\kappa} \cdot \boldsymbol{r}_\parallel - \kappa z),$$

$$\boldsymbol{\kappa} \equiv (\kappa_1, \kappa_2), \qquad \kappa \equiv (\kappa_1^2 + \kappa_2^2)^{\frac{1}{2}}, \tag{5.4.1}$$

the functions with factors $\exp(\kappa z)$ being eliminated by the requirement that ψ vanish as $z \to \infty$. On setting $z = 0$, (5.4.1) reduces to an ordinary Fourier representation whose inverse is given by (C.2):

$$u(\boldsymbol{r}_\parallel) \equiv \psi(x, y, 0) = \int \mathrm{d}^2\kappa\, b(\boldsymbol{\kappa}) \exp(i\boldsymbol{\kappa} \cdot \boldsymbol{r}_\parallel), \tag{5.4.2a}$$

$$b(\boldsymbol{\kappa}) = (2\pi)^{-2} \int \mathrm{d}^2 r_\parallel\, u(\boldsymbol{r}_\parallel) \exp(-i \mid \boldsymbol{\kappa} \cdot \boldsymbol{r}_\parallel). \tag{5.4.2b}$$

Substitution of the known function (5.4.2b) into (5.4.1) then yields ψ by quadratures.

This method is convenient if $u(\boldsymbol{r}_\parallel)$ and $b(\boldsymbol{\kappa})$ are relatively simple functions:

Exercise: Determine $\psi(\boldsymbol{r})$ when $u = \alpha \exp(-r_\parallel^2/a^2)$.

On the other hand, irrespective of whether u and b are simple or complicated, (5.4.1) and (5.4.2b) can be combined and rearranged to read

$$\psi(\boldsymbol{r}) = \int \mathrm{d}^2\kappa \left\{\frac{1}{(2\pi)^2} \int \mathrm{d}^2 r'_\parallel\, u(\boldsymbol{r}'_\parallel) \exp(-i\boldsymbol{\kappa} \cdot \boldsymbol{r}'_\parallel)\right\} \exp(i\boldsymbol{\kappa}\boldsymbol{r}_\parallel - \kappa z)$$

$$= \int \mathrm{d}^2 r'_\parallel\, u(\boldsymbol{r}'_\parallel) \left\{\frac{1}{(2\pi)^2} \int \mathrm{d}^2\kappa \exp[i\boldsymbol{\kappa} \cdot (\boldsymbol{r}_\parallel - \boldsymbol{r}'_\parallel) - \kappa z]\right\}; \tag{5.4.3}$$

by comparison with $\psi = g_\mathrm{D} = \int \mathrm{d}^2 r'_\parallel\, u(\boldsymbol{r}'_\parallel)\, \partial G_\mathrm{D}(\boldsymbol{r} \mid \boldsymbol{r}')/\partial z'$, the quantity within braces in the rightmost expression must evidently be the same as

$$\frac{\partial}{\partial z'} G_\mathrm{D}(\boldsymbol{r} \mid \boldsymbol{r}')|_{z'=0},$$

where G_D is the image Green's function (5.3.11).

Exercise: Verify this explicitly.

For a second example take the interior of a circle of radius a centred on the origin, with prescribed $\psi_S = \psi(a, \phi) \equiv u(\phi)$. We use the functions (4.3.3a, b), rejecting those with a-coefficients, since they are not harmonic at the origin. Thus we look for ψ in the form

$$\psi(\mathbf{r}) = \sum_{m=-\infty}^{\infty} b_m (r/a)^{|m|} \exp(im\phi), \tag{5.4.4a}$$

where, ψ being real, the b_m must satisfy the condition

$$b_{-m} = b_m^*. \tag{5.4.4b}$$

Exercise: Prove (5.4.4b) from (5.4.4a) by setting $\psi^* = \psi$.

With $r = a$ and inverting the resultant Fourier series, one obtains

$$u(\phi) = \sum_{m=-\infty}^{\infty} b_m \exp(im\phi), \tag{5.4.5a}$$

$$b_m = \frac{1}{2\pi}\int_{-\pi}^{\pi} \mathrm{d}\phi\, u(\phi) \exp(-im\phi). \tag{5.4.5b}$$

Again, this method is tailor-made for cases where in the series for u only a few terms are important. At the other extreme, the magic rule is probably preferable if $u(\phi)$ is discontinuous or even singular, as in the examples (5.3.31) and (5.3.32). For comparison, we apply (5.4.4, 5) to the problem where $u(\phi) = \alpha\,\delta(\phi)$, whose solution is already known from (5.3.32). Here (5.4.5b) entails $b_m = \alpha/2\pi$ for all m, whence (5.4.4a) yields

$$\begin{aligned}
\psi &= (\alpha/2\pi)\left\{1 + 2\,\mathrm{Re}\sum_{m=1}^{\infty} (r/a)^m \exp(im\phi)\right\} \\
&= (\alpha/2\pi)\,\mathrm{Re}\left\{-1 + 2\sum_{m=0}^{\infty} (r\exp(i\phi)/a)^m\right\} \\
&= \frac{\alpha}{2\pi}\mathrm{Re}\left\{-1 + \frac{2}{1 - r\exp(i\phi)/a}\right\} \\
&= \frac{\alpha}{2\pi}\mathrm{Re}\left\{\frac{1 + r\exp(i\phi)/a}{1 - r\exp(i\phi)/a}\right\} \\
&= \frac{\alpha}{2\pi}\mathrm{Re}\frac{(a - r\exp(-i\phi))(a + r\exp(i\phi))}{(a^2 + r^2 - 2ar\cos\phi)} \\
&= \frac{\alpha}{2\pi}\frac{(a^2 - r^2)}{(a^2 + r^2 - 2ar\cos\phi)},
\end{aligned} \tag{5.4.6}$$

which agrees with (5.3.32) as it should.

Problems

5.1 In dry air, an electric field greater than $3\times10^6\,\mathrm{V\,m^{-1}}$ causes sparking.

(i) How high can one raise the potential of a conducting sphere of radius 10^{-2} m before sparking occurs?
(ii) How high can one raise the charge per unit length of an infinitely long conducting cylinder of radius 10^{-2} m before sparking occurs?
(iii) What would be your reaction if in (ii) the words 'charge per unit length' were replaced by 'potential'?

(In SI units, the so-called permittivity of the vacuum is $\varepsilon_0\approx 8.85\times10^{-12}\,\mathrm{C^2\,kg^{-1}\,m^{-3}\,s^2}$.)

5.2 This problem (like some later ones) calls for numerical summations by computer, in order to bring home the interplay between analytic and numerical procedures, which is crucial for efficiency.

$G_0(\boldsymbol{r}\,|\,\boldsymbol{r}')$ is the Dirichlet Green's function for the interior of the square $0\leqslant x,\, y\leqslant\pi$.

(i) To an accuracy of $1:10^3$, evaluate G with $\boldsymbol{r}=(\pi/4,\,\pi/4)$, $\boldsymbol{r}'=(3\pi/4,\,3\pi/4)$, using the double-sum representation (5.3.4).
(ii) Do the same, but using the single-sum representation (5.3.9).
(iii) Compare the number of terms you had to sum, and the time taken, in the two calculations.

Hint: In both methods, it pays to insert the given values of $\boldsymbol{r}$ and $\boldsymbol{r}'$ as soon as possible. In (ii), exploit sinh, cosh, and the hyperbolic versions of the standard trigonometric formulae for $\sin(A+B)$, etc.

5.3 (i) Find the eigenvalues and the normalized eigenfunctions of $-\nabla^2$ in the unit square $(0\leqslant x,\, y\leqslant 1)$, subject to homogeneous Dirichlet conditions $(\psi=0)$ on the pair of opposite sides $x=0,\,1$, and to homogeneous Neumann conditions $(\partial_n\psi=0)$ on the other two sides. List the four lowest eigenvalues and all the corresponding eigenfunctions.
(ii) Using your results, write down the expression for the Green's function as a double sum.

Hint: Your end-result should read

$$G(x,y\,|\,x',y')=\frac{1}{\pi^2}\left\{\sum_{n=1}^{\infty}\frac{2}{n^2}\sin(n\pi x)\sin(n\pi x')\right.$$
$$\left.+\sum_{n=1}^{\infty}\sum_{m=1}^{\infty}\frac{4}{n^2+m^2}\sin(n\pi x)\sin(n\pi x')\cos(m\pi y)\cos(m\pi y')\right\}.$$

5.4 Under the same conditions as in Problem 5.3, construct the Green's function as a single sum:

$$G(\boldsymbol{r} \mid \boldsymbol{r}') = \sum_{n=1}^{\infty} \sin(n\pi x) f_n(y, x', y').$$

Hints: Use the representation $\delta(x - x') = 2\sum_{n=1}^{\infty} \sin(n\pi x)\sin(n\pi x')$ to show that the defining equation $\nabla^2 G = -\delta(\boldsymbol{r} - \boldsymbol{r}')$ entails

$$\left\{\frac{\partial^2 f_n}{\partial y^2} - n^2\pi^2 f_n\right\} = -2\sin(n\pi x')\,\delta(y - y'),$$

and solve this equation subject to the BCs on $y = 0, 1$.

Your end-result should read

$$G(\boldsymbol{r} \mid \boldsymbol{r}') = \frac{2}{\pi}\sum_{n=1}^{\infty} \frac{\sin(n\pi x)\sin(n\pi x')}{n \sinh(n\pi)} \cosh(n\pi y_<)\cosh(n\pi(1 - y_>)).$$

5.5 A spherical shell of radius a has its northern hemisphere at potential V, and its southern hemisphere at potential zero.

(i) Show (by reference to the corresponding problem with both hemispheres at V) that in the equatorial plane inside the sphere the electric field is everywhere normal to this plane.
(ii) Is this true outside?
(iii) Determine the electric field at the centre.

Hint: The magic rule will deliver the answer to (iii), but there are quicker ways.

5.6 As Problem 5.5, but for a circle.

5.7 A point charge Q is located at a distance $r' > a$ from the centre of an earthed conducting sphere of radius a.

(i) Calculate the attractive potential energy U and the force F between sphere and point charge.
(ii) Show that the total charge induced on the sphere is $-Qa/r'$.
(iii) Verify that, for $r'/a \gg 1$, $|F| \sim Q^2 a/4\pi r'^3$.

Hint: Recall the decomposition $G_D = G_0 + \chi$ from Section 5.1, and the polar expansion of G_D. The induced charge can be found by appeal to Gauss's theorem.

5.8 In the region $z \geqslant 0$, the function ψ obeys the Laplace equation. On the xy-plane, ψ vanishes except in the square $|x| \leqslant a$, $|y| \leqslant a$, where $\psi = 1$.

(i) Write down the expression for ψ given by the magic rule.

(ii) Evaluate $\psi(x, y, z)$ accurately up to and including terms of order $1/z^4$ in the region where $z \gg a, |x|, |y|$.

5.9 The $x = 0$ and $y = 0$ planes are maintained at zero potential, and a point charge $+Q$ is placed at $(x', y', 0)$. Adapt the method of images to find the potential $\psi(\boldsymbol{r})$, the force $\boldsymbol{F}$ on the charge, and the energy U (excluding the self-energy of the point charge).

Hint: Three image charges are needed. The end-results for U and $\boldsymbol{F}$ are best expressed in terms of plane polar coordinates; for instance,

$$U = \frac{Q^2}{16\pi r}\left\{\frac{1}{\sqrt{2}} - \frac{1}{\cos\phi} - \frac{1}{\sin\phi}\right\}.$$

5.10 Consider the *multiple-image series* for G_D given by (5.3.19).

(i) Verify that G_D obeys the defining equation $-\nabla^2 G_D = \delta(\boldsymbol{r} - \boldsymbol{r}')$, and that it vanishes when $z = 0$, $z = L$.
(ii) Verify that $G_D(\boldsymbol{r} \mid \boldsymbol{r}') = G_D(\boldsymbol{r}' \mid \boldsymbol{r})$.
(iii) Show that the energy of a point charge Q at $\boldsymbol{r}'$ (excluding its self-energy) may be expressed as

$$U(z) = \frac{1}{2}\frac{Q^2}{4\pi}\left\{-\frac{1}{2z} + \tfrac{1}{2}\sum_{N=1}^{\infty}\left[\frac{2}{NL} - \frac{1}{NL - z} - \frac{1}{NL + z}\right]\right\},$$

and evaluate $(8\pi L/Q^2)U(L/2)$ to 0.1%.

5.11 *A theorem of Gauss.* A source distribution $\rho(\boldsymbol{r})$ vanishes outside a region V bounded by the closed surface S; $\boldsymbol{r}_0$ is any point outside V, and $\boldsymbol{r}_i$ any point inside; G_0 is the usual unbounded-space Green's function, and G_D the Dirichlet Green's function for V.

(i) Show, by applying Green's theorem to $G_0(\boldsymbol{r}_0 \mid \boldsymbol{r}')$ and $G_D(\boldsymbol{r}_i \mid \boldsymbol{r}')$ over V, that

$$G_0(\boldsymbol{r}_0 \mid \boldsymbol{r}_i) = -\int_S \mathrm{d}S'\, G_0(\boldsymbol{r}_0 \mid \boldsymbol{r}')\, \partial'_n G_D(\boldsymbol{r}' \mid \boldsymbol{r}_i). \tag{1}$$

(ii) Use (1) to prove the theorem (due to Gauss) that one can find a surface source-distribution σ on S (source strength $\sigma(\boldsymbol{r}')\,\mathrm{d}S'$ on the element $\mathrm{d}S'$ situated at $\boldsymbol{r}'$), such that, at points outside V, the potential due to σ is the same as that due to ρ. Express σ in terms of ρ and of G_D.

5.12 *Proof that in a finite volume the spectrum of $-\nabla^2$ is discrete.* Consider the eigenvalue problem $-\nabla^2\phi_n = \lambda_n\phi_n$ in a finite volume V, under the BC $\phi_n = 0$ on the bounding surface S of V. The main object is to show that, corresponding to a finite range $\Lambda_1 \leqslant \lambda_n \leqslant \Lambda_2$, there are at most a finite number of linearly independent eigen*functions* ϕ_n. Take the ϕ_n as orthonormal.

(i) Show that

$$\phi_n(\boldsymbol{r}) = \lambda_n \int_V \mathrm{d}V' G_\mathrm{D}(\boldsymbol{r}\,|\,\boldsymbol{r}')\phi_n(\boldsymbol{r}'). \tag{1}$$

(ii) Define

$$J \equiv \int_V \mathrm{d}V \int_V \mathrm{d}V' \left\{ G_\mathrm{D}(\boldsymbol{r}\,|\,\boldsymbol{r}') - \sum_{(n)} \frac{\phi_n(\boldsymbol{r})\phi_n(\boldsymbol{r}')}{\lambda_n} \right\}^2 > 0, \tag{2}$$

where the restricted sum $\Sigma_{(n)}$ runs over any finite set of eigenfunctions chosen so that $\Lambda_1 \leqslant \lambda_n \leqslant \Lambda_2$. Show (by appeal to (1)) that

$$J = W - \sum_{(n)} 1/\lambda_n^2 > 0, \tag{3}$$

where $W \equiv \int_V \mathrm{d}V \int_V \mathrm{d}V' \, G_\mathrm{D}^2(\boldsymbol{r}\,|\,\boldsymbol{r}')$.

(iii) By appeal to the decomposition $G_\mathrm{D} = G_0 + \chi$, and to the harmonic nature of χ, show, separately in 3D, 2D, and 1D, that $W < \infty$.

(iv) If there were an infinity of eigenfunctions with λ_n in the given range, then the sum $\Sigma_{(n)} 1/\lambda_n^2$ could be increased indefinitely by including more and more terms. This would contradict (3). ■

(v) As a byproduct, show that the sum $\Sigma_n 1/\lambda_n^2$, extended now over *all* eigenfunctions, converges.

(vi) Verify this explicitly for rectangular domains in 1D, 2D, and 3D.

5.13 Prove the result stated in Problem 5.12 under NBCs.

5.14 (i) Write down $G_\mathrm{D}(\boldsymbol{r}\,|\,\boldsymbol{r}')$ for the semi-infinite 2D strip $0 \leqslant x \leqslant a$, $y \geqslant 0$, by taking the limit $b \to \infty$ of eqn (5.3.9).

(ii) By adapting the method described at the end of Section 6.5 below, show that

$$G_\mathrm{D}(\boldsymbol{r}\,|\,\boldsymbol{r}') = -\frac{1}{4\pi} \log \frac{[\sinh^2 \eta + \sin^2 \xi][\sinh^2 Y + \sin^2 X]}{[\sinh^2 Y + \sin^2 \xi][\sinh^2 \eta + \sin^2 X]},$$

where $\xi \equiv \frac{1}{2}(x - x')$, $X \equiv \frac{1}{2}(x + x')$, $\eta \equiv \frac{1}{2}(y_> - y_<)$, $Y \equiv \frac{1}{2}(y_> + y_<)$.

(iii) Use the magic rule to solve the Laplace equation in the strip, given the BCs $\psi(x, 0) = \alpha$, $\psi(0, y) = 0 = \psi(a, y)$. Explore the behaviour of $\psi(\boldsymbol{r})$ as $\boldsymbol{r} \to 0$.

6 Poisson's equation: III. Neumann problems

This chapter concentrates mainly on the differences from the Dirichlet problem; like Chapter 5, it draws heavily on Sections 2.4 and 4.5. The important but rather technical applications to spherical regions are given in Appendix H.

6.1 The peculiarities of the Neumann problem

Our problem is to solve Poisson's equation (5.1.1), $-\nabla^2\psi = \rho$, in a region V, subject to prescribed values of the outward-normal derivative $\partial_n\psi_S$ on the closed surface S bounding V. Unlike the input data ρ and ψ_S of the Dirichlet problem, those of the Neumann problem must satisfy a self-consistency condition, since Gauss's theorem implies

$$\int_V \mathrm{d}V\,\rho(\boldsymbol{r}) = -\int_V \mathrm{d}V\,\nabla^2\psi(\boldsymbol{r}) = -\int_S \mathrm{d}S\,\partial_n\psi_S. \tag{6.1.1}$$

We assume (6.1.1) is obeyed; otherwise there can be no solution.

It is clear that the solution to the Neumann problem contains an arbitrary additive constant, because both the equation and the BCs involve ψ only through grad ψ, which remains unaffected by such a constant.

Mathematically, this is a reflection of the fact, already known from Section 4.5, that the totally homogeneous Neumann problem (Laplace equation $\nabla^2\psi = 0$ plus $\partial_n\psi_S = 0$) always has the (technically speaking non-trivial because non-zero) solution $\psi =$ constant; in other words the operator $-\nabla^2$ under NBCs always has $\lambda_0 = 0$ as an eigenvalue, with $\phi_0 = 1/V^{\frac{1}{2}}$ as the normalized eigenfunction. (We revert to calling Neumann eigenvalues λ rather than μ.) Hence we are always in the *special case* of the Fredholm alternative explained in Section 2.4, and (6.1.1) is just the condition on the input data which ensures that we are in fact in the *special special case* allowing a solution to exist nevertheless. The additive constant left undetermined by the Neumann problem is just $C\phi_0 = C/V^{\frac{1}{2}}$ as expected from the theorem (2.4.11).

By the same argument as in (6.1.1) for ψ, Gauss's theorem makes it impossible for a Green's function obeying the standard differential eqn (5.1.2), $-\nabla^2 G(\boldsymbol{r}\,|\,\boldsymbol{r}') = \delta(\boldsymbol{r}-\boldsymbol{r}')$, to satisfy the homogeneous NBCs $\partial_n G = 0$ on S. Green's functions G_N obeying suitably modified BCs will be constructed in Sections 6.2 and 6.5 below: but first we consider the pseudo Green's function H already introduced in Section 2.4, which is probably the more useful.

6.2 The magic rule with the pseudo Green's function $H(\mathbf{r}\,|\,\mathbf{r}')$

In terms of the eigenfunctions and eigenvalues of $-\nabla^2$ under the NBCs $\partial_n \phi_{pS} = 0$ as discussed in Section 4.5, we define the pseudo Green's function

$$H(\mathbf{r}\,|\,\mathbf{r}') = \sum_p{}' \phi_p^*(\mathbf{r}')\phi_p(\mathbf{r})/\lambda_p. \tag{6.2.1}$$

The label p is used instead of n to avoid a clash with the normal derivative ∂_n. The prime on Σ_p' signifies that the sum omits the term with $p=0$, corresponding to $\lambda_0 = 0$ and $\phi_0 = 1/V^{\frac{1}{2}}$ (H is sometimes called 'a Green's function in a generalized sense'.)

By choosing the set $\{\phi_p\}$ to be real (cf. Appendix E), one can make $\phi_p^*(\mathbf{r}') = \phi_p(\mathbf{r}')$ in (6.2.1), which shows that H is symmetric:

$$H(\mathbf{r}\,|\,\mathbf{r}') = H(\mathbf{r}'\,|\,\mathbf{r}). \tag{6.2.2}$$

Acting on H with $-\nabla^2$ we obtain, as in (2.4.8–9),

$$\begin{aligned} -\nabla^2 H(\mathbf{r}\,|\,\mathbf{r}') &= \sum_p{}' \phi_p^*(\mathbf{r}')\phi_p(\mathbf{r}) \\ &= \sum_p \phi_p^*(\mathbf{r}')\phi_p(\mathbf{r}) - \phi_0^*(\mathbf{r}')\phi_0(\mathbf{r}), \\ -\nabla^2 H(\mathbf{r}\,|\,\mathbf{r}') &= \delta(\mathbf{r}-\mathbf{r}') - V^{-1}, \end{aligned} \tag{6.2.3}$$

where in the last step we have appealed to closure and to $\phi_0(\mathbf{r}) = V^{-\frac{1}{2}}$. Finally, because $\partial_n \phi_{pS} = 0$, H satisfies the homogeneous NBCs

$$\partial_n H(\mathbf{r}\,|\,\mathbf{r}') = 0 \quad \text{for } \mathbf{r} \text{ on } \quad S. \tag{6.2.4}$$

Notice that H is not a (true) Green's function, because the RHS of (6.2.3) is not simply a delta-function. However, this distinction disappears if V is infinite, e.g. for a halfspace or for the exterior of a sphere.

We are now in a position to derive the magic rule appropriate to H, by the same strategy as in Section 5.2, namely by applying Green's theorem to $H(\mathbf{r}'\,|\,\mathbf{r})$ and $\psi(\mathbf{r}')$. Since the argument is familiar we give it with minimal comment:

$$\int_V dV' \{\psi(\mathbf{r}')\nabla'^2 H(\mathbf{r}' \mid \mathbf{r}) - (\nabla'^2\psi(\mathbf{r}'))H(\mathbf{r}' \mid \mathbf{r})\}$$

$$= \int_V dV' \left\{\psi(\mathbf{r}')\left[-\delta(\mathbf{r}-\mathbf{r}') + \frac{1}{V}\right] + \rho(\mathbf{r}')H(\mathbf{r}' \mid \mathbf{r})\right\}$$

$$= -\psi(\mathbf{r}) + \langle\psi\rangle_V + \int_V dV'\, \rho(\mathbf{r}')H(\mathbf{r}' \mid \mathbf{r})$$

$$= \int_S dS' \{\psi(\mathbf{r}')\,\partial_n' H(\mathbf{r}' \mid \mathbf{r}) - (\partial_n'\psi(\mathbf{r}'))H(\mathbf{r}' \mid \mathbf{r})\}$$

$$= -\int_S dS' (\partial_n'\psi(\mathbf{r}'))H(\mathbf{r}' \mid \mathbf{r}). \tag{6.2.5}$$

Here we have used (6.2.3) for $\nabla^2 H$ and Poisson's equation for $\nabla^2\psi$; have defined the volume average of ψ by

$$\langle\psi\rangle_V \equiv \frac{1}{V}\int_V dV'\, \psi(\mathbf{r}'); \tag{6.2.6}$$

and have dropped a surface integral which vanishes by virtue of (6.2.4). Rearrangement of (6.2.5) yields the magic rule

$$\psi(\mathbf{r}) = \int_V dV'\, \rho(\mathbf{r}')H(\mathbf{r}' \mid \mathbf{r}) + \int_S dS' (\partial_n'\psi_S(\mathbf{r}'))H(\mathbf{r}' \mid \mathbf{r}) + \langle\psi\rangle_V, \tag{6.2.7}$$

which should be compared with the Dirichlet magic rule (5.2.4).

The standard Neumann data ρ and $\partial_n\psi_S$ are accommodated in the first two terms; note that by contrast to (5.2.4) the surface integral is now prefaced by a plus sign. The term $\langle\psi\rangle_V$ is just the additive constant which we already know to expect. To determine ψ uniquely, $\langle\psi\rangle_V$ must be supplied as an additional datum. However, it is irrelevant to grad ψ, which may be all that is physically required.

Exercise: By integrating both sides of (6.2.7) over V, verify directly that the third term on the right is indeed given by (6.2.6). Use the orthonormality of the ϕ_p plus the fact that $\phi_0 = V^{-\frac{1}{2}}$.

Again, it is not immediately obvious that, with ψ given by (6.2.7), $\partial_n\psi(\mathbf{r})$ tends to the prescribed values as $\mathbf{r}$ approaches the surface.

Exercise: Prove this along the lines of Section 7.1 below.

6.3 The magic rule with the Green's function

As explained in Section 6.1, no Green's function G_N obeying

$$-\nabla^2 G_N(\boldsymbol{r}\,|\,\boldsymbol{r}') = \delta(\boldsymbol{r}-\boldsymbol{r}') \tag{6.3.1}$$

can satisfy homogeneous NBCs everywhere on S, since Gauss's theorem entails

$$\int_S \mathrm{d}S\, \partial_n G_N(\boldsymbol{r}\,|\,\boldsymbol{r}') = -1. \tag{6.3.2}$$

If one wishes to work with G_N instead of H, one must therefore subject G_N to inhomogeneous NBCs; for these there is no universal prescription, and different problems might favour different choices, as illustrated in Section 6.5 below. But one popular choice is to make $\partial_n G_N$ constant on S; then (6.3.2) shows that the value of this constant must be $-1/A$, where A is the total area of S:

$$\partial_n G_N(\boldsymbol{r}\,|\,\boldsymbol{r}') = -1/A \quad \text{for} \quad \boldsymbol{r} \quad \text{on} \quad S. \tag{6.3.3}$$

If A is infinite (e.g. for a halfspace, or for the exterior problem of the sphere), then $1/A = 0$, and $\partial_n G_N$ does vanish on S. For all but obviously outlandish regions, V and A would both be finite or both infinite; if they are both infinite, then the equations and the BCs defining H and G_N are the same, whence H and G_N can differ at most by an additive constant $C(\boldsymbol{r}')$, independent of $\boldsymbol{r}$ but possibly a function of $\boldsymbol{r}'$.

The magic rule appropriate to G_N is derived by the same strategy as for (6.2.5–7), except that now we shall need the surface average of ψ defined by

$$\langle\psi\rangle_S \equiv \frac{1}{A}\int_S \mathrm{d}S'\, \psi_S(\boldsymbol{r}'). \tag{6.3.4}$$

One finds the magic rule

$$\begin{aligned}\psi(\boldsymbol{r}) &= \int_V \mathrm{d}V'\, \rho(\boldsymbol{r}')G_N(\boldsymbol{r}'\,|\,\boldsymbol{r}) + \int_S \mathrm{d}S'\, (\partial_n'\psi_S(\boldsymbol{r}'))G_N(\boldsymbol{r}'\,|\,\boldsymbol{r}) + \langle\psi\rangle_S \\ &\equiv f_N(\boldsymbol{r}) + g_N(\boldsymbol{r}) + \langle\psi\rangle_S,\end{aligned} \tag{6.3.5}$$

which should be compared with (6.2.7) and with the Dirichlet magic rule (5.2.4). Again, f_N and g_N stand for the volume and the surface integral respectively, regarded as explicit constructs from the data ρ and $\partial_n\psi_S$; $\langle\psi\rangle_S$ is the expected additive constant, and must be supplied as an additional datum. We have refrained from defining analogous functions f and g for the magic rule (6.2.7) simply to avoid a proliferation of symbols.

Exercises: (i) Derive (6.3.5). (ii) By integrating both sides over S, verify directly that the third term on the right is indeed given by (6.3.4).

Unless V and A are both infinite, H and G_N can differ very significantly; in particular, G_N can be quite hard to determine because, unlike H, it is not readily given by an eigenfunction expansion. It seems to be a matter of experience and judgement which to use for any particular problem.

The equations (6.3.1, 3) defining G_N do not force it to be symmetric in $\mathbf{r}$ and $\mathbf{r}'$; but it can always be made symmetric by a suitable choice of the otherwise arbitrary additive constant, call it $\gamma(\mathbf{r}')$, as we shall show. (By contrast, $H(\mathbf{r}\,|\,\mathbf{r}')$ is necessarily symmetric by virtue of its definition (6.2.1).)

The argument starts in the same way as the proof of the symmetry of the Dirichlet Green's function G_D asserted in Theorem (5.1.7), namely by applying Green's theorem to $\hat{G}_N(\mathbf{r}\,|\,\mathbf{r}')$ and $\hat{G}_N(\mathbf{r}\,|\,\mathbf{r}'')$, where $\hat{G}_N$ satisfies the eqn (6.3.1) and the boundary condition (6.3.3), but is not yet presumed to be symmetric. In view of the non-zero right-hand side of (6.3.3), one finds

$$\begin{aligned}\hat{G}_N(\mathbf{r}''\,|\,\mathbf{r}') - \hat{G}_N(\mathbf{r}'\,|\,\mathbf{r}'') &= \frac{1}{A}\int_S \mathrm{d}S\,\{\hat{G}_N(\mathbf{r}\,|\,\mathbf{r}') - \hat{G}_N(\mathbf{r}\,|\,\mathbf{r}'')\}\\ &\equiv \gamma(\mathbf{r}') - \gamma(\mathbf{r}''),\end{aligned} \tag{6.3.6}$$

where we have defined

$$\gamma(\mathbf{r}') \equiv (1/A)\int_S \mathrm{d}S\,\hat{G}_N(\mathbf{r}\,|\,\mathbf{r}'), \tag{6.3.7}$$

and $\gamma(\mathbf{r}'')$ similarly. Though (6.3.6) is compatible with symmetry, it does not require symmetry, because $\gamma(\mathbf{r}')$ could vary with $\mathbf{r}'$, making $\gamma(\mathbf{r}') - \gamma(\mathbf{r}'') \neq 0$. However, all we need do is to define a new function

$$G_N(\mathbf{r}\,|\,\mathbf{r}') \equiv \hat{G}_N(\mathbf{r}\,|\,\mathbf{r}') - \gamma(\mathbf{r}'); \tag{6.3.8}$$

since $\gamma(\mathbf{r}')$ depends only on $\mathbf{r}'$ but not on $\mathbf{r}$, G_N just like $\hat{G}_N$ satisfies the defining equations (6.3.1, 3). Thus we simply adopt G_N as the Neumann Green's function; then (6.3.6, 8) entails that

$$\begin{aligned}&\{G_N(\mathbf{r}''\,|\,\mathbf{r}') - G_N(\mathbf{r}'\,|\,\mathbf{r}'')\}\\ &\quad\equiv \{\hat{G}_N(\mathbf{r}''\,|\,\mathbf{r}') - \gamma(\mathbf{r}') - \hat{G}_N(\mathbf{r}'\,|\,\mathbf{r}'') + \gamma(\mathbf{r}'')\} = 0,\end{aligned}$$

whence G_N is indeed symmetric as desired. ■

Accordingly, we can define a Neumann Green's function G_N by the eqn (6.3.1), the BCs (6.3.3), *plus* the symmetry requirement

$$G_N(\mathbf{r}\,|\,\mathbf{r}') = G_N(\mathbf{r}'\,|\,\mathbf{r}). \tag{6.3.9}$$

The freedom to choose $\gamma(\mathbf{r}')$ in (6.3.8) will be exploited when we construct G_N for a circle in the next section.

6.4 Applications

For rectangular boundaries, the eigenfunction expansion (6.2.1) of H proceeds as in Section 5.3.2 with modifications that are obvious in principle but may need some care in practice. For instance, (5.3.3) is now replaced by

$$\phi_{00} = 1/\sqrt{(ab)}$$

$$\phi_{n0} = \sqrt{\frac{2}{a}}\cos\left(\frac{n\pi x}{a}\right)\sqrt{\frac{1}{b}}, \qquad \phi_{0m} = \sqrt{\frac{1}{a}}\sqrt{\frac{2}{b}}\cos\left(\frac{m\pi y}{b}\right),$$

$$\phi_{n,m} = \sqrt{\frac{2}{a}}\cos\left(\frac{n\pi x}{a}\right)\sqrt{\frac{2}{b}}\cos\left(\frac{m\pi y}{b}\right), \tag{6.4.1a}$$

$$\lambda_{n,m} = (n\pi/a)^2 + (m\pi/b)^2. \tag{6.4.1b}$$

The sum Σ' omits only the term with both $n = 0$ and $m = 0$; thus the analogue of (5.3.4) reads

$$H(\mathbf{r} \mid \mathbf{r}') = \frac{2}{ab}\sum_{n=1}^{\infty}\frac{\cos\left(\frac{n\pi x'}{a}\right)\cos\left(\frac{n\pi x}{a}\right)}{\left(\frac{n\pi}{a}\right)^2} + \frac{2}{ab}\sum_{m=1}^{\infty}\frac{\cos\left(\frac{m\pi y'}{b}\right)\cos\left(\frac{m\pi y}{b}\right)}{\left(\frac{m\pi}{b}\right)^2} + \frac{4}{ab}\sum_{n=1}^{\infty}\sum_{m=1}^{\infty}\frac{\cos\left(\frac{n\pi x'}{a}\right)\cos\left(\frac{n\pi x}{a}\right)\cos\left(\frac{m\pi y'}{b}\right)\cos\left(\frac{m\pi y}{b}\right)}{\left[\left(\frac{n\pi}{a}\right)^2 + \left(\frac{m\pi}{b}\right)^2\right]}. \tag{6.4.2}$$

The single-sum expansion analogous to (5.3.5–9) now starts from

$$H = f_0(y \mid y') + \sum_{n=1}^{\infty} \cos(n\pi x'/a)\cos(n\pi x/a) f_n(y \mid y'), \tag{6.4.3}$$

and in $-\nabla^2 H = \{\delta(\mathbf{r} - \mathbf{r}') - 1/ab\}$ uses the cosine representation (1.3.10) of $\delta(x - x')$ instead of the sine representation used in (5.3.6). The term $-1/ab$ on the right enters only the equation for the coefficient f_0 of the

x-independent term, which becomes

$$-\partial^2 f_0/\partial y^2 = (1/a)\,\delta(y-y') - 1/ab. \tag{6.4.4}$$

Integrating this subject to the BCs $\partial f_0/\partial y = 0$ at $y=0$, $y=b$, one needs two integration constants that are in fact functions of y'. The equation for the coefficient functions with $n \geqslant 1$ reads

$$-\partial^2 f_n/\partial y^2 + (n\pi/a)^2 f_n = (2/a)\,\delta(y-y'). \tag{6.4.5}$$

Eventually one finds

$$\begin{aligned} H(\boldsymbol{r}\,|\,\boldsymbol{r}') = &-\frac{1}{2a}\left[|y-y'| + (y+y') - \frac{1}{b}(y^2+y'^2)\right] \\ &+\frac{2}{\pi}\sum_{h=1}^{\infty} \frac{\cos(n\pi x'/a)\cos(n\pi x/a)}{n\sinh(n\pi b/a)} \\ &\times \cosh[n\pi y_</a]\cosh[n\pi(b-y_>)/a]. \end{aligned} \tag{6.4.6}$$

Exercise: Derive (6.4.6) in full.

For a halfspace, say for $z \geqslant 0$, the method of images works as easily with NBCs as with DBCs. Here, since both V and A are infinite, H and G_N are the same. Instead of (5.3.11) we now obtain

$$G_N(\boldsymbol{r}\,|\,\boldsymbol{r}') = G_0(\boldsymbol{r}\,|\,\boldsymbol{r}') + G_0(\boldsymbol{r}\,|\,\tilde{\boldsymbol{r}}'), \tag{6.4.7}$$

again with an image (but now a positive one) at $\boldsymbol{r}' = (x', y', -z')$. Because G_N is an even function of z, its z-derivative vanishes automatically on the plane $z=0$.

On the other hand, the Dirichlet multiple-image series (5.3.19) for a finite slab has no Neumann analogue. To enforce NBCs, all the images would have to be positive (those at $2NL - z'$ as well as those at $2NL + z'$); but then, as $|N| \to \infty$ with fixed $\boldsymbol{r}$ and $\boldsymbol{r}'$, one has $G_0^{(3)} \sim 1/4\pi NL$, whence the series diverges like $\Sigma^\infty 1/N$.

For spheres, the expressions for H and G_N are much more complicated than for G_D, and the most reasonable way to them is through their expansions in polar coordinates, which can eventually be summed in closed form. This procedure, though straightforward in principle, is quite lengthy, and we relegate both the method and the results to Appendix H. The same method works for circles, but G_N for a circle can also be obtained by educated guesswork. Since the guesswork illustrates some of the recently introduced ideas, we describe it here, independently of the more systematic polar expansion given in Appendix H.

To guess G_N for a circle from G_D for a circle (eqn (5.3.28)), one starts by observing the relation between the expressions for G_N and G_D for a halfspace, which are $G_0(\boldsymbol{r}\,|\,\boldsymbol{r}') \pm G_0(\boldsymbol{r}\,|\,\tilde{\boldsymbol{r}}')$, respectively. Regarding G_N

and G_D as functions of $\boldsymbol{r}$, one conjectures that the relation between them in a circle might be similar to their relation in a halfspace (but now with $\tilde{\boldsymbol{r}}' = a^2\boldsymbol{r}'/r'^2$). In fact this cannot be the whole story, because we know from (5.3.24) that $G_0(\boldsymbol{r} \mid a^2\boldsymbol{r}'/r'^2)$ is not symmetric in $\boldsymbol{r}$ and $\boldsymbol{r}'$. The obvious remedy is to make G_N symmetric by adding a suitable function of $\boldsymbol{r}'$ alone; for the interior problem, inspection will then reveal that the result indeed satisfies the BC. For the exterior problem, where the BC is different ($\partial_n G_N = 0$ instead of $\partial_n G_N = -1/A = -\frac{1}{4}\pi a^2$), one tries to accomodate this difference by adding a further symmetric function of $\boldsymbol{r}$ and $\boldsymbol{r}'$, which is harmonic outside the circle.

The image position $\tilde{\boldsymbol{r}}'$, and $\tilde{\boldsymbol{R}} = \boldsymbol{r} - \tilde{\boldsymbol{r}}'$, are defined as in Section 5.3.4; note in particular the expression (5.3.24) for the combination $r'R$, and the fact that it is symmetric.

Accordingly, in

$$G_D = (2\pi)^{-1}\{\log(a/R) - \log(a/\tilde{R}) + \log(r'/a)\}$$

as given by (5.3.28), we reverse the sign of the second term, replace the last term $\log(r'/a)$ by an as yet undetermined function of r' alone, and try the expression

$$G_{N,\text{int}} = \frac{1}{2\pi}\{\log(a/R) + \log(a/\tilde{R}) + \log\gamma(r')\}$$

$$= \frac{1}{2\pi}\log\left\{\frac{a^2\gamma(r')}{R\tilde{R}}\right\}. \tag{6.4.8}$$

In order to ensure symmetry without altering the dependence of G_N on $\boldsymbol{r}$, we choose $\gamma(r') = a/r'$, whence

$$G_{N,\text{int}}(\boldsymbol{r} \mid \boldsymbol{r}') = \frac{1}{2\pi}\log\left(\frac{a^3}{R\tilde{R}r'}\right)$$

$$= \frac{1}{2\pi}\log\left\{\frac{a^2}{[r^2 + r'^2 - 2\boldsymbol{r}\cdot\boldsymbol{r}']^{\frac{1}{2}}[r^2r'^2/a^2 + a^2 - 2\boldsymbol{r}\cdot\boldsymbol{r}']^{\frac{1}{2}}}\right\}. \tag{6.4.9}$$

Then, for use in the magic rule,

$$G_{N,\text{int}}(\boldsymbol{r} \mid \boldsymbol{r}')|_{r'=a} = \frac{1}{2\pi}\log\left\{\frac{a^2}{[r^2 + a^2 - 2\boldsymbol{r}\cdot\boldsymbol{r}']}\right\}$$

$$= \frac{1}{2\pi}\log\left\{\frac{a^2}{[r^2 + a^2 - 2ar\cos(\phi - \phi')]}\right\}. \tag{6.4.10}$$

Remarkably, explicit differentiation (or reference to Appendix H) confirms that the BC is indeed satisfied, whence (6.4.9) is indeed the correct interior Neumann Green's function.

Exercise: Verify that BC, noting the hint in the exercise just below (5.3.26).

For the exterior problem, one must add to the expression (6.4.9) a function of r and r' such that (i) it is harmonic when $r > a$ (so as to respect the differential equation); (ii) its $\partial/\partial r$ derivative cancels that of (6.4.9) at $r = a$ (so as to validate the exterior BC $\partial_r G_N|_{r=a} = 0$); (iii) it is symmetric in r and r'. Trial and error soon shows that the requisite addend is $(1/2\pi)\log(rr'/a^2)$, whence

$$G_{N,\mathrm{ext}}(\boldsymbol{r} \mid \boldsymbol{r}') = \frac{1}{2\pi}\log\left(\frac{arr'}{R\tilde{R}r'}\right)$$
$$= \frac{1}{2\pi}\log\left\{\frac{rr'}{[r^2 + r'^2 - 2\boldsymbol{r}\cdot\boldsymbol{r}']^{\frac{1}{2}}[r^2r'^2/a^2 + a^2 - 2\boldsymbol{r}\cdot\boldsymbol{r}']^{\frac{1}{2}}}\right\}. \tag{6.4.11}$$

Consequently

$$G_{N,\mathrm{ext}}(\boldsymbol{r} \mid \boldsymbol{r}')|_{r'=a} = \frac{1}{2\pi}\log\left\{\frac{ar}{[r^2 + a^2 - 2\boldsymbol{r}\cdot\boldsymbol{r}']}\right\}, \tag{6.4.12}$$

which should be compared with (6.4.10).

6.5 A different Green's function, and the reduction of dimensionality under Neumann boundary conditions

We tackle the 2D Neumann problem in the infinite strip ($0 \leqslant x \leqslant a$, $-\infty < y < \infty$). This special case is used (i) to show that useful Green's functions can be defined under BCs other than (6.3.3); and (ii) to demonstrate how the 1D Green's function can arise from this 2D problem (where no flux escapes through the sides of the strip), as an approximation for small enough strip width a. This is the third mechanism mentioned in Section 3.3 for reducing the effective dimensionality.

Consider a Green's function defined by the standard differential equation (6.3.1) under the NBCs

$$\partial\psi/\partial x = 0 \quad \text{at} \quad x = 0 \quad \text{and} \quad x = a. \tag{6.5.1}$$

We do not yet impose BCs as $y \to \pm\infty$, since our experience with unbounded space in Section 4.4 suggests that it is best to fit these to the physics by hindsight. We look for G_N in the single-sum form familiar from Sections 5.3.2 and 6.4, writing

$$G_N(\mathbf{r} \mid \mathbf{r}') = f_0(y \mid y') + \sum_{n=1}^{\infty} \cos(n\pi x'/a) \cos(n\pi x/a) f_n(y \mid y'), \qquad (6.5.2)$$

which satisfies the BCs (6.5.1) by construction, and anticipates the symmetry of G_N in x and x', known from the end of Section 6.3. This is the same form (6.4.3) in which we looked for the pseudo Green's function for a rectangle. On the right of (6.3.1) we write $\delta(\mathbf{r} - \mathbf{r}') = \delta(x - x')\,\delta(y - y')$, and for $\delta(x - x')$ we again substitute its cosine representation (1.3.10); inserting (6.5.2) on the left of (6.3.1) and equating coefficients, one obtains

$$-\partial^2 f_0/\partial y^2 = (1/a)\,\delta(y - y'), \qquad (6.5.3a)$$

$$-\partial^2 f_n/\partial y^2 + (n\pi/a)^2 f_n = (2/a)\,\delta(y - y'), \qquad n = 1, 2, 3, \ldots \qquad (6.5.3b)$$

(These are (of course) the same equations as (6.4.4, 5) but the f_n are subject now to different BCs.) Solve (6.5.3b) first. We require that f_n be bounded as $y \to \pm\infty$; then the methods of Chapter 2 lead directly to $f_n(y \mid y') = (1/\pi n) \exp(-n\pi\,|y - y'|/a)$.

Exercise: Derive this solution explicitly.

Next, solve (6.5.3a). Apart from the symbol for the independent variable, and from the factor $1/a$ on the right, this is the same as the equation (4.4.8) for the 1D Green's function in unbounded space. Imposing the same BCs at infinity as we did there, we obtain the appropriate analogue of (4.4.9c), namely $f_0 = (-1/2a)\,|y - y'|$. Accordingly,

$$G_N(\mathbf{r} \mid \mathbf{r}') = -\frac{1}{2a}|y - y'| + \sum_{n=1}^{\infty} \frac{1}{\pi n} \cos\left(\frac{n\pi x'}{a}\right) \cos\left(\frac{n\pi x}{a}\right) \exp(-n\pi\,|y - y'|/a) \qquad (6.5.4)$$

Notice that this is not the limit of the pseudo Green's function H as the strip length becomes infinite: by construction, H obeys the BC $\partial_n H = 0$ on S, whereas (6.5.4) entails $\lim_{y \to \pm\infty} (\partial G_N/\partial y) = (\mp)(1/2a) \neq 0$. By the same token, (6.5.4) is not the limit of the Neumann Green's function subject to the BCs (6.3.3) chosen previously, since for an infinitely long strip that function coincides with H.

Finally, consider G_N for fixed y, y' (with $y \neq y'$), in the limit as $a \to 0$, i.e. in a regime where $|y/a|$, $|y'/a|$, $|y - y'|/a \gg 1$. Then the sum in (6.5.4) becomes exponentially small compared to the first term; in other words one has $G_N \to -|y-y'|/2a = G_0^{(1)}(y \mid y')/a$. To appreciate the implications, consider the 2D Poisson problem $-\nabla^2\psi = \rho$, with $\partial_n \psi = 0$ on the sides of the strip. Using (6.5.4), the solution can be written

$$\psi(\mathbf{r}) = \int \mathrm{d}V'\, \rho(\mathbf{r}')G_N(\mathbf{r} \mid \mathbf{r}')$$
$$= \int_{-\infty}^{\infty} \mathrm{d}y' \int_0^a \mathrm{d}x'\, \rho(x', y')\Big\{-\frac{1}{2a}|y-y'| + \sum \cdots\Big\}.$$

As $a \to 0$ at fixed y, y', the sum becomes negligible, whence

$$\psi(\mathbf{r}) \to \int_{-\infty}^{\infty} \mathrm{d}y'\, \{-\tfrac{1}{2}|y-y'|\}\bar{\rho}(y') = \int_{-\infty}^{\infty} \mathrm{d}y'\, G_0^{(1)}(y \mid y')\bar{\rho}(y'), \qquad (6.5.5)$$

where $\bar{\rho}(y') \equiv (1/a)\int_0^a \mathrm{d}x'\, \rho(x', y')$ is the average of ρ across the strip. But (6.5.5) is precisely the solution of Poisson's equation in 1D with a source density $\bar{\rho}$; thus the 1D problem indeed emerges as an appropriate limit of a 2D problem.

Exercise: Explain why DBCs admit no such simple descent from 2D to 1D.

Remarkably, the sum in (6.5.4), call it F, can be found in closed form. Define $\xi = \pi x/a$, $\eta = \pi y/a$, and primed variables similarly. Writing

$$\cos(n\xi')\cos(n\xi) = \tfrac{1}{2}\{\cos[n(\xi+\xi')] + \cos[n(\xi-\xi')]\},$$

we have

$$F = \frac{1}{2\pi}\operatorname{Re}\sum_{n=1}^{\infty}\frac{1}{n}\{\exp n[-|\eta-\eta'| + i(\xi+\xi')] + \exp n[-|\eta-\eta'| + i(\xi-\xi')]\}.$$

In view of the expansion $-\log(1-z) = \sum_{n=1}^{\infty} z^n/n$, we identify z appropriately in each of the two terms, and find

$$F = -\frac{1}{2\pi}\operatorname{Re}\log\{[1-\exp(-|\eta-\eta'| + i(\xi+\xi'))][1-\exp(-|\eta-\eta'|$$
$$+ i(\xi-\xi'))]\}$$
$$= \frac{1}{2\pi}\operatorname{Re}\log\Big\{\exp\Big(-\frac{|\eta-\eta'| + i(\xi+\xi')}{2}\Big)2\sinh\Big(\frac{|\eta-\eta'| + i(\xi+\xi')}{2}\Big)$$
$$\times \exp\Big(-\frac{|\eta-\eta'| + i(\xi-\xi')}{2}\Big)2\sinh\Big(\frac{|\eta-\eta'| + i(\xi-\xi')}{2}\Big)\Big\}.$$

We write $\sinh(\alpha + i\beta) = \sinh\alpha\cos\beta + i\cosh\alpha\sin\beta$, whence $|\sinh(\alpha+i\beta)| = [\sinh^2\alpha + \sin^2\beta]^{\frac{1}{2}}$; note that only the modulus of $\{\cdots\}$ is relevant to the real part of log $\{\cdots\}$; and are finally led to

$$F = -\frac{1}{2\pi}\log\left\{4\exp(-|\eta-\eta'|)\left[\sinh^2\left(\frac{\eta-\eta'}{2}\right) + \sin^2\left(\frac{\xi+\xi'}{2}\right)\right]^{\frac{1}{2}}\left[\sinh^2\left(\frac{\eta-\eta'}{2}\right) + \sin^2\left(\frac{\xi-\xi'}{2}\right)\right]^{\frac{1}{2}}\right\}. \quad (6.5.6)$$

The first term $-|y-y'|/2a = -|\eta-\eta'|/2\pi$ in (6.5.4) cancels the component $-(2\pi)^{-1}\log\{\exp(-|\eta-\eta'|)\}$ of F as written in (6.5.6); however, when $|\eta-\eta'| \gg 1$, it is better to keep the two terms in (6.5.4) separate, because we know that the first term then dominates overall.

To illustrate the uses of the closed form (6.5.6), take the special case where x, x' are both on the median plane, $x = x' = a/2$; then F simplifies considerably, and one finds

$$G_N = \{-|y-y'|/2a - (2\pi)^{-1}\log[1-\exp(-2\pi\,|y-y'|/a]\}.$$

Hence the flow velocity in the median plane due to a unit point source at $(x', y') = (a/2, 0)$ is

$$-\frac{\partial G_N}{\partial y} = \varepsilon(y-y')\left\{\frac{1}{2a} + \frac{1}{a}\cdot\frac{1}{[\exp(2\pi\,|y-y'|/a) - 1]}\right\}.$$

Exercise: For the same strip, find the Dirichlet Green's function in closed form; compare $G_D(a/2, y \mid a/2, y')$ with G_N and compare the flow velocities in the median plane.

Problems

6.1 In the square $0 \leqslant x \leqslant \pi$, $0 \leqslant y \leqslant \pi$, the flow is source-free, and satisfies the following BCs: on the top and right-hand edges, $v_n = 0$; on the left-hand edge, $v_n = -1$ (inflow); on the bottom edge, $v_n = +1$ (outflow). By using the magic rule with the pseudo Green's function H, and the eigenfunction-expansion of H, determine the flow velocity everywhere. You may quote the Fourier series (A.5.6).

6.2 In the 3D halfspace $z \geqslant 0$, the velocity potential ψ obeys the Laplace equation; on the xy-plane, $-\partial_n \psi = u$ inside the circle of radius a centred on the origin, and vanishes outside the circle.

(i) Write down the magic rule for $\psi(r, \theta, \phi)$. Evaluate ψ and $\boldsymbol{v}(\boldsymbol{r}) = -\boldsymbol{\nabla}\psi = (v_r, v_\theta, v_\phi)$ as $r/a \to \infty$ at fixed θ, ϕ, keeping in $\boldsymbol{v}$ terms up to order $1/r^4$ inclusive.
(ii) Verify and explain the fact that the leading term in $\boldsymbol{v}$ (of order $1/r^2$) is what would result in unbounded space from a point source at the origin, of total strength $2\pi a^2 u$.

6.3 In magnetostatics (a hypothetical science), magnetic pole strength M plays the role that electric charge Q plays in electrostatics. The difference is that at a (super)conducting surface, the magnetic scalar potential obeys NBCs instead of DBCs, so that $\partial_n \psi_S = 0$. Thus, a pole and its image in a flat conducting surface have the same instead of opposite signs, whence they repel rather than attract.

In Problem 5.9, substitute M for Q, and solve the magnetic problem that results. Compare the magnitude and the direction of the force in the two cases. Verify that on the diagonal ($x' = y'$), $U_{\text{mag}}/U_{\text{el}} = -(5/3)M^2/Q^2$.

6.4 On the unit circle, the velocity-potential obeys the NBC $\partial_r \psi = 0$.

(i) Determine ψ, and thence $\boldsymbol{v} = -\text{grad}\,\psi$, when a point source of strength $+1$ is placed at $(r, \phi) = (1-, 0)$ just inside the enclosure, and a point source of strength -1 is placed at the centre. Verify that the tangential flow speed just inside the enclosure is

$$\frac{1}{2\pi}\cos(\phi/2),$$

while along the axis the velocity is $(1 + x)/2\pi x(1 - x)$.
(ii) What would be your reaction if you were asked the same question but without the source at the centre?

6.5 A point-dipole source of strength $\boldsymbol{p}$ is placed at the centre of a rigid spherical enclosure.

(i) Use the magic rule to determine the velocity potential ψ and the flow velocity

$$\boldsymbol{v} = -\nabla\psi = (v_r(r, \theta, \phi), v_\theta(r, \theta, \phi), v_\phi(r, \theta, \phi)).$$

(ii) The density of the fluid is σ. Calculate the total kinetic-energy difference ΔU between the flow in the enclosure (namely $\frac{1}{2}\sigma \int_v dV\, \boldsymbol{v}^2(\boldsymbol{r})$), and the flow that would result from the same source in unbounded space.

Hints: The source density is $\rho(\boldsymbol{r}) = -\boldsymbol{p} \cdot \nabla\, \delta(\boldsymbol{r})$. One could start from G_N as given in closed form by (H.4.18); but it is probably quicker to start from (H.4.16, 17) and (H.1.4), realizing that, with such a source, one can ignore components with $l > 1$ in the polar expansion of G_N. The result is $\Delta U = \sigma p^2/12\pi a^2$.

6.6 For magnetostatics as explained in Problem 6.3, consider the mutual potential energy U and the force between a point pole M and a superconducting sphere of radius a. This is the Neumann analogue of the Dirichlet problem 5.7. Recall that in the case of a flat instead of a spherical surface, the change from DBCs to NBCs simply reverses the sign of the effect. It is interesting to ask whether this remains true in the present problem. (We would expect it to remain true at least in the limiting case where $r' \equiv a + \xi$ with $\xi \ll a$, since a point source so placed would see the sphere as essentially flat. For instance, the electrostatic energy then indeed reduces to $(Q^2/4\pi)(1/4\xi)$.)

(i) Determine $U(r')$ and $F = -\nabla' U$. Does your result conform to the (tentative) expectations formulated above?
(ii) Verify that in the opposite limit $r/a \to \infty$ one has $U \sim M^2 a^3/16\pi r^4$, and compare this with the asymptotic behaviour of the electrostatic energy.

Hint: Equation (H.4.25) identifies $\chi(\boldsymbol{r} \mid \boldsymbol{r}')$. Notice that $\boldsymbol{r} \to \boldsymbol{r}'$ implies $\mu \to 1$ and $r \to r'$. In the argument of the logarithm, both numerator and denominator vanish at $\mu = 1$; one must use l'Hôpital's rule to implement $\mu \to 1$ before letting $r \to r'$.

6.7 A uniform flow field $\boldsymbol{v}_0 = \hat{\boldsymbol{z}} = -\text{grad}\, \psi_0$, $\psi_0 = -z$, is disturbed by a fixed rigid sphere of radius a centred on the origin.

(i) Show that the appropriate solution of the Laplace equation for $r \geqslant a$ can be found in the form $\psi = z + Az/r^3$, and determine the constant A from the BC $\partial\psi/\partial r = 0$ at $r = a$.
(ii) What is the maximum speed in the disturbed flow pattern, and where does it occur?

(iii) What is the difference between the total energies of the flows with and without the sphere?

6.8 One would expect that the solution $\psi = z + Az/r$ to Problem 6.7(i), or at least $\nabla\psi$, is reproduced by $QG_N(\boldsymbol{r} \mid \boldsymbol{r}')$ in the limit $Q \to \infty$, $r' \to \infty$, $Q/4\pi r'^2 = 1$. Verify that this is so, using either the polar decomposition or the closed form of the Neumann Green's function for the exterior of the sphere.

Hint: If you use the closed form, note the hint in Problem 6.6.

6.9 (i) By adapting the argument from Problem 5.11, prove that

$$G_0(\boldsymbol{r}_0 \mid \boldsymbol{r}_i) = \frac{1}{A}\int_S dS'\, G_0(\boldsymbol{r}_0 \mid \boldsymbol{r}') + \int_S dS'\, G_0(\boldsymbol{r}_0 \mid \boldsymbol{r}')\, \partial'_n G_N(\boldsymbol{r}_i \mid \boldsymbol{r}'), \quad (1)$$

where G_N is the Neumann Green's function for V, defined by (6.3.1, 3).

(ii) By acting on (1) with $\int_V dV_i\, \rho(\boldsymbol{r}_i)$, show that at points $\boldsymbol{r}_0$ outside V, the potential $\psi(\boldsymbol{r}_0)$ due to ρ inside V can be represented as

$$\psi(\boldsymbol{r}_0) = \frac{Q}{A}\int_S dS'\, G_0(\boldsymbol{r}_0 \mid \boldsymbol{r}') + \int_S dS'\, (\partial'_n G_0(\boldsymbol{r}_0 \mid \boldsymbol{r}')) \int_V dV_i\, G_N(\boldsymbol{r}_i \mid \boldsymbol{r}')\rho(\boldsymbol{r}_i), \quad (2)$$

where $Q \equiv \int_V dV_i\, \rho(\boldsymbol{r}_i)$ is the total charge.

(iii) Convince yourself that the RHS of (2) could be interpreted as follows. The first term is the potential due to a uniform surface source-density $\sigma = Q/A$ on S; the second term is the potential due to an outward-pointing dipole layer on S, of variable strength $\int_V dV_i\, G_N(\boldsymbol{r}_i \mid \boldsymbol{r}')\rho(\boldsymbol{r}_i)$.

6.10 By adapting the reasoning of Problem 6.9 to the pseudo Green's function, show that

$$G_0(\boldsymbol{r}_0 \mid \boldsymbol{r}_i) = \frac{1}{V}\int_V dV'\, G_0(\boldsymbol{r}_0 \mid \boldsymbol{r}') + \int_S dS'(\partial'_n G_0(\boldsymbol{r}_0 \mid \boldsymbol{r}'))H(\boldsymbol{r}_i \mid \boldsymbol{r}').$$

Deduce and interpret the representation

$$\psi(\boldsymbol{r}_0) = \frac{Q}{V}\int_V dV'\, G_0(\boldsymbol{r}_0 \mid \boldsymbol{r}') + \int_S dS'\, (\partial'_n G_0(\boldsymbol{r}_0 \mid \boldsymbol{r}')) \int_V dV_i\, \rho(\boldsymbol{r}_i)H(\boldsymbol{r}_i \mid \boldsymbol{r}').$$

7 Poisson's equation: IV. Some points of principle

The methods we use to solve linear partial differential equations skirt some points of mathematical principle which are nevertheless worth a glance even in a mundane approach like ours. Elsewhere we keep comments on such points to the barest minimum, but we raise some of them here in connection, somewhat arbitrarily, just with Poisson's equation. Very far from trying to be systematic, we aim only to convey some of the flavour of such problems.

Of this chapter, only Section 7.5, on the Kirchhoff representation, is used later in the book.

7.1 Existence of solutions: the self-consistency of the magic rule

Commenting on the magic rule in Section 5.2 (cf. comment (v)), we deferred the proof that the expression $f_D(\boldsymbol{r}) + g_D(\boldsymbol{r})$ in (5.2.4) indeed approaches the prescribed boundary-value $\psi_S(\boldsymbol{r})$ as the field point $\boldsymbol{r}$ approaches the surface S (we write this as $\boldsymbol{r} \to S$ for short). Since the volume integral $f_D(\boldsymbol{r}) \equiv \int_V dV'\, \rho(\boldsymbol{r}')G_D(\boldsymbol{r}' \mid \boldsymbol{r})$ vanishes as $\boldsymbol{r} \to S$ (because G_D does, by definition), we need consider only the surface integral $g_D(\boldsymbol{r}) \equiv -\int_S dS'\, \psi_S(\boldsymbol{r}')\, \partial'_n G_D(\boldsymbol{r}' \mid \boldsymbol{r})$, whose behaviour is evidently governed by that of $\partial'_n G_D$. Because the normal derivative ∂'_n is unambiguous only at points where the surface is smooth, we exclude corners and edges from the argument for simplicity. (Problem 5.14 displays in closed form one Green's function for a region with sharp corners.) We also exclude points or curves on S where the prescribed ψ_S is discontinuous or singular, though Section 7.2 deals with one such example explicitly.

Accordingly, we aim to show that $g_D \to \psi_S$ as $\boldsymbol{r} \to S$, provided that S is smooth and ψ_S continuous at the point in question. Without further loss of generality, we now choose this point as the origin; thus we must show that $g_D(\boldsymbol{r} \to 0) = \psi_S(0)$. Our essential tool is the image-in-the-plane Green's function from Section 5.3.3.

Since we are concerned only with points arbitrarily close to $\boldsymbol{r} = 0$, we can treat S as locally flat; thus we replace S by its tangent plane at $\boldsymbol{r} = 0$, choose this plane as the xy-plane, and choose the positive z-axis along the inward normal. The crucial observation is that, as the field point $\boldsymbol{r} \to 0$, the effects of other points of the surface (infinitely far away on the scale of r) become irrelevant, so that we can replace G_D by the explicit image Green's function (5.3.11); correspondingly, (5.3.17) yields

$$g_D(\mathbf{r}) = (2\pi)^{-1}\int \mathrm{d}S'\ \psi_S(\mathbf{r}')z/|\mathbf{r}-\mathbf{r}'|^3.$$

We can put $x = 0 = y$ immediately, whence

$$|\mathbf{r}-\mathbf{r}'| = [(x-x')^2+(y-y')^2+z^2]^{\frac{1}{2}} = [x'^2+y'^2+z^2]^{\frac{1}{2}},$$

and need to prove that

$$\lim_{\mathbf{r}\to 0} g_D(\mathbf{r}) = \lim_{z\to 0}\iint \mathrm{d}x'\,\mathrm{d}y'\ \psi_S(x', y', 0)\frac{1}{2\pi}\frac{z}{[x'^2+y'^2+z^2]^{\frac{3}{2}}}$$
$$= \psi_S(0, 0, 0). \tag{7.1.1}$$

In other words, we must establish

$$\lim_{z\to 0}\frac{1}{2\pi}\frac{z}{[x'^2+y'^2+z^2]^{\frac{1}{2}}} = \delta(x')\,\delta(y'). \tag{7.1.2}$$

Proof: (i) Unless $x' = 0$ and $y' = 0$, the expression on the left certainly vanishes when $z = 0$.

(ii) The integral of this expression with respect to $\mathrm{d}S'$ is unity over any domain including the origin. To see this, choose the domain as a circle of radius η centred on 0. The integration proceeds by a change to plane-polar variables ρ', ϕ', and by then defining a new radial variable σ by $\rho' = (x'^2+y'^2)^{\frac{1}{2}} \equiv z\sigma$. The integral is

$$\lim_{z\to 0}\int_0^{\eta} 2\pi\,\mathrm{d}\rho'\,\rho'\frac{1}{2\pi}\frac{z}{(\rho'^2+z^2)^{\frac{3}{2}}} = \lim_{z\to 0}\int_0^{\eta/z}\frac{\mathrm{d}\sigma\,\sigma}{(\sigma^2+1)^{\frac{3}{2}}}$$
$$= \int_0^{\infty}\mathrm{d}\sigma\,\sigma/(\sigma^2+1)^{\frac{3}{2}} = 1.$$ ■

Exercise: Prove that the magic rules (6.2.7), (6.3.5) for Neumann problems are likewise self-consistent.

Of course, in practice the magic rule is freely applied even to regions that do have corners and edges: for instance, we often deal with rectangular boundaries. While the mathematician might regard such BVPs as ill-defined at a corner, the physicist takes the explicit expression $\psi \equiv f_D + g_D$ as his solution; if he needs ψ near a corner, then he simply consults the expression $f_D + g_D$ in order to see what the solution is. The justification is that, on the one hand, the corners of real physical boundaries, like metal boxes, are not mathematically sharp; while, on the other hand, only in regions negligibly small by normal standards is there any appreciable difference between the solution calculated for the true shape with rounded corners, and the idealized solution calculated for idealized rectangular shapes, albeit ignoring any mathematical ambiguities at the corners.

7.2 Existence of solutions: discontinuous boundary values

To the Dirichlet BVP with $\rho = 0$, the magic rule (5.2.4) yields the solution $\psi = g_D(\boldsymbol{r}) = -\int_S dS'\, \psi_S(\boldsymbol{r}')\, \partial'_n G_D(\boldsymbol{r}' \mid \boldsymbol{r})$, provided only that the integral converges. The integral can easily converge even if the prescribed boundary function ψ_S is singular, challenging one to investigate how $g_D(\boldsymbol{r})$ behaves as $\boldsymbol{r}$ tends to such a surface singularity. In eqns (5.3.31, 32) we have already seen explicit solutions to extreme examples where ψ_S has an $(n-1)$-dimensional delta-function singularity on the boundary S of an n-dimensional region V.

More interesting is the case where ψ_S is simply discontinuous rather than infinite. We investigate the not untypical example of the 2D halfplane $x \geqslant 0$, whose boundary is the y-axis; we prescribe $\psi_S = 0$ on the positive y-axis, and $\psi_S = \alpha$ on the negative y-axis, as shown in Fig. 7.1. (Of course this can equally well be regarded as a 3D problem in the halfspace $x \geqslant 0$; then the boundary is the yz-plane, still with $\psi_S = \alpha$ for $y < 0$ and $\psi_S = 0$ for $y > 0$.) The value of ψ_S at (exactly) $y = 0$ is physically irrelevant and need not be defined, since it does not affect the integral g_D. The solution $g_D(\boldsymbol{r})$ is given at once by the subtended-angle construction (5.3.18). Inspection of Fig. 7.1 shows that the total angle

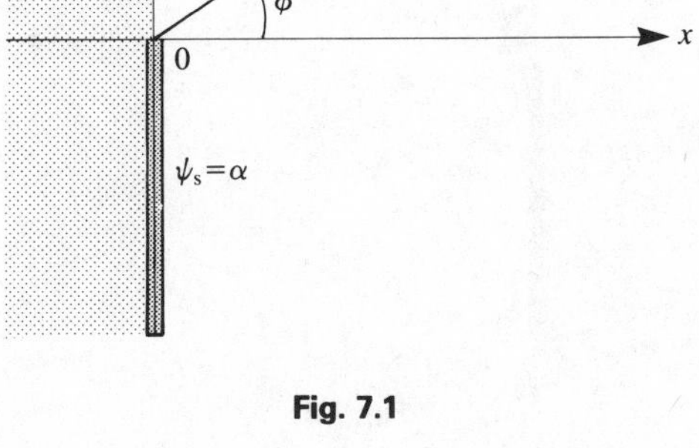

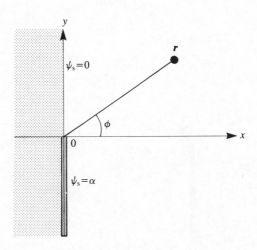

Fig. 7.1

subtended at the field point $\boldsymbol{r}$ by the negative y-axis is $\chi = (\pi/2 - \phi) = [\pi/2 - \tan^{-1}(y/x)]$; hence

$$g_D(\boldsymbol{r}) = \alpha(1/\pi)(\pi/2 - \phi). \tag{7.2.1}$$

Exercise: Derive (7.2.1) from (5.3.18) for the 3D problem.

To verify that $\nabla^2 g_D = 0$, simply write $\nabla^2 = r^{-1}(\partial/\partial r)(r^{-1}\,\partial/\partial r) + r^{-2}\,\partial^2/\partial\phi^2$. The BCs are verified as follows: when $x \to 0$ at fixed $y > 0$, one has $\phi \to \pi/2$, whence $g_D \to 0$; when $x \to 0$ at fixed $y < 0$, one has $\phi \to -\pi/2$, whence $g_D \to (\alpha/\pi)(\pi/2 + \pi/2) = \alpha$.

It remains to investigate $g_D(\boldsymbol{r} \to 0)$. This limit evidently depends on the path along which $\boldsymbol{r}$ approaches the origin. For instance, as $r \to 0$ at fixed ϕ, g_D is actually independent of r, and the limiting value equals this fixed value given directly by (7.2.1). One must simply accept the fact that the value of ψ *at* $r = 0$ is ill-defined in this sense. While such behaviour would be intolerable at any point *inside* the region V, on the boundary it is quite harmless. One need only remember that, physically, such a sharp discontinuity in ψ_S is an unrealizable idealization. In practice, we might have a narrow strip on the boundary across which ψ_S drops fast but continuously from α to 0. We might even have a modified boundary as shown in Fig. 7.2, with a cylindrical bulge of arbitrarily small but finite radius η, along which ψ_S varies precisely according to (7.2.1); then (7.2.1) itself persists unchanged throughout V.

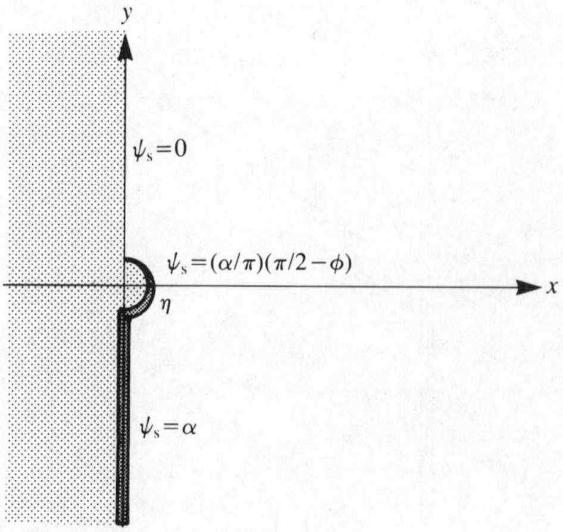

Fig. 7.2

By contrast to this view of the physics, prematurely ossified mathematical formulations often founder on such examples. For instance, suppose that, observing the admitted need for $\psi(\boldsymbol{r})$ to be well-defined everywhere within V, one had at some early stage of formulating the problem required of the solution ψ not only that $\nabla^2\psi = 0$ in V and that $\psi = \psi_S$ on S, but also that 'ψ be twice differentiable and continuous everywhere in V'. Thus stated, the problem has no solution, because ψ lacks a unique limit not only as $\boldsymbol{r} \to 0$ *along* S (which is merely another way of saying that ψ_S is discontinuous), but also as $\boldsymbol{r} \to 0$ *from within* V, as we have just seen. Thus, an unenlightened mathematical formulation might lead to the assertion that no solution exists, while an enlightened formulation (or a reformulation prodded by the physics) simply widens the class of functions to which, from the outset, any eventual solutions are required to belong. Experiences like these have made physicists wary of non-existence proofs; perhaps excessively so, as suggested by the cautionary example in Appendix I on variational arguments.

7.3 Uniqueness: the paradox of the ineffective constraints

Consider the Dirichlet problem of determining a harmonic function in V ($\nabla^2\psi = 0$), subject to prescribed boundary values ψ_S on S. Surprisingly, if one imposes the further constraint that at some selected point $\boldsymbol{r}_1$ within V the solution must assume an arbitrarily prescribed finite value α ($\psi(\boldsymbol{r}_1) = \alpha$), the solution at all other points ($\boldsymbol{r} \neq \boldsymbol{r}_1$) remains totally unaffected. We confine ourselves to illustrating this through a limiting procedure in two trivially soluble cases, one in 3D and one in 2D.

In 3D, consider the region V between two concentric spheres S_1 and S_2 with radii $a < b$ respectively, prescribing $\psi_{S_1} = \alpha$ and $\psi_{S_2} = 0$ (outer sphere earthed, inner sphere at fixed potential α). The (unique) harmonic function satisfying these BCs is $\psi_{00}(\boldsymbol{r})$ given by (4.3.6), where the coefficients a_{00} and b_{00} are determined by setting $\psi_{00}(a) = \alpha$ and $\psi_{00}(b) = 0$. Thus

$$\psi(\boldsymbol{r}) = \alpha \frac{a}{b-a}\left(\frac{b}{r} - 1\right), \tag{7.3.1}$$

as one verifies by inspection. (The component $1/r$ is admissible because V excludes the origin.)

Now let $a \to 0$. While $\psi(a) = \alpha$ remains fixed by construction, (7.3.1) shows explicitly that with fixed α, b, and $r \neq 0$, we have

$$\lim_{a \to 0} \psi(\boldsymbol{r}) = 0. \tag{7.3.2}$$

Thus, at any fixed $r \neq 0$, ψ tends to the same value (namely zero) which is (uniquely) appropriate to the solution of a different problem, where there is no inner sphere at all, i.e. where one prescribes $\psi_{S_2} = \psi(b) = 0$, but with no constraints at any points interior to S_2.

Accordingly, in the limit $a = 0$ we are left with a discontinuous solution given

by

$$\psi(r \neq 0) = 0, \qquad \psi(0) = \alpha. \tag{7.3.3}$$

In this example, $\mathbf{r}_1 = 0$. Such points are called 'removable singularities', presumably because the singularity of ψ at $\mathbf{r}_1$ can be 'removed' simply by redefining ψ at $\mathbf{r}_1$.

Physically, the result might strike one as paradoxical: for instance, if α is high enough and is maintained by connecting the inner sphere to a power supply, then anyone exploring the region V risks electrocution if he touches the origin, but receives no warning (through a gradually rising potential and field) as the origin is approached. On the other hand, the result is routine if the system is thought of as a charged capacitor. In order for the inner sphere to be at potential α, it must carry a charge $Q = C\alpha$, where $C = 4\pi ab/(b-a)$ is the capacity of the arrangement, while the outer sphere carries an equal and opposite charge $-C\alpha$. Since C and therefore Q vanish with a, in the limit $a = 0$ there are no charges and therefore no fields. The total energy $\frac{1}{2}\int_V dV\,(\nabla\psi)^2 = \frac{1}{2}C\alpha^2$ likewise vanishes in the limit $a \to 0$, which removes the apparent contradiction with the uniqueness theorem (cf. the remarks in Section 4.2).

Though we have discussed only a very special configuration, the conclusion is general. It is somewhat laborious but not too difficult to prove the following. In V, surround any denumerable set of points $\mathbf{r}_n$ by small surfaces S_n. On each S_n prescribe $\psi_{S_n} = \alpha_n$ arbitrarily. Let each S_n shrink to the point $\mathbf{r}_n$. In the limit, the solution, except at the points $\mathbf{r}_n$, is the same as if these surfaces and assignments had never been. (Unfortunately, textbooks do not seem to formulate the conclusion in just this way.)

In 2D we consider two concentric circles of radii $a < b$, with the same BCs as above. Then

$$\psi(\mathbf{r}) = \alpha \frac{\log(b/r)}{\log(b/a)}, \tag{7.3.4}$$

and again, for fixed α, b, $r \neq 0$, we have $\lim_{a \to 0} \psi = 0$, since $\log(b/a) \to \infty$.

Physically, one can think of a membrane stretched between two initially coplanar and concentric rings; the outer ring is held fixed, and the inner ring is raised a distance α vertically above the outer. Our result asserts that in the limit $a \to 0$, though the centre $r = 0$ is still raised a distance α above the plane of the fixed outer rim, yet everywhere else the membrane is perfectly flat. This is quite startling, even when one remembers that for a stretched membrane Poisson's equation is only an approximation valid when the slopes are small, which for finite a they obviously cannot be in the region next to the inner ring.

7.4 What is wrong with wrong boundary conditions?

7.4.1 Non-existence: Cauchy conditions on a closed surface

In Chapter 3 the BCs that make a problem 'well-posed' were simply quoted. In Section 4.2 we proved that any solution they yield is unique (up to an additive constant in Neumann problems), but we have not yet considered whether other

BCs might not be equally acceptable. Here, without trying to be in any way systematic, we show what goes wrong with the more obviously tempting alternatives. Recall that the requisite conditions are, in fact, Dirichlet or Neumann (but not both, i.e. not Cauchy) on a closed boundary. It will suffice in the following to think of the (homogeneous) Laplace equation with inhomogeneous BCs.

One possibility is easily eliminated, namely that of imposing Cauchy conditions (prescribing both ψ_S and $\partial_n\psi_S$) on a closed surface. Section 4.2 showed that ψ_S, alone, determines ψ uniquely throughout V, and that $\partial_n\psi_S$, alone, determines ψ uniquely up to an additive constant (provided that $\int_S dS\, \partial_n\psi_S = 0$, which we assume throughout the rest of this section). If ψ_S and $\partial_n\psi_S$ are *both* prescribed independently of each other, then the ψ's they determine will not in general agree, and the problem will have no solution. Hence Cauchy conditions on a closed surface are inconsistent in general. Consider for example the interior of a circle of radius a, and prescribe $\psi_S = \alpha\cos(m\phi)$, $\partial_n\psi_S = \partial\psi/\partial r|_{r=a} = (\beta/a)\cos(n\phi)$. In view of the harmonic functions listed in (4.3.3b), ψ_S demands that $\psi = \psi_1 \equiv \alpha(r/a)^m\cos(m\phi)$, while $\partial_n\psi_S$ demands that $\psi = \psi_2 \equiv (\beta/n)(r/a)^n\cos(n\phi)$. But $\psi_1 \neq \psi_2$ unless, by coincidence or conspiracy, $m = n$ and $\alpha = \beta$. While it could thus happen for special choices of ψ_S and $\partial_n\psi_S$ that the solutions they determine agree, such conditions on the input data are very stringent (incomparably more so than the Neumann self-consistency condition $\int_S dS\, \partial_n\psi_S = 0$), half of the data are then redundant, and the resulting problem is not a BVP in the usual sense.

Although the argument which has just eliminated Cauchy conditions seems straightforward, the conclusion appears very startling indeed if one approaches the Laplace BVP variationally, as sketched briefly in Appendix I.

7.4.2 Instability: Cauchy conditions on an open surface

Far more interesting is the attempt to prescribe Cauchy conditions on an open surface. For instance, in 3D we might prescribe both ψ and $\partial\psi/\partial z$ on the xy-plane. After all, in Section 3.5 we stated that precisely such Cauchy conditions, over part of an open hyperboundary, are appropriate to the wave equation regarded as an equation in hyperspace, where $\mathbf{r}$ and t are treated on an equal footing. However, the same attempt for elliptic rather than for hyperbolic equations goes badly wrong, even in the most favourable case where the equation of the open surface S, and the prescribed ψ_S and $\partial_n\psi_S$, are all at least piecewise analytic. ('Analytic' here means that they can be represented by convergent power series in their independent variables; 'piecewise' as regards ψ_S say means that ψ_S can be written as such a series everywhere on S except perhaps on some curves or points subdividing S into regions, with different series representing ψ_S in different regions.) Then, solutions under Cauchy BCs still exist in some finite regions adjacent to the surface; but these solutions become singular at some points within V, where, on physical grounds, no singularities can be tolerated. In the vicinity of the singularities the solutions are effectively *unstable*: small changes in the input data lead to arbitrarily large changes in the solutions. Of course it is again possible to find very special (highly constrained) input data from which acceptable solutions do follow, but again this is not a boundary-value problem in the usual sense.

The theory substantiating these remarks in general is quite involved. Sneddon (1957) for instance applies the method of characteristics. Morse and Feshbach (1953) Section 6.2 apply the theory of difference equations, which shows vividly how numerical integration starting from S can deliver a solution in a limited region, but fails eventually because of instability. Hadamard (1964) gives a penetrating critique of popular misconceptions surrounding the implications of the famous Cauchy–Kowalewska proof that a solution near S exists when the data are analytic.

Here we confine ourselves to an example which exhibits the crucial features. For simplicity we consider only the 2D Laplace equation in the halfplane $y \geqslant 0$, subject to Cauchy conditions on the x-axis, but with no conditions imposed at infinity. Thus the x-axis indeed constitutes an open boundary. This problem will be tackled in two ways: first by appeal to the special 2D method outlined at the end of Section 4.3, and then by an expansion in harmonic functions $\psi_k(\boldsymbol{r})$ of the type (4.3.2).

Along $y = 0$ we prescribe the *analytic* Cauchy conditions

$$\psi(x, 0) = h(x) \equiv \frac{1}{[a^2 + x^2]}, \qquad \psi_y(x, 0) = 0, \tag{7.4.1a, b}$$

where a is real and positive. The arguments for eqn (4.3.17) show that the function

$$\psi(x, y) = \operatorname{Re} h(x + iy) = \operatorname{Re} \frac{1}{[a^2 + (x + iy)^2]}, \tag{7.4.2}$$

is harmonic, at least for small enough y; inspection reveals that on the x-axis ψ reduces to the prescribed function $h(x)$; finally, $\partial\psi/\partial y = \operatorname{Re}\,(-2i)(x + iy)/[a^2 + (x + iy)^2]^2$ manifestly vanishes when $y = 0$, whence the second Cauchy condition (7.4.1b) is also obeyed. Hence ψ is the solution we seek. Explicitly,

$$\begin{aligned}\psi(x, y) &= \operatorname{Re} \frac{1}{[a^2 + x^2 - y^2 + 2ixy]} \\ &= \operatorname{Re} \frac{[a^2 + x^2 - y^2 - 2ixy]}{[(a^2 + x^2 - y^2)^2 + 4x^2y^2]} \equiv \chi(x, y; a),\end{aligned}$$

$$\psi(x, y) = \chi(x, y; a) = \frac{[a^2 + x^2 - y^2]}{[(a^2 + x^2 - y^2)^2 + 4x^2y^2]}. \tag{7.4.3}$$

The trouble with this function is that it is ill-defined at the point where the denominator vanishes, i.e. at $(x, y) = (0, a)$, and that it can diverge as this point is approached. To see this, note the behaviour of ψ as one moves up the y-axis:

$$\psi(0, y) = \frac{(a^2 - y^2)}{(a^2 - y^2)^2} = \frac{1}{(a^2 - y^2)}, \qquad \psi(0, y \to a) \to \infty. \tag{7.4.4a, b}$$

On the other hand, as the same point is approached horizontally, one has

$$\psi(x, a) = \frac{x^2}{(x^4 + 4a^2x^2)} = \frac{1}{(x^2 + 4a^2)}, \qquad \psi(x \to 0, a) = \frac{1}{4a^2}. \qquad (7.4.5a, b)$$

Exercise: Determine the behaviour of $\psi(x, y)$ as the point $(0, a)$ is approached along straight lines of finite slope.

Since we are looking for a solution in source-free space, such singular behaviour is physically unacceptable: the mathematical problem is 'ill-posed' in the sense discussed in Section 3.5.1.

The situation is even worse if the prescribed boundary function $h(x)$ is not analytic. Suppose for instance that $h(x)$ for $x > 0$ is given by $1/(a^2 + x^2)$ as above, while for $x < 0$ it is given by $1/(b^2 + x^2)$, with $b^2 \neq a^2$. Thus, $\psi_S(x, 0) = h(x)$ is discontinuous at the origin. Then just above the positive x-axis one has $\psi = \chi(x, y; a)$ as defined in (7.4.3), but the same arguments show that just above the negative x-axis one has $\psi = \chi(x, y; b) = (b^2 + x^2 - y^2)/[(b^2 + x^2 - y^2)^2 + 4x^2y^2]$. Since these two functions are different, we must choose a curve Γ in the upper halfplane (i.e. within V), extending from the origin to infinity, with $\psi = \chi(x, y; a)$ to the right of Γ and $\psi = \chi(x, y; b)$ to the left. Though the precise location of Γ is a matter of choice, along Γ ψ is unavoidably discontinuous (and thereby unstable); in particular, because one end of Γ is anchored to the origin, i.e. to a point of the boundary, it is now impossible to construct a solution that is physically acceptable throughout any strip of non-zero width adjacent to the boundary.

Exercise: Adapt these arguments to the Cauchy conditions $\psi(x, 0) = 0$, $\psi_y(x, 0) = 1/(a^2 + x^2)$.

It is instructive to reach the same conclusions by another method, namely by looking for ψ as a linear combination of the set of harmonic functions (4.3.2):

$$\psi(x, y) = \int_{-\infty}^{\infty} dk\, \{A(k) \exp(ikx - |k|\, y) + B(k) \exp(ikx + |k|\, y)\}. \qquad (7.4.6)$$

Notice that the B-term is now admissible: we are not imposing any conditions as $y \to \infty$, whence exponential increase cannot be ruled out *a priori*. The coefficients A and B are determined by fitting to the BCs (7.4.1):

$$\psi_y(x, 0) = 0 = \int_{-\infty}^{\infty} dk \exp(ikx) ik\{A(k) - B(k)\}$$

gives $A(k) = B(k)$; then the other BC

$$\psi(x, 0) = \frac{1}{a^2 + x^2} = \int_{-\infty}^{\infty} dk \exp(ikx)\{A(k) + B(k)\} = \int_{-\infty}^{\infty} dk \exp(ikx) 2A(k)$$

gives, on inversion,

$$A(k) = \frac{1}{2} \cdot \frac{1}{2\pi} \int_{-\infty}^{\infty} dx \frac{\exp(-ikx)}{a^2 + x^2} = \frac{1}{4\pi} \int_{-\infty}^{\infty} dx \frac{\cos(kx)}{a^2 + x^2},$$

$$B(k) = A(k) = \frac{1}{4a} \exp(-a|k|). \tag{7.4.7}$$

Finally, substituting (7.4.7) into (7.4.6), one finds

$$\psi(x, y) = \frac{1}{4a} \int_{-\infty}^{\infty} dk \, \{\exp[-a|k| + ikx - |k|y] + \exp[-a|k| + ikx + |k|y]\}$$

$$\psi(x, y) = \frac{1}{2a} \operatorname{Re} \int_0^{\infty} dk \, \{\exp[-k(a + y) + ikx] + \exp[-k(a - y) + ikx]\}. \tag{7.4.8}$$

The crucial question is whether these integrals converge. Since the first integrand contains the factor $\exp[-k(a+y)] \leqslant \exp[-ka]$, this term will converge for any (positive) value of y. By contrast, the second integrand contains the factor $\exp[-k(a-y)]$; when $y > a$ this exponential increases as $k \to \infty$, and the integral diverges. Thus we recognize, even without actually evaluating the integrals, that $\psi(x, y)$ must have a singularity at some point or points with $y = a$; hence ψ cannot possibly be a physically acceptable solution for *all* x except in the limited region $0 \leqslant y < a$.

Of course in the present case the integrals can be evaluated straightforwardly (for $y < a$):

$$\psi(x, y) = \frac{1}{2a} \operatorname{Re} \left\{ \frac{1}{a + y - ix} + \frac{1}{a - y - ix} \right\}$$

$$= \frac{1}{2a} \operatorname{Re} \left\{ \frac{a + y + ix}{(a + y)^2 + x^2} + \frac{a - y - ix}{(a - y)^2 + x^2} \right\},$$

which on simplifying indeed reduces to the known expression (7.4.3).

7.5 The Kirchhoff integral representation

As regards the Laplace and Poisson equations, this topic is interesting in principle, and helps to sharpen one's mathematical perception of the problems. But the approach comes into its own for the Helmholtz equation, where it is the basis of the theory of diffraction. It is in that context that Kirchhoff's name is normally attached to the representation.

To understand the physics, we shall need the results (4.4.25, 29) for the discontinuities of $\partial_n \psi$ and ψ across surfaces carrying source and dipole layers.

Up to this point, we have always considered BVPs in a given region V, bounded by S. Both source point $\boldsymbol{r}'$ and field point $\boldsymbol{r}$ were confined to V, and no meaning was sought for the resulting expressions, like the magic rules (5.2.4), (6.2.7), or (6.3.5), when either of these points lay outside V. This restriction follows from the nature of the case: for instance, when we constructed G_D as in (5.1.13) from the eigenfunctions of $-\nabla^2$, $G_D = \sum_n \lambda_n^{-1}\phi_n^*(\boldsymbol{r}')\phi_n(\boldsymbol{r})$, these functions ϕ_n were defined only for $\boldsymbol{r}$ within V or on S, and although it might have been possible to extend the definition to points beyond V (in 2D for instance by analytic continuation), there was no physical motivation for doing so. This will now change. The differences between the following argument and those in Sections 5.2, 6.2, and 6.3 constitute the point of the exercise, and should be watched carefully.

We now consider the solution $\psi(\boldsymbol{r})$ of Poisson's equation

$$-\nabla^2\psi(\boldsymbol{r}) = \rho(\boldsymbol{r}), \qquad \psi(\boldsymbol{r}\to\infty) = 0, \tag{7.5.1}$$

with $\rho(\boldsymbol{r})$ given, throughout all of space. When, presently, we introduce a limited region V bounded by a closed surface S, this will be done just as a mathematical device, and not in order to impose any mathematical pre-conditions or physical constraints at S. We stress again that ψ is now defined (and remains to be discovered) in all space.

The basic idea is to select a region V surrounded by a closed surface S, and to apply Green's theorem to $\psi(\boldsymbol{r})$ and to the *free-space* Green's function $G_0(\boldsymbol{r}\,|\,\boldsymbol{r}')$ which, like ψ, is defined for all $\boldsymbol{r}$ and $\boldsymbol{r}'$, and obeys $-\nabla^2 G_0(\boldsymbol{r}\,|\,\boldsymbol{r}') = \delta(\boldsymbol{r}-\boldsymbol{r}')$. Reasoning exactly as in eqns (5.2.1–3), but with G_0 in place of G_D, we find

$$\begin{aligned}
&-\int_V dV'\,\{\psi(\boldsymbol{r}')\nabla'^2 G_0(\boldsymbol{r}'\,|\,\boldsymbol{r}) - (\nabla'^2\psi(\boldsymbol{r}'))G_0(\boldsymbol{r}'\,|\,\boldsymbol{r})\}\\
&\quad = -\int_S dS'\,\{\psi(\boldsymbol{r}')\,\partial'_n G_0(\boldsymbol{r}'\,|\,\boldsymbol{r}) - (\partial'_n\psi(\boldsymbol{r}'))G_0(\boldsymbol{r}'\,|\,\boldsymbol{r})\}\\
&\quad = \int_V dV'\,\{\psi(\boldsymbol{r}')\,\delta(\boldsymbol{r}'-\boldsymbol{r}) - \rho(\boldsymbol{r}')G_0(\boldsymbol{r}'\,|\,\boldsymbol{r})\}.
\end{aligned} \tag{7.5.2}$$

The new feature is that it now makes sense to consider this equation with the field point $\boldsymbol{r}$ outside as well as inside V (though, as before, the volume integrals with respect to dV' run only over V).

On the right of (7.5.2), the first integral is zero if $\boldsymbol{r}$ is outside V (the argument of $\delta(\boldsymbol{r}'-\boldsymbol{r})$ does not then vanish anywhere in the integration region); if $\boldsymbol{r}$ is inside V, then this integral is $\psi(\boldsymbol{r})$. In the surface integral, neither term vanishes now that we are using G_0 rather than G_D or G_N. Rearrangement of the last equality in (7.5.2) yields what we call the

Kirchhoff representation:

$$f_K(\boldsymbol{r}) + g_K(\boldsymbol{r}) \equiv \int_V dV'\, \rho(\boldsymbol{r}')G_0(\boldsymbol{r}' \mid \boldsymbol{r})$$

$$- \int_S dS'\, \{\psi(\boldsymbol{r}')\, \partial_n' G_0(\boldsymbol{r}' \mid \boldsymbol{r}) - (\partial_n' \psi(\boldsymbol{r}'))G_0(\boldsymbol{r}' \mid \boldsymbol{r})\}$$

$$= \begin{cases} \psi(\boldsymbol{r}) & \text{if } \boldsymbol{r} \text{ is in } V; \\ 0, & \text{if } \boldsymbol{r} \text{ is not in } V. \end{cases} \qquad (7.5.3)$$

Here we have defined the volume integral f_K and the surface integral g_K by analogy with (5.2.4) say. Notice that only g_K but not f_K features ψ or $\partial_n \psi$. G_0 in 3D, 2D, and 1D is given by (4.4.4, 7, 9c).

(i) Equation (7.5.3) shows that the expression on its left is discontinuous when the field point $\boldsymbol{r}$ crosses S, even though this is far from obvious on mere inspection. Arguments like those in Section 4.4.5 show that the discontinuity stems wholly from the surface integral g_K.

(ii) *If* both ψ and $\partial_n \psi$ were known on S, then, from (7.5.3), we could evaluate $\psi(\boldsymbol{r})$ everywhere within S. (Recall that S could, for instance, be a sphere, and V the region inside the sphere; but equally, S could be the same sphere plus the surface at infinity, and V is then the region outside the sphere. In g_K all the derivatives ∂_n' are taken in the direction of the outward normal.) The derivation shows that, in any case of doubt, $\partial_n' \psi(\boldsymbol{r}')$ and $\psi(\boldsymbol{r}')$ are to be interpreted as the limits when $\boldsymbol{r}'$ tends to S from *within* V; this matters if our (artificially chosen) surface S happens to coincide with a physical source or dipole layer, on which we know from (4.4.25, 29) that $\partial_n \psi$ or ψ are discontinuous.

(iii) Of course, the snag is that in general we do *not* know both ψ and $\partial_n \psi$ until the problem has been solved. In other words, f_K and g_K, unlike $f_{D,N}$ and $g_{D,N}$ cannot be evaluated directly from the data. Therefore (7.5.3) is in no sense a solution, and its status is altogether different from that of the magic rules, which are solutions. This is why we call (7.5.3) an 'integral representation' instead: it simply *represents* the solution inside V in terms of the solution and of its normal derivatives on S. Outside V it does not represent the solution.

We recall the point emphasized in Section 7.4.1, that if we were to prescribe ψ and $\partial_n \psi$ *arbitrarily* on S, then they would in general be inconsistent with each other. In terms of the Kirchhoff representation this inconsistency shows up as follows. If one does insert arbitrarily chosen values of $\psi(\boldsymbol{r}')$ and $\partial_n' \psi(\boldsymbol{r}')$ into g_K, then the resulting expression $(f_K + g_K)$ does still solve Poisson's equation (7.5.1) inside V, as is easily verified by substituting (7.5.3) into (7.5.1) and appealing to $-\nabla^2 G_0 =$

$\delta(\boldsymbol{r}-\boldsymbol{r}')$; but as the field point $\boldsymbol{r}$ approaches S from within V, the expression does not, in general, tend to the prescribed values. Therefore, though it is a solution of *some* Poisson problem, it does not, in a manner of speaking, solve the problem we intended that it should solve.

(iv) A physical interpretation of (7.5.3) can be constructed as follows. The idea is simply to notice that $(f_K(\boldsymbol{r})+g_K(\boldsymbol{r}))$ solves *some* problem (one of whose demands is that $-\nabla^2\psi$ equals ρ inside V, and vanishes outside V), and to ask what precisely that problem is.

The volume integral $f_K(\boldsymbol{r})$ is just the potential due to the sources inside V. As already stated in (i), this term is continuous across S. To procure the jump of (f_K+g_K) from zero just outside to $\psi(\boldsymbol{r}')$ just inside S at the point $\boldsymbol{r}'$ on S, the surface element $\mathrm{d}S'$ must carry a dipole moment $\mathrm{d}\boldsymbol{p}'=-\mathrm{d}\boldsymbol{S}'\,\psi(\boldsymbol{r}')$, as we know from (4.4.29); the first part of g_K is precisely the potential due to this dipole layer. Lastly, to procure the jump of $\partial_n(f_K+g_K)$ from zero just outside to $\partial'_n\psi(\boldsymbol{r}')$ just inside S at the point $\boldsymbol{r}'$ on S, the surface element $\mathrm{d}S'$ must carry source strength $\mathrm{d}Q=\mathrm{d}S'\,\partial'_n\psi(\boldsymbol{r}')$, as we know from (4.4.25); the second part of g_K is precisely the potential due to this source layer.

Accordingly, the expression (f_K+g_K) featured in the Kirchhoff representation does not give the solution of the original problem; in view of (7.5.1) that solution is $\psi(\boldsymbol{r})=\int \mathrm{d}V'\,\rho(\boldsymbol{r}')G_0(\boldsymbol{r}'\mid\boldsymbol{r})$, where the integral extends over all space. Instead, the representation gives the solution of another problem, where the sources outside V have been removed, and where a dipole layer and a source layer have been placed on S, their strength determined, as described above, by the actual solution of the original problem and by its normal derivative at S. Then the solution of this new problem coincides with the solution of the original problem at all points inside S; but, unlike the solution of the original problem, that of the new problem vanishes identically outside S.

III Diffusion

Summary

Diffusion or heat equation:

$$L \equiv \left(\frac{\partial}{\partial t} - \mathrm{D}\nabla^2\right); \qquad L\psi(\boldsymbol{r}, t) = \rho(\boldsymbol{r}, t).$$

D is positive by the second law of thermodynamics.
BCs: either D or N (or mixed), on a closed surface, i.e. $\psi(\boldsymbol{r}, t)$ or $\partial_n\psi(\boldsymbol{r}, t)$ given for $\boldsymbol{r}$ on S *for all* $t \geq$ *starting time* t_0.

Initial conditions: $\psi(\boldsymbol{r}, t_0)$ throughout V.

Irreversibility: solutions *cannot* in general be found for all $t < t_0$.

Propagator: $LK(\boldsymbol{r}, t \mid \boldsymbol{r}', t') = 0$; homogeneous BCs on S (D or N (or mixed) respectively); initial condition: $K(\boldsymbol{r}, t \mid \boldsymbol{r}', t') = \delta(\boldsymbol{r} - \boldsymbol{r}')$.

- *Time-translation invariance:* $K(\boldsymbol{r}, t \mid \boldsymbol{r}', t') = K(\boldsymbol{r}, \boldsymbol{r}'; \tau)$, $\tau \equiv t - t'$.
- *Symmetry:* $K(\boldsymbol{r}, \boldsymbol{r}'; \tau) = K(\boldsymbol{r}', \boldsymbol{r}, \tau)$.
- *General solution of the homogeneous IVP:*

$$\psi(\boldsymbol{r}, t) = \int_V \mathrm{d}V'\, K(\boldsymbol{r}, t \mid \boldsymbol{r}', t_0)\psi(\boldsymbol{r}, t_0), \qquad (t > t_0).$$

Green's function: $LG(\boldsymbol{r}, t \mid \boldsymbol{r}', t') = \delta(t - t')\,\delta(\boldsymbol{r} - \boldsymbol{r}')$; homogeneous BCs; $G = 0$ for $t < t'$.

$$G(\boldsymbol{r}, t \mid \boldsymbol{r}', t') = \theta(t - t')K(\boldsymbol{r}, t \mid \boldsymbol{r}', t').$$

Magic rule (under any BCs):

$$\begin{aligned}\psi(\boldsymbol{r}, t) = &\int_{t_0}^{t} \mathrm{d}t' \int_V \mathrm{d}V'\, G(\boldsymbol{r}, t \mid \boldsymbol{r}', t')\rho(\boldsymbol{r}', t') \\ &+ D\int_{t_0}^{t} \mathrm{d}t' \int_S \mathrm{d}S'\, [G(\boldsymbol{r}, t \mid \boldsymbol{r}', t')\,\partial_n'\psi_S(\boldsymbol{r}', t') \\ &- (\partial_n' G(\boldsymbol{r}, t \mid \boldsymbol{r}', t'))\psi_S(\boldsymbol{r}', t')] \\ &+ \int_V \mathrm{d}V'\, G(\boldsymbol{r}, t \mid \boldsymbol{r}', t_0)\psi(\boldsymbol{r}', t_0), \qquad (t > t_0).\end{aligned}$$

Eigenfunction expansion: With $-\nabla^2\phi_p = \lambda_p\phi_p$, under homogeneous BCs on ϕ_p,

$$K(\boldsymbol{r}, t \mid \boldsymbol{r}', t') = \sum_p \phi_p^*(\boldsymbol{r}')\phi_p(r)\exp(-\lambda_p \mathrm{D}(t-t')).$$

In unbounded nD **space:**

$$K_0^{(n)} = \frac{1}{(4\pi \mathrm{D}\tau)^{n/2}}\exp(-R^2/4\mathrm{D}\tau); \qquad \boldsymbol{R} \equiv \boldsymbol{r} - \boldsymbol{r}', \qquad \tau \equiv t - t'$$

8 The diffusion equation: I. Unbounded space

Although in many ways Poisson's equation is closer to the wave equation than to the diffusion equation, we treat the diffusion equation first, because it is somewhat simpler, being only of first order in $\partial/\partial t$. The present chapter deals mainly with the important special case of unbounded space, where propagator and Green's function can be found in closed form. This makes it easier to develop a feel for what is new when time enters as one of the independent variables. The general theory, including the symmetries and the irreversibility of the diffusion equation, is reserved for Chapter 9.

From here on, we pay at most passing attention to finer points like those discussed for Poisson's equation in Chapter 7. Such aspects are treated, by, for example, Hadamard (1964) and Sobolev (1964).

8.1 Introduction and basic physics

The diffusion equation reads

$$L\psi \equiv \{\partial/\partial t - D\nabla^2\}\psi(\boldsymbol{r}, t) = \rho(\boldsymbol{r}, t). \tag{8.1.1}$$

In diffusion, ψ is the concentration (number per unit volume) of the particles, and ρ is the number injected into the system per unit volume per unit time. D is a parameter called the diffusion constant, having dimensions $[\mathrm{L}^2\mathrm{T}^{-1}]$.

We define also the diffusion current density $\boldsymbol{j}(\boldsymbol{r}, t)$, such that the (net) number of particles crossing an element of area $\delta\boldsymbol{S}$ per unit time is $\boldsymbol{j}\cdot\delta\boldsymbol{S}$. Then the (local) conservation law for particles reads

$$\frac{\partial\psi}{\partial t} = -\operatorname{div}\boldsymbol{j} + \rho. \tag{8.1.2}$$

To validate the diffusion equation, the conservation law, which is a universal truth, must be supplemented by Fick's law for $\boldsymbol{j}$, which is only an approximation:

$$\boldsymbol{j} = -D\,\operatorname{grad}\psi. \tag{8.1.3}$$

Laws like (8.1.3) often apply to currents arising from random (e.g. Brownian) motion. Substituting (8.1.3) into (8.1.2) one obtains (8.1.1). To summarize, the diffusion equation is simply the local conservation law for the density, appropriate in the special case where the current is governed by Fick's law.

The same equation arises in heat conduction, where it is often called the 'heat equation'. Let ψ be the temperature, c the heat capacity per unit volume (density times specific heat), h the heat-energy input per unit volume per unit time, and $\boldsymbol{j}$ the heat flux. Then the conservation of heat energy entails $c\,\partial\psi/\partial t = -\operatorname{div}\boldsymbol{j} + h$, while in many systems $\boldsymbol{j} = -\kappa \operatorname{grad}\psi$, κ being the thermal conductivity. Thus one obtains $\partial\psi/\partial t = -(\kappa/c)\nabla^2\psi + (h/c)$; accordingly, (8.1.1) covers heat conduction if we identify

$$D = \kappa/c, \qquad \rho = h/c. \tag{8.1.4}$$

We shall generally ignore such distinctions, calling D the conductivity and ρ the source distribution for short.

Since there is no deep physical reason why D should not depend on ψ, nor why $\boldsymbol{j}$ should not depend on higher derivatives of ψ, the diffusion equation is indeed only an approximation, though often a very good one. In this respect it contrasts with Poisson's equation in electrostatics, and with the wave equation in electromagnetism, which are exact laws of nature (at least *in vacuo,* and as long as quantum effects are negligible).

The random molecular-level motions underlying (8.1.3) are irreversible (in the thermodynamic sense); correspondingly, and in contrast to the wave equation, the diffusion equation works only forwards in time, as shown in Section 9.5. Specifically, the second law of thermodynamics implies that heat is conducted towards lower temperatures, and that particles diffuse towards lower concentrations; hence D is always positive:

$$D > 0. \tag{8.1.5}$$

Appendix D showed that for Brownian motion in a medium exerting a viscous drag $-\eta\boldsymbol{v}$, the diffusion constant is given by the Einstein relation $D = kT/\eta$.

Boundary conditions on ψ are dictated by the physics. For instance, in heat conduction, DBCs apply on boundaries whose temperature is prescribed by the experimenter; NBCs apply if he prescribes the heat flow across the boundary. In general, the prescribed values of ψ_S or of $\partial_n\psi_S$ vary with time, though in particular problems they may happen to be constants. When the value of ψ_S is enforced, the system itself adjusts the value of $\partial_n\psi_S$, and vice versa; thus, Cauchy BCs are not realizable physically. Mixed BCs are quite common; for physical examples where Churchill BCs are appropriate, see e.g. Carslaw and Jaeger (1959).

Integration of (8.1.2) over a (fixed) region V bounded by a (fixed) closed surface S yields a global (as opposed to a local) conservation law.

Noting that

$$\frac{d}{dt}\int_V dV\,\psi = \int_V dV\frac{\partial\psi}{\partial t},$$

we find

$$\frac{d}{dt}\int_V dV\,\psi = -\int_V dV\,\operatorname{div}\boldsymbol{j} + \int_V dV\,\rho,$$

$$\frac{d}{dt}\int_V dV\,\psi = -\int_S dS\,j_n + \int_V dV\,\rho \tag{8.1.5a}$$

$$= D\int_S dS\,\partial_n\psi + \int_V dV\,\rho. \tag{8.1.5b}$$

While (8.1.5a) is again a universal truth, (8.1.5b) depends on Fick's law. If ψ falls faster than $1/r^2$, one can allow S to recede to infinity; then

$$\frac{d}{dt}\int dV\,\psi = \int_V dV\,\rho. \tag{8.1.6}$$

If neither ρ nor any of the prescribed boundary values vary with time, then the diffusion equation has time-independent solutions, for which $\partial\psi/\partial t = 0$ and (8.1.1) reduces to Poisson's equation (apart from the scaling factor D):

$$-D\nabla^2\psi(\boldsymbol{r}) = \rho(\boldsymbol{r}). \tag{8.1.7}$$

This is appropriate to the physics only if ρ and the boundary values *never* change, and even then only if the initial values are suitably chosen. On the other hand, if say a source is switched on suddenly and then remains steady, the resulting disturbance will spread outwards, and ψ will certainly not be time-independent. Similarly, if we start in source-free space from a localized disturbance at time t_0, for instance from $\psi(\boldsymbol{r}, t_0) = \alpha\exp(-r^2/a^2)$, then ψ will steadily drop to zero with increasing time, and will not be time-independent even though there are no time-dependent sources. By contrast, (8.1.7) does apply for instance in an enclosure whose walls are kept at a constant temperature, if there are only steady sources inside. Examples are given in Sections 8.5 and 9.4.

In strictly static problems the techniques to be discussed in this and in the next chapter are superfluous, and Poisson's equation suffices. Often the static results are useful also as easily-obtained limits of time-varying solutions, approached asymptotically as transients die out. For instance, suppose that $\psi(\boldsymbol{r}, t_0) = 0$, and that a point source is suddenly switched on at $t = t_0$, remaining steady for ever after: $\rho(\boldsymbol{r}, t) = \theta(t - t_0)\alpha\,\delta(\boldsymbol{r})$. Though $\psi(\boldsymbol{r}, t)$ for $t > t_0$ must be worked for, at fixed $\boldsymbol{r}$ the solution at large times will in some cases approach the solution of (8.1.7) with $\rho = \alpha\,\delta(\mathbf{r})$; thus, in 3D, $\psi(\mathbf{r}, t \to \infty) = 1/4\pi Dr$, though Section 8.5 shows that no such thing happens in 1D or in 2D.

8.2 Simple-harmonic solutions

If sources and prescribed boundary values vary harmonically in time, then the diffusion equation has simple-harmonic solutions. (Do not confuse this usage of 'harmonic', referring to time, with 'harmonic' as a description of solutions of the Laplace equation, as in Section 4.3.) For the present we consider only harmonic solutions in unbounded source-free space. (It will appear later that some BVPs too can be solved by writing down such solutions, and then choosing their coefficients so as to satisfy the BCs.)

The 1D case displays all the essential features. Write

$$\psi(x, t) = \operatorname{Re} \Psi(x, t), \tag{8.2.1}$$

$$\Psi(x, t) = A \exp(i\delta) \exp(-i\omega t) f(x), \tag{8.2.2}$$

where $\omega > 0$ and A, δ are real constants. Substitute into (8.1.1), setting $\rho = 0$, to find

$$\frac{\mathrm{d}^2 f}{\mathrm{d}x^2} = -\frac{i\omega}{D} f, \qquad f = \exp\left\{\mp\left(-\frac{i\omega}{D}\right)^{\frac{1}{2}} x\right\}, \tag{8.2.3a,b}$$

where $(-i)^{\frac{1}{2}} \equiv (1 - i)/\sqrt{2}$. Thus the two independent solutions of (8.2.3a) are

$$f(x) = \exp\{\mp(1 - i)(\omega/2D)^{\frac{1}{2}} x\}. \tag{8.2.4}$$

Accordingly,

$$\Psi_\pm = A_\pm \exp(\mp(\omega/2D)^{\frac{1}{2}} x) \exp\{-i\omega t + i\,\delta_\pm \pm i(\omega/2D)^{\frac{1}{2}} x\}, \tag{8.2.5}$$

$$\psi_\pm(x, t) = A_\pm \exp(\mp(\omega/2D)^{\frac{1}{2}} x) \cos\{\omega t \mp (\omega/2D)^{\frac{1}{2}} x - \delta_\pm\}. \tag{8.2.6}$$

The most general solution is a linear combination of $\psi_\pm$, with arbitrary values of $A_\pm$ and of $\delta_\pm$. The upper (lower) signs describe waves travelling in the positive (negative) direction, and damped as they travel. One could define a wavenumber $\kappa \equiv (\omega/2D)^{\frac{1}{2}}$ by writing the cosine in (8.2.6) as $\cos\{\omega t \mp \kappa x - \delta_\pm\}$. The waves are strongly dispersive, because the phase velocity $v_{\mathrm{p}} \equiv \omega/\kappa = 2D\kappa = (2D\omega)^{\frac{1}{2}}$ is far from independent of κ; the group velocity is $v_{\mathrm{g}} \equiv \mathrm{d}\omega/\mathrm{d}\kappa = 2v_{\mathrm{p}}$. However, these 'velocities' have only limited physical significance, because the damping is very strong: in travelling one 'wavelength' (i.e. as x increases by $\lambda \equiv 2\pi/\kappa = 2\pi(2D/\omega)^{\frac{1}{2}}$, the amplitude drops by the factor $\exp(-2\pi) \approx 1.87 \times 10^{-3}$.

The uses of (8.2.6) are illustrated by a system where the temperature of the yz-plane is made to vary according to $\psi(0, t) = \alpha \cos(\omega t)$. On physical grounds, we impose the boundary condition $\psi(|x| \to \infty, t) = 0$. For $x \geqslant 0$, say, this eliminates ψ_-. Equating $\psi_+(0, t)$ to $\alpha \cos(\omega t)$, we find by inspection that $A_+ = \alpha$, $\delta_+ = 0$, whence

$$\psi(x, t) = \alpha \exp(-(\omega/2D)^{\frac{1}{2}}x) \cos\{\omega t - (\omega/2D)^{\frac{1}{2}}x\}, \qquad (x \geqslant 0). \quad (8.2.7)$$

For $x \leqslant 0$, one simply changes $x \to -x$; in other words (8.2.7) holds for all x if we replace x by $|x|$. The power per unit area that must be supplied to the yz-plane in order to maintain this temperature variation is clearly

$$W = -2D\psi_x(0+, t) = \alpha 2(\omega D)^{\frac{1}{2}} \cos(\omega t + \pi/4);$$

thus $\psi(0, t)$ follows $W(t)$ with a phase lag of $\pi/4$.

Exercise: Determine $\psi(x, t)$ if $W = \beta \sin(\omega t)$.

Because the diffusion equation is linear, and because any arbitrary time-variation at $x = 0$ can be represented as a Fourier integral, say as

$$\psi(0, t) = \int_0^\infty d\omega \, \{\alpha(\omega) \cos(\omega t) + \beta(\omega) \sin(\omega t)\}, \qquad (8.2.8)$$

it is possible in principle to solve such BVPs with any prescribed $\psi(0, t)$ (or $\psi_x(0, t)$) by constructing an appropriate linear combination of the solutions (8.2.6). We simply write, for $x \geqslant 0$,

$$\psi(x, t) = \int_0^\infty d\omega \, A(\omega) \cos\{\omega t - (\omega/2D)^{\frac{1}{2}}x - \delta(\omega)\}, \qquad (8.2.9)$$

whence

$$\psi(x, 0) = \int_0^\infty d\omega \, A(\omega)\{\cos(\omega t) \cos\delta + \sin(\omega t) \sin\delta\}. \qquad (8.2.10)$$

Equating coefficients of $\cos(\omega t)$ and $\sin(\omega t)$ under the integrals in (8.2.8, 9) we find $A = (\alpha^2 + \beta^2)^{\frac{1}{2}}$, $\tan\delta = \beta/\alpha$, which determines (8.2.9). Whether this is better than to use the Green's function and the magic rule developed below depends purely on which integrals are easier to evaluate.

It is clear from (8.2.3) that in effect we have been dealing with the 1D Helmholtz equation $(\nabla^2 + k^2)\psi = 0$, but with pure imaginary $k^2 = i\omega/D$. This relation between the diffusion and the Helmholtz equations applies equally in 2D and 3D, and indicates how one might proceed there. For instance, in a 3D polar expansion $\psi_{lm}(\mathbf{r}, t) = \exp(-i\omega t) f_{lm}(r) Y_{lm}(\Omega)$, the f_{lm} are spherical Bessel functions of complex argument; this opens an alternative approach to the example treated by the normal-mode expansion in Section 9.4.2 below. Similarly, in a 2D expansion $\psi_m(r, t) = \exp(-i\omega t) f_m(r) \exp(im\phi)$, the f_m are ordinary Bessel functions of complex argument.

8.3 The initial-value problem and the propagator

8.3.1 The propagator in 1D

The diffusion equation has the unique feature that its propagator in unbounded space exhibits no important formal differences between 3D, 2D, and 1D; hence we start in 1D. Further, we start with the homogeneous (source-free) equation, where the only inhomogeneity in the problem is the initial function $\psi(x, t_0) \equiv a(x)$, prescribed for all x at some starting time t_0. Defining the differential operator

$$L \equiv \partial/\partial t - D\nabla^2 \tag{8.3.1}$$

as in (8.1.1), our problem is to determine $\psi(x, t)$ for all $t > t_0$, given

$$L\psi = 0, \qquad \psi(x, t_0) = a(x). \tag{8.3.2a, b}$$

To this end we define the *propagator* K_0 by

$$LK_0(x, t \mid x', t') = 0, \qquad \text{for } t > t', \tag{8.3.3a}$$

$$K_0(x, t' \mid x', t') = \delta(x - x'), \qquad \text{(at equal times)}, \tag{8.3.3b}$$

$$K_0 \to 0 \qquad \text{as } |x| \to \infty. \tag{8.3.3c}$$

Note that K_0 is not a Green's function, since it satisfies the homogeneous equation (8.3.3a), without delta-functions on the right. The suffix 0 specifies that K_0 is defined in unbounded space, i.e. subject to the BC (8.3.3c). When necessary, the dimensionality will be indicated by a superscript: then our present function would be written $K_0^{(1)}$.

Once K_0 is known, the problem (8.3.2) is solved:

$$\psi(x, t) = \int_{-\infty}^{\infty} dx' \, K_0(x, t \mid x', t_0) a(x') \equiv h_0(x, t), \tag{8.3.4}$$

where h_0 stands for the integral regarded as an explicit construct from the data a.

Proof: (i) The differential equation (8.3.2a) is satisfies because

$$L\psi = L \int dx' \, K_0 \psi(x') = \int dx' \, \{LK_0\} \psi(x') = 0,$$

where (8.3.3a) has been used in the last step.

(ii) The initial condition (8.3.2b) is satisfied, because

$$\psi(x, t_0) = \int dx' \, K_0(x, t_0 \mid x', t_0) a(x') = \int dx' \, \delta(x - x') a(x') = a(x),$$

where (8.3.3b) has been used in the last step. ■

K_0 is best determined as a Fourier integral with respect to its x-dependence. Try

$$K_0(x, t \mid x', t') = \int_{-\infty}^{\infty} dk\, A(k, t, x', t') \exp(ikx). \tag{8.3.5}$$

(By contrast, Appendix F.5 determines K_0 (or rather the related Green's function G_0) through a double Fourier representation, with respect to its dependence on t as well as on x.) Substitution into (8.3.3a) yields the time-dependence of A:

$$LK_0 = \int dk\, L(A \exp(ikx)) = \int dk \left[\frac{\partial A}{\partial t} + Dk^2 A\right] \exp(ikx) = 0,$$

$$\frac{\partial A}{\partial t} = -Dk^2 A, \qquad A = B(k, x', t') \exp(-Dk^2 t). \tag{8.3.6a, b}$$

Substitute (8.3.6b) into (8.3.5), set $t = t'$, appeal to (8.3.3b), and use the standard Fourier representation of $\delta(x - x')$:

$$\int dk \exp(-Dk^2 t') B(k, x', t') \exp(ikx)$$

$$= \delta(x - x') = \frac{1}{2\pi} \int dk \exp(ik(x - x')).$$

Equating coefficients of $\exp(ikx)$, we find

$$B(k, x', t') = \frac{1}{2\pi} \exp(Dk^2 t') \exp(-ikx'),$$

whence

$$A(k, t, x', t') = \frac{1}{2\pi} \exp(-D(t - t')k^2) \exp(-ikx'), \tag{8.3.7}$$

$$K_0(x, t \mid x', t') = \frac{1}{2\pi} \int_{-\infty}^{\infty} dk \exp(-D(t - t')k^2) \exp(ik(x - x')). \tag{8.3.8}$$

Clearly, K_0 depends only on the combinations

$$\xi \equiv x - x', \qquad \tau \equiv t - t'. \tag{8.3.9}$$

(In later chapters, $x - x'$ will generally be denoted by X rather than ξ.) For brevity we shall often write

$$K_0(x, t \mid x', t') \equiv K_0(\xi, \tau), \tag{8.3.10}$$

$$K_0(\xi, \tau) = \frac{1}{2\pi} \int_{-\infty}^{\infty} dk \exp(-D\tau k^2 + ik\xi). \tag{8.3.11}$$

The integral in (8.3.11) is in fact standard: since $\exp(-Dk^2\tau)$ is even in k, only the even part $\cos(k\xi)$ of $\exp(ik\xi)$ contributes, whence

$$K_0(\xi, \tau) = \frac{1}{2\pi}\int_{-\infty}^{\infty} \mathrm{d}k \exp(-D\tau k^2)\cos(k\xi)$$

$$= \frac{1}{2\pi}2\int_0^{\infty} \mathrm{d}k \exp(-D\tau k^2)\cos(k\xi)$$

$$= \frac{1}{2\pi}2\frac{\pi^{\frac{1}{2}}}{2(D\tau)^{\frac{1}{2}}}\exp(-\xi^2/4D\tau),$$

$$K_0(\xi, \tau) = \left(\frac{1}{4\pi D\tau}\right)^{\frac{1}{2}}\exp(-\xi^2/4D\tau). \tag{8.3.12}$$

Proof: To evaluate K_0 from first principles, one starts by completing the square in the exponent in (8.3.11):

$$-D\tau[k^2 - ik\xi/D\tau] = -D\tau[(k - i\xi/2D\tau)^2 + \xi^2/4D^2\tau^2];$$

$$K_0(\xi, \tau) = \frac{\exp(-\xi^2/4D\tau)}{2\pi}\int_{-\infty}^{\infty} \mathrm{d}k \exp(-D\tau(k - i\xi/2D\tau)^2. \tag{8.3.13}$$

Change the integration variable to $\kappa \equiv (k - i\xi/2D\tau)$; then the integral becomes $\int_C \mathrm{d}\kappa \exp(-D\tau\kappa^2)$, where the integration contour C runs parallel to the real axis, above it as shown in Fig. 8.1 if $\xi < 0$, and below it if $\xi > 0$. In either case the contour may be shifted to run along the real axis. (Those unacquainted with the theory of functions of a complex variable should just shut their eyes to the fact that κ is complex: all that the above argument asserts is that this makes no difference here.) One

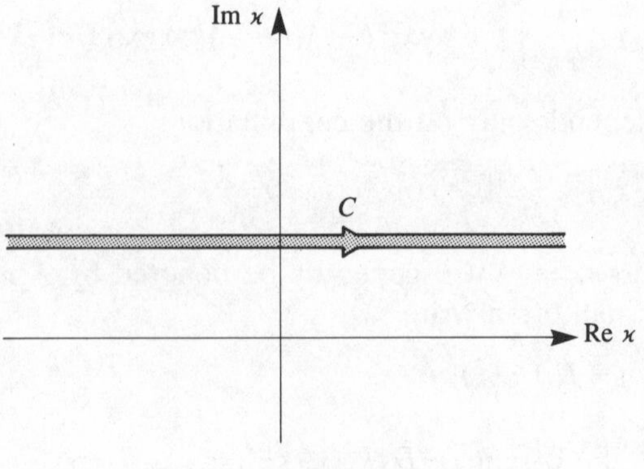

Fig. 8.1

obtains the familiar Gaussian integral (see Appendix A.7)

$$\int_C = \int_{-\infty}^{\infty} d\kappa \exp(-D\tau\kappa^2) = (\pi/D\tau)^{\frac{1}{2}}, \tag{8.3.14}$$

whence (8.3.12) follows. ∎

Restoring our customary notation $\psi(x', t_0)$ instead of $a(x')$ in (8.3.4), we see that the general inhomogeneous IVP is solved by either of the equivalent expressions

$$\psi(x, t) = \int_{-\infty}^{\infty} dx' \frac{\exp\{-(x-x')^2/4D(t-t_0)\}}{[4\pi D(t-t_0)]^{\frac{1}{2}}} \psi(x', t_0), \tag{8.3.15a}$$

$$\psi(x, t) = \int_{-\infty}^{\infty} d\xi \frac{\exp\{-\xi^2/4D(t-t_0)\}}{[4\pi D(t-t_0)]^{\frac{1}{2}}} \psi(x-\xi, t_0). \tag{8.3.15b}$$

The second is called Laplace's solution: it stresses the extent to which $\psi(x, t)$ is affected by initial values at points a distance ξ from x.

Two general features merit attention.

(i) *Evolution in time smoothes* ψ. In (8.3.15a), the x-dependence resides wholly in the exponent. This entails that for any $t > t_0$, $\psi(x, t)$ is continuous and differentiable n times with respect to x, provided only that

$$\int_{-\infty}^{\infty} dx' \, |x'|^n \exp\{-(x-x')^2/4D(t-t')\}\psi(x', t_0) < \infty,$$

which is so for any n if $\psi(x', t_0)$ rises no faster than exponentially as $|x'| \to \infty$: a condition that will certainly be satisfied by any $\psi(x', t_0)$ met in practice. Consequently, any discontinuities in the initial ψ or in any of its derivatives disappear after an arbitrarily short time lapse. (For an example, see Section 8.3.4.)

Smoothing with time is very much a peculiarity of the diffusion equation (i.e. of parabolic equations); we shall see in Chapters 11 and 12 (note especially Section 12.2.4) that no such thing occurs with the wave equation (i.e. with hyperbolic equations).

(ii) *Asymptotic behaviour.* It is clear from the exponent $-(x-x')^2/4D(t-t')$ in K_0 that the limits $|x| \to \infty$ and $t \to \infty$ are incompatible, since the exponent diverges in the first case and vanishes in the second. Here we are interested only in the limit $t \to \infty$ at fixed x. Suppose that $\psi(x', t_0)$ is appreciable only in a limited range $|x'| \lesssim a$; we consider on the basis of (8.3.15a) how $\psi(x, t)$ behaves at fixed x once t is large enough to ensure that $D(t-t_0) \gg (|x|+a)^2$. Then the exponent is small wherever $\psi(x', t_0)$ is appreciable, whence the exponential is approximately unity. Therefore

$$\psi(x, t\to\infty) \sim \frac{1}{[4\pi D(t-t_0)]^{\frac{1}{2}}} \int_{-\infty}^{\infty} dx'\, \psi(x', t_0) = \frac{N}{[4\pi D(t-t_0)]^{\frac{1}{2}}}, \tag{8.3.16}$$

where

$$N = \int_{-\infty}^{\infty} dx\, \psi(x, t_0) = \int_{-\infty}^{\infty} dx\, \psi(x, t)$$

is the total number of particles, which in the present source-free case is independent of time, as we know from (8.1.6).

At first sight (8.3.16) seems paradoxical, because the RHS is independent of x, and therefore, in particular, not integrable from $x = -\infty$ to $x = +\infty$. The reason is that, although (8.3.16) does apply at any fixed x, nevertheless for any fixed t it fails at large enough $|x|$, where the assumed inequality $D(t - t') \gg x^2$ fails. In fact, most particles are located at points where (8.3.16) does not apply.

8.3.2. Comments

(i) If ξ and τ are regarded as new independent variables replacing x and t (while x', t' continue to be regarded as parameters), then the operator L and the definition (8.3.3) of K_0 read

$$LK_0 = (\partial/\partial\tau - D\, \partial^2/\partial\xi^2)K_0 = 0, \tag{8.3.17a}$$

$$K_0(\xi, 0) = \delta(\xi), \tag{8.3.17b}$$

$$K_0(|\xi|\to\infty, \tau) = 0. \tag{8.3.17c}$$

Even ahead of an explicit solution, we could have foreseen that K_0 depends on x, x', t, t' only through ξ and τ, from the fact that only ξ and τ occur in these defining relations. The dependence on t and t' through τ alone is a special case of 'invariance under time-translation', discussed more generally in Section 9.2.

(ii) It is easy to verify by explicit differentiation that K_0 as given by (8.3.12) indeed satisfies (8.3 17a).

Exercise: Verify this.

That the initial condition (8.3.17b) is satisfied can be seen by setting $4D\tau \equiv \varepsilon^2$; then $\tau\to 0$ implies $\varepsilon\to 0$, and K_0 turns into the standard Gaussian representation of the delta-function:

$$\lim_{\varepsilon\to 0} (\varepsilon\pi^{\frac{1}{2}})^{-1} \exp(-\xi^2/\varepsilon^2) = \delta(\xi).$$

(iii) K_0 asserts that the initial delta-function peak at $\tau = 0$ broadens at finite τ into a Gaussian error curve having root-mean-square deviation $(2D\tau)^{\frac{1}{2}}$. The area under the curve is unity independently of τ, as demanded by the global conservation law (8.1.6):

$$\frac{\mathrm{d}}{\mathrm{d}\tau}\int_{-\infty}^{\infty} \mathrm{d}\xi\, K_0(\xi, \tau) = 0,$$

$$\int_{-\infty}^{\infty} \mathrm{d}\xi\, K_0(\xi, \tau) = \int_{-\infty}^{\infty} \mathrm{d}\xi\, K_0(\xi, 0) = \int_{-\infty}^{\infty} \mathrm{d}\xi\, \delta(\xi) = 1.$$

(iv) Regarded as a function of τ at fixed ξ, K_0 has an essential singularity at $\tau = 0$, because the exponential diverges faster than any power of $1/\tau$. Evidently $K_0(\xi, \tau)$ makes no physical sense for negative τ: (a) it is complex on account of the prefactor $\tau^{-\frac{1}{2}}$, and (b) it diverges like $\exp(\xi^2/4D\,|\tau|)$, so that for reasonable $\psi(x, t_0)$ the integrals (8.3.15) would diverge. This feature reflects the irreversibility of the diffusion equation, discussed further in Section 9.5.

(v) The diffusion equation implies infinitely fast propagation. On the one hand, $K(\xi, 0)$ vanishes for all $\xi \neq 0$; thus, at $\tau = 0$, all the particles described by K_0 are at $\xi = 0$. On the other hand, for any $\tau > 0$ however small, K_0 is non-zero for all $|\xi|$ however large. Hence some of the particles must have had, or must immediately have acquired, arbitrarily high velocities; in any case, it is clear that signals of some kind have travelled from $\xi = 0$ arbitrarily fast. This is wrong on at least two counts. First, according to the special theory of relativity, no signals can travel faster than light. Second, it is wrong quite irrespective of relativity, because we know from continuum mechanics that no small-amplitude signals in a medium are likely to travel faster than sound; in practice it is this second limitation which is operative. The point is that Fick's law (8.1.3), and consequently the diffusion equation, fail over very small times or equivalently at very high frequencies, because as regards the response of the diffusing particles to an applied force, or to the fluctuating Langevin force, they allow only for viscous resistance, but not for the inertia of the particles. If both effects are allowed for, the diffusion equation is replaced by the 'equation of telegraphy'

$$\left\{\frac{\partial}{\partial t} + D\left[\frac{1}{u^2}\frac{\partial^2}{\partial t^2} - \nabla^2\right]\right\}\psi = \rho(\mathbf{r}, t), \qquad (u = \text{constant}),$$

whose propagator is free of such paradoxes (see Morse and Feshbach 1953, Section 7.4). However, in most diffusion and heat-conduction problems, the solutions (8.3.15) constructed from (8.3.12) describe the interesting physics adequately.

8.3.3 The propagator in nD

In nD, define the n-component vector $\boldsymbol{R} = \boldsymbol{r} - \boldsymbol{r}'$ as before, represent $\delta(\boldsymbol{R}) = (2\pi)^{-n} \int \mathrm{d}^n k \exp(i\boldsymbol{k} \cdot \boldsymbol{R})$ as in (1.4.3), and look for K_0 in the form $\int \mathrm{d}^n k\, A(\boldsymbol{k}, t, \boldsymbol{r}', t') \exp(i\boldsymbol{k} \cdot \boldsymbol{r})$ analogous to (8.3.5). The same argument that gave (8.3.8) now leads to

$$K_0^{(n)}(\boldsymbol{r}, t \mid \boldsymbol{r}', t') = \int \frac{\mathrm{d}^n k}{(2\pi)^n} \exp\{-D\tau k^2 + i\boldsymbol{k} \cdot \boldsymbol{R}\}$$
$$\equiv K_0^{(n)}(\boldsymbol{R}, \tau). \tag{8.3.18}$$

Since

$$\begin{aligned}\exp\{-D\tau k^2 + i\boldsymbol{k} \cdot \boldsymbol{R}\} &= \exp\{-D\tau(k_1^2 + \cdots + k_n^2) \\ &\qquad + i(k_1 R_1 + \cdots + k_n R_n)\} \\ &= \exp\{-D\tau k_1^2 + ik_1 R_1\} \\ &\qquad \times \cdots \times \exp\{-D\tau k_n^2 + ik_n R_n\},\end{aligned}$$

we can write $K_0^{(n)}$ as the product

$$\begin{aligned}K_0^{(n)}(\boldsymbol{R}, \tau) &= \prod_{j=1}^{n} \left\{\int_{-\infty}^{\infty} \frac{\mathrm{d}k_j}{2\pi} \exp(-D\tau k_j^2 + ik_j R_j)\right\} \\ &= \prod_{j=1}^{n} \left\{\frac{\exp(-R_j^2/4D\tau)}{(4\pi D\tau)^{\frac{1}{2}}}\right\} = \frac{\exp\{-\sum_{j=1}^{n} R_j^2/4D\tau\}}{(4\pi D\tau)^{n/2}},\end{aligned}$$

$$K_0^{(n)}(\boldsymbol{R}, \tau) = \frac{1}{(4\pi D\tau)^{n/2}} \exp(-R^2/4D\tau). \tag{8.3.19}$$

The solution of the IVP in nD is

$$\psi(\boldsymbol{r}, t) = \int \mathrm{d}V'\, K_0^{(n)}(\boldsymbol{R}, t - t_0)\psi(\boldsymbol{r}', t_0) \equiv h_0(\boldsymbol{r}, t), \tag{8.3.20}$$

where h_0 stands for the integral regarded as an explicit construct from the data $\psi(\boldsymbol{r}', t_0)$.

Exercise: Verify (8.3.20) explicitly.

The large-time behaviour at fixed $\boldsymbol{r}$ now becomes proportional to $t^{-n/2}$, which generalizes (8.3.16).

It is a peculiarity of the diffusion equation that its propagator $K_0^{(n)}$ depends on the dimensionality n in such a simple and explicit manner. (Unfortunately this is far from being the case for the wave equation.) However, the close formal resemblance between the $K_0^{(n)}$ for different n does not preclude drastic differences of physical behaviour as described by the solutions of superficially similar problems in 1D, 2D, and 3D; witness the example in Section 8.5 below.

8.3.4 Example: diffusion from a hollow spherical shell

We determine the solution of the homogeneous 3D diffusion equation for $t \geqslant 0$, given the initial concentration $\psi(\boldsymbol{r}, 0) = \lambda\, \delta(r - a)$ at $t = t_0 = 0$. In particular we determine the maximum value reached by the concentration at the centre.

By (8.3.20, 19),

$$\psi(\boldsymbol{r}, t) = \int \mathrm{d}^3 r' \frac{\exp\{-|\boldsymbol{r} - \boldsymbol{r}'|^2/4Dt\}}{(4\pi Dt)^{\frac{3}{2}}} \lambda\, \delta(r' - a). \tag{i}$$

Following the standard procedure described in Appendix F.1 (albeit for exponential rather than Gaussian integrands), we choose the polar axis along $\boldsymbol{r}$; then $|\boldsymbol{r} - \boldsymbol{r}'|^2 = r^2 + r'^2 - 2rr' \cos\theta'$, whence

$$\begin{aligned}\psi(\boldsymbol{r}, t) &= \frac{\lambda}{(4\pi Dt)^{\frac{3}{2}}} \int_0^\infty \mathrm{d}r'\, r'^2\, \delta(r' - a) \\ &\quad \times \int \mathrm{d}\Omega' \exp\left\{-\frac{r^2 + r'^2 - 2rr'\cos\theta'}{4Dt}\right\} \\ &= \frac{\lambda a^2}{(4\pi Dt)^{\frac{3}{2}}} \exp\left\{-\frac{r^2 + a^2}{4Dt}\right\} 2\pi \int_{-1}^{1} \mathrm{d}\cos\theta' \exp\left\{\frac{2ar\cos\theta'}{4Dt}\right\} \\ &= \frac{\lambda a^2}{(4\pi Dt)^{\frac{3}{2}}} \exp\left\{-\frac{r^2 + a^2}{4Dt}\right\} 2\pi \frac{4Dt}{2ar} 2\sinh\left\{\frac{ar}{2Dt}\right\},\end{aligned}$$

$$\psi(\boldsymbol{r}, t) = \frac{\lambda}{(\pi Dt)^{\frac{1}{2}}} \frac{a}{r} \sinh\left\{\frac{ar}{2Dt}\right\} \exp\left\{-\frac{r^2 + a^2}{4Dt}\right\}. \tag{ii}$$

Exercise: Verify explicitly that, as $t \to \infty$, (ii) behaves like $t^{-\frac{3}{2}}$.

As $r \to 0$, we have $(a/r)\sinh(ar/2Dt) \to a^2/2Dt$, which entails

$$\psi(0, t) = \frac{\lambda a^2}{\pi^{\frac{1}{2}}(Dt)^{\frac{3}{2}}} \exp\{-a^2/4Dt\}. \tag{iii}$$

This vanishes at $Dt = 0$, and again as $Dt \to \infty$; it peaks when $\mathrm{d}\psi(0, t)/\mathrm{d}(Dt) = 0$, which yields $Dt = Dt^* = a^2/6$. Hence the maximum concentration at the origin is

$$\psi(0, t^*) \equiv \psi_{\max} = (\lambda/a)(6^{\frac{3}{2}}/\pi^{\frac{1}{2}}) \exp(-3/2).$$

Notice that $\psi_{\max}$ is independent of D; D merely determines the time scale (e.g. the value of t^*).

Exercise: Instead of evaluating the integral (i) from scratch (as above), derive (ii) by exploiting the fact that ψ is given by a convolution. Represent ψ as a 3D Fourier integral, and use the 3D version of Theorem (C.5). One needs to evaluate only the Fourier transform of $\psi(\boldsymbol{r}, 0)$ (which is somewhat easier than the integral in (i)); then one appeals to the standard integrals

$$\int d\Omega \exp(i\boldsymbol{k}\cdot\boldsymbol{r}) = 4\pi \sin(kr)/(kr)$$

(see Appendix F.1) and

$$\int_0^\infty dx \exp(-x^2/b^2)\cos(kx) = (\pi^{\frac{1}{2}}b/2)\exp(-k^2b^2/4)$$

(see eqns (8.3.12)).

8.4 The inhomogeneous equation and the Green's function

8.4.1 The Green's function

In the absence of boundaries, the most general problem is to solve (8.1.1) with given $\rho(\boldsymbol{r}, t)$ and, as in Section 8.3, with given initial values $\psi(\boldsymbol{r}, t_0) = a(\boldsymbol{r})$. The Green's function designed to deliver this solution is defined by

$$LG_0 \equiv (\partial/\partial t - D\nabla^2)G_0(\boldsymbol{r}, t \mid \boldsymbol{r}', t') = \delta(t-t')\,\delta(\boldsymbol{r}-\boldsymbol{r}'), \tag{8.4.1a}$$

$$G_0 = 0 \quad \text{for} \quad t < t', \tag{8.4.1b}$$

$$G_0 \to 0 \quad \text{as} \quad r \to \infty. \tag{8.4.1c}$$

The physical reason for choosing the initial condition (8.4.1b) will appear later. By the argument applied to K_0 at the start of Section 8.3.2, G_0 too depends only on $\boldsymbol{R}$ and τ, since its definition (8.4.1) can be written

$$(\partial/\partial t - D\nabla_R^2)G_0(\boldsymbol{R}, \tau) = \delta(\tau)\,\delta(\boldsymbol{R}), \tag{8.4.2a}$$

$$G_0 = 0 \quad \text{for} \quad \tau < 0, \tag{8.4.2b}$$

$$G_0 \to 0 \quad \text{as} \quad R \to \infty. \tag{8.4.2c}$$

In fact K_0 fully determines G_0:

$$G_0(\boldsymbol{R}, \tau) = \theta(\tau)K_0(\boldsymbol{R}, \tau). \tag{8.4.3}$$

Proof: We show that θK_0 satisfies the defining relations (8.4.2). The initial condition (8.4.2b) is satisfied by virtue of the step-function. The boundary condition (8.4.2c) is satisfied because K_0 satisfies it (cf. the nD version of (8.3.17c)). To verify the differential equation (8.4.2a), one acts on θK_0 with L:

$$(\partial/\partial\tau - D\nabla_R^2)\theta(\tau)K_0(\boldsymbol{R},\tau) = \frac{\partial\theta}{\partial\tau}K_0 + \theta(\tau)(\partial/\partial\tau - D\nabla_R^2)K_0$$

$$= \delta(\tau)K_0(\boldsymbol{R},\tau) = \delta(\tau)K_0(\boldsymbol{R},0) = \delta(\tau)\,\delta(\boldsymbol{R}), \tag{8.4.4}$$

where we have appealed in turn to the product rule for the action of $\partial/\partial\tau$, and to the nD version of (8.3.17a, b). ■

When $\tau > 0$, G_0 and K_0 coincide; this leads to a more explicit form of the initial condition on G_0, obtained by taking the limit of (8.4.3) as $\tau \to 0+$:

$$\lim_{\tau\to 0+} G_0(\boldsymbol{R},\tau) = \lim_{\tau\to 0+} K_0(\boldsymbol{R},\tau) = K_0(\boldsymbol{R},0) = \delta(\boldsymbol{R}). \tag{8.4.5a}$$

The same conclusion follows from integrating (8.4.2a) with respect to τ from $0-$ to $0+$ (at fixed $\boldsymbol{R}$). By the principles of Section 2.1.3, only the term $\partial G_0/\partial\tau$ contributes on the left:

$$\int_{0-}^{0+} d\tau \frac{\partial G_0}{\partial\tau} = G_0(\boldsymbol{R},0+) - G_0(\boldsymbol{R},0-)$$

$$= G_0(\boldsymbol{R},0+) = \int_{0-}^{0+} d\tau\,\delta(\tau)\,\delta(\boldsymbol{R}) = \delta(\boldsymbol{R}), \tag{8.4.5b}$$

where the last step relies on (8.4.2b).

For the general theory in Chapter 9 it will be essential that the proof of (8.4.3) applies provided the boundary conditions on K_0 and G_0 are the same, but irrespective of what these conditions actually are. Thus the corresponding relation $G(\boldsymbol{r},t\,|\,\boldsymbol{r}',t') = \theta(t-t')K(\boldsymbol{r},t\,|\,\boldsymbol{r}',t')$ persists for finite region V, where the unbounded-space conditions (8.3.3c) and (8.4.1c) are replaced by homogeneous DBCs or NBCs on the surface S bounding V. Similarly, the initial conditions (8.4.5) persist in finite regions:

$$\lim_{t\to t'+} G(\boldsymbol{r},t\,|\,\boldsymbol{r}',t') = \lim_{t\to t'+} K(\boldsymbol{r},t\,|\,\boldsymbol{r}',t')$$

$$= K(\boldsymbol{r},t'\,|\,\boldsymbol{r}',t') = \delta(\boldsymbol{r}-\boldsymbol{r}').$$

Although the crucial relation (8.4.3) may seem to have been pulled out of a hat, in Appendix F.5 we derive it straightforwardly, i.e. without any preliminary guesswork, through an $(n+1)$-fold Fourier integral representation of G_0. More to the point, it is suggested by the following physical argument.

The function G describes the distribution resulting from the injection of one particle (or of unit amount of heat) at time t' at the source point

$\boldsymbol{r}'$, no particles having been present previously. On the other hand, K describes the distribution resulting from one particle known to have been at $\boldsymbol{r}'$ at time t'. Clearly, for times $t > t'$, these two distributions are the same: hence the relation $G = \theta(t - t')K$.

8.4.2 The magic rule in unbounded space

Suppose that there were no particles at time $t = -\infty$, and that a prescribed source distribution ρ has been operating since. Then, simply because the diffusion equation is linear, we have

$$\psi(\boldsymbol{r}, t) = \int_{-\infty}^{t+} \mathrm{d}t' \int \mathrm{d}V' \, G_0(\boldsymbol{r}, t \mid \boldsymbol{r}', t')\rho(\boldsymbol{r}', t'). \tag{8.4.6}$$

In other words, $\psi(\boldsymbol{r}, t)$ is just the linear combination of the contributions from all the elementary sources $\rho(\boldsymbol{r}', t') \, \mathrm{d}V' \, \mathrm{d}t'$ that have ever acted. The upper limit $t+$ means that, whenever this makes any difference, we evaluate the integral up to $t' = t + \varepsilon$, and then take the limit $\varepsilon \to 0+$.

Proof:

$$L\psi = \int_{-\infty}^{t+} \mathrm{d}t' \int \mathrm{d}V' \, [LG_0]\rho(\boldsymbol{r}', t')$$

$$= \int_{-\infty}^{t+} \mathrm{d}t' \int \mathrm{d}V \, \delta(t - t') \, \delta(\boldsymbol{r} - \boldsymbol{r}')\rho(\boldsymbol{r}', t') = \rho(\boldsymbol{r}, t). \qquad \blacksquare$$

In fact, except for some delicate arguments like this proof, the $t+$ prescription is not required, and generally we shall write the upper limit simply as t instead of $t+$.

For instance, if no sources act before some initial time t_0 (or if no account need be taken of them, because their effects are subsumed into a known initial distribution $\psi(\boldsymbol{r}, t_0)$ which is catered for separately), then in this more relaxed notation (8.4.6) may be written

$$\psi(\boldsymbol{r}, t) = \int_{t_0}^{t} \mathrm{d}t' \int \mathrm{d}V' \, G_0(\boldsymbol{r}, t \mid \boldsymbol{r}', t')\rho(\boldsymbol{r}', t')$$

$$= \int_{t_0}^{t} \mathrm{d}t' \int \mathrm{d}V' \, K_0(\boldsymbol{r}, t \mid \boldsymbol{r}', t')\rho(\boldsymbol{r}', t') \equiv f_0(\boldsymbol{r}, t). \tag{8.4.7}$$

G_0 can be replaced by K_0 as in the second step, because, once it is safe to replace the upper limit $t+$ by t, we have $t \geqslant t'$, whence $G_0 = K_0$. As before, f_0 stands for the integrals regarded as explicit constructs from the data ρ.

The general inhomogeneous problem can now be solved immediately, by combining the solutions (8.3.20) of the homogeneous equation for given $\psi(\boldsymbol{r}, t_0)$ with the solution (8.4.7) of the inhomogeneous equation with given ρ. This yields the magic rule in the form appropriate to unbounded space:

$$\psi(\mathbf{r}, t) = f_0(\mathbf{r}, t) + h_0(\mathbf{r}, t)$$
$$= \int_{t_0}^{t} \mathrm{d}t' \int \mathrm{d}V'\, G_0(\mathbf{r}, t \mid \mathbf{r}', t')\rho(\mathbf{r}', t')$$
$$+ \int \mathrm{d}V'\, G_0(\mathbf{r}, t \mid \mathbf{r}', t_0)\psi(\mathbf{r}', t_0), \tag{8.4.8}$$

where G_0 may be replaced by K_0 in either term or in both.

Proof: (i) $L(f_0 + h_0) = \rho$, because $Lf_0 = \rho$ as proved just above, while $Lh_0 = 0$, as proved for (8.3.4).

(ii) The initial conditions are satisfied, because $h_0(\mathbf{r}, t_0) = \psi(\mathbf{r}, t_0)$ as proved for (8.3.4), while $f_0(\mathbf{r}, t_0) = 0$, because, as $t \to t_0$, the t' integration region shrinks to zero. In that case, an integral can remain non-zero only if the integrand contains a delta-function; but G_0 itself, unlike $\partial G_0/\partial t$, contains no delta-functions. ■

The integrands of f_0 and h_0 can be rewritten in Laplace's form (8.3.15b) as regards their dependence on $\mathbf{r}$ and $\mathbf{r}'$. The integrand of f_0 can be rearranged analogously as regards its dependence on t and t':

$$f_0(\mathbf{r}, t) = \int_{t_0}^{t} \mathrm{d}t' \int \mathrm{d}V' \frac{\exp\{-|\mathbf{r} - \mathbf{r}'|^2/4D(t-t')\}}{[4\pi D(t-t')]^{n/2}} \rho(\mathbf{r}', t'), \tag{8.4.9a}$$

$$= \int_{0}^{t-t_0} \mathrm{d}\tau \int \mathrm{d}V' \frac{\exp\{-|\mathbf{r} - \mathbf{r}'|^2/4D\tau\}}{[4\pi D\tau]^{n/2}} \rho(\mathbf{r}', t-\tau). \tag{8.4.9b}$$

Examples of these formulae appear elsewhere. Appendix J shows how, in the limit $t_0 \to -\infty$, an expression similar to (8.4.9b) with a simple-harmonic point source reproduces the simple-harmonic solutions of Section 8.2 (cf. also Section 9.3.2). A more interesting example of f_0 is worked through in the next section. An example of h_0 was worked through in Section 8.3.4.

8.5 Some remarkable consequences of the dimensionality of space

8.5.1 Introduction

To illustrate the magic rule (8.4.8) we consider the temperature $\psi(\mathbf{r}, t)$ at times $t \geqslant 0$, for the case that $\psi(\mathbf{r}, 0) = 0$, and that a point source

$$\rho(\mathbf{r}, t) = W\theta(t)\,\delta(\mathbf{r}) \tag{8.5.1}$$

is switched on at $t = 0$, remaining steady thereafter. Thus, only the integral f_0 contributes. We solve the problem in 1D, 2D, and 3D. At the outset, one might reasonably expect that as $t \to \infty$ for fixed $\mathbf{r}$, $\psi(\mathbf{r}, t)$ should approach the steady-state solutions $\psi_{\text{stat}}(\mathbf{r})$ appropriate to a point source $\rho(\mathbf{r}) = W\,\delta(\mathbf{r})$ that has been active every since time $-\infty$. Accordingly we shall pay special attention to $\psi(\mathbf{r}, t)$ in the approximation appropriate to large t.

For reference we quote the steady-state solutions (cf. Section 4.4.2):

$$\psi^{(1)}_{\text{stat}} = A - \frac{W}{2D} r, \tag{8.5.2}$$

$$\psi^{(2)}_{\text{stat}} = A - \frac{W}{2\pi D} \log r, \tag{8.5.3}$$

$$\psi^{(3)}_{\text{stat}} = \frac{W}{D}\frac{1}{4\pi r}, \tag{8.5.4}$$

where the A are arbitrary constants. In 1D, $r \equiv |x|$.

According to the magic rule, $\psi = f_0$ is given by (8.4.7) with $t_0 = 0$ and with ρ from (8.5.1):

$$\psi^{(n)}(\mathbf{r}, t) = \int_0^t \mathrm{d}\tau \int \mathrm{d}V' \frac{\exp\{-|\mathbf{r} - \mathbf{r}'|^2/4D\tau\}}{(4\pi D\tau)^{n/2}} W\,\delta(\mathbf{r}')$$

$$= W \int_0^t \mathrm{d}\tau \exp\{-r^2/4D\tau\}/(4\pi D\tau)^{n/2}. \tag{8.5.5}$$

Change the integration variable to ξ, where $\xi^2 = r^2/4D\tau$, $\tau = r^2/4D\xi^2$, and define

$$\alpha \equiv (r^2/4Dt)^{\frac{1}{2}}; \tag{8.5.6}$$

this yields

$$\psi^{(n)}(\mathbf{r}, t) = W \int_{(r^2/4Dt)^{\frac{1}{2}}}^{\infty} \frac{r^2}{4D} \frac{2\,\mathrm{d}\xi}{\xi^3} \exp(-\xi^2) \frac{1}{(\pi r^2/\xi^2)^{n/2}},$$

$$\psi^{(n)}(\mathbf{r}, t) = \frac{W}{2D}\left(\frac{1}{\pi}\right)^{n/2} r^{2-n} \int_\alpha^\infty \mathrm{d}\xi \frac{\exp(-\xi^2)}{\xi^{3-n}}. \tag{8.5.7}$$

Since we are interested mainly in $t \to \infty$, we think of α as small.

8.5.2 One dimension

Setting $n = 1$ in (8.5.7), we find

$$\psi^{(1)}(r, t) = \frac{W}{2D} \frac{r}{\pi^{\frac{1}{2}}} J_2(\alpha), \tag{8.5.8}$$

$$J_2(\alpha) \equiv \int_\alpha^\infty \mathrm{d}\xi \exp(-\xi^2)/\xi^2. \tag{8.5.9}$$

To determine the behaviour of J_2 (which is related to the error function) as $\alpha \to 0$, we integrate by parts:

$$J_2(\alpha) = -\frac{1}{\xi}\exp(-\xi^2)\Big|_\alpha^\infty - \int_\alpha^\infty d\xi\left(-\frac{1}{\xi}\right)(-2\xi\exp(-\xi^2))$$
$$= \frac{1}{\alpha}\exp(-\alpha^2) - 2\int_\alpha^\infty d\xi \exp(-\xi^2)$$
$$= \frac{1}{\alpha}\exp(-\alpha^2) - 2\int_0^\infty d\xi\exp(-\xi^2) + 2\int_0^\alpha d\xi\exp(-\xi^2). \tag{8.5.10}$$

The first term is

$$\alpha^{-1}(1-\alpha^2+\alpha^4/2!+\cdots) = \alpha^{-1} - \alpha + O(\alpha^3).$$

The middle term, where the integral is standard, is $(-2)(\frac{1}{2}\pi^{\frac{1}{2}}) = -\pi^{\frac{1}{2}}$. In the third term, the integrand can be expanded in powers of ξ^2, giving

$$2\int_0^\alpha d\xi(1-\xi^2/2!+\cdots) = 2\alpha + O(\alpha^3).$$

Thus the three terms combine into

$$J_2(\alpha) = \frac{1}{\alpha} - \pi^{\frac{1}{2}} + \alpha + O(\alpha^3). \tag{8.5.11}$$

Substituting into (8.5.8), using (8.5.6), and restoring $r = |x|$, we find

$$\psi^{(1)}(x, t\to\infty) = \frac{W}{2D}\frac{|x|}{\pi^{\frac{1}{2}}}\left\{\left(\frac{4Dt}{x^2}\right)^{\frac{1}{2}} - \pi^{\frac{1}{2}} + \left(\frac{x^2}{4Dt}\right)^{\frac{1}{2}} + \cdots\right\}$$
$$= W\left(\frac{t}{D\pi}\right)^{\frac{1}{2}} - \frac{W}{2D}|x| + \frac{W}{2D}\frac{x^2}{(4\pi Dt)^{\frac{1}{2}}} + \cdots. \tag{8.5.12}$$

Remarkably, the first term depends only on t but not on x, while the second term depends only on x but not on t. The first term dominates, and shows that at fixed x the temperature goes on rising indefinitely: no steady state is ever approached, not even asymptotically. Thus $\psi^{(1)}_{\text{stat}}(x)$ by itself would certainly be quite insufficient to predict $\psi^{(1)}(x, t\to\infty)$. Hindsight shows that $\psi^{(1)}_{\text{stat}}$ tallies with the leading x-dependent term (the second term in (8.5.12), while the constant A there has in fact turned into an indefinitely increasing function of time, $A = W(t/D\pi)^{\frac{1}{2}}$.

On the other hand, the heat flux, being proportional to grad ψ, is dominated by the second term in (8.5.12). For $x > 0$ we have

$$j(x, t\to\infty) = -D\frac{\partial\psi}{\partial x} \approx \frac{W}{2D} + \frac{W}{2D}\left(\frac{x^2}{\pi Dt}\right)^{\frac{1}{2}} + \cdots, \tag{8.5.13}$$

where the leading term does reproduce the steady-state flux $j_{\text{stat}} = -D\,\partial\psi_{\text{stat}}/\partial x$.

One could describe these results by saying that one-dimensional space is so cramped that the heat from our point source simply cannot disperse fast enough for a steady temperature to be approached at any given point. From this point of view it becomes interesting to investigate the 2D and 3D versions of the same problem.

8.5.3 Two dimensions

Setting $n = 2$ in (8.5.7) one obtains

$$\psi^{(2)}(\mathbf{r}, t) = \frac{W}{2D}\frac{1}{\pi}J_1(\alpha), \tag{8.5.14}$$

$$J_1(\alpha) \equiv \int_\alpha^\infty d\xi \exp(-\xi^2)/\xi. \tag{8.5.15}$$

J_1 is related not to the error function but to the exponential integral. Changing the integration variable to $\eta \equiv \xi^2$, and integrating by parts, we find

$$\begin{aligned}
J_1(\alpha) &= \frac{1}{2}\int_{\alpha^2}^\infty d\eta \exp(-\eta)/\eta \\
&= \frac{1}{2}\left\{\log\eta\cdot\exp(-\eta)\Big|_{\alpha^2}^{\infty} - \int_{\alpha^2}^\infty d\eta\,(\log\eta)(-\exp(-\eta))\right\} \\
&= \frac{1}{2}\left\{-\exp(-\alpha^2)\log\alpha^2 + \int_{\alpha^2}^\infty d\eta\,(\log\eta)\exp(-\eta)\right\} \\
&= \frac{1}{2}\left\{-\exp(-\alpha^2)\log\alpha^2 + \int_0^\infty d\eta\,(\log\eta)\exp(-\eta)\right. \\
&\qquad \left. - \int_0^{\alpha^2} d\eta\,(\log\eta)\exp(-\eta)\right\}.
\end{aligned} \tag{8.5.16}$$

As $\alpha \to 0$, we approximate the first term by $-\frac{1}{2}\log\alpha^2$. The integral in the middle term is just a number, and in fact equals $-\gamma$, where $\gamma \approx 0.577$ is Euler's constant. The third term vanishes as $\alpha \to 0$. Thus

$$J_1(\alpha) = -\tfrac{1}{2}[\log\alpha^2 + \gamma] + (\text{terms that vanish as } \alpha \to 0) \tag{8.5.17}$$

Accordingly,

$$\psi^{(2)}(\mathbf{r}, t \to \infty) = \frac{W}{2D}\frac{1}{2\pi}\left\{\log\left(\frac{4Dt}{r^2}\right) - \gamma + \cdots\right\} \tag{8.5.18a}$$

$$= \frac{W}{4\pi D}[\log(4Dt) - \gamma] - \frac{W}{2\pi D}\log r + \cdots, \tag{8.5.18b}$$

which should be compared with $\psi_{stat}^{(2)}$ in (8.5.3). As in 1D, the dominant r-dependent term tallies with ψ_{stat}, but the constant A has again become an indefinitely rising function of time. The rise at fixed r is now only logarithmic in t, i.e. much slower than the $t^{\frac{1}{2}}$-proportional rise in 1D.

8.5.4 Three dimensions

Setting $n = 3$ in (8.5.7) one obtains

$$\psi^{(3)}(\mathbf{r}, t) = \frac{W}{2D}\left(\frac{1}{\pi}\right)^{\frac{3}{2}} \frac{1}{r} J_0(\alpha), \tag{8.5.19}$$

$$J_0(\alpha) \equiv \int_\alpha^\infty \mathrm{d}\xi \exp(-\xi^2); \tag{8.5.20}$$

$$J_0(\alpha) = \int_0^\infty \mathrm{d}\xi \exp(-\xi^2) - \int_0^\alpha \mathrm{d}\xi\,(1 - \xi^2 + \xi^4/2! + \cdots)$$

$$= \tfrac{1}{2}\pi^{\frac{1}{2}} - \alpha + O(\alpha^3); \tag{8.5.21}$$

$$\psi^{(3)}(\mathbf{r}, t \to \infty) = \frac{W}{2D}\frac{1}{\pi^{\frac{3}{2}}}\frac{1}{r}\left\{\tfrac{1}{2}\pi^{\frac{1}{2}} - \left(\frac{r^2}{4Dt}\right)^{\frac{1}{2}} + \cdots\right\}$$

$$= \frac{W}{4\pi D r}\left\{1 - \left(\frac{r^2}{\pi D t}\right)^{\frac{1}{2}} + \cdots\right\}. \tag{8.5.22}$$

Thus three dimensions, unlike two or one, are spacious enough to allow heat from our point source to disperse as fast as is necessary to realize the naive expectation that $\lim_{t\to\infty} \psi(\mathbf{r}, t) = \psi_{stat}(\mathbf{r})$.

Exercise: Pretending that the dimensionality can be varied continuously, show that $\lim_{t\to\infty} \psi^{(n)}(\mathbf{r}, t)$ exists if and only if $n > 2$.

8.5.5 Dimensional analysis

It is an interesting question to what extent one could have foreseen the discrepancies between $\psi_{stat}(\mathbf{r})$ and $\psi(\mathbf{r}, t \to \infty)$ on purely dimensional grounds. If terms independent of $\mathbf{r}$, like the constants A in (8.5.2, 3), are to emerge from the integral f_0, then they must do so in the form

$$A^{(n)} = \beta \cdot W \cdot D^a \cdot t^b, \tag{8.5.23}$$

where β is a pure number, and where proportionality to W follows automatically from the linearity of the problem. The concentration ψ represents number per unit volume; since A is an addend in ψ, its dimensions are $[A] = [\psi] = L^{-n}$. In (8.5.1), the source density ρ represents number per unit volume per unit time, $[\rho] = L^{-n}T^{-1}$, while $[\delta(\mathbf{r})] = L^{-n}$. Hence $[\rho] = L^{-n}T^{-1} = [W][\delta(\mathbf{r})] = [W]L^{-n}$, which yields

$[W] = T^{-1}$. Finally, $[D] = L^2T^{-1}$. Thus, matching dimensions in (8.5.23) leads to

$$L^{-n} = T^{-1}(L^2T^{-1})^a T^b \Rightarrow a = -n/2, \qquad b = (1 - n/2).$$

Substituting into (8.5 23), we obtain

$$A^{(n)} = \beta \cdot W \cdot D^{-n/2} \cdot t^{1-n/2}, \tag{8.5.24}$$

$$A^{(1)} = \beta W (t/D)^{\frac{1}{2}}, \qquad A^{(2)} = \beta W D^{-1}. \tag{8.5.25a,b}$$

In 1D, it is evident from (8.5.25a) that the 'constant' $A^{(1)}$ must in fact be proportional to $t^{\frac{1}{2}}$, which is just what we found in (8.5.12). By contrast, in 2D a genuine (time-independent) constant $A^{(2)}$ would be admissible as far as (8.5.25b) is concerned; in other words, purely dimensional reasoning fails to give us any inkling of the actual logarithmically intertwined dependence of the leading terms on t and r through their dimensionless combination Dt/r^2.

Problems

8.1 Solve $(\partial/\partial t - D\,\partial^2/\partial x^2)\psi = 0$, given the initial conditions

(i) $\psi(x, 0) = x$;
(ii) $\psi(x, 0) = \cos(kx)$;
(iii) $\psi(x, 0) = \frac{1}{2}\varepsilon(x)$.

Hint: In (i, ii) use common sense first, and then check that the magic rule gives the same result albeit with more effort.

8.2 The temperature ψ in an unbounded medium of conductivity D and of unit heat capacity per unit volume obeys the diffusion equation $(\partial/\partial t - D\nabla^2)\psi = \rho$, and is given in spherical polar coordinates by

$$\psi(\boldsymbol{r}, t) = A\cos(\omega t)\exp(-\lambda r).$$

(i) What is the heat $\rho(\boldsymbol{r}, t)$ generated per unit volume per unit time?
(ii) What is the total heat $W(t) = \int \mathrm{d}V\rho$ generated per unit time in the system?
(iii) Verify that your expression for W vanishes in the steady state where $\omega = 0$. Show without reference to any explicit expression for W that this must be so, by considering only the total outflow of heat through a sphere of arbitrarily large radius.

8.3 As in Problem 8.2, but with

$$\psi(\boldsymbol{r}, t) = B\cos(\omega t)\exp(-\lambda r)/r.$$

(i) Show that the heat-source density $\rho(\boldsymbol{r}, t)$ is of the form $\rho = f(t)\exp(-\lambda r)/r + g(t)\,\delta(\boldsymbol{r})$, and determine f and g. (Verify that there were no $\delta(\boldsymbol{r})$-proportional sources in Problem 8.2.)
(ii) In the steady state $\omega = 0$, show that $W = \int \mathrm{d}V\,\rho$ vanishes.
(iii) Verify that this must be so, along the lines of Problem 8.2 (iii).

Notice that, if you had overlooked the $\delta(\boldsymbol{r})$-proportional component of ρ in (i), then there would be a contradiction between your conclusions from (ii) and (iii).

(*Note:* Problems 8.4–8.7 form a progression.)

8.4 *Solutions by integration.* Because the propagator K_0 obeys $LK_0 = 0$, and because $[L, \partial/\partial x] = 0 = [L, \partial/\partial t]$, it is obvious that $LK_{0x} = 0$ and $LK_{0t} = 0$ (see eqn (4.3.7)). It is natural to ask whether other solutions can be generated by integrating rather than differentiating K_0.

(i) Show that $\alpha(x, t, a) \equiv \int_a^x \mathrm{d}x'\, K_0(x', t)$ obeys $L\alpha = 0$ if and only if

K_{0x} vanishes at $x = a$, i.e. if and only if $a = 0$ or $a = \infty$. This yields two solutions A and B of the homogeneous equation, which are not however linearly independent:

$$A(x, t) \equiv \int_0^x \mathrm{d}x' \, K_0(x', t), \qquad B(x, t) \equiv \int_x^\infty \mathrm{d}x' \, K_0(x', t).$$

Verify that $A + B = 1/2$.

(ii) Show that

$$A(x, t) = \pi^{-\frac{1}{2}} \int_0^{x/(4Dt)^{\frac{1}{2}}} \mathrm{d}\xi \, \exp(-\xi^2) \equiv 2 \operatorname{erf}\left(\frac{x}{(4Dt)^{\frac{1}{2}}}\right).$$

Here erf is the standard error function.

(iii) Obviously one should ask what IVP $A(x, t)$ is the solution of. Show that $A(x, 0) = \frac{1}{2}\varepsilon(x)$, and verify that $A(x, t)$ indeed agrees with the solution of Problem 8.1(iii).

8.5 Consider the function

$$\gamma(x, t, b) \equiv \int_b^t \mathrm{d}t' \, K_0(x, t').$$

(i) Show that $L\gamma = K_0(x, b)$. Therefore no choice of b can make $L\gamma$ vanish for all x. But for $b = 0$ we see that $C(x, t) \equiv \int_0^t \mathrm{d}t' \, K_0(x, t')$ obeys the *inhomogeneous* equation

$$LC(x, t) = \delta(x), \qquad (t > 0).$$

(ii) Show that

$$C(x, t) = \frac{|x|}{2D\pi^{\frac{1}{2}}} \int_{|x|/(4Dt)^{\frac{1}{2}}}^\infty \frac{\mathrm{d}\xi}{\xi^2} \exp(-\xi^2).$$

(Notice that C is an even function of x from its definition, because K_0 is. The modulus signs merely reflect this property.)

(iii) What problem is $C(x, t)$ the solution of? Show that

$$\lim_{t \to 0+} C(x, t) = 0 \qquad (\text{for all } x \neq 0);$$

$$\lim_{x \to 0} C(x, t) = 0 \qquad (\text{for all } t > 0).$$

8.6 The IVP $L\psi = 0$, $\psi(x, 0) = \frac{1}{2}\psi_0 \cdot \varepsilon(x)$ is solved by $\psi = \psi_0 \cdot A(x, t)$, as shown in Problems 8.1(iii) and 8.4. When restricted to the halfspace $x \geqslant 0$, ψ obviously solves the problem $L\psi = 0$, $\psi(x, 0) = \frac{1}{2}\psi_0$, but subject to the BC $\psi(0, t) = 0$ for $t > 0$. (This is a special case of a general class of BVPs explored in Section 9.4.4 by the method of images.)

(i) Show that the flux across the origin is

$$j(0, t) \equiv -D\psi_x(0, t) = -\psi_0(D/4\pi t)^{\frac{1}{2}}.$$

(ii) Show that the total transfer across the fixed point x up to time t is

$$Q(x, t) \equiv \int_0^t \mathrm{d}t\, j(x, t) = -\psi_0 DB(x, t).$$

8.7 *Kelvin's estimate of the age of the Earth.* Let a flat Earth be represented by the halfspace $x \geqslant 0$. Assume that it solidified at a temperature ψ_0 (the melting-point of rocks) at time $t = 0$, and that its surface temperature immediately dropped to zero (approximately that of outer space). Let c be the heat-capacity per unit volume, and κ the thermal conductivity, so that $D = \kappa/c$.

(i) Show that at time t, the heat flux out across the surface is

$$j_{\text{heat}}(0, t) = c\psi_0(D/\pi t)^{\frac{1}{2}}.$$

(ii) The measured value of j_{heat} now, averaged over the surface of the Earth, is $0.06\ \text{W m}^{-2}$. Take numerical values roughly appropriate to SiO_2, namely $\psi_0 \approx 2000°\text{K}$; $\kappa \approx 2\ \text{W m}^{-1}\,\text{K}^{-1}$; (heat capacity per kg) $\approx 10^3\ \text{J K}^{-1}$; density $\approx 2.7 \times 10^3\ \text{kg m}^{-3}$. Estimate the age of the Earth. (The result is of the order 6×10^7 years. Geologists know that the true age of the Earth is close to 5×10^9 years.)
(iii) Now take into account the fact that the Earth is a sphere of radius $R \approx 6.4 \times 10^6$ m, having finite surface area $4\pi R^2$ and finite volume $4\pi R^3/3$. Continuing nevertheless to use the estimates of heat flux from a flat surface as above, estimate what fraction of the initial heat content (on solidification) has been lost.

8.8 In the 1D switched-on source problem of Sections 8.5.1, 2, determine

(i) the temperature of the heater, i.e. $\psi(0, t)$ as a function of time;
(ii) the temperature distribution $\psi(x, t)$ at small x. Explain over what range of x your estimate applies for given t.

Hint: No calculations should be needed beyond those in Section 8.5.

8.9 *Phase-controlled line heater.* An infinitely long line source extends along the z-axis. It is simple-harmonic in time, but its phase can be controlled:

$$\rho(\mathbf{r}, t) = Q \cos(\omega t + \phi(z))\, \delta(x)\, \delta(y).$$

(i) How should $\phi(z)$ be chosen so as to maximize the amplitude of the temperature oscillations at a distance r from the axis?
(ii) Verify that, with this optimal choice of phase,

$$\psi(r, t) = \frac{Q}{2\pi D} \cos(\omega t) \int_1^\infty \frac{d\xi}{(\xi^2 - 1)^{\frac{1}{2}}} \exp\left[-\xi\left(\frac{\omega r^2}{2D}\right)^{\frac{1}{2}}\right]$$

$$\equiv \frac{Q}{2\pi D} \cos(\omega t) K_0\left[\left(\frac{\omega r^2}{2D}\right)^{\frac{1}{2}}\right].$$

(iii) For large argument, the Bessel function K_0 behaves like $K_0(\zeta \to \infty) \sim (\pi/2\zeta)^{\frac{1}{2}} \exp(-\zeta)$. If the phase is changed to $\phi = 0$, show that, in the asymptotic region $r \gg (2D/\omega)^{\frac{1}{2}}$, the amplitude drops only by a factor $2^{-\frac{1}{4}}$, but that the temperature now oscillates proportionately to $\cos(\omega t - \delta)$, where $\delta = \pi/4 + r(\omega/2D)^{\frac{1}{2}}$.

Hint: Use a complex representation $\cos(\omega t + \phi) = \text{Re} \exp[i(\omega t + \phi)]$ and start with the magic rule; $\psi(r, t)$ will then emerge as an integral over z, whose integrand can be written down at once by careful comparison with eqns (J.1, 7).

9 The diffusion equation: II. General theory

Here, as throughout this book, we deal only with regions V that are fixed in space, so that their bounding surfaces S are stationary. Unbounded space as considered in Chapter 8 is a special limiting case, with S arbitrarily far away in all directions. For the diffusion equation this limit is straightforward, because all diffusive effects originating from a finite region automatically vanish at infinity. (By contrast, we shall see later that the asymptotics of the wave equation are far more subtle.)

9.1 Boundary conditions, initial conditions, and uniqueness

It was asserted in Section 3.54 that the problem of solving the diffusion equation is 'well-posed' under ICs prescribing $\psi(\boldsymbol{r}, t_0)$ throughout V at the starting time t_0, plus BCs prescribing either $\psi_S(\boldsymbol{r}, t')$ or $\partial_n\psi_S(\boldsymbol{r}, t')$ (but not both) everywhere on the closed surface S bounding V, and for all $t_0 \leqslant t' \leqslant t$. (See Appendix G about our nomenclature for initial and for boundary values.) Under mixed BCs, ψ_S is prescribed over part of S and $\partial_n\psi_S$ over the rest; purely for simplicity we consider only cases where on any given part of S one prescribes either ψ (for all t') or $\partial_n\psi$ (for all t').

We show that under such conditions the solution is unique. (Existence and stability will be taken for granted.) The argument can be compared with the uniqueness proof for Poisson's equation in Section 4.2; the basic idea is similar, but the details are rather different, not simply because of the extra variable (the time), but because the structure of the equation (in hyperspace) is different, i.e. parabolic rather than elliptic.

Proof: Consider two functions ψ_1 and ψ_2, both of which satisfy all the requirements of the problem. Thus $L\psi_1 = \rho$, $L\psi_2 = \rho$, $\psi_1(\boldsymbol{r}, t_0) = \psi(\boldsymbol{r}, t_0) = \psi_2(\boldsymbol{r}, t_0)$; and, when $\boldsymbol{r}$ is on S, either $\psi_1(\boldsymbol{r}, t) = \psi_S(\boldsymbol{r}, t) = \psi_2(\boldsymbol{r}, t)$, or $\partial_n\psi_1(\boldsymbol{r}, t) = \partial_n\psi_S(\boldsymbol{r}, t) = \partial_n\psi_2(\boldsymbol{r}, t)$. We establish uniqueness by showing that the difference

$$\phi(\boldsymbol{r}, t) \equiv \psi_1(\boldsymbol{r}, t) - \psi_2(\boldsymbol{r}, t) \tag{9.1.1}$$

vanishes always and everywhere. Observe that ϕ obeys the *homogeneous* equation

$$L\phi = (\partial/\partial t - D\nabla^2)\phi = 0, \tag{9.1.2}$$

since $L\psi_1 - L\psi_2 = \rho - \rho = 0$, and that it satisfies *homogeneous* ICs and BCs, since $\phi(\boldsymbol{r}, t_0) = \psi_1(\boldsymbol{r}, t_0) - \psi_2(\boldsymbol{r}, t_0) = 0$, and $\phi_S(\boldsymbol{r}, t) = \psi_{1S}(\boldsymbol{r}, t) -$

$\psi_{2S}(\mathbf{r}, t) = 0$, or similarly for the normal derivative. The crucial idea is to consider the integral

$$J \equiv \int_{t_0}^{T} \mathrm{d}t \int_V \mathrm{d}V\, \phi(\partial/\partial t - D\nabla^2)\phi = 0, \tag{9.1.3}$$

where T is any time later than t_0. J vanishes by (9.1.2).

The first term on the left is

$$J_1 \equiv \int_{t_0}^{T} \mathrm{d}t \int_V \mathrm{d}V \frac{1}{2}\frac{\partial}{\partial t}\phi^2$$
$$= \frac{1}{2}\int_{t_0}^{T} \mathrm{d}t \frac{\mathrm{d}}{\mathrm{d}t}\int_V \mathrm{d}V\, \phi^2(\mathbf{r}, t) = \frac{1}{2}\int_V \mathrm{d}V\, \phi^2(\mathbf{r}, T), \tag{9.1.4}$$

since the contribution from the lower limit of the time integral vanishes because $\phi(\mathbf{r}, t_0) = 0$. Notice that J_1 cannot be negative.

In the second term on the left of (9.1.3), we recast the space integral:

$$\int_V \mathrm{d}V\, \phi\nabla^2\phi = \int_V \mathrm{d}V\, \{\mathrm{div}\,(\phi\, \mathrm{grad}\, \phi) - (\mathrm{grad}\, \phi)^2\}$$
$$= \int_S \mathrm{d}S\, \phi\, \partial_n\phi - \int_V \mathrm{d}V\, (\mathrm{grad}\, \phi)^2 = -\int_V \mathrm{d}V\, (\mathrm{grad}\, \phi)^2;$$

the surface integral vanishes because either ϕ_S or $\partial_n\phi_S$ vanishes (for all t). Thus the second term in J is

$$J_2 = \int_{t_0}^{T} \mathrm{d}t\, D \int_V \mathrm{d}V\, (\mathrm{grad}\, \phi)^2. \tag{9.1.5}$$

Notice that J_2, like J_1, is non-negative.

Finally we combine J_1 and J_2:

$$J = J_1 + J_2 = \frac{1}{2}\int_V \mathrm{d}V\, \phi^2(\mathbf{r}, T) + D\int_{t_0}^{T} \mathrm{d}t \int_V \mathrm{d}V\, (\mathrm{grad}\, \phi(\mathbf{r}, t))^2 = 0. \tag{9.1.6}$$

Since both integrals are non-negative, they cannot cancel each other, and each must vanish separately. By the same token, since the integrands are non-negative, they must vanish for all values of their arguments. In particular, $\phi(\mathbf{r}, T)$ must vanish everywhere in V for all times $T > t_0$. ■

9.2 Propagator and Green's function

9.2.1 Definitions

The propagator K is defined by

$$LK \equiv (\partial/\partial t - D\nabla^2)K(\mathbf{r}, t\,|\,\mathbf{r}', t') = 0 \quad \text{for} \quad t > t', \tag{9.2.1a}$$

$$K(\mathbf{r}, t'\,|\,\mathbf{r}', t') = \delta(\mathbf{r} - \mathbf{r}'), \quad \text{(equal times)}, \tag{9.2.1b}$$

plus the *homogeneous* version of the BCs imposed on ψ in the actual problem:

$$K(\mathbf{r}, t \mid \mathbf{r}', t') = 0 \quad \text{or} \quad \partial_n K(\mathbf{r}, t \mid \mathbf{r}', t') = 0, \quad \text{for } \mathbf{r} \text{ on } S. \tag{9.2.1c}$$

As a rule we do not now introduce suffices D or N to label the two cases explicitly, especially because mixed BCs are much more common with the diffusion than with Poisson's equation; i.e. K might vanish on parts of S while $\partial_n K$ vanishes on the other parts.

The Green's function G is defined by

$$LG(\mathbf{r}, t \mid \mathbf{r}', t') = \delta(t - t')\, \delta(\mathbf{r} - \mathbf{r}'), \tag{9.2.2a}$$

$$G(\mathbf{r}, t \mid \mathbf{r}', t') = 0 \quad \text{for} \quad t < t', \tag{9.2.2b}$$

$$G(\mathbf{r}, t \mid \mathbf{r}', t') = 0 \quad \text{or} \quad \partial_n G(\mathbf{r}, t \mid \mathbf{r}', t') = 0, \quad \text{for } \mathbf{r} \text{ on } S, \tag{9.2.2c}$$

with the same BCs for G as for K.

These defining relations should be compared with those in Sections 8.3, 4 for unbounded space. By the argument of Section 8.4.1,

$$G(\mathbf{r}, t \mid \mathbf{r}', t') = \theta(t - t') K(\mathbf{r}, t \mid \mathbf{r}', t'). \tag{9.2.3}$$

To establish the magic rule which solves the most general diffusion problem, one needs the *reciprocal equation* obeyed by G regarded as a function of $\mathbf{r}'$ and t'. This in turn depends on the symmetry properties of K and G, which are interesting in their own right; the reader is asked to tolerate the delay while we study them before proceeding to the magic rule. The symmetry properties are non-trivial for the obvious reason that, on account of the BCs, K, unlike K_0, depends on $\mathbf{r}$ and $\mathbf{r}'$ separately, and not only through $\mathbf{R} = \mathbf{r} - \mathbf{r}'$.

9.2.2 Invariance under time-translation

Provided only that the boundary S of the region V is stationary, i.e. provided that V never changes with time, the propagator K depends on t and t' only through the combination $\tau \equiv t - t'$. In this respect K is like the propagator K_0 for unbounded space (Section 8.3.2), by contrast to the fact that, on $\mathbf{r}$ and $\mathbf{r}'$, K depends individually, and not just through $\mathbf{R} = \mathbf{r} - \mathbf{r}'$. The same remarks apply to the Green's functions. In other words, K and G are invariant under the transformation $t \to t + a$, $t' \to t' + a$, which increments ('translates') all times by the same arbitrary amount a, and which in consequence is called 'invariance under time-translation'.

Later, it will be equally important that K and G for the wave equation are likewise invariant under time-translation, the proof being essentially the same as for diffusion (Section 10.3.1).

Theorem: K and G are functions of t and t' only through $\tau \equiv t - t'$; hence they can be written as $K(\boldsymbol{r}, \boldsymbol{r}'; \tau)$ and $G(\boldsymbol{r}, \boldsymbol{r}'; \tau)$ whenever convenient.

Proof: In the relations (9.2.1) defining K, change the independent variables from $(\boldsymbol{r}, t)$ to $(\boldsymbol{r}, \tau)$. The only effect of this on L is to change $\partial/\partial t$ into $\partial/\partial\tau$; hence the defining relations (9.2.1a, b) become

$$(\partial/\partial\tau - D\nabla^2)K = 0 \quad \text{for} \quad \tau \geqslant 0, \tag{9.2.4a}$$

$$K = \delta(\boldsymbol{r} - \boldsymbol{r}') \quad \text{for} \quad \tau = 0, \tag{9.2.4b}$$

while (9.2.1c) is unaffected (because, by assumption, the boundary S is stationary). Since the defining relations involve only $\boldsymbol{r}, \boldsymbol{r}', \tau$, the solution K is a function only of these variables. Since (9.2.3) reads

$$G = \theta(\tau)K, \tag{9.2.5}$$

the same is true of G. ■

Corollary: Since $\tau = (t - t')$ is unaffected by (invariant under) the replacements $t \to -t'$, $t' \to -t$, we have

$$K(\boldsymbol{r}, t \mid \boldsymbol{r}', t') = K(\boldsymbol{r}, -t' \mid \boldsymbol{r}', -t). \tag{9.2.6}$$

In view of (9.2.5), it is clear that every symmetry relation for K that leaves τ unchanged entails a corresponding symmetry relation for G.

9.2.3 Symmetry and reciprocity

Theorem:

$$K(\boldsymbol{r}, t \mid \boldsymbol{r}', t') = K(\boldsymbol{r}', t \mid \boldsymbol{r}, t'). \tag{9.2.7}$$

We call this the *symmetry relation* (in the narrow sense).

Proof: Start by repeating (9.2.1a,b):

$$(\partial/\partial t - D\nabla^2)K(\boldsymbol{r}, t \mid \boldsymbol{r}', t') = 0, \qquad K(\boldsymbol{r}, t' \mid \boldsymbol{r}', t') = \delta(\boldsymbol{r} - \boldsymbol{r}'). \tag{9.2.8a,b}$$

In (9.2.8a), write $(\boldsymbol{r}'', t'')$ instead of $(\boldsymbol{r}', t')$, and then rearrange by appeal to (9.2.6):

$$(\partial/\partial t - D\nabla^2)K(\boldsymbol{r}, t \mid \boldsymbol{r}'', t'') = (\partial/\partial t - D\nabla^2)K(\boldsymbol{r}, -t'' \mid \boldsymbol{r}'', -t) = 0.$$

Change the names of the two time-variables as follows: $t \to -t$ (hence also $\partial/\partial t \to -\partial/\partial t$), $t'' \to -t''$. This yields

$$(-\partial/\partial t - D\nabla^2)K(\boldsymbol{r}, t'' \mid \boldsymbol{r}'', t) = 0, \qquad K(\boldsymbol{r}, t'' \mid \boldsymbol{r}'', t'') = \delta(\boldsymbol{r} - \boldsymbol{r}''). \quad (9.2.9\text{a,b})$$

Multiply (9.2.8a) by $K(\boldsymbol{r}, t'' \mid \boldsymbol{r}'', t)$ and (9.2.9a) by $K(\boldsymbol{r}, t \mid \boldsymbol{r}', t')$, take the difference, and integrate it $\int_{t_1}^{t_2} \mathrm{d}t \int_V \mathrm{d}V \ldots$, where t_1 and t_2 are arbitrary:

$$\int_{t_1}^{t_2} \mathrm{d}t \int_V \mathrm{d}V \left\{ \left[K(\boldsymbol{r}, t'' \mid \boldsymbol{r}'', t) \frac{\partial}{\partial t} K(\boldsymbol{r}, t \mid \boldsymbol{r}', t') \right.\right.$$
$$\left. + \frac{\partial K(\boldsymbol{r}, t'' \mid \boldsymbol{r}'', t)}{\partial t} K(\boldsymbol{r}, t \mid \boldsymbol{r}', t') \right] - D[K(\boldsymbol{r}, t'' \mid \boldsymbol{r}'', t)\nabla^2 K(\boldsymbol{r}, t \mid \boldsymbol{r}', t')$$
$$\left. - (\nabla^2 K(\boldsymbol{r}, t'' \mid \boldsymbol{r}'', t))K(\boldsymbol{r}, t \mid \boldsymbol{r}', t')] \right\} = 0. \qquad (9.2.10)$$

The contents of the second pair of square brackets are integrated with respect to volume, using Green's theorem, which yields

$$\int_{t_1}^{t_2} \mathrm{d}t \int_S \mathrm{d}S \, [K(\boldsymbol{r}, t'' \mid \boldsymbol{r}'', t) \, \partial_n K(\boldsymbol{r}, t \mid \boldsymbol{r}', t')$$
$$- (\partial_n K(\boldsymbol{r}, t'' \mid \boldsymbol{r}'', t))K(\boldsymbol{r}, t \mid \boldsymbol{r}', t')] = 0;$$

the surface integral vanishes because, by virtue of the BCs, either K or $\partial_n K$ vanishes everywhere on S.

Accordingly, the contents of the first pair of square brackets in (9.2.10) must also integrate to zero. Inspection reveals that these contents are just the total time-derivative of the product of the two propagators. In an obvious shorthand,

$$\int_{t_1}^{t_2} \mathrm{d}t \int_V \mathrm{d}V \frac{\partial}{\partial t}(KK) = \int_{t_1}^{t_2} \mathrm{d}t \frac{\mathrm{d}}{\mathrm{d}t} \int_V \mathrm{d}V \, KK$$
$$= \int_V \mathrm{d}V \, KK\big|_{t=t_2} - \int_V \mathrm{d}V \, KK\big|_{t=t_1} = 0.$$

In other words the volume integral $\int_V \mathrm{d}V \, K(\boldsymbol{r}, t'' \mid \boldsymbol{r}'', t)K(\boldsymbol{r}, t \mid \boldsymbol{r}', t')$ is independent of t.

Finally we equate the values of this integral at $t = t''$ and at $t = t'$, determining them by exploiting the delta-functions on the right of (9.2.9b) and (9.2.8b):

$$\int_V \mathrm{d}V \, K(\boldsymbol{r}, t'' \mid \boldsymbol{r}'', t'')K(\boldsymbol{r}, t'' \mid \boldsymbol{r}', t')$$
$$= \int_V \mathrm{d}V \, \delta(\boldsymbol{r} - \boldsymbol{r}'')K(\boldsymbol{r}, t'' \mid \boldsymbol{r}', t') = K(\boldsymbol{r}'', t'' \mid \boldsymbol{r}', t')$$

$$= \int_V \mathrm{d}V\, K(\boldsymbol{r}, t'' \mid \boldsymbol{r}'', t')K(\boldsymbol{r}, t' \mid \boldsymbol{r}', t')$$

$$= \int_V \mathrm{d}V\, K(\boldsymbol{r}, t'' \mid \boldsymbol{r}'', t')\, \delta(\boldsymbol{r} - \boldsymbol{r}') = K(\boldsymbol{r}', t'' \mid \boldsymbol{r}'', t'),$$

$$K(\boldsymbol{r}'', t'' \mid \boldsymbol{r}', t') = K(\boldsymbol{r}', t'' \mid \boldsymbol{r}'', t'),$$

which is precisely (9.2.7) except for the names of the variables. ■

Time-translation invariance (9.2.6) and symmetry (9.2.7) together imply the *reciprocity relation*

$$K(\boldsymbol{r}, t \mid \boldsymbol{r}', t') = K(\boldsymbol{r}', -t' \mid \boldsymbol{r}, -t), \tag{9.2.11}$$

where both the primed variables have moved to the left of the upright.

By virtue of $G(\boldsymbol{r}, \boldsymbol{r}'; \tau) = \theta(\tau)K(\boldsymbol{r}, \boldsymbol{r}'; \tau)$ as in (9.2.5), the time-translation, symmetry, and reciprocity relations for K entail the corresponding relations for G, namely

$$G(\boldsymbol{r}, t \mid \boldsymbol{r}', t') = G(\boldsymbol{r}, -t' \mid \boldsymbol{r}', -t), \tag{9.2.12}$$

$$= G(\boldsymbol{r}', t \mid \boldsymbol{r}, t'), \tag{9.2.13}$$

$$= G(\boldsymbol{r}', -t' \mid \boldsymbol{r}, -t). \tag{9.2.14}$$

Evidently any two of these imply the third. In unbounded space, all are obeyed trivially, because K_0 and G_0 depend on $\boldsymbol{r}$ and $\boldsymbol{r}'$ only through $|\boldsymbol{R}| = |\boldsymbol{r} - \boldsymbol{r}'|$, which is automatically invariant under $\boldsymbol{r} \rightleftarrows \boldsymbol{r}'$. The symmetry relation (9.2.13) should be compared with the symmetry relation $G(\boldsymbol{r} \mid \boldsymbol{r}') = G(\boldsymbol{r}' \mid \boldsymbol{r})$ appropriate to Poisson's equation (cf. eqn (5.1.7) and Section 6.3). Remarkably, (9.2.13) implies that if a given amount of heat is introduced instantaneously at $\boldsymbol{r}$, then after a time lapse τ it produces at $\boldsymbol{r}'$ the same effect as one would produce at $\boldsymbol{r}$, after the same time lapse, by introducing the same amount of heat at $\boldsymbol{r}'$; this equality applies however asymmetrically $\boldsymbol{r}$ and $\boldsymbol{r}'$ are situated relative to the boundary, and under any homogeneous boundary conditions.

Lastly, the symmetry relations show that K and G obey the same boundary conditions as functions of $\boldsymbol{r}'$ as of $\boldsymbol{r}$; for instance, if $G(\boldsymbol{r}, t \mid \boldsymbol{r}', t')$ vanishes when $\boldsymbol{r}$ is on S, then it vanishes when $\boldsymbol{r}'$ is on S.

9.2.4 The reciprocal equation

To deduce the differential equation for $G(\boldsymbol{r}, t \mid \boldsymbol{r}', t')$ with respect to $\boldsymbol{r}', t'$ as the independent variables, one starts from the original differential equation (9.2.2a) and the reciprocity relation (9.2.14):

$$(\partial/\partial t - D\nabla^2)G(\boldsymbol{r}, t \mid \boldsymbol{r}', t') = (\partial/\partial t - D\nabla^2)G(\boldsymbol{r}', -t' \mid \boldsymbol{r}, -t)$$
$$= \delta(t - t')\, \delta(\boldsymbol{r} - \boldsymbol{r}').$$

To put the second equality into standard form, change the names of the variables as follows: $t \rightleftharpoons -t'$ (hence also $\partial/\partial t \rightleftharpoons -\partial/\partial t'$); $\boldsymbol{r} \rightleftharpoons \boldsymbol{r}'$ (hence also $\nabla^2 \rightleftharpoons \nabla'^2$). This yields

$$(-\partial/\partial t' - D\nabla'^2)G(\boldsymbol{r}, t \mid \boldsymbol{r}', t') = \delta(-t+t')\,\delta(\boldsymbol{r}'-\boldsymbol{r})$$
$$= \delta(t-t')\,\delta(\boldsymbol{r}-\boldsymbol{r}') \qquad (9.2.15)$$

which we call the *reciprocal equation*; often it is called the *adjoint* equation. (The last step in (9.2.15) is legitimate because we shall need only the weak definition of the delta-function, under which it is an even function.)

The boundary conditions for the reciprocal equation are the same as those appropriate for the unprimed variables, as pointed out at the end of Section 9.2.3. The initial conditions are also the same, namely $G = 0$ for $\tau = t - t' < 0$.

Exercises: (i) Prove this. (ii) Determine the reciprocal equation for K, together with the appropriate boundary and initial conditions.

Finally we compare the operators featuring in the reciprocal and in the original equations. Here one cares only about their form, but not about the symbols, primed or unprimed, for the independent variables. Evidently the operator $\tilde{L} \equiv (-\partial/\partial t - D\nabla^2)$ on the left of the reciprocal equation (9.2.15) is different in form from the operator $L \equiv (\partial/\partial t - D\nabla^2)$ on the left of the original eqn (9.2.2a), because the time derivatives enter with opposite signs. This manifests yet again the irreversibility of the diffusion equation (cf. Section 9.5). For instance, we shall find in Section 10.3 that the Green's function for the wave equation, which *is* reversible, satisfies the same reciprocity *relation* (9.2.14) as does G for the diffusion equation, but that the form of the reciprocal *equation* agrees (rather than disagrees, as here) with that of the original wave equation.

In mathematics, the operator $\tilde{L}$ is called the *adjoint* of L; since $\tilde{L} \neq L$, L is not self-adjoint. Here we do not pursue the deeper mathematical implications of the adjoint relationship; an excellent account (beyond the scope of this book) is given by Sobolev (1964), and another, along somewhat unconventional lines, by Lanczos (1961). Very briefly, the linear operator L, subject to homogeneous boundary conditions on the closed hypersurface of a region in the hyperspace spanned by $\boldsymbol{r}$ and t jointly, is *not Hermitean.* Hence the set $\{\Phi_n(\boldsymbol{r}, t)\}$ of its eigenfunctions, even if not empty, is certainly not complete. In this L is unlike the Laplace operator $-\nabla^2$ in ordinary space, and equally unlike the wave operator $\left(\frac{1}{c^2}\frac{\partial^2}{\partial t^2} - \nabla^2\right)$ in hyperspace. Accordingly, there can be no question of expanding K, or G, or the solutions $\psi(\boldsymbol{r}, t)$ of the diffusion equation, in terms of such a set of eigenfunctions. All that we can do, as in the next subsection, is to expand ψ, as regards its $\boldsymbol{r}$-dependence, in terms of the eigenfunctions of $-\nabla^2$ in ordinary space, assigning an appropriate time-dependent factor to each term. From a point of view in hyperspace, this is analogous *not* to the expansion of a solution of say the 2D Poisson equation in eigenfunctions of $-\nabla^2 = -(\partial^2/\partial x^2 + \partial^2/\partial y^2)$, but rather to its single-sum expansion in terms of the eigenfunctions of $-\partial^2/\partial x^2$, with the y-dependence of each term determined separately, as in Section 5.3.2.

9.2.5 Expansion in eigenfunctions of $-\nabla^2$

For simplicity, let us consider the diffusion equation with either DBCs everywhere on S, or with NBCs everywhere, so that the propagator and the Green's function obey homogeneous DBCs or NBCs everywhere (and for all t). As in Sections 4.5 and 5.3.2, let λ_p and ϕ_p be the eigenvalues and eigenfunctions of $-\nabla^2$ under these BCs. From our experience so far it is easy to guess, and in any case it is easy to verify, that the propagator may be written

$$K(\boldsymbol{r}, t \mid \boldsymbol{r}', t') = \sum_p \phi_p^*(\boldsymbol{r}')\phi_p(\boldsymbol{r}) \exp(-\lambda_p D(t-t')). \qquad (9.2.16)$$

(As explained in Section 4.5.2, the method that gave K_0 in Section 8.3 is just a special case of this with plane waves $(2\pi)^{-n/2} \exp(i\boldsymbol{k}\cdot\boldsymbol{r})$ for ϕ_p and $\int d^n k$ for Σ_p, appropriately to unbounded space.)

Time-translation invariance and its Corollary (9.2.6) are visible from (9.2.16) by inspection. With the set $\{\phi_p(\boldsymbol{r})\}$ chosen as real (cf. Appendix E), the symmetry and reciprocity relations (9.2.7, 11) are likewise manifest. The remark at the end of Section 5.1 bears repeating here.

When the volume V is finite, the allowed values of λ_p are discrete; in that case it is clear from (9.2.16) that the behaviour of K as $t \to \infty$, i.e. for $\lambda D\tau \gg 1$, is determined by the smallest eigenvalue λ_p; all other terms will be exponentially small by comparison.

Example: Consider a 1D system, e.g. a sufficiently thin conducting rod with insulated sides (cf. Section 9.4.5 below). Under DBCs at both ends, say at $x = 0$ and $x = L$, we have from (9.2.3, 16)

$$G_D(x, t \mid x', t') = \theta(t-t')\frac{2}{L}\sum_{n=1}^{\infty} \sin\left(\frac{n\pi x'}{L}\right)\sin\left(\frac{n\pi x}{L}\right) \times \exp\left\{-D\left(\frac{n\pi}{L}\right)^2(t-t')\right\}. \qquad (9.2.17)$$

Under NBCs at both ends,

$$G_N(x, t \mid x', t') = \theta(t-t')\frac{2}{L}\left\{\frac{1}{2} + \sum_{n=1}^{\infty} \cos\left(\frac{n\pi x'}{L}\right)\cos\left(\frac{n\pi x}{L}\right) \times \exp\left\{-D\left(\frac{n\pi}{L}\right)^2(t-t')\right\}\right\}. \qquad (9.2.18)$$

Suppose finally that the temperature at $x = 0$ is prescribed as a function of time, while the end at $x = L$ is insulated. Then G must be expanded in terms of eigenfunctions of $-\mathrm{d}^2/\mathrm{d}x^2$ that obey $\phi(0) = 0$ and $\phi_x(L) = 0$.

Exercises: (i) Determine G under these (mixed) BCs. (ii) Compare its behaviour as $t \to \infty$ with that of (9.2.17, 18).

It is clear from the exponentials that the series (9.2.17, 18) converge rapidly unless $D(t - t')/L^2$ is small (see Problem 9.1). When convergence is inconveniently slow, i.e. for very short time-lapses, a useful alternative is provided by the multiple-image series discussed at the end of Section 9.4.4 (see also Problem 9.9).

9.3 The magic rule

9.3.1 Derivation

We now have all the tools for solving the most general diffusion problem, namely

$$(\partial/\partial t' - D\nabla'^2)\psi(\mathbf{r}', t') = \rho(\mathbf{r}', t'), \tag{9.3.1}$$

under ICs prescribing $\psi(\mathbf{r}', t_0)$, and inhomogeneous BCs on S. (Primed and unprimed variables are being chosen by hindsight. This is not essential, but ensures that the magic rule emerges in convenient form. Otherwise one would have to relabel variables at a late stage.) We need also the reciprocal equation (9.2.15) for G:

$$(-\partial/\partial t' - D\nabla'^2)G(\mathbf{r}, t \mid \mathbf{r}', t') = \delta(t - t')\,\delta(\mathbf{r} - \mathbf{r}'). \tag{9.3.2}$$

The reasoning is reminiscent of the uniqueness proof. Multiply (9.3.1) by $G(\mathbf{r}, t \mid \mathbf{r}', t')$, and (9.3.2) by $\psi(\mathbf{r}', t')$; subtract; integrate with respect to $\mathrm{d}V'$ over V, and with respect to t' from t_0 to $t+$, where $t_+ \equiv t + \varepsilon$ as in Section 8.4.2. On the left one obtains

$$\int_{t_0}^{t+} \mathrm{d}t' \int_V \mathrm{d}V' \left\{ \frac{\partial}{\partial t'}[G(\mathbf{r}, t \mid \mathbf{r}', t')\psi(\mathbf{r}', t')] - D[G\nabla'^2\psi(\mathbf{r}', t') - (\nabla'^2 G)\psi(\mathbf{r}', t')] \right\}$$

$$= \left\{ \left[\int_V \mathrm{d}V'\, G(\mathbf{r}, t \mid \mathbf{r}', t')\psi(\mathbf{r}', t') \right|_{t'=t_0}^{t'=t+} - \int_{t_0}^{t+} \mathrm{d}t' \int_S \mathrm{d}S'\, [G\,\partial_n'\psi_S(\mathbf{r}', t') - (\partial_n' G)\psi_S(\mathbf{r}', t')] \right\}.$$

In the first term, the upper limit vanishes because $t' = t+ > t$, so that G vanishes. In the second term, we can replace $t+$ by t. Thus, on the left, our operations on (9.3.1, 2) yield

$$\left\{-\int_V dV'\, G(\mathbf{r}, t \mid \mathbf{r}', t_0)\psi(\mathbf{r}', t_0)\right.$$
$$\left. - D\int_{t_0}^{t} dt' \int_S dS'\, [G\, \partial_n' \psi_S(\mathbf{r}', t') - (\partial_n' G)\psi_S(\mathbf{r}', t')]\right\}. \tag{9.3.3a}$$

On the right, our operations yield

$$\int_{t_0}^{t+} dt' \int_V dV'\, \{G(\mathbf{r}, t \mid \mathbf{r}', t')\rho(\mathbf{r}', t') - \psi(\mathbf{r}', t')\, \delta(t-t')\, \delta(\mathbf{r}-\mathbf{r}')\}$$
$$= \int_{t_0}^{t} dt' \int_V dV'\, G(\mathbf{r}, t \mid \mathbf{r}', t')\rho(\mathbf{r}', t') - \psi(\mathbf{r}, t), \tag{9.3.3b}$$

where in the remaining integral one has once again replaced $t+$ by t. Equating (9.3.3a) to (9.3.3b), and rearranging, we obtain the magic rule:

$$\psi(\mathbf{r}, t) = \int_{t_0}^{t} dt' \int_V dV'\, G(\mathbf{r}, t \mid \mathbf{r}', t')\rho(\mathbf{r}', t')$$
$$+ D\int_{t_0}^{t} dt' \int_S dS'\, [G(\mathbf{r}, t \mid \mathbf{r}', t')\, \partial_n' \psi_S(\mathbf{r}', t')$$
$$- (\partial_n' G(\mathbf{r}, t \mid \mathbf{r}', t'))\psi_S(\mathbf{r}', t')]$$
$$+ \int_V dV'\, G(\mathbf{r}, t \mid \mathbf{r}', t_0)\psi(\mathbf{r}', t_0)$$
$$\equiv f(\mathbf{r}, t) + g(\mathbf{r}, t) + h(\mathbf{r}, t). \tag{9.3.4}$$

As before, f, g, h are regarded as explicit constructs from the data ρ, ψ_S or $\partial_n \psi_S$, and $\psi(\mathbf{r}, t_0)$. They obey the equations

$$Lf = \rho, \qquad Lg = 0, \qquad Lh = 0. \tag{9.3.5a,b,c}$$

Exercise: Verify this.

9.3.2 Comments

(i) At this stage, provided $t > t_0$ is understood, G in (9.3.4) may be replaced by K.

(ii) Recall that ∂_n is the outward-normal derivative.

(iii) The integral f propagates the effects of sources acting between the starting time t_0 and the time t at which $\psi(\mathbf{r}, t)$ is required.

(iv) The integral g propagates the effects of the inhomogeneous boundary values prescribed between t_0 and t. Recall that at any one point $\boldsymbol{r}'$ on S, either $\psi_S(\boldsymbol{r}', t')$ or $\partial'_n\psi_S(\boldsymbol{r}', t')$ is given for all $t_0 \leqslant t' \leqslant t$. If $\psi_S(\boldsymbol{r}', t')$ is given, then, from its definition, G is zero there, and therefore the value of $\partial'_n\psi_S(\boldsymbol{r}', t')$ is irrelevant, because in (9.3.4) it is multiplied by G. Conversely, if $\partial'_n\psi_S$ is given, then $\partial'_n G$ is zero there, and ψ_S at that point is irrelevant to g. This is precisely what happened with Poisson's equation: boundary values that are not prescribed are not needed in the magic rule.

In some problems, either DBCs or NBCs are imposed everywhere on S; then one or other of the two surface integrals constituting g vanishes altogether.

(v) The integral h propagates the effects of the initial conditions.

(vi) There is no reason why the lower limits should not be $t_0 = -\infty$. Normally in such cases the initial values $\psi(\boldsymbol{r}, -\infty)$ vanish, so that h is absent, and everything that happens later is determined either by time-varying sources (through f) or by time-varying boundary values (through g). A somewhat perverse application of this idea to the simple-harmonic solutions of Section 8.2 is worked through in Appendix J.

(vii) K and G involve the time only through the product $D\tau = D(t - t')$, as one can see explicitly from the eigenfunction expansion (9.2.16). In h, G is the only source of time-dependence; hence the only role of D is to determine the time-scale, as illustrated in Sections 8.3.4 and 9.4.3. In the special case where ρ and ψ_S or $\partial_n\psi_S$ are likewise time-independent for $t > t_0$, the same is true of any transient components of f and g; this is illustrated for g in Section 9.4.1 below. However, in the general case where sources or prescribed boundary values vary non-trivially with time, their time-variation influences that of ψ just as strongly as do transients evolving at rates governed wholly by D.

(viii) All questions can be answered by the magic rule. But it is not always the quickest way to the answer. In particular, if there are no volume sources (so that f vanishes), it is sometimes easier to expand $\psi(\boldsymbol{r}, t)$ directly in eigenfunctions of $-\nabla^2$ with appropriately time-varying coefficients, i.e. to bypass h and g, rather than to use the same basic idea in order to obtain G first, as we did above (compare this alternative strategy with that of Section 5.4).

(ix) Notice that the magic rule always works; nothing can get in the way of the explicit construction culminating in (9.3.4). In particular, purely NBCs do not now (as they did for Poisson's equation) require special treatment. They are, however, distinguished by the feature that the zero eigenvalue of $-\nabla^2$ (cf. Section 4.5) admits to G a constant term V^{-1}, which remains non-zero as $t \to \infty$ (cf. Section 9.2.5).

(x) Section 7.1 demonstrated the self-consistency of the magic rule for Poisson's equation. We shall not give the corresponding argument for (9.3.4), and merely point out what would need proving.

To verify the initial conditions, one shows that in the limit $t \to t_0+$, f and g vanish, while h tends to $\psi(\boldsymbol{r}, t_0)$. This is quite straightforward: one replaces G by K and appeals to the initial condition on K, i.e. effectively one uses (8.4.5a), which is equally valid in bounded regions. (See also the proof of (8.4.8).)

To verify the BCs, say DBCs, one shows first that as $\boldsymbol{r}$ tends to a point $\boldsymbol{r}_1$ on S, f and h vanish. Then, writing $g = -\int_{t_0}^{t+} dt' \{D \int_S dS' (\partial_n' G)\psi_S(\boldsymbol{r}', t')\}$, one must show that as $\boldsymbol{r}$ tends to $\boldsymbol{r}_1$, $\{D \int_S \cdots\}$ tends to $\delta(t'-t)\psi_S$. The argument requires care, and at intermediate stages the result is not at all obvious. The subtleties echo through some of the *prima facie* paradoxical expressions encountered in the examples of Sections 9.4.2 and 9.4.4 below.

9.4 Examples

9.4.1 Prescribed sources: the integral f

An effectively 1D rod, $0 \leqslant x \leqslant \pi$, is insulated both along its sides and at its ends, so that $\psi_x(0, t) = 0 = \psi_x(\pi, t)$. Initially, the temperature is zero: $\psi(x, 0) = 0$. At $t = 0$, a point source of heat is switched on at the midpoint, and then remains constant: $\rho(x, t) = \theta(t)\alpha\,\delta(x - \pi/2)$. We determine $\psi(x, t)$, and particularly its asymptotic behaviour as $t \to \infty$.

The Neumann Green's function is given by (9.2.18):

$$G(x, t \mid x', t') = \theta(t-t')\frac{2}{\pi}\left\{\tfrac{1}{2} + \sum_{n=1}^{\infty} \cos(nx')\cos(nx)\exp(-Dn^2(t-t'))\right\}. \quad (9.4.1)$$

Since h and g vanish, ψ is given by f alone:

$$\begin{aligned}\psi(x, t) &= f(x, t)\\ &= \int_0^t dt' \int_0^\pi dx' \frac{2}{\pi}\left\{\tfrac{1}{2} + \sum_{n=1}^{\infty} \cos(nx')\cos(nx)\exp(-Dn^2(t-t'))\right\}\\ &\quad \times \alpha\,\delta(x' - \pi/2)\\ &= \frac{2\alpha}{\pi}\int_0^t dt' \left\{\tfrac{1}{2} + \sum_{n=1}^{\infty} \cos(n\pi/2)\cos(nx)\exp(-Dn^2(t-t'))\right\}.\end{aligned}$$

When n is odd, $\cos(n\pi/2)$ vanishes; when $n = 2m$ is even,

$\cos(n\pi/2) = \cos(m\pi) = (-1)^m$; thus

$$\psi = \frac{2\alpha}{\pi}\int_0^t dt' \left\{\tfrac{1}{2} + \sum_{m=1}^{\infty} (-1)^m \cos(m2x) \exp(-4Dm^2(t-t'))\right\},$$

$$\psi(x,t) = \frac{\alpha t}{\pi} + \frac{\alpha}{2\pi D}\sum_{m=1}^{\infty}(-1)^m \cos(m2x)\frac{(1-\exp(-4Dm^2 t))}{m^2}. \tag{9.4.2}$$

Since $\int_0^\pi dx \cos(2mx) = 0$, only the term $\alpha t/\pi$ contributes to the average value of ψ, and so to the total heat content of the rod:

$$\int_0^\pi dx\, \psi(x,t) = \int_0^\pi dx\, \alpha t/\pi = \alpha t$$

is the total heat injected up to time t, as it should be.

When $Dt \gg 1$, the exponentials contribute negligibly to the Fourier coefficients; accordingly

$$\psi(x, t\to\infty) \approx \alpha t/\pi + (\alpha/2\pi D)S(x), \tag{9.4.3a}$$

$$S(x) \equiv \sum_{m=1}^{\infty}(-1)^m \frac{1}{m^2}\cos(2mx). \tag{9.4.3b}$$

To identify $S(x)$ one starts from the standard Fourier series (see Appendix A.5)

$$\cos\theta - \frac{1}{2^2}\cos 2\theta + \frac{1}{3^2}\cos 3\theta - \cdots = [\pi^2/12 - \theta^2/4], \quad -\pi \leqslant \theta \leqslant \pi. \tag{9.4.4}$$

This must now be handled with a watchful eye on the range of the independent variable. Set $\theta = 2x$, $x = \theta/2$, obtaining in the first place

$$S(x) = -[\pi^2/12 - x^2], \qquad -\pi/2 \leqslant x \leqslant \pi/2.$$

This range of x overlaps our desired range $0 \leqslant x \leqslant \pi$ only partially, i.e. only when $0 \leqslant x \leqslant \pi/2$. To get around this, we introduce the distance ξ from the midpoint, $x \equiv (\pi/2 + \xi)$, because we know from the symmetry of the problem that ψ must be an even function of ξ, i.e. a function of $|\xi|$ only. Accordingly, in the legitimate overlap region $0 \leqslant x \leqslant \pi/2$, where $-\pi/2 \leqslant \xi \leqslant 0$, we have

$$S = -[\pi^2/12 - (\pi/2 + \xi)^2] = [\pi^2/6 + \pi\xi + \xi^2]$$
$$= [\pi^2/6 - \pi|\xi| + \xi^2] = [\pi^2/6 - \pi|x - \pi/2| + (x - \pi/2)^2].$$

In its final form this result now applies over the entire physical system

$0 \leqslant x \leqslant \pi$. Thus

$$\psi(x, t \gg 1/D) \approx \frac{\alpha t}{\pi} + \frac{\alpha}{2\pi D}[\pi^2/6 - \pi\,|x - \pi/2| + (x - \pi/2)^2],$$
$$0 \leqslant x \leqslant \pi. \tag{9.4.5}$$

Exercises: (i) Sketch (9.4.5) as a function of x, and verify explicitly that it obeys the diffusion equation and the boundary conditions. Explain why, nevertheless, (9.4.5) is not the solution to the problem. (ii) Solve the corresponding problem with $\rho = \alpha\theta(t)\sin(\omega t)$.

9.4.2 Prescribed boundary values: the integral g

The temperature ψ satisfies the homogeneous heat equation between concentric spheres of radii $a < b$. At $t = 0$, when ψ vanishes everywhere, the temperature of the inner sphere begins to rise linearly with time, while the temperature of the outer sphere is maintained at zero. Thus $\psi(a, t) = \alpha\theta(t)t$, $\psi(b, t) = 0$. We determine $\psi(\mathbf{r}, t)$.

Since f and h vanish, $\psi = g$, as defined in (9.3.4). We require the Dirichlet Green's function vanishing at $r = a, b$. Because the problem is isotropic, only the Ω-independent terms, call them Γ, need be kept in the eigenfunction expansion (9.2.16):

$$\phi_n = \frac{1}{r}\left(\frac{2}{c}\right)^{\frac{1}{2}} \sin[n\pi(r-a)/c]Y_{00}, \qquad \lambda_n = (n\pi/c)^2, \tag{9.4.6a,b}$$

where we set $b - a = c$ for short. Since $Y^*_{00}(\Omega')Y_{00}(\Omega) = 1/4\pi$, we can write (with $K \to G \to \Gamma$)

$$\Gamma(\mathbf{r}, t \,|\, \mathbf{r}', t') = \frac{1}{4\pi r r'}\frac{2}{c}\sum_{n=1}^{\infty} \sin\left[\frac{n\pi(r'-a)}{c}\right]\sin\left[\frac{n\pi(r-a)}{c}\right]$$
$$\times \exp(-D(n\pi/c)^2(t-t')). \tag{9.4.7}$$

Only the inner sphere contributes to the integral; by isotropy, this integral is just $4\pi a^2$ times the integrand $-(\partial'_n\Gamma)\psi_S = \partial\Gamma/\partial r'|_{r'=a}\alpha\theta(t')t'$. (The first integral in g vanishes, as explained in Section 9.3.2, comment (iv).) But

$$\left.\frac{\partial\Gamma}{\partial r'}\right|_{r'=a} = \frac{1}{4\pi r a}\frac{2}{c}\sum_{n=1}^{\infty}\frac{n\pi}{c}\sin\left[\frac{n\pi(r-a)}{c}\right]\exp(-D(n\pi/c)^2(t-t')), \tag{9.4.8}$$

because, if $\partial/\partial r'$ acts on the prefactor $1/r'$ in (9.4.7), then the factors $\sin[n\pi(r'-a)/c]$ remain in $\partial\Gamma/\partial r'$, and vanish at $r' = a$.

Accordingly,

$$\psi(\mathbf{r}, t) = g(\mathbf{r}, t) = -D \int_0^t \mathrm{d}t' \int_S \mathrm{d}S' \, \partial'_n \Gamma|_{r'=a} \psi_S$$

$$= D \int_0^t \mathrm{d}t' \, 4\pi a^2 \frac{1}{4\pi r a} \frac{2}{c} \sum_{n=1}^{\infty} \frac{n\pi}{c} \sin\left[\frac{n\pi(r-a)}{c}\right]$$

$$\times \exp\left(-D(n\pi/c)^2(t-t')\right) \cdot \alpha t'$$

$$= \frac{\alpha D}{r} \frac{2\pi a}{c^2} \sum_{n=1}^{\infty} n \sin\left[\frac{n\pi(r-a)}{c}\right] \int_0^t \mathrm{d}t' \, t' \exp\left(-D(n\pi/c)^2(t-t')\right),$$

$$\psi(\mathbf{r}, t) = \alpha \frac{2a}{\pi r} \sum_{n=1}^{\infty} \frac{1}{n} \sin\left[\frac{n\pi(r-a)}{c}\right] \left\{ t - \frac{1}{(n\pi D/c)} + \frac{\exp\left(-D(n\pi/c)^2 t\right)}{(n\pi D/c)} \right\}. \tag{9.4.9}$$

Since every term in the series vanishes at $r = a$, it seems as if the BC $\psi(a, t) = \alpha t$ were violated. However, the first sum, namely (see Appendix A.5)

$$\sum_{n=1}^{\infty} \frac{1}{n} \sin(n\theta) = \tfrac{1}{2}(\pi - \theta), \qquad (0 < \theta < 2\pi),$$

fails to converge uniformly, whence the limit $\theta \to 0$ may not be taken under the summation sign. Instead, writing $\theta \equiv \pi(r-a)/(b-a)$, we see that the explicitly t-proportional part of (9.4.9) is just $(\alpha a/r)t(b-r)/(b-a)$, which does, as it should, reduce to αt when $r \to a$. The other two series do converge uniformly, and do vanish at $r = a$.

Exercise: Solve the corresponding problem with $\psi(a, t) = \beta\theta(t)$.

9.4.3 Prescribed initial values: the integral h

A flat square slab is thermally insulated except at its edges, which are kept at zero temperature. The initial temperature $\psi(\mathbf{r}, t_0) \equiv a(\mathbf{r})$ is non-negative. We determine the temperature distribution $\psi(\mathbf{r}, t)$ as $t \to \infty$, and show that only its absolute normalization but not its shape depend on $a(\mathbf{r})$.

Since f and g vanish, $\psi = h$. We require the 2D Dirichlet propagator. Taking the slab as the region $0 \le x, y \le \pi$,

$$K(\mathbf{r}, t \,|\, \mathbf{r}', t') = \left(\frac{2}{\pi}\right)^2 \sum_{n=1}^{\infty} \sum_{m=1}^{\infty} \sin(nx') \sin(nx) \sin(my') \sin(my)$$

$$\times \exp\{-D(n^2 + m^2)(t - t')\}, \tag{9.4.10}$$

$$\psi(\mathbf{r}, t) = h(\mathbf{r}, t) = \int_0^{\pi} dx' \int_0^{\pi} dy'\, K(\mathbf{r}, t \mid \mathbf{r}', t_0) a(\mathbf{r}')$$
$$= \sum_{n=1}^{\infty} \sum_{m=1}^{\infty} \sin(nx) \sin(my) \exp\{-D(n^2+m^2)(t-t_0)\} \left(\frac{2}{\pi}\right)^2$$
$$\times \int_0^{\pi} dx' \int_0^{\pi} dy' \sin(nx') \sin(mx') a(x', y'). \tag{9.4.11}$$

The final temperature, i.e. $\lim_{t\to\infty} \psi$, is obviously zero. When $D(t-t_0)$ is large enough, only the term with $n = 1 = m$ is appreciable, whence

$$\psi(\mathbf{r}, t\to\infty) \approx \sin(x) \sin(y) \exp(-2D(t-t_0))$$
$$\times \left(\frac{2}{\pi}\right)^2 \int_0^{\pi} dx' \int_0^{\pi} dy' \sin(x') \sin(y') a(x', y'). \tag{9.4.12}$$

Because a is non-negative, the integrand is non-negative, and the integral cannot vanish; in other words, this leading term cannot cancel accidentally.

The leading term alone will be a good approximation to ψ if it is much larger than the next term, which it will be if

$$\exp[-2D(t-t_0)]/\exp[-5D(t-t_0)] = \exp[3D(t-t_0)] \gg 1.$$

For instance, this ratio exceeds 10^2 as soon as $(t-t_0) \geqslant 1.53/D$.

9.4.4 The method of images

In a halfspace, propagators and Green's functions can be written down as soon as one thinks of the method of images (cf. Sections 5.3.3 and 6.4). In the 1D halfspace $x \geqslant 0$, under D and NBCs, respectively, at $x = 0$, and requiring $\psi(x\to\infty, t) = 0$, one has

$$G_{D,N}(x, t \mid x', t') = \frac{\theta(\tau)}{(4\pi D\tau)^{\frac{1}{2}}} \{\exp(-(x-x')^2/4D\tau)$$
$$\mp \exp(-(x+x')^2/4D\tau)\}. \tag{9.4.13}$$

Exercises: (i) Verify that this expression satisfies all the requirements. (ii) Write down G_D and G_N for a halfspace in 2D and in 3D.

As an example, consider a Dirichlet problem, with $\psi(x, 0) = 0$, $\rho = 0$, and $\psi(0, t) \equiv a(t)$. Then $f = 0 = h$, and we have $\psi = g$, for which one needs

$$-\partial_n' G_D|_{x'=0} = \left.\frac{\partial G_D}{\partial x'}\right|_{x'=0} = \frac{\theta(\tau)}{(4\pi D\tau)^{\frac{1}{2}}} \cdot \frac{2x}{4D\tau} \exp(-x^2/4D\tau) \cdot 2;$$

this yields

$$\psi(x, t) = \frac{x}{(4\pi D)^{\frac{1}{2}}} \int_{t_0}^{t} \mathrm{d}t' \frac{1}{(t-t')^{\frac{3}{2}}} \exp(-x^2/4D(t-t'))a(t') \tag{9.4.14a}$$

$$= \frac{x}{(4\pi D)^{\frac{1}{2}}} \int_{0}^{(t-t_0)} \mathrm{d}\tau \frac{1}{\tau^{\frac{3}{2}}} \exp(-x^2/4D\tau)a(t-\tau). \tag{9.4.14b}$$

In the corresponding Neumann problem where $\psi_x(0, t) \equiv b(t)$ is prescribed, we need

$$G_N(x, t \mid 0, t') = 2\theta(\tau) \exp(-x^2/4D\tau)/(4\pi D\tau)^{\frac{1}{2}},$$

whence

$$\psi(x, t) = -\left(\frac{D}{\pi}\right)^{\frac{1}{2}} \int_{0}^{(t-t_0)} \mathrm{d}\tau \frac{1}{\tau^{\frac{1}{2}}} \exp(-x^2/4D\tau)b(t-\tau). \tag{9.4.15}$$

(i) It may not be easy to evaluate integrals like (9.4.14, 15) except numerically, or in limiting cases, even for quite simple data $a(t)$ or $b(t)$.

(ii) One would expect to recover the simple-harmonic solutions of Section 8.2 by making $a(t)$ or $b(t)$ simple-harmonic, and then letting $t_0 \to -\infty$, which makes the upper limits in (9.4.14b, 15) positive infinite (and independent of t). The requisite manipulations are given in Appendix J.

(iii) There is an apparent paradox in (9.4.14). As $x \to 0$, the RHS must approach the datum $a(t)$; yet the factor x suggests that in that limit the RHS vanishes. A similar paradox is implicit in (9.4.15): the integral depends on x only through x^2, i.e. it is an even function of x, so that its derivative at $x = 0$ might seem to vanish; yet in fact it must equal the datum $b(t)$. These paradoxes are akin to the point raised at the end of Section 9.4.2 above.

Exercise: Resolve the paradoxes by paying attention to the convergence of the integrals. Recall that what matters physically is not the value of these expressions *at* $x = 0$, but their limits as $x \to 0+$.

The method of images extends to the slab $0 \leqslant x \leqslant L$ just as it did for Poisson's equation (Section 5.3.3). In 1D we have the *multiple-image series*

$$G_{D,N}^{(1)}(x, t \mid x', t') =$$

$$\sum_{N=-\infty}^{\infty} \{G_0^{(1)}(x, t \mid 2NL + x', t') \mp G_0^{(1)}(x, t \mid 2NL - x', t')\}, \tag{9.4.16}$$

where the upper (lower) sign applies to G_D (G_N).

Exercise: Verify by direct substitution that (9.4.16) obeys the defining differential equation and the BCs, and that it has the symmetry properties (9.2.12–14).

The multiple-image series are useful especially when $D(t-t')/L^2$ is small, so that the normal-mode expansions (9.2.17, 18) converge very slowly. These alternative representations are linked by Poisson's summation formula (Appendix C), as follows. Consider $G_D^{(1)}$:

$$G_D(x, x'; \tau) = \frac{\theta(\tau)}{(4\pi D\tau)^{\frac{1}{2}}} \sum_{N=-\infty}^{\infty} \{\exp[-(x-x'-2NL)^2/4D\tau] - \exp[-(x+x'-2NL)^2/4D\tau]\}. \quad (9.4.17)$$

We apply Poisson's formula (C.13) to the positive and to the negative images separately. Identify

$$F(\xi) = (4\pi D\tau)^{-\frac{1}{2}} \exp[-(x \mp \lambda' - \xi)^2/4D\tau]$$

in turn, and set $\lambda = 2L$; then (C.13) yields

$$G_D = \frac{\theta(\tau)}{(4\pi D\tau)^{\frac{1}{2}}} \sum_{N=-\infty}^{\infty} \frac{1}{2L} \int_{-\infty}^{\infty} d\xi \exp(2\pi i N\xi/2L)\{\exp[-(x-x'-\xi)^2/4D\tau] - \exp[-(x+x'-\xi)^2/4D\tau]\}.$$

In the two terms, change the integration variable to $\eta = (\xi - x \pm x')$ respectively; one finds

$$G_D = \frac{\theta(\tau)}{(4\pi D\tau)^{\frac{1}{2}}} \frac{1}{2L} \sum_{N=-\infty}^{\infty} \{\exp[i\pi N(x-x')/L] - \exp[i\pi N(x+x')/L]\} \times \int_{-\infty}^{\infty} d\eta \exp(-\eta^2/4D\tau)\cos(\eta\pi N/L)$$

$$= \frac{\theta(\tau)}{(4\pi D\tau)^{\frac{1}{2}}} \frac{1}{2L} (4\pi D\tau)^{\frac{1}{2}} \sum_{N=-\infty}^{\infty} \exp(-\pi^2 N^2 D\tau/L^2) \times \{\cos[\pi N(x-x')/L] - \cos[\pi N(x+x')/L]\},$$

$$G_D = \theta(\tau)\frac{2}{L} \sum_{N=1}^{\infty} \exp(-\pi^2 N^2 D\tau/L^2) \sin(N\pi x/L)\sin(N\pi x'/L), \quad (9.4.18)$$

which correctly reproduces the normal-mode expansion (9.2.17).

Exercise: Apply the same reasoning to G_N in (9.4.16), and reproduce (9.2.18).

Applications of the multiple-image series are called for in Problem 9.9, and may be advantageous in Problem 9.1.

9.4.5 The reduction of dimensionality under Neumann boundary conditions

In a strip much longer than it is wide, and under NBCs on its long sides, the 2D diffusion equation mimics 1D behaviour except over very small time intervals. Similarly, the effective dimensionality can be reduced from 3 to 2 or from 3 to 1. The mechanism (see Section 3.3, point (iii)) should be compared with that of Section 6.5 for Poisson's equation. An example makes this clear.

Consider the rectangle $0 \leqslant x \leqslant a$, $0 \leqslant y \leqslant b$. Then

$$K_N^{(2)}(\boldsymbol{r}, t \mid \boldsymbol{r}', t') = \frac{1}{ab}$$
$$+ \frac{2}{ab} \sum_{n=1}^{\infty} \cos\left(\frac{n\pi x'}{a}\right) \cos\left(\frac{n\pi x}{a}\right) \exp\left[-D(n\pi/a)^2 \tau\right]$$
$$+ \frac{2}{ab} \sum_{m=1}^{\infty} \cos\left(\frac{m\pi y'}{b}\right) \cos\left(\frac{m\pi y}{b}\right) \exp\left[-D(m\pi/b)^2 \tau\right]$$
$$+ \frac{4}{ab} \sum_{n=1}^{\infty} \sum_{m=1}^{\infty} \cos\left(\frac{n\pi x'}{a}\right) \cos\left(\frac{n\pi x}{a}\right) \cos\left(\frac{m\pi y'}{b}\right) \cos\left(\frac{m\pi y}{b}\right)$$
$$\times \exp\left\{-D\left[\left(\frac{n\pi}{a}\right)^2 + \left(\frac{m\pi}{b}\right)^2\right]\tau\right\}. \tag{9.4.19}$$

(Compare this with the Poisson pseudo Green's function in eqn (6.4.2), which lacks merely the $\boldsymbol{r}$-independent zero-eigenvalue contribution $1/ab$.) Now suppose $a \ll b$, i.e. $1/a^2 \gg 1/b^2$, and that $D\tau/a^2 \gg 1$ while $D\tau/b^2 \leqslant O(1)$; in other words, $a^2 \ll D\tau \leqslant b^2$. Then, in K, all the terms with $n \geqslant 1$ become negligible, whence

$$K_N^{(2)}(x, y, t \mid x', y', t') \approx \frac{1}{a}\left\{\frac{1}{b}\right.$$
$$\left. + \frac{2}{b} \sum_{m=1}^{\infty} \cos\left(\frac{m\pi y'}{b}\right) \cos\left(\frac{m\pi y}{b}\right) \exp\left[-D(m\pi/b)^2 \tau\right]\right\}$$
$$= \frac{1}{a} K_N^{(1)}(y, t \mid y', t'). \tag{9.4.20}$$

For instance, the integral $h^{(2)}$ becomes

$$h^{(2)}(x, y, t) \equiv \int_0^a \int_0^b \mathrm{d}x'\, \mathrm{d}y'\, K_N^{(2)}(x, y, t \mid x', y', t') \psi(x', y', t_0)$$
$$\approx \int_0^b \mathrm{d}y'\, K^{(1)}(y, t \mid y', t_0) \bar{\psi}(y', t_0) = \bar{h}^{(1)}(y, t), \tag{9.4.21}$$

where $\bar{\psi}(y', t_0) \equiv a^{-1} \int_0^a dx' \psi(x', y', t_0)$ is the average of $\psi(\mathbf{r}', t_0)$ across the narrow dimension of the strip, and $\bar{h}^{(1)}$ is the corresponding 1D integral. The approximation (9.4.21) stems from the obvious fact that in a long narrow strip with insulated sides temperature differences across the strip equalize much faster than they can change along the strip.

9.5 Irreversibility

What we call reversibility in an equation is more properly known as invariance under time-reversal. Irreversibility is simply the lack of such invariance. The irreversibility of the diffusion equation is best seen by contrast with the reversibility of the wave equation; what is in question is time-evolution governed by the homogeneous equation, under time independent boundary conditions, i.e. the integrals h in the magic rules.

The formal statement of invariance under time-reversal in the *wave equation* is as follows.

If $\psi(\mathbf{r}, t)$ satisfies

$$\left(\frac{1}{c^2}\frac{\partial^2}{\partial t^2} - \nabla^2\right)\psi(\mathbf{r}, t) = 0, \tag{9.5.1}$$

then the different but related function

$$\psi^T(\mathbf{r}, t) \equiv \psi(\mathbf{r}, -t) \tag{9.5.2}$$

satisfies the same equation, i.e.

$$\left(\frac{1}{c^2}\frac{\partial^2}{\partial t^2} - \nabla^2\right)\psi^T(\mathbf{r}, t) = 0.$$

(The superscript T stands for 'time-reversed'.)

Proof: In (9.5.1) change variables $t \to -t$, whence $\partial/\partial t \to -\partial/\partial t$ and $\partial^2/\partial t^2 \to \partial^2/\partial t^2$. This yields

$$\left(\frac{1}{c^2}\frac{\partial^2}{\partial(-t)^2} - \nabla^2\right)\psi(\mathbf{r}, -t) = \left(\frac{1}{c^2}\frac{\partial^2}{\partial t^2} - \nabla^2\right)\psi^T(\mathbf{r}, t) = 0. \quad \blacksquare \tag{9.5.3}$$

To see the physical implications, suppose we film a process where $\psi(\mathbf{r}, t)$ develops in accordance with (9.5.1). Run the film backwards, and ask: does the process shown, described by $\psi^T(\mathbf{r}, t)$, satisfy the same governing equation? If the answer is yes, we say that the equation is invariant under time-reversal. For the wave equation, (9.5.3) shows that the answer is yes. In other words, one cannot tell, simply from watching, whether the film is running forwards or backwards. If $\psi(\mathbf{r}, t)$ shows an initial state $\psi(\mathbf{r}, 0)$ evolving, then $\psi^T(\mathbf{r}, t)$ shows a process evolving *into* $\psi(\mathbf{r}, 0)$. In particular, any possible initial state is also a possible final state of some evolution governed by the homogeneous wave equation.

Contrast this with the *diffusion equation*. If $\psi(\mathbf{r}, t)$ satisfies

$$(\partial/\partial t - D\nabla^2)\psi(\mathbf{r}, t) = 0, \tag{9.5.4}$$

i.e. $L\psi = 0$, then $\psi^T(\mathbf{r}, t)$ is governed by a *different* equation, namely by

$$\left(\frac{\partial}{\partial(-t)} - D\nabla^2\right)\psi(\mathbf{r}, -t) = \left(-\frac{\partial}{\partial t} - D\nabla^2\right)\psi^T(\mathbf{r}, t) = 0, \tag{9.5.5}$$

(i.e. $\tilde{L}\psi^T = 0$, where $\tilde{L}$ is in fact the adjoint operator from Section 9.2.3). Accordingly, a film run backwards shows a process ψ^T not satisfying the diffusion equation, and one could tell, simply from watching, that the film is running backwards.

In particular, while all states are possible as initial states, there are states that are not possible as final states of any evolution process governed by the homogeneous diffusion equation. (Such states can be prepared, say at $t = 0$, only by suitably arranged sources $\rho(\mathbf{r}, t)$ acting at times $t \leqslant 0$, i.e. through some time evolution governed by the *inhomogeneous* equation.)

For instance, under homogeneous DBCs at $x = 0$ and $x = \pi$, $\psi = \sin(n\pi x/L)\exp\{-D(n\pi/L)^2 t\}$ describes a possible process, but $\psi^T = \sin(n\pi x/L)\exp\{D(n\pi/L)^2 t\}$ does not. More dramatically, functions with discontinuities, like $\varepsilon(x)$ and $\theta(x)$, are possible initially but not finally. This merely paraphrases the smoothing property of time-evolution (cf. Section 8.3.1, comment (i)).

It follows automatically that under the homogeneous diffusion equation, while prediction is always possible, *retrodiction* is possible only under special restrictions on the present state, and only over limited excursions into the past. For instance, suppose we are given $\psi(x, 0) = \alpha\exp\{-(x/a)^2\}$, and are asked what the distribution was at earlier times. This is what is meant by retrodiction. Comparison with the propagator $K_0(x, \tau) = (4\pi D\tau)^{-\frac{1}{2}}\exp\{-x^2/4D\tau\}$ shows that $\psi(x, 0)$ has evolved from a distribution $\lambda\delta(x)$ at $t = T < 0$, where $T = a^2/4D$ and $\lambda = 4\pi DT\alpha$. Thus the question can be answered only for times earlier by less than T ('less than T ago'). As regards times more than T ago, the answer is: there exists no distribution from which $\psi(x, 0)$ could have arisen by undisturbed diffusion over such times. This is signalled by the essential singularity of K at $\tau = 0$, and by the unphysical properties of $K_0(x, \tau)$ for negative τ, as discussed in Section 8.3.2, comment (iv).

Reversing the sign of t in the propagator has the same effect as reversing the sign of D, which would contravene the second law of thermodynamics, as discussed in Section 8.1. Thus, the diffusion equation is irreversible for the same reasons that thermodynamics is; the fundamental physical problem of accounting for such irreversibility from first principles is discussed briefly at the end of Appendix D.

9.6 The Schroedinger equation

We confine ourselves to some very brief comments on the free-particle Schroedinger equation

$$\left(i\hbar\frac{\partial}{\partial t}+\frac{\hbar^2}{2m}\nabla^2\right)\psi(\boldsymbol{r},t)=0, \tag{9.6.1}$$

which is always homogeneous.

(i) As already mentioned in Section 3.2, the replacements

$$t\to it, \qquad D\to\hbar/2m \tag{9.6.2}$$

turn the diffusion into the Schroedinger equation. Both are parabolic, and require the same boundary and initial conditions in order to be 'well-posed' (see Section 3.5.1).

(ii) The replacements (9.6.2) turn the diffusion propagator $(4\pi D\tau)^{-\frac{3}{2}}$ $\exp\{-R^2/4D\tau\}$ into the Schroedinger propagator

$$K(\boldsymbol{R},\tau)=\left(\frac{m}{2\pi\hbar\tau}\right)^{\frac{3}{2}}\exp(-3\pi i/4)\exp(imR^2/2\hbar\tau), \tag{9.6.3}$$

which obeys (9.6.1) and the initial condition $K(\boldsymbol{R},0)=\delta(\boldsymbol{R})$. However, this formal correspondence masks very significant differences, both mathematical and physical.

(iii) In quantum mechanics, $\psi(\boldsymbol{r},t)$ is non-trivially complex (Appendix E explains what this means). Even if $\psi(\boldsymbol{r},t)$ should happen to be real at a particular time t, $\psi(\boldsymbol{r},t+\delta t)\equiv\psi+\delta\psi$ will be complex because (9.6.1) implies that $\delta\psi\approx i(\hbar/2m)\nabla^2\psi\,\delta t$ is pure imaginary.

The wavefunction ψ is not itself observable, though we can calculate observable distributions only through ψ. Examples are the probability density $|\psi|^2$ and the probability-current density $\boldsymbol{j}=-(i\hbar/2m)(\psi^*\boldsymbol{\nabla}\psi-(\boldsymbol{\nabla}\psi^*)\psi)$. Notice that these are quadratic in ψ, in sharp contrast to the diffusion equation, where ψ itself is the density that (in the absence of sources) is conserved.

(iv) The Schroedinger equation is reversible. By contrast to (9.5.2), one has $\psi^T(\boldsymbol{r},t)\equiv\psi^*(\boldsymbol{r},-t)$, which satisfies the same equation (9.6.1) as does ψ itself, and describes the time-reversed process (film running backwards). For instance, $\boldsymbol{j}^T$ constructed from ψ^T is the negative of $\boldsymbol{j}$ constructed from ψ.

Exercise: Verify these statements.

Correspondingly, unlike the diffusion propagator, (9.6.3) does make physical sense for negative τ, in spite of the essential singularity at $\tau=0$.

Problems

9.1 *DBCs in 1D*. The temperature $\psi(x, t)$ obeys the homogeneous diffusion equation for $0 \leqslant x \leqslant \pi$, and vanishes at $t = 0$. For $t > 0$ it obeys the BCs $\psi(0, t) = A$, $\psi(\pi, t) = 0$; in other words, at time zero the temperature of the left-hand end is suddenly raised from 0 to the subsequently constant value A, while the right-hand end is maintained at 0.

(i) Use the eigenfunction expansion of the propagator, and the magic rule, to show that

$$\psi(x, t) = \frac{2A}{\pi} \sum_{n=1}^{\infty} \frac{1}{n} \sin(nx)(1 - \exp(-n^2 Dt)).$$

The series $(2/\pi) \sum_1^\infty (1/n) \sin(nx) = (1 - x/\pi)$, $(0 < x < 2\pi)$, is standard. The series $S(x, t) \equiv (2/\pi) \sum_1^\infty (1/n) \sin(nx) \exp(-n^2 Dt)$ is the centrepiece of the problem.
(ii) Check that, as $t \to \infty$, your expression for ψ reproduces the eventually expected steady-state distribution $\psi(x, \infty) = A(1 - x/\pi)$.
(iii) Set $D = 1$, $A = 1$, and to an accuracy of $1{:}10^3$ determine the midpoint temperature $\psi(\pi/2, t)$ at $t = 0.01$, 0.1, 1, 10, 100. How many terms do you need to sum in each case?

Hint: For small t, it is much better to transform the series for $S(\pi/2, t)$ by Poissons' formula (Appendix C). You will then need an integral from Appendix A, and the asymptotic expression given there for the error function.

9.2 *NBCs in 1D*. As in Problem 9.1, but with the BCs $\psi_x(0, t) = -B$, $\psi_x(\pi, t) = 0$ for $t > 0$, where B is a constant.

(i) Determine $\psi(x, t)$.
(ii) Verify that

$$\psi(x, t \to \infty) \approx B\left\{Dt/\pi + (2/\pi) \sum_1^\infty (1/n^2) \cos(nx)\right\}.$$

(iii) Verify that this approximate asymptotic expression obeys the diffusion equation and the BCs exactly, and discuss it briefly in the light of the physics prescribed by the BCs.

Hint: You may quote the Fourier series (A.5.6).

9.3 *Mixed BCs in 1D*. As in Problems 9.1 and 9.2, but with the BCs $\psi_x(0, t) = -B$, $\psi(\pi, t) = 0$ for $t > 0$; in other words the left-hand end suffers a steady heat input, while the right-hand end is kept at zero temperature.

(i) Construct the appropriate propagator K, from the eigenfunctions ϕ_n of $-\partial^2/\partial x^2$, subject to the BCs $\phi_{nx}(0) = 0 = \phi_n(\pi)$, and from the corresponding eigenvalues.
(ii) Use K and the magic rule to determine $\psi(x, t)$.
(iii) Verify that

$$\psi(x, \infty) = B\frac{8}{\pi}\sum_{n=0}^{\infty}\frac{\cos[(2n+1)x/2]}{(2n+1)^2}.$$

Is this compatible with the eventually-expected steady-state solution $B(\pi - x)$?

9.4 *Radioactive heating.* The region $0 \leqslant x \leqslant L$ is at zero temperature at $t = 0$, and the temperature satisfies the BCs $\psi(0, t) = 0 = \psi(L, t)$ for all t. At $t = 0$, radioactive material is deposited uniformly throughout the region, producing a source distribution $\rho(x, t) = Q\exp(-\Lambda t)$.

(i) Use the magic rule and the appropriate eigenfunction expansion of the Green's function to write down a series for the midpoint temperature $\psi(L/2, t)$ for $t > 0$. It is interesting to ask what is the maximum value of $\psi(L/2, t)$ and at what time t^* it is reached.
(ii) Explain why the analogous question under NBCs would be trivial.
(iii) If the radioactive decay is fast, $\Lambda \gg D\pi^2/L^2$, show that at times $t \gtrsim 1/\Lambda$ your exact expression is reasonably approximated by

$$\psi(L/2, t) \approx \frac{Q}{\Lambda}\frac{4}{\pi}\sum_{n=0}^{\infty}\frac{(-1)^n}{(2n+1)}\exp[-D\pi^2(2n+1)^2 t/L^2].$$

(iv) If the decay is slow, $\Lambda \ll D\pi^2/L^2$, show that at times $t \gtrsim L^2/D\pi^2$ your exact expression is reasonably approximated by

$$\psi(L/2, t) \approx Q\frac{4L^2}{D\pi^3}\exp(-\Lambda t)\sum_{n=0}^{\infty}\frac{(-1)^n}{(2n+1)^3}.$$

(The sum equals $\pi^3/32$.)
(v) If $\Lambda = D\pi^2/L^2$, then, in terms of $\theta \equiv \Lambda t$, your expression should reduce to

$$\psi(L/2, t) = \frac{Q}{\Lambda}\frac{4}{\pi}\exp(-\theta)\Big\{\theta + \sum_{m=1}^{\infty}\frac{(-1)^m}{2m(2m+1)(2m+2)} \times[1 - \exp(-4m+1)\theta)]\Big\}.$$

This peaks at θ^*. Determine θ^* and $(\Lambda/Q)\psi(L/2, t^*)$ numerically to an accuracy of 1%. (Notice that the sum now amounts only to a fairly small correction.)

9.5 For $x \geqslant 0$, ψ satisfies the homogeneous diffusion equation, and for longer than anyone can remember has satisfied the BC $\partial\psi/\partial x = -B\cos(\omega t)$ at $x = 0$.

(i) Determine $\psi(x, t)$ by adapting the approach of Section 8.2.
(ii) Tackle the same problem the hard way, by starting from the magic rule and using an approach like that of Appendix J.

Hint: In (ii) you might meet the integral

$$\int_0^\infty d\xi \frac{1}{\xi^2} \exp[-(\xi - \lambda/\xi)^2].$$

Changing the integration variable from ξ to $\eta = V/\xi$ reduces this to the one discussed in Appendix J.

9.6 *Method of images.* In a long column of solvent, of unit cross-sectional area, the concentration obeys the homogeneous diffusion equation for $x \geqslant 0$, and vanishes at $t = 0$. For $t > 0$, it obeys the BC $\psi(0, t) = A$.

(i) Calculate $\psi(x, t)$.
(ii) Calculate the total amount of solute $N(t) = \int_0^\infty dx\, \psi(x, t)$ in solution at time t.
(iii) Calculate dN/dt.
(iv) Show that $-D\psi_x(x, t) = A(D/\pi t)^{\frac{1}{2}} \exp(-x^2/4Dt)$, and verify that $-D\psi_x(0, t) = A(D/\pi t)^{\frac{1}{2}}$ agrees with the result of (iii).

9.7 *Exterior problem of the sphere.* The surface of a sphere of radius b is suddenly raised to temperature A at time $t = 0$, and is then maintained at this temperature. At $t = 0$, the outside temperature is zero.

(i) Find $\psi(r, t)$ for $t > 0$, $r > b$, by writing it as $\psi(r, t) = \phi(r, t)/r$; the resulting equation and BC for $\phi(r, t)$ then define an effectively 1D problem, solvable by our by-now-familiar 1D methods. Verify that your solution can be put into the form

$$\psi(r, t) = \frac{Ab}{r} \frac{2}{\pi^{\frac{1}{2}}} \int_{(r-b)/(4Dt)^{\frac{1}{2}}}^{\infty} d\xi \exp(-\xi^2).$$

(ii) Consider the limit $b \to 0$ at fixed A, r, t, and verify that in this limit ψ vanishes (see Section 7.3).

9.8 *Interior problem of the sphere.* Consider the isotropic solution of the homogeneous diffusion equation inside a sphere of radius b. Write down the isotropic part $\gamma(r, t \mid r', t')$ of the Dirichlet Green's function, once from first principles, and then by taking the limit $a \rightarrow 0$ in the example of Section 9.4.2. (The fact that this limit gives the correct result is remarkable in itself; it echoes Section 7.3 and Problem 9.7 above.)

9.9 *Fourier series or multiple images*? In the example of Section 9.4.1, set $D = 1$, $\alpha = 1$, and to an accuracy of $1:10^3$ evaluate the endpoint temperature $\psi(0, t)$ for $t = 0.01$, 0.1, 1, 10.

Hint: For small t, it is sensible either to use the multiple-image form (9.4.16) of the Neumann Green's function from the start, or else to transform the series for $\psi(0, t)$ by Poisson's formula (Appendix C). For large t, the normal-mode expansion (9.4.1) is more convenient. One should convince oneself of this by trying, just once, to use the less sensible approach (say the normal-mode form for $t = 0.01$, and the multiple-image form for $t = 10$), and comparing the number of terms needed for the requisite accuracy. For intermediate values of t, either form should serve.

IV Waves

Summary

Wave equation:

$$\Box^2 \equiv \left(\frac{1}{c^2}\frac{\partial^2}{\partial t^2} - \nabla^2\right); \qquad \Box^2\psi(\boldsymbol{r}, t) = \rho(\boldsymbol{r}, t)$$

BCs: as for diffusion, i.e. either D or N (or mixed) on a closed surface S, for all $t \geqslant t_0$.

Initial conditions: $\psi(\boldsymbol{r}, t_0)$ and $\psi_t(\boldsymbol{r}, t_0)$ throughout V.

For sound waves: density of fluid $\equiv \sigma$; velocity potential $= \psi$, $(\boldsymbol{v} = -\nabla\psi)$; excess pressure $\Delta p = \sigma\psi_t$.
Density and flux of available energy: H and $\boldsymbol{N}$,

$$H = \tfrac{1}{2}\sigma\left\{\frac{1}{c^2}\psi_t^2 + (\nabla\psi)^2\right\}, \qquad \boldsymbol{N} = -\sigma\psi_t\nabla\psi;$$

$$\frac{\partial H}{\partial t} + \operatorname{div}\boldsymbol{N} = \sigma\rho\psi_t.$$

Propagator: $\Box^2 K(\boldsymbol{r}, t \mid \boldsymbol{r}', t') = 0$; homogeneous BCs;

$$\lim_{t\to t'+} K(\boldsymbol{r}, t \mid \boldsymbol{r}', t') = 0,$$

$$\lim_{t\to t'+} K_t(\boldsymbol{r}, t \mid \boldsymbol{r}', t') = c^2\,\delta(\boldsymbol{r} - \boldsymbol{r}').$$

Symmetries:

$$K(\boldsymbol{r}, t \mid \boldsymbol{r}', t') = K(\boldsymbol{r}, \boldsymbol{r}'; \tau \equiv t - t');$$
$$K(\boldsymbol{r}, \boldsymbol{r}'; \tau) = K(\boldsymbol{r}', \boldsymbol{r}; \tau).$$

Green's function: $\Box^2 G(\boldsymbol{r}, t \mid \boldsymbol{r}', t') = \delta(t - t')\,\delta(\boldsymbol{r} - \boldsymbol{r}')$; homogeneous BCs; $G = 0$ for $t < t'$.

$$G(\boldsymbol{r}, t \mid \boldsymbol{r}', t') = \theta(t - t')K(\boldsymbol{r}, t \mid \boldsymbol{r}', t').$$

Magic rule:

$$\psi(\mathbf{r}, t) = \int_{t_0}^{t+} \mathrm{d}t' \int_V \mathrm{d}V\,' G(\mathbf{r}, t \,|\, \mathbf{r}', t')\rho(\mathbf{r}', t')$$

$$+ \int_{t_0}^{t+} \mathrm{d}t' \int_S \mathrm{d}S' \, [G(\mathbf{r}, t \,|\, \mathbf{r}', t') \, \partial_n' \psi_S(\mathbf{r}', t')$$

$$- (\partial_n' G(\mathbf{r}, t \,|\, \mathbf{r}', t'))\psi_S(\mathbf{r}', t')]$$

$$+ \int_V \mathrm{d}V' \frac{1}{c^2} G(\mathbf{r}, t \,|\, \mathbf{r}', t_0)\psi_{t_0}(\mathbf{r}', t_0)$$

$$+ \frac{\partial}{\partial t} \int_V \mathrm{d}V' \frac{1}{c^2} G(\mathbf{r}, t \,|\, \mathbf{r}', t_0)\psi(\mathbf{r}', t_0).$$

Eigenfunction expansion: with $-\nabla^2 \phi_p = \lambda_p \phi_p$ plus homogeneous BCs; $\omega_p^2 \equiv c^2 \lambda_p$:

$$K(\mathbf{r}, t \,|\, \mathbf{r}', t') = c^2 \left\{ \frac{1}{V}(t - t') + \sum_p{}' \phi_p^*(\mathbf{r}')\phi_p(\mathbf{r}) \frac{\sin [\omega_p(t - t')]}{\omega_p} \right\}$$

where the first term is present only if $\lambda_0 = 0$ is an eigenvalue.

In unbounded space: $\boldsymbol{R} \equiv \boldsymbol{r} - \boldsymbol{r}'$, $\tau \equiv t - t'$:

$$K_0(R, \tau) = c(2\pi)^{-n} \int \mathrm{d}^n k \, \frac{\sin (kc\tau)}{k} \exp (i\boldsymbol{k} \cdot \boldsymbol{R})$$

$$G_0^{(3)} = \theta(\tau) \frac{\delta(\tau - R/c)}{4\pi R}; \qquad G_0^{(2)} = \theta(\tau) \frac{1}{2\pi} \frac{\theta(\tau - R/c)}{[\tau^2 - R^2/c^2]^{\frac{1}{2}}};$$

$$G_0^{(1)} = \theta(\tau) \frac{c}{2} \theta(\tau - |X|/c).$$

Poisson's solution: in 3D unbounded space

$$\psi(\mathbf{r}, t) = (t - t_0) M_{\mathbf{r}, c(t - t_0)} \{\psi_t(\mathbf{r}', t_0)\} + \frac{\partial}{\partial t} [(t - t_0) M_{\mathbf{r}, c(t - t_0)} \{\psi(\mathbf{r}', t_0)\}]$$

where

$$M_{\mathbf{r}, R}\{f\} \equiv \left\{ \begin{matrix} \text{average of } f \text{ over a sphere} \\ \text{with centre } \boldsymbol{r} \text{ and radius } R \end{matrix} \right\}$$

10 The wave equation: I. General theory

Diffusion has already acquainted us with several of the characteristics attending time as one of the independent variables; general comments will now be reserved mainly for features peculiar to waves, stemming ultimately from the appearance of $\partial^2/\partial t^2$ instead of $\partial/\partial t$. Here we start from the general theory (bounded regions), largely because the behaviour of waves in unbounded space is so various, and so spectacularly different in different dimensions, that it is better tackled separately in the next chapter, having first disposed of some generalities in this one. Although, in principle, unbounded space is a special limiting case of bounded regions, we shall see that for waves the limit is more delicate than for diffusion. In this respect as in some others the wave equation is closer mathematically to Poisson's than to the diffusion equation.

From previous chapters, we shall need especially the ideas on time-translation invariance (Section 9.2.2) and reversibility (Section 9.5). Propagators and Green's functions are again central, but differ very significantly from those we have met before.

Only scalar waves will be considered. Electromagnetic waves are discussed in many excellent texts: the favourites of the present writer are Feynman *et al.* (1964), Landau and Lifshitz (1975), Stratton (1941), Panofsky and Phillips (1955), and Jackson (1962). Though electromagnetic waves are more important in physics, they are also far more complicated in their details, while scalar waves provide perfectly fair yet incomparably more accessible analogues to those wave properties that do not depend critically on polarization or on the conservation of electric charge (i.e. of source strength). Nevertheless, our emphasis and choice of examples are slanted towards making these analogies with electromagnetism as vivid as possible.

We shall go into relatively little detail about waves in bounded regions, because they are well covered in several elementary texts. A clear introduction to waves in general is given by Coulson and Jeffrey (1977).

We must stress that we shall confine ourselves strictly to the linear wave equation (10.1.1) and to its mathematical consequences, even though it applies to acoustic and to elastic waves only as a first approximation (cf. Appendix K). Thus, without further apology, we shall investigate singular solutions which certainly violate the physical assumptions under which such scalar wave equations are derived in the first place. Consequently, such solutions can be applied to sound waves only with a pinch of salt. By contrast, in classical electromagnetism their vectorial analogues are exact, and the relevance of singular solutions there is limited only by practical limitations on emitters and receivers.

Finally, though we shall consider energy propagation and the power output of radiation sources, we lack the space to discuss momentum flux and the forces exerted on bounding surfaces.

10.1 Introduction

Waves are encountered so widely in elementary physics that a few introductory comments should suffice.

(i) We shall consider only scalar waves, i.e. only waves governed by the equation

$$\Box^2\psi(\mathbf{r}, t) \equiv \left(\frac{1}{c^2}\frac{\partial^2}{\partial t^2} - \nabla^2\right)\psi(\mathbf{r}, t) = \rho(\mathbf{r}, t), \tag{10.1.1}$$

where the wave function ψ and the source density ρ are scalar quantities. We write $\Box^2$ instead of L for the wave operator, following common practice. The parameter c has the dimensions of velocity.

(ii) Generally we shall envisage sound waves, i.e. small-amplitude density waves in an ideal (non-viscous) compressible fluid. Their physics, leading to the expressions quoted below, is outlined in Appendix K. For sound, ψ is the velocity potential, related to the fluid velocity $\mathbf{v}$ and to the excess pressure Δp by

$$\mathbf{v} = -\nabla\psi, \qquad \Delta p = \sigma\frac{\partial\psi}{\partial t}, \tag{10.1.2a,b}$$

where σ is the equilibrium density of the fluid (called σ_0 in Appendix K), and $c^2 \equiv \partial p/\partial\sigma$ is the bulk modulus, taken as a constant. As in the steady flow of an incompressible fluid (Section 4.1), $\rho(\mathbf{r}, t)\,\delta V\,\delta t$ is the volume of fluid injected into the (geometric) volume element δV in time δt.

(iii) Evidently, NBCs apply on a surface S if one prescribes the normal velocity $v_{nS} = -\partial_n\psi_S$ across S (see Appendix G for the notation). Often such conditions are enforced by moving an impenetrable bounding surface with velocity v_n along its normal. DBCs are enforced by prescribing the excess pressure exerted across S; this prescribes $\partial\psi_S(\mathbf{r}, t)/\partial t$, and hence $\psi_S(\mathbf{r}, t)$ (by integration with respect to time, starting from equilibrium where ψ_S may be taken as zero). For instance, homogeneous DBCs apply to sound waves at an open end of a pipe (in an approximation where no excess pressure can be supported by whatever lies outside), while homogeneous NBCs apply at a closed end.

(iv) The so-called acoustic energy density H, the acoustic energy flux $\boldsymbol{N}$, and the local conservation law connecting them, are

$$H = \tfrac{1}{2}\sigma\left\{\frac{1}{c^2}\left(\frac{\partial\psi}{\partial t}\right)^2 + (\boldsymbol{\nabla}\psi)^2\right\}, \tag{10.1.3}$$

$$\boldsymbol{N} = -\sigma\frac{\partial\psi}{\partial t}\boldsymbol{\nabla}\psi = \Delta p\boldsymbol{v}, \tag{10.1.4}$$

$$\frac{\partial H}{\partial t} + \operatorname{div}\boldsymbol{N} = \sigma\rho\frac{\partial\psi}{\partial t} = \rho\Delta p. \tag{10.1.5}$$

Appendix K derives (10.1.5), and explains that H is the density not of the total energy but of the so-called available energy. However, we shall not explicitly qualify H and $\boldsymbol{N}$ as 'available' or 'acoustic', partly for brevity, and partly because their electromagnetic analogues *in vacuo* are indeed the total energy density and the Poynting vector. Note that the second term on the right of (10.1.3) is the kinetic-energy density, while the first term stems from the potential-energy changes on compression.

The expression on the right of (10.1.5) is the rate at which the sources ρ must do work per unit volume against the excess pressure. Integration over V yields the rate of change of the overall energy of the system:

$$\frac{\mathrm{d}H}{\mathrm{d}t} \equiv \frac{\mathrm{d}}{\mathrm{d}t}\int_V \mathrm{d}V\,H(\boldsymbol{r},t) = \int_V \mathrm{d}V\frac{\partial H}{\partial t} = \int_V \mathrm{d}V\left\{\rho\sigma\frac{\partial\psi}{\partial t} - \operatorname{div}\boldsymbol{N}\right\}, \tag{10.1.6a}$$

$$\frac{\mathrm{d}H}{\mathrm{d}t} = \int_V \mathrm{d}V\,\rho\sigma\frac{\partial\psi}{\partial t} - \int_S \mathrm{d}SN_n$$
$$\equiv W_V(t) + W_S(t). \tag{10.1.6b}$$

The term W_V is the overall power output of the sources ρ. The other term W_S can be viewed in two ways. Obviously, its integrand $\mathrm{d}SN_n$ is the power outflow across $\mathrm{d}S$; equivalently, $-\mathrm{d}SN_n$ is the rate at which energy must be supplied across $\mathrm{d}S$ in order to enforce any inhomogeneous BCs on $\mathrm{d}S$. Notice that W_S vanishes under homogeneous BCs, whether Dirichlet (ψ_S and hence $\partial\psi_S/\partial t$ zero), or Neumann ($\partial_n\psi_S$ zero): in either case N_n vanishes on S.

(v) For sound, the directly observable quantities are the derivatives $\partial\psi/\partial t = \Delta p/\sigma$ and grad $\psi = -\boldsymbol{v}$, rather than ψ itself. (Compare the roles of fields and potentials in classical electromagnetism.)

(vi) By contrast, the wave equation, say in 1D, is equally appropriate if ψ is the y-displacement of an elastic string stretched along the x-axis, so that ψ itself is observable. (The equation then applies provided $|\partial\psi/\partial x| \ll 1$: see Section 4.1.) In this case $\rho\,\delta x$ is the transverse external force applied to the element δx of the string. With T the equilibrium tension and σ the mass per unit length (taken as constant), one has $c = (T/\sigma)^{\frac{1}{2}}$, and

$$H = \tfrac{1}{2}\sigma\left(\frac{\partial\psi}{\partial t}\right)^2 + \tfrac{1}{2}T\left(\frac{\partial\psi}{\partial x}\right)^2 = \tfrac{1}{2}\sigma c^2\left\{\frac{1}{c^2}\left(\frac{\partial\psi}{\partial t}\right)^2 + (\boldsymbol{\nabla}\psi)^2\right\}, \tag{10.1.7}$$

$$\boldsymbol{N} = -T\frac{\partial\psi}{\partial t}\frac{\partial\psi}{\partial x} = -\sigma c^2\frac{\partial\psi}{\partial t}\frac{\partial\psi}{\partial x}. \tag{10.1.8}$$

These expressions should be compared with (10.1.3, 4); notice that the kinetic-energy density now depends on the t rather than the x derivative. The energy balance is again described by (10.1.5). DBCs now apply when the displacement at the ends is prescribed; NBCs apply if, at an endpoint, one prescribes the applied transverse force to be balanced by the string. The 2D analogues of these expressions apply to the transverse vibrations of a stretched membrane.

Exercise: Derive (10.1.7, 8).

(vii) The wave equation reduces to Poisson's equation when there is no time-dependence, so that, in particular, $\rho(\boldsymbol{r}, t) = \rho(\boldsymbol{r})$, admitting the possibility that $\partial^2\psi/\partial t^2 = 0$. More interesting is the so-called *non-retarded* or *non-relativistic* limit where, formally, we let $c \to \infty$. The wave equation again reduces to Poisson's equation, even if the sources are time-dependent. Then one has

$$-\nabla^2\psi(\boldsymbol{r}, t) = \rho(\boldsymbol{r}, t), \qquad (c \to \infty), \tag{10.1.9}$$

so that t is demoted from independent variable to a mere parameter. In other words, $\psi(\boldsymbol{r}, t)$ at a field point $\boldsymbol{r}$ now follows any time-dependent changes of $\rho(\boldsymbol{r}', t)$ at any source point $\boldsymbol{r}'$ instantaneously, irrespective of the distance $|\boldsymbol{r} - \boldsymbol{r}'|$: into the *relation* between the functions $\psi(\boldsymbol{r}, t)$ and $\rho(\boldsymbol{r}, t)$, i.e. into the Poisson Green's functions, time does not enter at all.

It is fascinating to ask under what conditions such non-retarded expressions are adequate *approximations* to what really happens. On dimensional grounds one would guess that they might apply when $L/cT \ll 1$, where L is the largest relevant length in the problem, and T is the shortest relevant time-scale. Roughly speaking, this guess turns out to be correct whenever L and T can be identified with confidence. For instance, if a 3D point source oscillates in strength with frequency $f = 1/T$, and if we require ψ at a distance L from the source, then the non-retarded expressions apply provided $L/cT = fL/c = L/\lambda \ll 1$, where λ is the wavelength of (plane) waves having frequency f. Section 12.4 gives an example. On the other hand, Section 11.4 points out that in 2D and 1D the non-retarded (i.e. $c \to \infty$) limit for the Green's functions (as distinct from the wave equations) cannot be implemented straightforwardly.

(viii) The unmistakable relativistic features of the wave operator $\square^2$ are exploited in Appendix O.

(ix) The homogeneous wave equation under homogeneous BCs can be solved by separating the variables t and $\boldsymbol{r}$, which yields the normal-mode solutions $F_n(t)\phi_n(\boldsymbol{r})$. Here the ϕ_n are the eigenfunctions of $-\nabla^2$ (Section 4.5); thus

$$-\nabla^2\phi_n = k_n^2\phi_n \equiv (\omega_n/c)^2\phi_n. \tag{10.1.10}$$

For convenience, the ϕ_n are always normed in the region V, whose volume is likewise written as V. Thus $\int_V \mathrm{d}V\, \phi_{n'}^*\phi_n = \delta_{n'n}$. Recall that the k_n^2 are discretely spaced when V is finite. The F_n obey the ordinary differential equation $\mathrm{d}^2F_n/\mathrm{d}t^2 = \omega_n^2F_n$. Under NBCs, the lowest eigenvalue, labelled by $n = 0$, is $\omega_0^2 = 0$, and $\phi_0 = V^{-\frac{1}{2}}$. Under any other BCs, no ω_n^2 vanishes. In this notation, $F_0(t) = A_0t + \Delta_0$, while for $n \neq 0$, $F_n(t) = a_n \sin[\omega_nt + \delta_n]$, with arbitrary A_0, a_n, Δ_0, δ_n. Accordingly, the full normal-mode functions can be written

$$\psi_0(\boldsymbol{r}, t) = (A_0t + \Delta_0)\phi_0(\boldsymbol{r}) = (A_0t + \Delta_0)V^{-\frac{1}{2}},$$
$$\psi_n(\boldsymbol{r}, t) = a_n \sin[\omega_nt + \delta_n]\phi_n(\boldsymbol{r}), \qquad n \neq 0. \tag{10.1.11}$$

(Note that A_0 and Δ_0 have different dimensions from a_n, δ_n; thus $[A_0] = [a_n]T^{-1}$. We shall write sums over all normal modes except $n = 0$ as $\sum_n'$; thus

$$\sum_n{}' \cdots \equiv \sum_{n(\neq 0)} \cdots. \tag{10.1.12}$$

Except under NBCs, there is no $n = 0$ term, and $\sum_n' = \sum_n$.

(x) As discussed in Section 3.2, the wave equation admits simple-harmonic solutions $\psi(\boldsymbol{r}, t) = \psi_\omega(\boldsymbol{r}) \exp(-i\omega t)$, provided any prescribed boundary values, and the source density $\rho = \rho_\omega(\boldsymbol{r}) \exp(-i\omega t)$, are themselves simple harmonic. Then ψ_ω obeys the associated inhomogeneous Helmholtz equation

$$-(\nabla^2 + k^2)\psi_\omega(\boldsymbol{r}) = \rho_\omega(\boldsymbol{r}), \tag{10.1.13}$$

where $k^2 \equiv \omega^2/c^2$.

For simplicity, consider a system driven by sources ρ as above, and subject to homogeneous BCs, like a resonator with rigid walls. From Sections 2.3, 4 we recall the Fredholm alternatives for solutions of (10.1.13) (see especially Table 2.1). In the general case, where $\omega \neq \omega_n$, there is a unique solution; in the general special case, where $\omega = \omega_n$ and

$\int_V dV\,\phi_n^*\rho_\omega \neq 0$, there is no solution; in the special special case, where $\omega = \omega_n$ and $\int_V dV\,\phi_n^*\rho_\omega = 0$, (10.1.13) has solutions, which are undetermined up to an addend constant $\times\, \phi_n(\mathbf{r})$. The physics underlying these alternatives will become clearer from the example in Section 10.5.2, which illustrates the fact that the wave equation always has a unique solution, though even in response to a harmonic driving term this solution need not always be harmonic.

(xi) The wave operator $\Box^2$ unlike the diffusion operator $(\partial/\partial t - D\nabla^2)$ is self-adjoint in hyperspace, but we shall make no direct use of this property. Indirectly, it does make the uniqueness proof more transparent, and entails that the reciprocal equation for the Green's function is the same as the original equation (cf. Section 10.3.3 below).

10.2 Boundary conditions, initial conditions, and uniqueness

Section 3.5.5 asserted that the problem of solving the wave equation is 'well-posed' under the same BCs conditions as for diffusion (D or N or mixed, but not Cauchy), while the initial conditions at $t = t_0$ must now prescribe both $\psi(\mathbf{r}, t_0)$ and $\psi_t(\mathbf{r}, t_0)$ throughout V (Cauchy ICs). We proceed to show that under such conditions the solution is unique; the strategy is broadly the same as in Section 9.1 for diffusion.

Proof: Consider two functions ψ_1 and ψ_2, both satisfying all the requirements. Then their difference $\phi(\mathbf{r}, t) \equiv \psi_1(\mathbf{r}, t) - \psi_2(\mathbf{r}, t)$ obeys the homogeneous wave equation, the homogeneous version of the BCs on ψ or $\partial_n\psi$, and the homogeneous Cauchy ICs $\phi(\mathbf{r}, t_0) = 0 = \phi_t(\mathbf{r}, t_0)$. We show that $\phi(\mathbf{r}, t)$ vanishes at all $\mathbf{r}$ and for all $t > t_0$.

The crucial idea is to work with the time-integral $J = \int_{t_0}^{t} dt' \int_V dV\, dH(\mathbf{r}, t')/dt'$ of the conservation law (10.1.6), where

$$H = \tfrac{1}{2}\sigma\left\{\left(\frac{\partial\phi}{\partial t}\right)^2 + (\nabla\phi)^2\right\}$$

is the energy density, and $\mathbf{N} = -\sigma(\partial\phi/\partial t)\nabla\phi$ the energy flux corresponding to ϕ as in (10.1.3–5). (The factor σ is irrelevant here, but we carry it to avoid a clash of notations.) In this case $\rho = 0$ (because ϕ obeys the homogeneous equation), and N_n vanishes on S (because ϕ obeys homogeneous BCs); therefore W_V and W_S both vanish, and (10.1.6) entails

$$J \equiv \int_{t_0}^{t} dt' \frac{d}{dt'} H = H(t) - H(t_0) = H(t) = 0. \qquad (10.2.1)$$

The second step follows because $H(t_0)$ vanishes by virtue of the homogeneous ICs on ϕ; the last step follows because the RHS of (10.1.6) vanishes, as explained above. Thus

$$H(t) = \tfrac{1}{2}\sigma \int_V \mathrm{d}V \left\{ \left(\frac{\partial \phi}{\partial t} \right)^2 + (\nabla \phi)^2 \right\} = 0; \tag{10.2.2}$$

since the integrand is non-negative, the integral can vanish only if the integrand vanishes for all $\boldsymbol{r}$. Therefore ϕ is a constant independent of $\boldsymbol{r}$ and of t, which in fact vanishes by continuity since it vanishes at $t = t_0$. ∎

To avoid confusion, one should state explicitly that the wave equation *may* be solvable, though perhaps not uniquely, under hyperboundary conditions different from those adopted above. For instance, instead of $\psi(\boldsymbol{r}, t_0)$ and $\psi_t(\boldsymbol{r}, t_0)$, one could try to prescribe $\psi(\boldsymbol{r}, t_0)$ and $\psi(\boldsymbol{r}, t_1)$, and then to determine $\psi(\boldsymbol{r}, t)$ for $t_0 \leqslant t \leqslant t_1$. However, a solution of (10.1.1) satisfying these requirements exists only under further conditions on the data (i.e. on $\psi(\boldsymbol{r}, t_0)$, $\psi(\boldsymbol{r}, t_1)$, $\rho(\boldsymbol{r}, t)$, and on the prescribed boundary values); hence the problem is not 'well-posed', because it cannot be solved for arbitrary data. It is similar to the worked example of Section 2.5 in that, from a point of view in hyperspace, it constitutes a pure BVP (with a closed hyperboundary) rather than an IVP (with an open hyperboundary).

10.3 Propagator and Green's function

The programme is to define the propagator K; determine its symmetries; expand it in normal modes; define the Green's function G; and finally to establish the reciprocal equation for G, which is needed in the magic rule. The reasoning follows the same lines as did Section 9.2 for diffusion; consequently we proceed quite briskly. The reader is asked to bear with this formalism, pending applications once the magic rule has been derived in Section 10.4. Meanwhile we anticipate that K solves the homogeneous problem with prescribed initial values, while G solves the general inhomogeneous problem with prescribed volume and surface sources in addition.

10.3.1 The propagator and its symmetries

The propagator K is defined by

$$\square^2 K(\boldsymbol{r}, t \mid \boldsymbol{r}', t') \equiv \left(\frac{1}{c^2} \frac{\partial^2}{\partial t^2} - \nabla^2 \right) K(\boldsymbol{r}, t \mid \boldsymbol{r}', t') = 0, \tag{10.3.1a}$$

$$\lim_{t \to t'+} K(\boldsymbol{r}, t \mid \boldsymbol{r}', t') = 0, \tag{10.3.1b}$$

$$\lim_{t \to t'+} K_t(\boldsymbol{r}, t \mid \boldsymbol{r}', t') = c^2\, \delta(\boldsymbol{r} - \boldsymbol{r}'), \tag{10.3.1c}$$

plus the *homogeneous* version of the BCs to be imposed on ψ:

$$K(\boldsymbol{r}, t \mid \boldsymbol{r}', t') = 0, \quad \text{or} \quad \partial_n K(\boldsymbol{r}, t \mid \boldsymbol{r}', t') = 0 \quad \text{for} \quad \boldsymbol{r} \text{ on } S. \tag{10.3.1d}$$

The initial conditions (10.3.1b, c) are chosen by hindsight, and will be motivated by the magic rule in Section 10.4. The factor c^2 on the right of (10.3.1c) likewise stems from hindsight. Often but not always one can ignore the limiting prescriptions on the left of (10.3.1b, c), and simply set $t = t'$.

The dimensions of K in nD follow from (10.3.1c): $T^{-1}[K] = (LT^{-1})^2 L^{-n}$, whence

$$[K] = L^{2-n} T^{-1}. \tag{10.3.2}$$

By exactly the same argument as in Section 9.2.2, K is invariant under *time-translation*, and therefore depends on t and t' only through the combination $t - t'$: whenever convenient we can write

$$K(\boldsymbol{r}, t \mid \boldsymbol{r}', t') = K(\boldsymbol{r}, \boldsymbol{r}'; \tau), \qquad \tau \equiv t - t'. \tag{10.3.3}$$

Corollaries:

$$K(\boldsymbol{r}, t \mid \boldsymbol{r}', t') = K(\boldsymbol{r}, -t' \mid \boldsymbol{r}, -t), \tag{10.3.4}$$

$$K_t(\boldsymbol{r}, t \mid \boldsymbol{r}', t') = -K_{t'}(\boldsymbol{r}, t \mid \boldsymbol{r}', t'). \tag{10.3.5}$$

In view of (10.3.3) the defining relations (10.3.1) can be rewritten as

$$\left(\frac{1}{c^2}\frac{\partial^2}{\partial \tau^2} - \nabla^2\right) K(\boldsymbol{r}, \boldsymbol{r}'; \tau) = 0, \tag{10.3.6a}$$

$$\lim_{\tau \to 0+} K(\boldsymbol{r}, \boldsymbol{r}'; \tau) = 0, \qquad \lim_{\tau \to 0+} K_\tau(\boldsymbol{r}, \boldsymbol{r}'; \tau) = c^2\, \delta(\boldsymbol{r} - \boldsymbol{r}'), \tag{10.3.6b,c}$$

$$K(\boldsymbol{r}, \boldsymbol{r}'; \tau) = 0, \quad \text{or} \quad \partial_n K(\boldsymbol{r}, \boldsymbol{r}'; \tau) = 0 \quad \text{for} \quad \boldsymbol{r} \text{ on } S. \tag{10.3.6d}$$

As discussed in Section 9.5, the homogeneous wave equation (10.3.6a) is *invariant under the time-reversal transformation* $\tau \to -\tau$. Regarding (10.3.6a) for the moment as an equation that governs the τ-dependence of K at fixed $\boldsymbol{r}$ *and* $\boldsymbol{r}'$, its invariance under $\tau \to -\tau$ implies that its two linearly independent solutions may be chosen as respectively even and odd in τ.

Exercise: Prove this explicitly. The proof is like the argument in quantum mechanics about the parity of energy eigenstates in an even potential.

The initial conditions (10.3.6b, c) then show that K is the odd solution,

whence

$$K(\mathbf{r}, \mathbf{r}'; -\tau) = -K(\mathbf{r}, \mathbf{r}'; \tau), \quad K_\tau(\mathbf{r}, \mathbf{r}'; -\tau) = K_\tau(\mathbf{r}, \mathbf{r}', \tau). \quad (10.3.7\text{a,b})$$

In terms of t and t', (10.3.7a) entails

$$K(\mathbf{r}, t \mid \mathbf{r}', t') = -K(\mathbf{r}, t' \mid \mathbf{r}, t). \quad (10.3.8)$$

By sharp contrast to the diffusion equation, K is perfectly well-defined for negative as well as for positive τ ($t' < t$ as well as $t > t'$).

As a function of $\mathbf{r}$ and $\mathbf{r}'$, K is *symmetric* (in the narrow sense):

$$K(\mathbf{r}, t \mid \mathbf{r}', t') = K(\mathbf{r}', t \mid \mathbf{r}, t'). \quad (10.3.9)$$

Proof: The argument resembles that for diffusion in Section 9.2.3, but is rather simpler. Start by rewriting the original differential equation (10.3.1a) with the parameters $(\mathbf{r}', t')$ renamed $(\mathbf{r}'', t'')$:

$$\left(\frac{1}{c^2}\frac{\partial^2}{\partial t^2} - \nabla^2\right)K(\mathbf{r}, t \mid \mathbf{r}'', t'') = 0.$$

Multiply this by $K(\mathbf{r}, t \mid \mathbf{r}', t')$; multiply (10.3.1a) by $K(\mathbf{r}, t \mid \mathbf{r}'', t'')$; take the difference; and integrate $\int_{t_1}^{t_2} \mathrm{d}t \int_V \mathrm{d}V \cdots$. The two time-derivatives in the integrand combine into a perfect time-derivative, while the two terms with Laplaceans turn into a surface integral by virtue of Green's theorem. Thus the result of our operations reads

$$\begin{aligned}\frac{1}{c^2}\int_{t_1}^{t_2} \mathrm{d}t \frac{\mathrm{d}}{\mathrm{d}t}\int_V \mathrm{d}V\, &[K(\mathbf{r}, t \mid \mathbf{r}'', t'')K_t(\mathbf{r}, t \mid \mathbf{r}', t') \\ &- K_t(\mathbf{r}, t \mid \mathbf{r}'', t'')K(\mathbf{r}, t \mid \mathbf{r}', t')] \\ &- \int_{t_1}^{t_2} \mathrm{d}t \int_S \mathrm{d}S\, [K(\mathbf{r}, t \mid \mathbf{r}'', t'')\, \partial_n K(\mathbf{r}, t \mid \mathbf{r}', t') \\ &- (\partial_n K(\mathbf{r}, t \mid \mathbf{r}'', t''))K(\mathbf{r}, t \mid \mathbf{r}', t')] = 0.\end{aligned}$$

The surface integral vanishes by virtue of the BCs, and one is left with

$$\int_V \mathrm{d}V\, [K(\mathbf{r}, t \mid \mathbf{r}'', t'')K_t(\mathbf{r}, t \mid \mathbf{r}', t') - K_t(\mathbf{r}, t \mid \mathbf{r}'', t'')K(\mathbf{r}, t \mid \mathbf{r}', t')]\Big|_{t=t_1}^{t=t_2} = 0.$$

Now choose $t_2 = t'$, $t_1 = t''$, and appeal to the ICs (10.3.1,b, c). Only the first term contributes to the upper limit, and only the second to the lower limit:

$$\begin{aligned}&\int_V \mathrm{d}V\, K(\mathbf{r}, t' \mid \mathbf{r}'', t'')\, \delta(\mathbf{r} - \mathbf{r}') + \int_V \mathrm{d}V\, \delta(\mathbf{r} - \mathbf{r}'')K(\mathbf{r}, t'' \mid \mathbf{r}', t') \\ &\quad = K(\mathbf{r}', t' \mid \mathbf{r}'', t'') + K(\mathbf{r}'', t'' \mid \mathbf{r}', t') = 0.\end{aligned}$$

Finally, we rewrite the second term by appeal to reversibility in the form (10.3.8):

$$K(\mathbf{r}', t' \mid \mathbf{r}'', t'') - K(\mathbf{r}'', t' \mid \mathbf{r}', t'') = 0,$$

which is precisely (10.3.9) except for the names of the variables. ■

Exercise: Prove (10.3.9) without explicit appeal to reversibility.

The corollary (10.3.4) of time-translation invariance, plus symmetry (10.3.9), imply the *reciprocity relation*

$$K(\mathbf{r}, t \mid \mathbf{r}', t') = K(\mathbf{r}, -t' \mid \mathbf{r}, -t), \tag{10.3.10}$$

where all the variables have crossed the upright.

10.3.2 Expansion of the propagator in normal modes

The propagator regarded as a function of $\mathbf{r}$ can naturally be expressed as a linear combination of the complete set of eigenfunctions ϕ_n of $-\nabla^2$ under the appropriate homogeneous BCs; the coefficients then depend on t, $\mathbf{r}'$, t'.

Exercise: Determine the coefficients from first principles; the result must agree with (10.3.13) below.

Since K obeys the homogeneous wave equation, this amounts to an expansion of K, regarded as a function both of $\mathbf{r}$ and of t, in terms of the normal modes (10.1.11). Accordingly, relying on time-translation invariance, we can write

$$\begin{aligned} K(\mathbf{r}, t \mid \mathbf{r}', t') = {} & [A_0(t-t') + \Delta_0]\phi_0 \\ & + \sum_n{}' a_n \sin[\omega_n(t-t') + \delta_n]\phi_n(\mathbf{r}). \end{aligned} \tag{10.3.11}$$

This obeys the wave equation because each term obeys it individually, and satisfies the BCs because the ϕ_n do. The notation Σ' was explained in (10.1.12); we recall that the term with $n = 0$ is present only under pure NBCs, and that $\phi_0 = 1/V^{\frac{1}{2}}$. It remains only to satisfy the ICs, by choosing A_0, a_n, Δ_0, δ_n so as to validate

$$K(\mathbf{r}, t' \mid \mathbf{r}', t') = \left\{\Delta_0\phi_0 + \sum_n{}' a_n \sin[\delta_n]\phi_n(\mathbf{r})\right\} = 0, \tag{10.3.12a}$$

$$\begin{aligned} K_t(\mathbf{r}, t \mid \mathbf{r}', t')\big|_{t=t'} &= \left\{A_0\phi_0 + \sum_n{}' a_n\omega_n \cos[\delta_n]\phi_n(\mathbf{r})\right\} \\ &= c^2\,\delta(\mathbf{r}-\mathbf{r}') = c^2 \sum_n \phi_n^*(\mathbf{r}')\phi_n(\mathbf{r}). \end{aligned} \tag{10.3.12b}$$

The first condition implies $\Delta_0=0$ and $\delta_n=0$ for all n, whence $\cos[\delta_n]=1$. Then, on equating coefficients of $\phi_n(\mathbf{r})$ in the second condition, we find $A_0=c^2\phi_0=c^2V^{-\frac{1}{2}}$, and $a_n=c^2\phi_n^*(\mathbf{r}')/\omega_n$ for $n>0$. Substitution into (10.3.11) yields

$$K(\mathbf{r},t\,|\,\mathbf{r}',t')=c^2\left\{\frac{1}{V}(t-t')+\sum_n{}'\,\phi_n^*(\mathbf{r}')\phi_n(\mathbf{r})\frac{\sin[\omega_n(t-t')]}{\omega_n}\right\}, \tag{10.3.13a}$$

$$K_t(\mathbf{r},t\,|\,\mathbf{r}',t')=c^2\sum_n\phi_n^*(\mathbf{r}')\phi_n(\mathbf{r})\cos[\omega_n(t-t')]. \tag{10.3.13b}$$

Notice that the $n=0$ term, if present, takes exactly the form one could have guessed naively from the $n>0$ terms in view of l'Hôpital's rule, by writing

$$\lim_{n\to 0}\frac{\sin[\omega_n(t-t')]}{\omega_n}=(t-t').$$

We call this guess naive because, in a finite volume, ω_n cannot be varied continuously as a function of n.

If the set $\{\phi_n\}$ is chosen as real (cf. Appendix E), then the expansion (10.3.13) manifests the symmetry and reciprocity relations explicitly.

10.3.3 The Green's function

The Green's function is defined by the inhomogeneous equation

$$\left(\frac{1}{c^2}\frac{\partial^2}{\partial t^2}-\nabla^2\right)G(\mathbf{r},t\,|\,\mathbf{r}',t')=\delta(t-t')\,\delta(\mathbf{r}-\mathbf{r}'), \tag{10.3.14a}$$

the same homogeneous BCs as for K:

$$G(\mathbf{r},t\,|\,\mathbf{r}',t')=0,\quad\text{or}\quad\partial_n G(\mathbf{r},t\,|\,\mathbf{r}',t')=0\quad\text{for}\quad\mathbf{r}\text{ on }S, \tag{10.3.14b}$$

and the IC (again chosen by hindsight)

$$G(\mathbf{r},t\,|\,\mathbf{r}',t')=0\quad\text{for}\quad t<t'. \tag{10.3.14c}$$

By integrating (10.3.14a) with respect to t from $t'-$ to $t'+$, at fixed $\mathbf{r}$, $\mathbf{r}'$, and using (10.3.14c) to eliminate the lower-limit contributions on the left, we can replace (10.3.14c) with the more explicit equal-time conditions

$$\lim_{t\to t'+}G(\mathbf{r},t\,|\,\mathbf{r}',t')=0,\qquad\lim_{t\to t'+}G_t(\mathbf{r},t\,|\,\mathbf{r}',t')=c^2\,\delta(\mathbf{r}-\mathbf{r}'). \tag{10.3.14d,e}$$

Evidently, G describes the effects at $(\boldsymbol{r}, t)$ of injecting unit volume of fluid at $(\boldsymbol{r}', t')$ into a system that is everywhere quiescent before time t'.

As for diffusion, G is related to K by

$$G(\boldsymbol{r}, t \mid \boldsymbol{r}', t') = \theta(t - t')K(\boldsymbol{r}, t \mid \boldsymbol{r}', t'). \tag{10.3.15}$$

Proof: (i) The BCs (10.3.14b) are satisfied because K satisfies them. (ii) The IC (10.3.14c) is satisfied by virtue of the step-function. (iii) To see that the differential equation (10.3.14a) is satisfied, act on θK with $\square^2$, performing the two time-differentiations successively:

$$\left(\frac{1}{c^2}\frac{\partial^2}{\partial t^2} - \nabla^2\right)\theta K = \frac{1}{c^2}\frac{\partial}{\partial t}\left[\frac{\partial \theta}{\partial t}K + \theta\frac{\partial K}{\partial t}\right] - \theta\nabla^2 K.$$

Now

$$(\partial\theta/\partial t)K = \delta(t - t')K(\boldsymbol{r}, t \mid \boldsymbol{r}', t') = \delta(t - t')K(\boldsymbol{r}, t' \mid \boldsymbol{r}', t') = 0;$$

this term vanishes by virtue of the IC (10.3.1d) on K, and we drop it. Performing the second time-differentiation, we find

$$\begin{aligned}\square^2(\theta K) &= \frac{1}{c^2}\frac{\partial\theta}{\partial t}\frac{\partial K}{\partial t} + \theta\cdot\left(\frac{1}{c^2}\frac{\partial^2}{\partial t^2} - \nabla^2\right)K \\ &= \frac{1}{c^2}\delta(t - t')K_t(\boldsymbol{r}, t \mid \boldsymbol{r}', t') = \delta(t - t')\,\delta(\boldsymbol{r} - \boldsymbol{r}').\end{aligned}$$

In the second step we have dropped $\theta\square^2 K$, which vanishes by (10.3.1a); in the last step we have used (10.3.1c) ■

It is instructive to see how the same conclusion is reached even if one fails to drop $(\partial\theta/\partial t)K$ so early in the argument. Then, differentiating routinely, one obtains

$$\begin{aligned}\square^2(\theta K) &= \theta\square^2 K + \frac{2}{c^2}\frac{\partial\theta}{\partial t}\frac{\partial K}{\partial t} + \frac{1}{c^2}\frac{\partial^2\theta}{\partial t^2}K \\ &= \frac{1}{c^2}\left\{2\,\delta(t - t')\frac{\partial K}{\partial t} + \delta'(t - t')K\right\}.\end{aligned} \tag{10.3.16}$$

The problem is to simplify the last term. The factors δ and δ' imply that this whole expression will, ultimately, be multiplied by some well-behaved test function $F(t)$, and integrated over t (see Chapter 1). Recalling the properties of δ', we have

$$\begin{aligned}\delta'(t - t')KF &= -\delta(t - t')\frac{\partial}{\partial t}[KF] \\ &= -\delta(t - t')[K_t F + KF_t] \\ &= -\delta(t - t')K_t F,\end{aligned}$$

where $\delta(t-t')K$ vanishes as above. Since this relation holds for arbitrary test functions F, it amounts to

$$\delta'(t-t')K = -\delta(t-t')K_{t'}. \tag{10.3.17}$$

Accordingly, the last term on the right of (10.3.16) cancels half the first term, and we recover the expected relation

$$\Box^2(\theta K) = \delta(t-t')K_t/c^2 = \delta(t-t')\,\delta(\boldsymbol{r}-\boldsymbol{r}'). \qquad \blacksquare$$

The symmetries of G follow from those of K through (10.3.15). Time-translation invariance, symmetry, and reciprocity entail

$$G(\boldsymbol{r}, t \mid \boldsymbol{r}', t') = G(\boldsymbol{r}, \boldsymbol{r}'; \tau) = \theta(\tau)K(\boldsymbol{r}, \boldsymbol{r}'; \tau), \tag{10.3.18a}$$

$$G(\boldsymbol{r}, t \mid \boldsymbol{r}', t') = G(\boldsymbol{r}, -t' \mid \boldsymbol{r}, -t), \tag{10.3.18b}$$

$$G_t(\boldsymbol{r}, t \mid \boldsymbol{r}', t') = -G_{t'}(\boldsymbol{r}, t \mid \boldsymbol{r}', t'), \tag{10.3.18c}$$

$$G(\boldsymbol{r}, t \mid \boldsymbol{r}', t') = G(\boldsymbol{r}', t \mid \boldsymbol{r}, t'), \tag{10.3.19}$$

$$= G(\boldsymbol{r}', -t' \mid \boldsymbol{r}, -t). \tag{10.3.20}$$

The physical significance of the symmetry relation (10.3.19) is the same as was described for diffusion near the end of Section 9.2.3, and is just as remarkable.

By contrast to these symmetries shared by K and G, for G (unlike K) the initial condition (10.3.14c) introduces an obvious and fundamental distinction between t and t' (between τ and $-\tau$), so that the time-reversal invariance of the differential equation (10.3.14a) has no immediate consequences for G. We shall comment on this further in Section 10.6.

While K determines G through (10.3.15), the fact that K is odd in τ allows it to be recovered from G through

$$\begin{aligned} K(\boldsymbol{r}, \boldsymbol{r}'; \tau) &= \varepsilon(\tau)G(\boldsymbol{r}, \boldsymbol{r}'; |\tau|) \\ &= G(\boldsymbol{r}, \boldsymbol{r}'; \tau) - G(\boldsymbol{r}, \boldsymbol{r}'; -\tau). \end{aligned} \tag{10.3.21}$$

The normal-mode expansion (10.3.13) entails

$$\begin{aligned} G(\boldsymbol{r}, t \mid \boldsymbol{r}', t') &= \theta(t-t')K(\boldsymbol{r}, t \mid \boldsymbol{r}', t') \\ &= \theta(t-t')c^2\left\{\frac{1}{V}(t-t') + \sum_n{}' \phi_n^*(\boldsymbol{r}')\phi_n(\boldsymbol{r})\frac{\sin[\omega_n(t-t')]}{\omega_n}\right\}. \end{aligned} \tag{10.3.22}$$

Notice that strictly speaking this expansion of G in eigenfunctions of $-\nabla^2$ is no longer an expansion in normal modes, because the expression on the right depends on t through the factor $\theta(t-t')$ as well as through the normal-mode functions $\phi_n(\boldsymbol{r})\sin[\omega_n(t-t')]$. In fact, G, not being a solution of the homogeneous wave equation, contains all frequencies, and not (like K) merely the eigenfrequencies ω_n. (For unbounded space, this distinction is illustrated explicitly by comparing the n-fold Fourier representation in Section 11.2.2 with the $(n+1)$-fold representation in Appendix F.4.)

In the magic rule (10.4.4–7), $\partial G/\partial t$ will feature as prominently as G. Differentiating (10.3.15), $\partial/\partial t$ need not act on the step-function, because $(\partial\theta/\partial t)K = \delta(t-t')K$, and K vanishes at equal times. Thus

$$\begin{aligned} G_t(\mathbf{r}, t \mid \mathbf{r}', t') &= \theta(t-t')K_t(\mathbf{r}, t \mid \mathbf{r}', t') \\ &= \theta(t-t')c^2 \sum_n \phi_n^*(\mathbf{r}')\phi_n(\mathbf{r}) \cos[\omega_n(t-t')], \end{aligned} \tag{10.3.23}$$

where (as in (10.3.13b)) the sum now includes the term $n = 0$, if it exists.

Finally, to derive the magic rule we shall need the *reciprocal equation* for G regarded as a function of $(\mathbf{r}', t')$. To find it, substitute the reciprocity relation (10.3.20) into the differential equation (10.3.14a); rename the variables $\mathbf{r} \rightleftharpoons \mathbf{r}'$, $t \rightleftharpoons -t'$ (whence $\partial^2/\partial t^2 \to \partial^2/\partial t'^2$), and use the fact that $\delta(\mathbf{r}-\mathbf{r}')$ and $\delta(t-t')$ are even functions of their arguments (under the weak definition, which is all that is needed here). This yields

$$\left(\frac{1}{c^2}\frac{\partial^2}{\partial t^2} - \nabla^2\right)G(\mathbf{r}', -t' \mid \mathbf{r}, -t) = \delta(t-t')\,\delta(\mathbf{r}-\mathbf{r}')$$

$$\left(\frac{1}{c^2}\frac{\partial^2}{\partial t'^2} - \nabla'^2\right)G(\mathbf{r}, t \mid \mathbf{r}', t') = \delta(t-t')\,\delta(\mathbf{r}-\mathbf{r}') \tag{10.3.24}$$

Notice that the operator on the left of the reciprocal equation (10.3.24) is just $\Box'^2$, i.e. it has the same form as the original wave operator (see the comments in Section 9.2.4, and Section 10.1, point (xi)).

10.4 The magic rule

We are now in a position to solve the most general 'well-posed' problem, as described in Section 10.2 above. The argument should be compared with Section 9.3.1.

10.4.1 Derivation

One starts from the wave equation for ψ, and from the reciprocal equation for G, written as

$$\left(\frac{1}{c^2}\frac{\partial^2}{\partial t'^2} - \nabla'^2\right)\psi(\mathbf{r}', t') = \rho(\mathbf{r}', t'), \tag{10.4.1}$$

$$\left(\frac{1}{c^2}\frac{\partial^2}{\partial t'^2} - \nabla'^2\right)G(\mathbf{r}, t \mid \mathbf{r}', t') = \delta(t-t')\,\delta(\mathbf{r}-\mathbf{r}'). \tag{10.4.2}$$

Recall once again that on the closed surface S of the region V, G obeys the homogeneous version of the BCs actually prescribed for ψ. In the now familiar way, we cross-multiply; take the difference; integrate with respect to $d^n r'$ over V; and integrate with respect to t' from t_0 to t_+, where, as always, $t+ \equiv t + \varepsilon$, with the limit $\varepsilon \to 0+$ to be taken at the end of the calculation. These operations yield

$$\int_{t_0}^{t+} dt' \int_V dV' \left\{ \frac{1}{c^2} \frac{\partial}{\partial t'} \left[G \frac{\partial \psi(\mathbf{r}', t')}{\partial t'} - \frac{\partial G}{\partial t'} \psi(\mathbf{r}', t') \right] \right.$$
$$\left. - [G \nabla'^2 \psi(\mathbf{r}', t') - (\nabla'^2 G) \psi(\mathbf{r}', t')] \right\}$$
$$= \int_{t_0}^{t+} dt' \int_V dV' \, [G\rho(\mathbf{r}', t') - \psi(\mathbf{r}', t') \, \delta(t - t') \, \delta(\mathbf{r} - \mathbf{r}')]$$
$$= \int_{t_0}^{t+} dt' \int_V dV' \, G(\mathbf{r}, t \mid \mathbf{r}', t') \rho(\mathbf{r}', t') - \psi(\mathbf{r}, t). \tag{10.4.3}$$

In the leftmost expression, integrate the t'-derivative with respect to t', and notice that the upper-limit contribution vanishes because $t' = t+$ entails $t' > t$, when $G(\mathbf{r}, t \mid \mathbf{r}', t')$ vanishes by the initial condition (10.3.14c). The volume integral over the two terms involving Laplaceans turns into a surface integral by Green's theorem. Then the leftmost expression becomes

$$-\frac{1}{c^2} \int_V dV' \left[G(\mathbf{r}, t \mid \mathbf{r}', t_0) \dot{\psi}_{t_0}(\mathbf{r}', t_0) - \frac{\partial G(\mathbf{r}, t \mid \mathbf{r}', t_0)}{\partial t_0} \psi(\mathbf{r}', t_0) \right]$$
$$- \int_{t_0}^{t+} dt' \int_S dS' \, [G(\mathbf{r}, t \mid \mathbf{r}', t') \, \partial_n' \psi(\mathbf{r}', t')$$
$$- (\partial_n' G(\mathbf{r}, t \mid \mathbf{r}', t')) \psi(\mathbf{r}', t')].$$

Equating this to the rightmost expression in (10.4.3), and rearranging, we obtain the *magic rule*

$$\psi(\mathbf{r}, t) = f(\mathbf{r}, t) + g(\mathbf{r}, t) + h(\mathbf{r}, t), \tag{10.4.4}$$

where

$$f(\mathbf{r}, t) \equiv \int_{t_0}^{t+} dt' \int_V dV' \, G(\mathbf{r}, t \mid \mathbf{r}', t') \rho(\mathbf{r}', t'), \tag{10.4.5}$$

$$g(\mathbf{r}, t) = \int_{t_0}^{t+} dt' \int_S dS' \, [G(\mathbf{r}, t \mid \mathbf{r}', t') \, \partial_n' \psi_S(\mathbf{r}', t')$$
$$- (\partial_n' G(\mathbf{r}, t \mid \mathbf{r}', t')) \psi_S(\mathbf{r}', t')], \tag{10.4.6}$$

$$h(\mathbf{r}, t) \equiv h_1(\mathbf{r}, t) + h_2(\mathbf{r}, t), \tag{10.4.7a}$$

$$h_1(\mathbf{r}, t) = \int_V \mathrm{d}V' \frac{1}{c^2} G(\mathbf{r}, t \mid \mathbf{r}', t_0) \psi_{t_0}(\mathbf{r}', t_0), \tag{10.4.7b}$$

$$h_2(\mathbf{r}, t) = -\int_V \mathrm{d}V' \frac{1}{c^2} G_{t_0}(\mathbf{r}, t \mid \mathbf{r}', t_0) \psi(\mathbf{r}', t_0), \tag{10.4.7c}$$

$$= \int_V \mathrm{d}V' \frac{1}{c^2} G_t(\mathbf{r}, t \mid \mathbf{r}', t_0) \psi(\mathbf{r}', t_0), \tag{10.4.7d}$$

$$= \frac{\partial}{\partial t} \int_V \mathrm{d}V' \frac{1}{c^2} G(\mathbf{r}, t \mid \mathbf{r}', t_0) \psi(\mathbf{r}', t_0). \tag{10.4.7e}$$

Observe from (10.4.7b, e) that if in h_2 one replaces the data $\psi(\mathbf{r}', t_0)$ by the data $\psi_{t_0}(\mathbf{r}', t_0)$, the result is just $\partial h_1/\partial t$.

The versions (10.4.7c, d) of h_2 are equivalent by virtue of time-translation invariance (see (10.3.18c)). The version (10.4.7e) then follows by taking $\partial/\partial t$ in $G_t \equiv \partial G/\partial t$ outside the volume integral. In 2D, $\partial G(\mathbf{r}, t \mid \mathbf{r}', t')/\partial t$ is ill-defined at $t = t'$, and the version (10.4.7e) may then be preferable. Similarly, when $\partial f/\partial t$ occurs in 2D applications, it may be safer not to take $\partial/\partial t$ under the integral. These points are discussed a little further in Sections 11.2.2 and 11.5 *à propos* of G in unbounded space, where the problem is the same. We need not be concerned with it in the present chapter, and fortunately it does not afflict 3D and 1D.

The functions f, g, h_1, h_2 are regarded as explicit constructs from the data ρ, $\partial'_n \psi_S$ or ψ_S, $\psi(\mathbf{r}, t_0)$, $\psi_t(\mathbf{r}, t_0)$. Taken separately, they obey

$$\Box^2 f = \rho, \qquad \Box^2 g = 0, \qquad \Box^2 h_1 = 0, \qquad \Box^2 h_2 0, \tag{10.4.8a,b,c,d}$$

$$f(\mathbf{r}, t_0) = 0, \qquad f_{t_0}(\mathbf{r}, t_0) = 0, \tag{10.4.9a}$$

$$g(\mathbf{r}, t_0) = 0, \qquad g_{t_0}(\mathbf{r}, t_0) = 0, \tag{10.4.9b}$$

$$h_1(\mathbf{r}, t_0) = \psi(\mathbf{r}, t_0) \equiv \beta(\mathbf{r}), \qquad h_{1t_0}(\mathbf{r}, t_0) = 0, \tag{10.4.9c}$$

$$h_2(\mathbf{r}, t_0) = 0, \qquad h_{2t_0}(\mathbf{r}, t_0) = \psi_{t_0}(\mathbf{r}, t_0) \equiv \alpha(\mathbf{r}). \tag{10.4.9d}$$

Under DBCs

$$f(\mathbf{r}, t) = 0, \qquad g(\mathbf{r}, t) = \psi_S(\mathbf{r}, t), \qquad h(\mathbf{r}, t) = 0, \qquad \mathbf{r} \text{ on } S\text{: DBCs.} \tag{10.4.10a,b,c}$$

Thus f solves the inhomogeneous equation with homogeneous BCs and homogeneous ICs; g solves the homogeneous equation with inhomogeneous BCs and homogeneous ICs; h solves the homogeneous equation with homogeneous BCs but with inhomogeneous ICs. By linearity, the solution of the general problem is the sum of these functions.

Under different BCs, one has the appropriately modified version of (10.4.10).

Exercise: Verify (10.4.8, 9, 10a, 10c) explicitly. To verify the BCs (10.4.10b) on g is harder: see Section 9.3.2, comment (x).

10.4.2 Comments

The comments that follow might become more lifelike in the light of the examples in Section 10.5 and in Chapters 11 and 12.

(i) In the derivation of the magic rule, taking $t+$ as the upper limit is just a device, more convenient than constructing an appropriate strong definition of $\delta(t-t')$ on the right of (10.4.2). Except in self-consistency (existence) checks like the exercise set above, it is safe to write the upper limit on the t' integrals as t rather than $t+$. Once this has been done, and provided $t > t_0$ is understood, G may be replaced by K.

(ii–iv) These comments from Section 9.3.2 apply word for word.

(v) Under D, N, or mixed BCs (i.e. under all except Churchill BCs), only one of the two terms contributes to the integrand of g at any given time and at any given point of S; but both the terms h_1 and h_2 contribute to h, since ψ and ψ_t are prescribed independently at t_0. Perhaps it bears repeating that no differentiations need be performed to obtain $\psi_0(\mathbf{r}', t_0) \equiv \alpha(\mathbf{r}')$ in the integrand of h_1: the function $\alpha(\mathbf{r}')$ is data, and must be supplied as part of the specification of the problem.

(vi) The lower limit t_0 can be $-\infty$.

(vii) On replacing G with K as represented by its normal-mode expansion, we observe that $h(\mathbf{r}, t)$ as a function of t contains only terms oscillating with the normal-mode frequencies ω_n. By contrast, the time-variation of f and g depends crucially on the time-variation of the prescribed sources and boundary values, as illustrated by the example on resonant forcing in Section 10.5.2.

(viii) The construction leading to (10.4.4–7) always works, so that for waves (as for diffusion) the magic rule answers all questions, though not always the shortest way (cf. Section 9.3.2, comment (viii)). NBCs, which caused major upheavals with Poisson's equation, are peculiar only in that they admit $\omega_0 = 0$, and the corresponding linearly-increasing term $c^2(t-t')/V$ in the normal-mode expansion (10.3.13a) of K.

(ix) Comparing h_1 with f, we observe that evolution from an initial distribution $\psi(\mathbf{r}, t_0) = 0$, $\psi_{t_0}(\mathbf{r}, t_0) = \alpha(\mathbf{r})$ is identical to evolution from a synchronously-flashed source distribution $\rho(\mathbf{r}, t) = \delta(t-t_0)\alpha(\mathbf{r})/c^2$. Conversely, if one knows h_1, in the sense of being able to write down the solution of the homogeneous equation with initially zero ψ but *arbitrary* initial ψ_t, then one can construct the solution f of the inhomogeneous equation with arbitrary sources ρ. This connection is sometimes called 'Duhamel's principle' (see e.g. Courant and Hilbert 1962).

In our approach, the information content of this 'principle' is exhausted by the connection $G = \theta(\tau)K$. The point is simply that K suffices to solve the homogeneous equation with initial values, while G is needed to solve the inhomogeneous equation; but that K in fact determines G. A similar 'principle' holds for the diffusion equation.

(x) In the integral h_2 the magic rule requires the initial values $\psi(\boldsymbol{r}, t_0)$, and under DBCs it requires in g the boundary values $\psi_S(\boldsymbol{r}, t)$. These data are natural in elasticity, where ψ itself is an observable quantity; but they can be awkward to ascertain in acoustics, where one is likely to have information about the observables $\partial\psi/\partial t = \Delta p/\sigma$ and $\boldsymbol{v} = -\nabla\psi$, rather than about ψ itself. Some ways of sidestepping such problems with h_2 will be discussed in Section 11.5 below. Here we merely state, without proof, how for bounded regions they can be tackled head-on. Given $\partial\psi_S(\boldsymbol{r}, t)/\partial t$, integration with respect to t at fixed $\boldsymbol{r}$ (on S) yields

$$\psi_S(\boldsymbol{r}, t) = \psi_S(\boldsymbol{r}, t_0) + \int_{t_0}^{t} dt'\, \partial\psi_S(\boldsymbol{r}, t')/\partial t';$$

and it may be shown that one must set $\psi_S(\boldsymbol{r}, t_0) = 0$, in order to guarantee that g will not generate observable effects without physical causes. A simple integration of this kind occurs in Section 10.5.2.

Similarly, given $\nabla\psi(\boldsymbol{r}, t_0)$, one has

$$\psi(\boldsymbol{r}, t_0) = \psi(\boldsymbol{r}_1, t_0) + \int_{\boldsymbol{r}_1}^{\boldsymbol{r}} d\boldsymbol{r}' \cdot \nabla'\psi(\boldsymbol{r}', t_0)$$

where the line integral runs along any path from any fixed point $\boldsymbol{r}_1$. In fact we choose $\boldsymbol{r}_1$ on the boundary S, where, for consistency with the assertion just made about $\psi_S(\boldsymbol{r}, t_0)$, one has $\psi(\boldsymbol{r}_1, t_0) = 0$.

10.5 Examples

10.5.1 The propagator in rectangular regions

Consider the 2D region $0 \leqslant x \leqslant a$, $0 \leqslant y \leqslant b$, under DBCs, as for Poisson's equation in Section 5.3.2. Labelling the normal modes as in (5.3.3) (but with $k_{mn}^2 = \omega_{mn}^2/c^2$ instead of λ_{mn}), the expansion (10.3.13) for the propagator reads

$$K(\boldsymbol{r}, \boldsymbol{r}'; \tau) = c^2 \frac{4}{ab} \sum_{n=1}^{\infty} \sum_{m=1}^{\infty} \sin\left(\frac{n\pi x'}{a}\right) \sin\left(\frac{n\pi x}{a}\right)$$

$$\times \sin\left(\frac{m\pi y'}{b}\right) \sin\left(\frac{m\pi y}{b}\right) \sin(\omega_{mm}\tau)/\omega_{mn}, \tag{10.5.1a}$$

$$\omega_{mn} = c[(n\pi/a)^2 + (m\pi/b)^2]^{\frac{1}{2}}. \tag{10.5.1b}$$

Because the ω_{mn} are generally incommensurable, i.e. because the ratios $\omega_{mn}/\omega_{m'n'}$ are not in general rational numbers, K is not a periodic function of $\tau = t - t'$: there is no period T such that $K(\boldsymbol{r}, \boldsymbol{r}'; \tau + T) = K(\boldsymbol{r}, \boldsymbol{r}'; \tau)$ for all $\boldsymbol{r}$, $\boldsymbol{r}'$, and for all τ. Consequently the same is true for the integrals h in the magic rule, except for very special ICs.

Repeated use of the trigonometric sum and difference formulae turns each term of K into a sum of plane waves. The product of the four sines may be written schematically as

$$\frac{1}{4}\left[\cos\left(\frac{n\pi(x-x')}{a}\right) - \cos\left(\frac{n\pi(x+x')}{a}\right)\right]\left[\cos\left(\frac{m\pi(y-y')}{b}\right)\right.$$

$$\left. - \cos\left(\frac{m\pi(y+y')}{b}\right)\right] \sin(\omega_{nm}\tau)$$

$$= \frac{1}{8}\left[\cos\left(\frac{n\pi(x-x')}{a} + \frac{m\pi(y-y')}{b}\right)\right.$$

$$\left. + \cos\left(\frac{n\pi(x-x')}{a} - \frac{m\pi(y-y')}{b}\right) + \cdots\right] \sin(\omega_{nm}\tau)$$

$$= \frac{1}{16}\left\{\sin\left(\omega_{nm}\tau + \frac{n\pi(x-x')}{a} + \frac{m\pi(y-y')}{b}\right)\right.$$

$$\left. + \sin\left(\omega_{nm}\tau - \frac{n\pi(x-x')}{a} - \frac{m\pi(y-y')}{b}\right) + \cdots\right\}, \tag{10.5.2}$$

where in each step only the first product on the left has been written out in full on the right. Sums over such expressions might reasonably be described as repeated-reflection series.

By contrast to 2D and to 3D, in 1D the frequencies $\omega_n = n\pi c/a$ are all integer multiples of the fundamental ω_1. Hence the motion described by K, or (irrespective of the initial conditions) by h, is automatically periodic with period $T = 2\pi/\omega_1 = 2a/c$, the time for a signal moving with speed c to traverse the system twice.

Further, the fact that $\omega_n = n\pi/a$ is proportional to n turns the 1D repeated-reflection series into ordinary Fourier series. For instance, the

1D analogue of (10.5.1, 2) for $0 \leqslant x \leqslant a$ reads

$$K(x, x'; \tau) = c^2 \frac{2}{a} \sum_{n=1}^{\infty} \sin\left(\frac{n\pi x'}{a}\right) \sin\left(\frac{n\pi x}{a}\right) \frac{\sin(n\pi c\tau/a)}{(n\pi c/a)} \tag{10.5.3a}$$

$$K = \frac{c}{2\pi} \sum_{n=1}^{\infty} \frac{1}{n} \left\{ \sin\left[\frac{n\pi}{a}(c\tau + x - x')\right] - \sin\left[\frac{n\pi}{a}(c\tau + x + x')\right] + \sin\left[\frac{n\pi}{a}(c\tau - x + x')\right] - \sin\left[\frac{n\pi}{a}(c\tau - x - x')\right] \right\}. \tag{10.5.3b}$$

This can be re-expressed in terms of the 1D unbounded-space propagator $K_0^{(1)}$, eqn (11.2.15) below. First, the Poisson summation formula (C.12) is applied to the four terms of (10.5.3) in turn (after appeal to

$$\sum_{n=1}^{\infty} \sin(\alpha n)/n = \frac{1}{2}\left\{-\alpha + \sum_{n=-\infty}^{\infty} \sin(\alpha n)/n\right\}),$$

yielding

$$K = \frac{c}{4\pi} \sum_{N=-\infty}^{\infty} \int_{-\infty}^{\infty} \frac{d\xi}{\xi} \cos(2\pi N\xi)\{\sin[\xi(c\tau + x' - x)\pi/a]$$
$$+ \sin[\xi(c\tau - x' + x)\pi/a] - \sin[\xi(c\tau + x' + x)\pi/a]$$
$$- \sin[\xi(c\tau - x' - x)\pi/c]\}.$$

The formula

$$\sin(A)\cos(B) = (\sin(A + B) + \sin(A - B))/2$$

turns these into Dirichlet integrals (B.1, 5); some manipulation of the result guided by hindsight (e.g. replacing some N by $-N$ as one may under $\sum_{N=-\infty}^{\infty}$) eventually produces

$$K = \frac{c}{4} \sum_{N=-\infty}^{\infty} \{\varepsilon[c\tau - x + (2Na + x')] + \varepsilon[c\tau + x - (2Na + x')]$$
$$- \varepsilon[c\tau - x + (2Na - x')] - \varepsilon[c\tau + x - (2Na - x')]\},$$

$$K = \sum_{N=-\infty}^{\infty} \{K_0^{(1)}(x, t \mid 2Na + x', t') - K_0^{(1)}(x, t \mid 2Na - x', t')\}, \tag{10.5.4}$$

in close analogy to the multiple-image series (5.3.19) for Poisson's equation and (9.4.16) for diffusion. However, it is incomparably more convenient to establish the multiple-image representations (in any number of dimensions) by direct verification, as at the end of Section 12.8.1 below.

10.5.2 Resonance

The following example is typical of the response of systems driven at a frequency close to the frequency of one of their normal modes. The general pattern is much the same whether the driving term is a source distribution $\rho(\mathbf{r}, t)$ or, as here, an inhomogeneous boundary value $\psi_S(\mathbf{r}, t)$. Incidentally, we illustrate the prescription given in Section 10.4.2 (point (x)) for reconstituting, from the prescribed values of $\partial\psi_S/\partial t$, the data ψ_S required in the magic rule.

Consider an effectively 1D resonator of length a, $0 \leqslant x \leqslant a$, open at both ends, so that DBCs apply (Section 10.1, point (iii)), and the normal modes are given by

$$\phi_m(2/a)^{\frac{1}{2}} \sin(m\pi x/a), \qquad \omega_m = m\pi c/a.$$

The resonator is quiescent at $t=0$; for $t \geqslant 0$, excess pressure $\Delta p_S = Q\sin(\omega t)$ is applied at $x=0$. We determine $\Delta p(x,t) = \sigma\,\partial\psi(x,t)/\partial t$, paying special attention to the case where the driving frequency ω is very close to one of the normal-mode frequencies, say $\omega \approx \omega_r \equiv r\pi c/a$.

The solution is given by the integral g in the magic rule (10.4.4–7); under DBCs, only the second term of g contributes. G is replaced by K (Section 10.4.2, point (i)), and for K we use the normal-mode expansion (10.5.3a). In 1D, the surface integration in G reduces to evaluating the integrand of g at the end-point $x=0$, where $\partial'_n = -\partial/\partial x'$. For the requisite input we write

$$\psi(0,t) = \int_0^t \mathrm{d}t' \frac{\partial\psi(0,t')}{\partial t'} = \int_0^t \mathrm{d}t' \frac{Q}{\sigma}\sin(\omega t'),$$

$$\psi(0,t) = \frac{Q}{\sigma\omega}(1-\cos(\omega t)). \tag{10.5.5}$$

Accordingly,

$$\psi(x,t) = g(x,t) = -\int_0^t \mathrm{d}t'\left(-\frac{\partial}{\partial x'}\right)\sum_{n=1}^{\infty} c^2\frac{2}{a}\sin\left(\frac{n\pi x'}{a}\right)\sin\left(\frac{n\pi x}{a}\right)\bigg|_{x'=0}$$

$$\times\frac{\sin[\omega_n(t-t')]}{(n\pi c/a)}\frac{Q}{\sigma\omega}(1-\cos(\omega t')),$$

$$\psi(x,t) = \frac{2cQ}{a\sigma\omega}\sum_{n=1}^{\infty}\sin\left(\frac{n\pi x}{a}\right)\alpha_n(t), \tag{10.5.6a}$$

$$\alpha_n(t) \equiv \int_0^t \mathrm{d}t' \sin[\omega_n(t-t')](1-\cos(\omega t'))$$

$$= \left\{\frac{1}{\omega_n}[1-\cos(\omega_n t)] - \frac{2\omega_n}{(\omega_n^2-\omega^2)}[\cos(\omega t)-\cos(\omega_n t)]\right\}. \tag{10.5.6b}$$

Acting on ψ (i.e. on the $\alpha_n(t)$) with $\sigma\,\partial/\partial t$, we obtain Δp:

$$\Delta p(x,t) = Q\frac{2c}{a\omega}\sum_{n=1}^{\infty}\sin\left(\frac{n\pi x}{a}\right)\beta_n(t), \tag{10.5.7a}$$

$$\beta_n(t) = \left\{\sin(\omega_n t) + \frac{2\omega_n}{(\omega_n^2 - \omega^2)}[\omega \sin(\omega t) - \omega_n \sin(\omega_n t)]\right\}. \tag{10.5.7b}$$

For most driving frequencies, the differences $|\omega - \omega_n|$ are all comparable to ω, and all the $|\beta_n(t)|$ have maxima of order no greater than unity. However, defining

$$\omega = \omega_r + \delta \tag{10.5.8}$$

we see that, if $|\delta| \ll \omega_r$, then $|\beta_r(t)|$ rises to order $\omega_r/\delta \gg 1$. Under such conditions, the one near-resonant term with $n = r$ will for most purposes dominate the sum. We drop all other terms, and approximate β_r to leading order in $1/\delta$, as follows.

(i) Drop the component $\sin(\omega_r t)$.
(ii) In the remaining component, set $2\omega_r/(\omega_r^2 - \omega^2) \approx -1/\delta$, and $\omega \sin(\omega t) \approx \omega_r \sin(\omega t) = \omega_r \sin[(\omega_r + \delta)t]$.

(We make no approximation in the argument of $\sin(\omega t)$, because this would induce errors accumulating indefinitely with time. By contrast, replacing the coefficient ω of $\sin(\omega t)$ by ω_r merely introduces a small constant error in the phase of $\beta_r(t)$.)

Accordingly, one finds

$$\begin{aligned}\beta_r(t) &\approx -(\omega_r)/\delta)\{\sin(\omega t) - \sin(\omega_r t)\} \\ &= (-\omega_r/\delta) 2 \sin[\tfrac{1}{2}(\omega - \omega_r)t] \cos[\tfrac{1}{2}(\omega + \omega_r)t] \\ &= (-2\omega_r/\delta) \sin[\tfrac{1}{2}\delta t] \cos[(\omega_r + \tfrac{1}{2}\delta)t],\end{aligned} \tag{10.5.9}$$

whence finally

$$\Delta p(x, t) \approx Q\frac{(4c/a)}{\delta} \sin(\tfrac{1}{2}\delta t) \cos[(\omega_r + \tfrac{1}{2}\delta)t] \sin(r\pi x/a). \tag{10.5.10}$$

While the factor $\cos[(\omega_r + \frac{1}{2}\delta)t] = \cos[(\omega - \frac{1}{2}\delta)t]$ of the response (10.5.10) oscillates nearly with the driving frequency, its coefficient $-Q(4c/a\delta)\sin(\frac{1}{2}\delta t)$ can be viewed as an amplitude varying slowly (with angular frequency $\delta/2$) between the large limits $\pm Q(4c/a\delta)$. The position-dependent factor $\sin(r\pi x/a)$ is just that of the near-resonant normal mode.

Exactly on resonance, i.e. as $\delta \to 0$, l'Hôpital's rule yields $\lim_{\delta \to 0}(1/\delta)\sin(\frac{1}{2}\delta t) = t/2$, whence

$$\Delta p(x, t) = -Q\frac{2ct}{a}\cos(\omega t)\sin(r\pi x/a), \qquad (\omega = \omega_r), \tag{10.5.11}$$

so that the amplitude grows indefinitely with time instead of oscillating slowly between large limits. (Of course, in an actual system the amplitude would be held within finite bounds by dissipation, which the wave equation leaves out of account.) As already anticipated in Section 3.5.6, the linearly increasing instead of harmonic response to a harmonic driving term with $\omega = \omega_r$ is the reason for the failure of the Helmholtz equation (in the general special case), since the Helmholtz equation assumes from the outset that the motion is harmonic (see also Problem 2.7).

It will cause no surprise by now that in the series (10.5.6a) each term vanishes as $x \to 0$, whereas $\psi(0, t)$ itself is non-zero (see for instance Section 9.4.2). *A fortiori*, the near-resonance approximation (10.5.10) violates the boundary condition: in practice this means that, in order to validate the approximation, one needs smaller δ for smaller x.

Finally, and independently of the resonance approximation, it is worth considering how the result (10.5.7) emerges if the problem is approached from first principles rather than through the magic rule, namely by expanding $\psi(x, t) = \sum a_n(t) \sin(n\pi x/a)$, and then solving the differential equations for the $a_n(t)$. The term in (10.5.7b) that contains $\sin(\omega t)$ obviously enters through the particular integral. The terms containing $\sin(\omega_n t)$ enter through the complementary function, their coefficients determined by the initial conditions $a_n(0) = 0 = \dot{a}_n(0)$. The first-principles approach has the advantage that it dispenses with the whole apparatus of propagators and Green's functions. It has the disadvantage that one must set up and solve the appropriate differential equations from scratch for every such problem, in contrast to the magic rule, which, as we have just seen, requires merely the evaluation of certain preselected integrals.

10.5.3 The power output of volume sources

Consider a region where a source distribution $\rho(\mathbf{r}, t)$ has acted after initial quiescence ($\psi = 0 = \psi_t$ as $t \to -\infty$). Beautiful and very useful formulae can be established for the total energy $H(t) = \int_V dV H(\mathbf{r}, t)$, and for the total power output of the sources, $W \equiv dH/dt$. For simplicity we admit only homogeneous BCs, so that, in the language of eqn (10.1.6), $W_S = 0$, and the volume sources ρ are the only mechanism capable of changing H. We deduce the general formulae first, and then apply them to the special but quite common case of a separable source distribution described by (10.5.17) below.

Obviously one could determine $H(t)$ first, by substituting $\psi = f(\mathbf{r}, t)$ from the magic rule into H, eqn (10.1.3), and integrating over V. But probably one gains more insight by starting with W instead. This illustrates one of the basic facts of field theory, namely that the energy of a field derives from work done by its sources against the fields (in this case against $\Delta p = \sigma \psi_t$) which the sources themselves have generated at earlier times.

Accordingly, we start from

$$W(t) = W_V(t) = \sigma \int_V dV\, \rho(\mathbf{r}, t)\, \partial\psi(\mathbf{r}, t)/\partial t$$

$$= \sigma \int_V dV\, \rho(\mathbf{r}, t) \frac{\partial}{\partial t} \int_{-\infty}^{t} dt' \int_V dV'\, K(\mathbf{r}, t \mid \mathbf{r}', t')\rho(\mathbf{r}', t'), \qquad (10.5.12)$$

$$W(t) = \sigma \int_V dV \int_V dV'\, \rho(\mathbf{r}, t) \int_{-\infty}^{t} dt'\, K_t(\mathbf{r}, t \mid \mathbf{r}', t')\rho(\mathbf{r}', t'). \qquad (10.5.13a)$$

(The derivative $\partial/\partial t$ acting on the upper limit of $\int_{-\infty}^{t} dt' \cdots$ in (10.5.12) yields no contribution to (10.5.13a), because the resulting factor $K(\mathbf{r}, t \mid \mathbf{r}', t)$ has $t = t'$ and vanishes by virtue of the initial conditions on K.) We can now write $K_t = -K_{t'}$, and integrate by parts with respect to t'. The integrated term vanishes for the reason just cited, and one finds the alternative form

$$W(t) = \sigma \int_V dV \int_V dV'\, \rho(\mathbf{r}, t) \int_{-\infty}^{t} dt'\, K(\mathbf{r}, t \mid \mathbf{r}', t')\rho_{t'}(\mathbf{r}', t'). \qquad (10.5.13b)$$

This form is often useful because it stresses the time-variation of the sources. A shortcut to it is pointed out in the exercise at the end of Section 11.5 below.

The total energy at time T can now be expressed as

$$H(T) = \int_{-\infty}^{T} dt\, W(t) = \sigma \int_V dV \int_V dV' \int_{-\infty}^{T} dt \int_{-\infty}^{t} dt'$$
$$\times \rho(\mathbf{r}, t) K_t(\mathbf{r}, t \mid \mathbf{r}', t')\rho(\mathbf{r}', t'). \qquad (10.5.14)$$

The repeated time integrals are simplified by observing that the integrand is unaffected by interchanging t and t', because K is odd and K_t therefore even in $(t - t')$. Hence the integral over the unshaded half of the square $-\infty \leq t,\ t' \leq T$ shown in Fig. 10.1 is equal to the integral over the shaded half, which is the region actually prescribed in (10.5.14). Accordingly, extending the integral over the whole square and multiplying by $\frac{1}{2}$, one obtains the symmetric expression

$$H(T) = \tfrac{1}{2}\sigma \int_V dV \int_V dV' \int_{-\infty}^{T} dt \int_{-\infty}^{T} dt'\, \rho(\mathbf{r}, t) K_t(\mathbf{r}, t \mid \mathbf{r}', t')\rho(\mathbf{r}', t'). \qquad (10.5.15)$$

Substituting for K_t its normal-mode expansion (10.3.23), and writing $\cos[\omega_n(t - t')] = \mathrm{Re}\exp[-i\omega_n(t - t')]$, we find

$$H(T) = \tfrac{1}{2}\sigma c^2\, \mathrm{Re} \sum_n \int_V dV \int_V dV' \int_{-\infty}^{T} dt \int_{-\infty}^{T} dt'\, \rho(\mathbf{r}, t)$$
$$\times \phi_n(\mathbf{r}) \exp[-i\omega_n(t - t')]\rho(\mathbf{r}', t')\phi_n^*(\mathbf{r}').$$

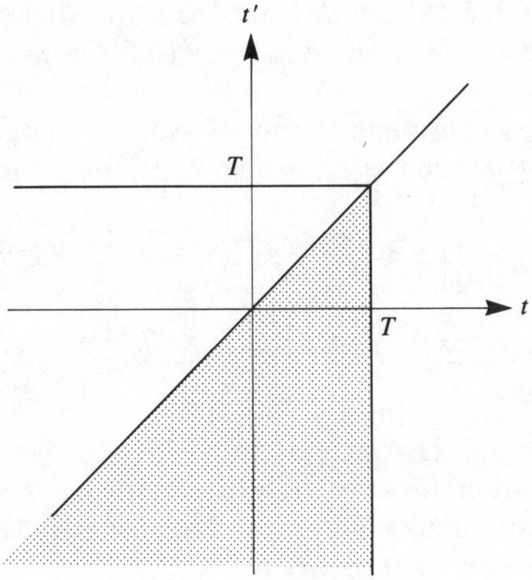

Fig. 10.1

Since the exponential factors into $\exp(-i\omega_n t)\exp(i\omega_n t')$, the entire integral factors; the result is explicitly real, so that the real-part prescription becomes irrelevant, and we are left with the remarkably elegant conclusion

$$H(T) = \tfrac{1}{2}\sigma c^2 \sum_n \left| \int_V \mathrm{d}V \int_{-\infty}^{T} \mathrm{d}t\, \rho(\boldsymbol{r}, t)\phi_n(\boldsymbol{r}) \exp(-i\omega_n t) \right|^2. \tag{10.5.16}$$

Up to this point the argument applies to arbitrary source distributions ρ.

Let us now specialize to the factored form

$$\rho(\boldsymbol{r}, t) = \alpha(\boldsymbol{r})F(t). \tag{10.5.17}$$

Then (10.5.16) reduces to

$$H(T) = \tfrac{1}{2}\sigma c^2 \sum_n |\alpha_n|^2 \left| \int_{-\infty}^{T} \mathrm{d}t\, F(t) \exp(-i\omega_n t) \right|^2, \tag{10.5.18a}$$

$$\alpha_n \equiv \int_V \mathrm{d}V\, \alpha(\boldsymbol{r})\phi_n(\boldsymbol{r}). \tag{10.5.18b}$$

The power can now be expressed in various equivalent forms, for instance as

$$W(t) = \frac{\mathrm{d}H}{\mathrm{d}t} = \sigma c^2 \sum_n |\alpha_n|^2 F(t) \int_{-\infty}^{t} \mathrm{d}t' \cos[\omega_n(t - t')]F(t'). \tag{10.5.19}$$

Exercise: Verify (10.5.19) either from (10.5.18), or by introducing the normal-mode expansion of K_t directly into (10.5.13a).

As an illustration, consider a harmonic source switched on at $t=0$, described by $F(t)=\theta(t)\sin(\omega t)$. Then straightforward integration yields

$$H(T)=\tfrac{1}{2}\sigma c^2\sum_n|\alpha_n|^2\left\{\left[\frac{1-\cos[(\omega_n+\omega)T]}{\omega_n+\omega}-\frac{1-\cos[(\omega_n-\omega)T]}{\omega_n-\omega}\right]^2\right.$$
$$\left.+\left[\frac{\sin[(\omega_n+\omega)T]}{\omega_n+\omega}-\frac{\sin[(\omega_n-\omega)T]}{\omega_n-\omega}\right]^2\right\}. \qquad (10.5.20)$$

Near resonance, where $\omega=\omega_r+\delta$ say, with $|\delta|\ll\omega_r$, it may be an excellent approximation to keep only the term $n=r$, and only those components that have denominators δ. Then one finds, for the source distribution $\rho=\theta(t)\alpha(\boldsymbol{r})\sin[(\omega_r+\delta)t]$,

$$H(T)\approx\tfrac{1}{2}\sigma c^2|\alpha_r|^2\left[\frac{\sin(\frac{1}{2}\delta T)}{\delta}\right]^2. \qquad (10.5.21)$$

While (10.5.16, 18) show that H is, naturally, non-negative, it is perfectly possible for W to be negative as well as positive. For instance, the near-resonance approximation (10.5.21) implies

$$W(t)=\mathrm{d}H/\mathrm{d}t\approx\tfrac{1}{4}\sigma c^2|\alpha_r|^2\sin(\delta t)/\delta, \qquad (10.5.22)$$

which oscillates with T unless $\delta=0$.

Exercise: The last equation implies that $W(t\to 0)\sim\frac{1}{4}\sigma c^2|\alpha_r|^2 t$, i.e. that $W(t)$ vanishes linearly with t. By contrast, $F(t)=\sin(\omega t)$ implies $F(t\to 0)\sim\omega t$, and substitution directly into the exact result (10.5.19) yields

$$W(t\to 0)\sim\sigma c^2\sum_n|\alpha_n|^2\,\omega t\int_0^t \mathrm{d}t'\,\omega t'=\tfrac{1}{2}\sigma c^2\omega^2 t^3\sum_n|\alpha_n|^2,$$

suggesting that $W(t)$ vanishes like t^3. Resolve this apparent contradiction.

All the above results are readily adapted to unbounded space by replacing $\phi_n(\boldsymbol{r})\to(2\pi)^{-n/2}\exp(i\boldsymbol{k}\cdot\boldsymbol{r})$, $\omega_n\to ck$, and $\sum_n\to\int\mathrm{d}^n k$, as explained in Section 4.5.2. In particular, (10.5.18) yields

$$H(t)=\tfrac{1}{2}\sigma c^2\int\mathrm{d}^n k\,|\tilde{\alpha}(\boldsymbol{k})|^2\left|\int_{-\infty}^{T}\mathrm{d}t\,F(t)\exp(-ickt)\right|^2, \qquad (10.5.23a)$$

$$\tilde{\alpha}(\boldsymbol{k})\equiv(2\pi)^{-n/2}\int\mathrm{d}V\,\alpha(\boldsymbol{r})\exp(\mathrm{i}\mathbf{k}\cdot\mathbf{r}). \qquad (10.5.23b)$$

Similarly, (10.5.19) yields

$$W(T) = \sigma c^2 \int \mathrm{d}^n k \, |\tilde{\alpha}(\boldsymbol{k})|^2 F(t) \int_{-\infty}^{T} \mathrm{d}t \cos[ck(t-t')]F(t'). \tag{10.5.24}$$

Such expressions can offer useful shortcuts, for instance to some of the answers in the case histories of Section 11.3 below.

10.6 Retrodiction

The homogeneous wave equation $\Box^2\psi = 0$ is invariant under time-reversal, as shown in Section 9.5, which explained the physical significance of such invariance, and what one means by retrodiction. Section 10.3.2 shows how time-reversal invariance makes the wave propagator $K(\boldsymbol{r}, \boldsymbol{r}'; \tau)$ odd in τ, and thus automatically well-defined for negative τ. Consequently, as regards waves (unlike diffusion), any preference for increasing (rather than decreasing) values of the time variable is a matter of physical choice rather than of mathematical necessity.

For instance, under homogeneous BCs, the equation $\Box^2\psi = 0$ plus 'final conditions' prescribing $\psi(\boldsymbol{r}, t_0)$ and $\psi_{t_0}(\boldsymbol{r}, t_0)$ fully determine $\psi(\boldsymbol{r}, t)$ for all $t \leqslant t_0$:

$$\psi(\boldsymbol{r}, t) = h(\boldsymbol{r}, t) = \int_V \mathrm{d}V' \frac{1}{c^2} \{K(\boldsymbol{r}, t \mid \boldsymbol{r}', t_0)\psi_{t_0}(\boldsymbol{r}', t_0) - K_{t_0}(\boldsymbol{r}, t \mid \boldsymbol{r}', t_0)\psi(\boldsymbol{r}', t_0)\}. \tag{10.6.1}$$

Similarly, the proof in Section 10.2 demonstrates the uniqueness of the solutions of $\Box^2\psi = \rho$ for $t \leqslant t_0$, given the final values $\psi(\boldsymbol{r}, t_0)$ and $\psi_{t_0}(\boldsymbol{r}, t_0)$, plus the data ρ, and ψ_S or $\partial_n\psi_S$, for $t \leqslant t_0$. A retrodictive magic rule for this problem can be constructed straightforwardly along the lines of Section 10.4, using the advanced Green's function $G_\mathrm{a} \equiv \theta(t_0 - t)K$ instead of the retarded Green's function $G_\mathrm{r} \equiv \theta(t - t_0)K$ that we have employed hitherto.

The reason why less attention is paid to retrodiction than to prediction is that in general the requisite input information about the future is not available, and that one seldom needs to reconstruct the past (even though the requisite data are known, referring as they do to the present). But one does resort to such calculations to determine what sources or what initial disturbances must be deployed now, in order to secure specific configurations ψ and ψ_t at some future time. All this contrasts sharply with diffusion, where, as we saw, K makes no sense for $t < t_0$, which makes it impossible to define a useful advanced Green's function even if one should wish to.

An example of retrodiction appears in Problem 12.3.

Problems

10.1 A pulsed source

$$\rho(x, t) = \alpha\, \delta(x - \pi/2)\left(\frac{2}{\pi^{\frac{1}{2}} T^3}\right) t \exp(-t^2/T^2)$$

acts in a 1D system, $\theta \leqslant x \leqslant \pi$, governed by

$$\Box^2 \psi = \rho, \quad \psi(x, -\infty) = 0 = \psi_t(x, -\infty), \quad \psi(x, t) = 0 = \psi(\pi, t).$$

(i) Calculate $\psi(x, t)$ in the asymptotic regime $t \gg T$. (Notice that $\psi(x, t \to \infty)$ fails to vanish even though $\int_{-\infty}^{\infty} \mathrm{d}t\, \rho(x, t) = 0$.)
(ii) Consider the limit $T \to 0$ and use it to check the result from (i).

Hint: In (i), use the magic rule and the normal-mode expansion of G, writing

$$\sin[\omega_n(t - t')] = \sin(\omega_n t') \cos(\omega_n t') - \cos(\omega_n t) \sin(\omega_n t')$$

before integrating over t'. In (ii), identify $\lim_{T \to 0} \rho(x, t)$ by reference to the Gaussian representation of $\delta(t)$.

10.2 In the region $0 \leqslant x, y \leqslant \pi$, the transverse displacement ψ of a stretched membrane, clamped at its edges, obeys the homogeneous wave equation, and the initial conditions

$$\psi(\boldsymbol{r}, 0) = x(\pi - x)y(\pi - y), \qquad \psi_t(\boldsymbol{r}, 0) = 0.$$

Take $c = 1$.

(i) Use the magic rule and the normal-mode expansion of the propagator to write down expressions for $\psi(\boldsymbol{r}, t)$ at $t \geqslant 0$.
(ii) To an absolute accuracy of 0.01, evaluate ψ at the centre, for times from $t = 0$ to 4π, choosing a reasonable step size.
(iii) Examine your results carefully for periodicity (see Section 10.5.1).

Hint:

$$\int \mathrm{d}\theta\, \theta \sin\theta = \sin\theta - \theta\cos\theta,$$

$$\int \mathrm{d}\theta\, \theta^2 \sin\theta = 2\theta\sin\theta - (\theta^2 - 2)\cos\theta.$$

10.3 In the region $0 \leqslant x \leqslant \pi$, $t \geqslant 0$, use the magic rule and the normal-mode expansion of the Green's function to find the function $\psi(x, t)$ defined by

$$\Box^2 \psi = 0, \qquad \psi_x(0, t) = -1/\pi = \psi_x(\pi, t),$$

$$\psi(x, 0) = (1 - x/\pi), \qquad \psi_t(x, 0) = 0.$$

Does the result surprise you?

Hint: You may quote the Fourier series (A.5.4).

10.4 *Adiabatic switching.* In the region $0 \leqslant x \leqslant \pi$, the function $\psi(x, t)$ is defined by

$$\square^2\psi = 0, \qquad \psi(x, -\infty) = 0 = \psi_t(x, -\infty),$$
$$\psi(0, t) = \{\theta(-t)\exp(t/T) + \theta(t)\}, \qquad \psi(\pi, t) = 0.$$

(i) Find $\psi(x, t)$ for $t \leqslant 0$ and for $t \geqslant 0$.
(ii) In each case, discuss the limit $T \to \infty$ at fixed t.

Hint: You may quote the Fourier series (A.5.5).

10.5 In a region V of finite volume, ψ obeys $\square^2\psi = 0$. On the boundary, either ψ vanishes everywhere, or $\partial_n\psi$ vanishes everywhere. The ICs are $\psi(\boldsymbol{r}, 0) = 0 = \psi_t(\boldsymbol{r}, 0)$. The source density is a single flash:

$$\rho(\boldsymbol{r}, t) = Q\,\delta(\boldsymbol{r} - \boldsymbol{r}_0)\,\delta(t - t_0), \qquad (\boldsymbol{r}_0 \text{ in } V,\ t_0 > 0).$$

Show that, in any number of dimensions, the energy needed to drive this source is

(i) formally independent of $\boldsymbol{r}_0$; and
(ii) divergent.

Hint: Use Section 10.5.3 plus the closure property.

10.6 A string of length L and mass σ per unit length is stretched under tension between two fixed points. At time $t = -\infty$ it is undisplaced and at rest. A transverse force $\rho(x, t)$ per unit length is applied according to

$$\rho(x, t) = \frac{Q}{(t^2 + T^2)}\,\delta(x - L/2).$$

(i) Calculate the total energy in the limit $t/T \to \infty$; call this H.
(ii) Check your expression dimensionally.
(iii) With $Q = AT^p$, discuss the behaviour of H in the limit $T \to 0$ at fixed A. Show that $\lim_{T\to 0} H$ is finite only if $p = \frac{3}{2}$, in which case it equals $\pi\sigma c A^2/4$.

10.7 Derive the expressions analogous to (10.5.16, 18) for the case where ψ obeys the homogeneous wave equation, but is subject

(i) to inhomogeneous DBCs;
(ii) to inhomogeneous NBCs.

10.8 *Resonance excitation with narrow-band and broad-band pulses.* A string is stretched between fixed supports at $x=0$ and $x=L$. Take $\sigma=1=c$. The system is quiescent at $t=-\infty$. A transverse force $\rho(x,t)$ per unit length is applied, where

$$\rho(x,t)=\alpha\,\delta(x-L/2)\int_{-\infty}^{\infty}\mathrm{d}\Omega\left[\frac{\Gamma/\pi}{(\Omega-\Omega_0)^2+\Gamma^2}\right]\cos(\Omega t)$$
$$=\alpha\,\delta(x-L/2)\exp(-\Gamma\,|t|)\cos(\Omega_0 t).$$

Consider only $\Omega_0\gg\pi/L$ and $\Omega_0\gg\Gamma$, but both $\Gamma\ll\pi/L$ (narrow-wave-band pulse), and $\Gamma\gg\pi/L$ (broad-wave-band pulse).

(i) Show that

$$\psi(x,t\to\infty)\sim\alpha\cdot 2\sum_{n=0}^{\infty}\sin[(2n+1)\pi x/L]\sin[(2n+1)\pi t/L]$$
$$\times\frac{1}{(2n+1)}\left[\frac{\Gamma/\pi}{(\Omega_0-(2n+1)\pi/L)^2+\Gamma^2}+\frac{\Gamma/\pi}{(\Omega_0+(2n+1)\pi/L)^2+\Gamma^2}\right]. \tag{1}$$

(ii) Using eqn (10.5.18), show that in the limit $t\to\infty$ the energy H of the string is

$$H(\Omega_0,\Gamma)=\alpha^2\frac{\pi^2}{L}\sum_{n=0}^{\infty}\left[\frac{\Gamma/\pi}{(\Omega_0-(2n+1)\pi/2)^2+\Gamma^2}\right.$$
$$\left.+\frac{\Gamma/\pi}{(\Omega_0+(2n+1)\pi/L)^2+\Gamma^2}\right]^2. \tag{2}$$

When $\Omega_0\gg\pi/L$, we can and now do neglect the second term in the square brackets.

(iii) *Narrow-band excitation.* As Ω_0 approaches one of the odd-numbered normal-mode frequencies, say as $\Omega_0\to(2\nu+1)\pi/L$, H is dominated by the single term $n=\nu$ in the sum. Sketch $H(\Omega_0,\Gamma)$ as a function of Ω_0 in this frequency range, and show that

$$H_{\max}\approx\alpha^2/L\Gamma^2. \tag{3}$$

(iv) *Broad-band excitation.* In this case, many normal modes contribute comparably with each other to the sum: roughly speaking, those having frequencies $(2n+1)\pi/L$ within the range $\Omega_0\pm\Gamma$. Then it becomes a reasonable approximation to replace $\sum_{n=0}^{\infty}$ by $\int_{-\infty}^{\infty}\mathrm{d}n$. Show that this yields $H(\Omega_0,\Gamma)\approx\alpha^2/8\Gamma$. This result is independent of Ω_0, and should be compared with (3).

10.9 An acoustic resonator, initially quiescent, experiences a burst of noise represented by sources $\rho(\mathbf{r}, t) = \alpha(\mathbf{r})F(t)$. The function F varies randomly, with the correlation

$$\langle F(t)F(t')\rangle = C(t)\,\delta(t-t').$$

Show that the average (i.e. the expected) final energy of the resonator is

$$\langle H\rangle = \tfrac{1}{2}\sigma c^2 \int_V \mathrm{d}V\,\alpha^2(\mathbf{r}) \int_{-\infty}^{\infty} \mathrm{d}t\,C(t).$$

10.10 Generalizing Problem 10.9 to sources subject only to

$$\langle \rho(\mathbf{r}, t)\rho(\mathbf{r}', t')\rangle = g(\mathbf{r}, \mathbf{r}')C(t)\,\delta(t-t'),$$

show that, as $t \to \infty$,

$$\langle H\rangle = \tfrac{1}{2}\sigma c^2 \int_V \mathrm{d}V\,g(\mathbf{r}, \mathbf{r}) \int_{-\infty}^{\infty} \mathrm{d}t\,C(t).$$

11 The wave equation: II. Unbounded space

11.1 The boundary conditions at infinity, and the magic rule

For Poisson's equation and the diffusion equation little new physics is needed in order to consider unbounded space as the limit of a bounded region whose bounding surface recedes to infinity. At least this is so in 3D. For instance, the potential $\phi(\boldsymbol{r})$ due to a point source at $\boldsymbol{r}'$ is little affected by homogeneous BCs over a closed surface S whose every point is much further from both $\boldsymbol{r}$ and $\boldsymbol{r}'$ than $\boldsymbol{r}$ and $\boldsymbol{r}'$ are from each other. The same is true for the diffusion equation. But for the wave equation such limits are more subtle, and are inextricably linked with the way the system evolves in time. The basic reason is that in 3D the wave equation has solutions that vanish only like $1/r$ as $r \to \infty$. Physically, these carry away or bring in finite amounts of energy, and it is impossible to distinguish between the two cases by BCs referring only to position but not to time. Mathematically, decrease proportional only to $1/r$ is too slow for mere inspection to reveal whether surface integrals like g in the magic rule (10.4.4–7) vanish in the appropriate limit.

On physical grounds, we require that, at spatial infinity, only an outward-bound disturbance result from sources or from initial values that are effectively localized in a finite region of space. Mathematical conditions that guarantee this are sometimes called *causal* BCs; in the special simple-harmonic case (e.g. with the Helmholtz equation as treated in Chapter 13) they are also called *outgoing-wave* or *Sommerfeld radiation conditions*. For simplicity we adopt the following condition: if the data, namely $\psi(\boldsymbol{r}, t_0)$, $\psi_t(\boldsymbol{r}, t_0)$, and $\rho(\boldsymbol{r}, t')$ for $t_0 \leqslant t' \leqslant t$, all vanish beyond a given limited region of space, then $\psi(\boldsymbol{r}, t)$ vanishes for $|\boldsymbol{r}|$ greater than some finite distance, which may and generally does increase with t. (It makes no significant difference if, instead of requiring first the data and then as a consequence the solution to vanish exactly beyond certain distances, we required merely that they decrease fast enough (say faster than any inverse power of $|\boldsymbol{r}|$) beyond some such distance.) Mathematically, this is a very strong condition (equally so in either form), and at the outset one might well wonder whether solutions obeying it exist at all. But we shall see (from the Green's functions and the magic rule) that physically relevant solutions in fact do obey the condition explicitly and quasi-automatically. Hence there would be little point (and much lost labour) in starting from some mathematical BCs that might appear less restrictive at first sight.

To establish the magic rule, we need the propagator K_0 and the Green's function G_0. As before, the suffix 0 identifies functions appropriate in unbounded space, and $\int dV \cdots$ now runs over all $\boldsymbol{r}$. The propagator is defined as in Section 10.3.1, except that now it is required to vanish as $r \to \infty$:

$$\Box^2 K_0(\boldsymbol{r}, t \mid \boldsymbol{r}', t') = 0, \tag{11.1.1a}$$

$$\lim_{t \to t'+} K_0(\boldsymbol{r}, t \mid \boldsymbol{r}', t') = 0, \qquad \lim_{t \to t'+} K_{0t}(\boldsymbol{r}, t \mid \boldsymbol{r}', t') = c^2\, \delta(\boldsymbol{r} - \boldsymbol{r}'), \tag{11.1.1b,c}$$

$$K_0 \to 0 \quad \text{as} \quad r \to \infty. \tag{11.1.1d}$$

The Green's function is defined correspondingly (see Section 10.3.3):

$$\Box^2 G_0(\boldsymbol{r}, t \mid \boldsymbol{r}', t') = \delta(t - t')\, \delta(\boldsymbol{r} - \boldsymbol{r}'), \tag{11.1.2a}$$

$$G_0(\boldsymbol{r}, t \mid \boldsymbol{r}', t') = 0 \quad \text{for} \quad t < t', \tag{11.1.2b}$$

$$G_0 \to 0 \quad \text{as} \quad r \to \infty. \tag{11.1.2c}$$

The IC (11.1.2b) again entails the explicit equal-time conditions (10.3.14d, e), and again G_0 is related to K_0 by

$$G_0(\boldsymbol{r}, t \mid \boldsymbol{r}', t') = \theta(t - t') K_0(\boldsymbol{r}, t \mid \boldsymbol{r}', t'). \tag{11.1.3}$$

All the symmetries (10.3.3–5), (10.3.7–10), and (10.3.18–20) remain in force. Moreover, K_0 and G_0 (unlike K and G) depend on $\boldsymbol{r}$ and $\boldsymbol{r}'$ only through the combination $\boldsymbol{R} = \boldsymbol{r} - \boldsymbol{r}'$, and in fact only through $R \equiv |\boldsymbol{R}|$, whence we can write

$$K_0(\boldsymbol{r}, t \mid \boldsymbol{r}', t') = K_0(R, \tau), \tag{11.1.4}$$

$$G_0(\boldsymbol{r}, t \mid \boldsymbol{r}', t') = G_0(R, \tau) = \theta(\tau) K_0(R, \tau). \tag{11.1.5a,b}$$

Proof: The reasoning is essentially the same as in the proof of time-translation invariance (Sections 9.2.2 and 10.3.1). We change the independent variables from $(\boldsymbol{r}, t)$ to $(\boldsymbol{R}, \tau)$, noting that this entails $\partial/\partial t = \partial/\partial \tau$, $\nabla_{\boldsymbol{r}} = \nabla_{\boldsymbol{R}}$, whence $\Box^2 = \left(\dfrac{1}{c^2}\dfrac{\partial^2}{\partial \tau^2} - \nabla_{\boldsymbol{R}}^2\right)$. Then one can see that all the defining relations (11.1.1, 2) are expressible in terms only of $\boldsymbol{R}$, τ, ∇_R, $\partial/\partial \tau$; therefore the solutions K_0, G_0 likewise can be expressed in terms only of $\boldsymbol{R}$ and of τ. (The only reason why this argument fails in bounded regions is that BCs *at finite* $\boldsymbol{r}$ do introduce a dependence on $\boldsymbol{r}$ independently of $\boldsymbol{r}'$.) ■

The magic rule is now established exactly as in Section 10.4, except that the surface integral g never appears, because by virtue of our BCs on ψ it vanishes when the surface S recedes to infinity. Accordingly, the magic rule for unbounded space reads

$$\psi(\boldsymbol{r}, t) = f_0(\boldsymbol{r}, t) + h_0(\boldsymbol{r}, t) = f_0(\boldsymbol{r}, t) + h_{10}(\boldsymbol{r}, t) + h_{20}(\boldsymbol{r}, t). \qquad (11.1.6)$$

The integrals f_0 and $h_0 = h_{10} + h_{20}$ are defined exactly as in (10.4.5, 7), but with G_0 instead of G. Taken separately, they satisfy (10.4.8, 9), and vanish as $r \to \infty$ because G_0 then vanishes.

To exploit the magic rule it remains to find K_0 and then G_0 from their defining relations. This is done in the next section.

Note that our BCs exclude not only waves arriving from infinitely distant sources, but also standing waves such as might be produced by reflectors suitably positioned at infinity. This is one reason why we cannot represent unbounded space simply as an infinitely large enclosure: the standing waves (normal modes) characteristic of finite enclosures would persist in the limit, and furnish mathematically quite legitimate solutions of a kind that we do not now wish to consider.

On the other hand, even though our formalism is designed for systems where all sources and all initial disturbances are well localized, it is flexible enough to cover physical situations where it is appropriate to relax some of these conditions, either tacitly or explicitly, in ways that are usually obvious from the context. For instance, in Problem 11.6 one can address an infinitely extended initial disturbance without any ado. More important, in Chapter 13 on diffraction it will be obvious how a signal arriving from a very (but finitely) distant point source can be idealized into an incoming plane wave which, formally, extends to infinity at all times. (The same idealization underlies much of scattering theory, which is not treated in this book.)

11.2 The calculation of $K_0^{(n)}$ and $G_0^{(n)}$

11.2.1 Preview of methods

The most straightforward strategy (Section 11.2.2) represents $K_0^{(n)}(\boldsymbol{R}, \tau)$, regarded as a function of $\boldsymbol{R}$ in nD space, by an n-fold Fourier integral, choosing the Fourier coefficients as appropriate functions of τ. This method works equally well in 3D, 2D, and 1D; in view of Section 4.5.2, it is the unbounded-space version of the normal-mode expansion (10.3.13), and as such the analogue of the method used for diffusion in Section 8.3. The differences between different dimensionalities appear only when the basic integral (11.2.1), common in form to all n, is evaluated explicitly. This is easy in 3D; in 1D it reduces to the Dirichlet integral; but in 2D it needs some otherwise rather obscure properties of Bessel functions.

The Green's function is again given by $G_0 = \theta(\tau)K_0$, by the same argument as in Section 10.3.3. Alternatively, this connection follows directly from the defining relations applied to the $(n+1)$-fold Fourier representation of G_0, regarded as a function of both $\boldsymbol{R}$ and τ, as in Appendix F.4.

Section 11.2.3 obtains $K_0^{(2)}$ and $K_0^{(1)}$ from $K_0^{(3)}$ by embedding, as was done for Poisson's equation in Section 4.4.3. This method needs to evaluate only one Fourier integral (the easy one in 3D), and dispenses with Bessel functions.

An alternative method (Appendix N) avoids Fourier integrals altogether, by adapting d'Alembert's solution of the homogeneous wave equation. It works easily in 1D; in 3D it requires only a little more analysis, but does require one to take very seriously the fact that the initial conditions on K are prescribed as limits $\tau \to 0+$ (rather than directly at $\tau = 0$). In 2D, d'Alembert's method does not apply, and leaves $K_0^{(2)}$ to be found by embedding.

Finally, Appendix O reverts to the primitive Fourier representation (11.2.1), but from the point of view of its Lorentz transformation properties. Once these are understood, they can serve to abbreviate the arguments very considerably.

11.2.2 Determination of $K_0^{(n)}$ and $G_0^{(n)}$ as n-fold Fourier integrals

The basic idea, as in Section 8.3, is to write $K_0(\boldsymbol{R}, \tau)$ as a Fourier integral with respect to its dependence on $\boldsymbol{R}$, with coefficients chosen so that the homogeneous wave equation is obeyed independently by each Fourier component. In the last step, the coefficients are then made to depend on $\boldsymbol{k}$ so as to satisfy the initial conditions $\lim_{\tau \to 0+} K(\boldsymbol{R}, \tau) = 0$ and $\lim_{\tau \to 0+} K_\tau(\boldsymbol{R}, \tau) = c^2\,\delta(\boldsymbol{R})$. In fact the Fourier coefficients can be written down immediately by adapting the normal-mode expansion (10.3.13) to unbounded space, according to the prescription of Section 4.5.2:

$$K_0(\boldsymbol{R}, \tau) = c^2 \int \frac{\mathrm{d}^n k}{(2\pi)^n} \frac{\sin[\omega(\boldsymbol{k})\tau]}{\omega(\boldsymbol{k})} \exp(i\boldsymbol{k} \cdot \boldsymbol{R}).$$

With $\omega(\boldsymbol{k}) = c\,|\boldsymbol{k}| \equiv ck$, this satisfies the homogeneous wave equation, and yields our basic result

$$K_0(\boldsymbol{R}, \tau) = \frac{c}{(2\pi)^n} \int \mathrm{d}^n k \, \frac{\sin(|\boldsymbol{k}|\, c\tau)}{|\boldsymbol{k}|} \exp(i\boldsymbol{k} \cdot \boldsymbol{R}). \tag{11.2.1}$$

The modulus signs on $|k|$ are, strictly speaking, redundant; we have inserted them because, later, their absence can invite confusion in 1D.

Exercise: Verify explicitly that (11.2.1) satisfies all the defining relations quoted above.

The integral in (11.2.1) now merely needs evaluating for $n = 3, 2, 1$. Ahead of that, Appendix O shows that K_0 and G_0 depend only on the Lorentz-invariant combination $\lambda^2 \equiv (c^2\tau^2 - R^2)$, and on the sign of τ when $\lambda^2 \geqslant 0$, and that they vanish when $\lambda^2 < 0$. This last is called the causality property.

(i) *Three dimensions.* The integral (11.2.1) has the form (F.1.1) considered in Appendix F.1, with $f(k) = c \sin(|k|\,c\tau)/(2\pi)^3\,|k|$. Accordingly, (F.1.4) yields

$$K_0^{(3)} = \frac{4\pi}{R}\int_0^\infty \mathrm{d}k\, k \frac{c}{(2\pi)^3}\frac{\sin(|k|\,c\tau)}{|k|}\sin(kR)$$

$$= \frac{c}{(2\pi)^3}\frac{4\pi}{R}\int_0^\infty \mathrm{d}k \sin(kc\tau)\sin(kR). \tag{11.2.2}$$

Because the integrand is even in k, we replace $\int_0^\infty \mathrm{d}k \cdots$ by $\frac{1}{2}\int_{-\infty}^\infty \mathrm{d}k \ldots$. Using the trigonometric sum and difference formulae, one finds

$$\int_0^\infty \mathrm{d}k \sin(kc\tau)\sin(kR) = \frac{1}{4}\int_{-\infty}^\infty \mathrm{d}k\,\{\cos[k(c\tau - R)] - \cos[k(c\tau + R)]\}$$

$$= \tfrac{1}{4}2\pi\{\delta(c\tau - R) - \delta(c\tau + R)\}$$

$$= \frac{\pi}{2c}\{\delta(\tau - R/c) - \delta(\tau + R/c)\}. \tag{11.2.3}$$

Substitution into (11.2.2) yields

$$K_0^{(3)}(\boldsymbol{R}, \tau) = \frac{1}{4\pi R}\{\delta(\tau - R/c) - \delta(\tau + R/c)\}. \tag{11.2.4}$$

Recall that R is non-negative by definition.

The delta-function has entered this expression through its Fourier representation, which defines it weakly, making it an even function of its argument. We can then verify that (11.2.4) is indeed an odd function of τ, because $\tau \to -\tau$ induces

$$\{\delta(\tau - R/c) - \delta(\tau + R/c)\} \to \delta(-\tau - R/c) - \delta(-\tau + R/c)\}$$
$$= \{\delta(\tau + R/c) - \delta(\tau - R/c)\} = -\{\delta(\tau - R/c) - \delta(\tau + R/c)\}.$$

Thus an equivalent to (11.2.4) featuring the sign function $\varepsilon(x)$ reads

$$K_0^{(3)}(\boldsymbol{R}, \tau) = \varepsilon(\tau)\, \delta(|\tau| - R/c)/4\pi R. \tag{11.2.5}$$

In order to verify explicitly that (11.2.4, 5) satisfy the ICs one needs to be surprisingly careful with the limit $\tau \to 0+$; too naive an approach can easily produce incorrect numerical factors. This point is dealt with in Appendix L; the lemmas proved there are needed again when $K_0^{(3)}$ is calculated in Appendix N by d'Alembert's method.

Multiplication of (11.2.5) by $\theta(\tau)$ yields G_0. Since $\theta(\tau)\varepsilon(\tau) = \theta(\tau)$, we find

$$\begin{aligned} G_0^{(3)}(\boldsymbol{R}, \tau) &= \theta(\tau)\, \delta(\tau - R/c)/4\pi R \\ &= \theta(\tau)\, \delta(c\tau - R)c/4\pi R. \end{aligned} \tag{11.2.6}$$

This follows also from (11.2.4), if one accepts that $\theta(\tau)\, \delta(\tau + R/c)$ is effectively zero, because $(\tau + R/c)$ cannot vanish when τ (like R) is positive. To make this argument foolproof it should be backed by the lemmas from Appendix L.

Because

$$\delta(\tau^2 - R^2/c^2) = \frac{1}{2(R/c)}\{\delta(\tau - R/c) + \delta(\tau + R/c)\},$$

and because the second delta-function when multiplied by $\theta(\tau)$ is zero (as argued just above), $G_0^{(3)}$ has the alternative form

$$\begin{aligned} G_0^{(3)}(\boldsymbol{R}, \tau) &= \theta(\tau)\frac{1}{2\pi c}\delta(\tau^2 - R^2/c^2) \\ &= \theta(\tau)\frac{c}{2\pi}\delta(c^2\tau^2 - R^2). \end{aligned} \tag{11.2.7}$$

Generally (11.2.7) is less convenient than (11.2.6) in calculations, but may be more so in general arguments. Notice that the step-function in (11.2.6) is actually redundant: if τ were negative, the argument $(\tau - R/c)$ could not vanish, and the delta-function would. By contrast, in (1.2.7) the step-function is essential.

From G_0 in either of its forms (11.2.6, 7) one can recover K_0 through the general rule (10.3.21):

$$K_0(\boldsymbol{R}, \tau) = G_0(\boldsymbol{R}, \tau) - G_0(\boldsymbol{R}, -\tau). \tag{11.2.8}$$

Exercise: Derive the relation $K_0^{(3)} = \varepsilon(\tau)\, \delta(\tau^2 - R^2/c^2)/2\pi c$ directly, without detouring via $G_0^{(3)}$.

In some problems one needs the expansions of $K_0^{(3)}$ or $G_0^{(3)}$ (regarded as functions of $\boldsymbol{r}$ with given $\boldsymbol{r}'$) in the spherical harmonics $Y_{lm}(\Omega)$, and especially the isotropic term of this expansion. Appendix M considers the expansion briefly, while the radial function of the isotropic term is derived in Appendix N.3 as an application of d'Alembert's method.

Finally, the versions (10.4.7c, d) of the integral h_{20} in the magic rule feature the time-derivative

$$-\partial G_0(\boldsymbol{r}, t \mid \boldsymbol{r}', t_0)/\partial t_0 = \partial G_0/\partial t = \partial G_0(\boldsymbol{R}, \tau)/\partial \tau$$

for $\tau > 0$, which is best obtained from (11.2.6):

$$K_{0\tau}^{(3)}(\boldsymbol{R}, \tau) = G_{0\tau}^{(3)}(\boldsymbol{R}, \tau) = \frac{\delta'(\tau - R/c)}{4\pi R} = \frac{c^2\,\delta'(c\tau - R)}{4\pi R}, \qquad (\tau > 0). \tag{11.2.9}$$

(ii) *Two dimensions.* For $n = 2$, the basic integral (11.2.1) involves Bessel functions in a manner unpleasing to most eyes. Using polar coordinates, one finds

$$K_0^{(2)}(\boldsymbol{R}, \tau) = \frac{c}{(2\pi)^2} \int_0^\infty \mathrm{d}k\, k \frac{\sin(kc\tau)}{k} \int_0^{2\pi} \mathrm{d}\phi \exp(ikR\cos\phi)$$

$$= \frac{c}{(2\pi)^2} \int_0^\infty \mathrm{d}k \sin(kc\tau) 2\pi J_0(kR), \tag{11.2.10}$$

$$K_0^{(2)}(\boldsymbol{R}, \tau) = \varepsilon(\tau) \frac{1}{2\pi} \frac{\theta(|\tau| - R/c)}{(\tau^2 - R^2/c^2)^{\frac{1}{2}}}, \tag{11.2.11}$$

$$G_0^{(2)}(\boldsymbol{R}, \tau) = \theta(\tau) \frac{1}{2\pi} \frac{\theta(\tau - R/c)}{(\tau^2 - R^2/c^2)^{\frac{1}{2}}}. \tag{11.2.12}$$

The last integral in (11.2.10) is standard, though seldom encountered elsewhere. As in 3D, the step function in (11.2.12) is redundant. In (11.2.11, 12) one could replace the final step function by $\theta(\tau^2 - R^2/c^2)$.

Because the denominators vanish, the discontinuities of $K_0^{(2)}$ and $G_0^{(2)}$ at $\tau = R/c$ are infinite, and the attempt to calculate the time derivative of (11.2.12) say can easily elicit partial nonsense at that point, namely

$$K_{0\tau}^{(2)}(\boldsymbol{R}, \tau) = G_{0\tau}^{(2)}(\boldsymbol{R}, \tau) = \frac{1}{2\pi} \left\{ \frac{\delta(\tau - R/c)}{(\tau^2 - R^2/c^2)^{\frac{1}{2}}} - \theta(\tau - R/c) \frac{\tau}{(\tau^2 - R^2/c^2)^{\frac{3}{2}}} \right\}, \qquad (\tau > 0). \tag{11.2.13}$$

(This is the awkwardness anticipated earlier below eqns (10.4.7).)

Neither term on the right of (11.2.13) is integrable on its own up to or across the singularity, though away from $\tau = R/c$ the expression is of course well-defined, and consists of the second term alone. By contrast, the first term is meaningless. Indeed the only meaning that $G_{0\tau}^{(2)}$ possesses at $\tau = R/c$ is that integration across this point reproduces $G_0^{(2)}$ itself (merely by definition). Hence it is safer to avoid quoting formulae that feature $G_{0\tau}^{(2)}$ in any position where it still needs to be manipulated: for instance, in 2D the initial-value integral h_{20} is better written as (10.4.7e) rather than as (10.4.7c, d). On the other hand, when $G_{0\tau}^{(2)}$ enters as an end-result (necessarily of an idealized problem, as in Section 11.5 below), it signifies merely the singularity manifest directly from $G_0^{(2)}$.

Fortunately, even in 2D correct conclusions are reached through formal integrations by parts featuring $G_{0\tau}$ at intermediate stages (as in Section 11.4 below), though we offer no general proof of this assertion. A careful treatment of some important and representative cases is given by Whitham (1974, Sections 7.4, 7.6).

(iii) *One dimension.* With $n = 1$ and $X \equiv x - x'$, the basic integral (11.2.1) reads

$$K_0^{(1)}(X, \tau) = \frac{c}{2\pi} \int_{-\infty}^{\infty} \mathrm{d}k \, \frac{\sin(|k| \, c\tau)}{|k|} [\cos(kX) + i \sin(kX)].$$

The modulus signs may be dropped (together), and the imaginary part integrates to zero because its integrand is odd in k. Standard trigonometric formulae yield

$$K_0^{(1)}(X, \tau) = \frac{c}{2\pi} \int_{-\infty}^{\infty} \mathrm{d}k \, \frac{1}{2k} \Big\{ \sin[k(c\tau - X)] + \sin[k(c\tau + X)] \Big\}. \quad (11.2.14)$$

These are Dirichlet integrals (see Appendix B); since

$$\int_{-\infty}^{\infty} \mathrm{d}k \, \sin(ak)/k = \pi\varepsilon(a),$$

we identify a in turn with $(c\tau - X)$ and $(c\tau + X)$, and find

$$K_0^{(1)}(X, \tau) = \frac{c}{2} \cdot \frac{1}{2} \{\varepsilon(\tau - X/c) + \varepsilon(\tau + X/c)\}. \quad (11.2.15)$$

This can be simplified as follows, thinking of $K_0^{(1)}$ as a function of X for fixed $\tau > 0$. When $X/c < -\tau$, the first ε is $+1$, the second is -1, and $K_0^{(1)}$ vanishes. When $-\tau < X/c < \tau$, the first ε is still $+1$, the second is also $+1$, and $K_0^{(1)} = \frac{1}{2}c$. When $X/c > \tau$, the first ε is -1, the second is $+1$, and $K_0^{(1)}$ again vanishes. Thus, for $\tau > 0$, we have $K_0^{(1)} = \frac{c}{2} \theta(\tau - |X|/c)$. Since

K_0 is odd in τ, either sign of τ is catered for by writing

$$K_0^{(1)}(X, \tau) = \varepsilon(\tau)\frac{c}{2}\theta(|\tau| - |X|/c), \tag{11.2.16a}$$

$$= \varepsilon(\tau)\frac{c}{2}\theta(\tau^2 - X^2/c^2); \tag{11.2.16b}$$

$$G_0^{(1)}(X, \tau) = \theta(\tau)\frac{c}{2}\tfrac{1}{2}\{\varepsilon(\tau - X/c) + \varepsilon(\tau + X/c)\}, \tag{11.2.17a}$$

$$= \theta(\tau)\frac{c}{2}\theta(\tau - |X|/c). \tag{11.2.17b}$$

The versions (11.2.15) and (11.2.17a) are often convenient in calculations, because they are expressed in terms of X rather than of $|X|$ or of X^2.

Exercises: (i) Verify (11.2.16a) from (11.2.15) by sketching $\varepsilon(\tau - X/c)$, $\varepsilon(\tau + X/c)$, and their sum, as functions of X for fixed τ first for $\tau > 0$ and then for $\tau < 0$. (ii) Verify explicitly that the expressions we have found for $K_0^{(n)}$, $(n = 3, 2, 1)$ have the correct dimensions as prescribed by (10.3.2).

The time-derivative is obtained, unproblematically, from (11.2.17b):

$$K_{0\tau}^{(1)}(\boldsymbol{R}, \tau) = G_{0\tau}^{(1)}(\boldsymbol{R}, \tau) = \frac{c}{2}\delta(\tau - |X|/c), \qquad (\tau > 0). \tag{11.2.18}$$

11.2.3 Embedding

One can obtain $G_0^{(2)}$ from $G_0^{(3)}$ by reasoning as in Section 3.3, and as we have already done for Poisson's equation in Section 4.4.3. One regards $G_0^{(2)}(\boldsymbol{R}, \tau)$ as the signal ψ, in 3D, from an infinitely long line source along the z-axis, $\rho(\boldsymbol{R}, \tau) = \delta(\tau)\,\delta(X)\,\delta(Y)$, of unit total output per unit length, flashing on and off instantaneously at $\tau = 0$. Then ψ is just the sum of the signals from all the 3D point sources which (loosely speaking) constitute this line source. Since ψ solves $\square^2\psi = \rho$, and since by symmetry it is clearly independent of Z, it in fact solves

$$\left(\frac{1}{c^2}\frac{\partial^2}{\partial t^2} - \frac{\partial^2}{\partial X^2} - \frac{\partial^2}{\partial Y^2}\right) = \delta(\tau)\,\delta(X)\,\delta(Y),$$

which identifies it as the solution $G_0^{(2)}$ that we seek. Accordingly

$$G_0^{(2)}(X, Y; \tau) = \int_{-\infty}^{\infty} \mathrm{d}Z\, G_0^{(3)}(X, Y, Z; \tau). \tag{11.2.19}$$

It may be reassuring to check explicitly that the integral on the right, call it J, indeed satisfies the conditions defining $G_0^{(2)}$. The asymptotic and initial conditions are met on inspection. To verify the differential equation $\square^{(2)^2} J = \delta(\tau)\,\delta^{(2)}(\boldsymbol{R})$ (where superscripts specify dimensionality), we take the 2D wave operator under the integral, turn it into $\square^{(3)^2}$ by subtracting $\partial^2/\partial Z^2$, and add $\partial^2/\partial Z^2$ to compensate:

$$\begin{aligned}\square^{(2)^2} J &= \int_{-\infty}^{\infty} \mathrm{d}Z\, \square^{(2)^2} G_0^{(3)} \\ &= \int_{-\infty}^{\infty} \mathrm{d}Z \left\{\square^{(3)^2} + \frac{\partial^2}{\partial Z^2}\right\} G_0^{(3)}(X, Y, Z; \tau).\end{aligned}$$

The second (compensating) term integrates to zero:

$$\int_{-\infty}^{\infty} \mathrm{d}Z\, \partial^2 G_0^{(3)}/\partial Z^2 = \partial G_0^{(3)}/\partial Z\big|_{Z\to\infty} - \partial G_0^{(3)}/\partial Z\big|_{Z\to-\infty} = 0;$$

the derivatives vanish because $Z \to \infty$ entails $R \to \infty$ and thereby $R > c\tau$. The first term supplies what we want:

$$\begin{aligned}\int_{-\infty}^{\infty} \mathrm{d}Z\, \square^{(3)^2} G_0^{(3)} &= \int_{-\infty}^{\infty} \mathrm{d}Z\, \delta(\tau)\,\delta(X)\,\delta(Y)\,\delta(Z) \\ &= \delta(\tau)\,\delta(X)\,\delta(Y) \int_{-\infty}^{\infty} \mathrm{d}Z\, \delta(Z) \\ &= \delta(\tau)\,\delta^{(2)}(\boldsymbol{R}).\end{aligned}$$

■

To evaluate (11.2.19) we introduce polar coordinates $(R_\parallel, \phi)$ in the (X, Y)-plane, so that $R^2 = (R_\parallel^2 + Z^2)$. Then, using the form (11.2.7) of $G_0^{(3)}$, we find

$$G_0^{(3)} = \theta(\tau) \frac{c}{2\pi} \int_{-\infty}^{\infty} \mathrm{d}Z\, \delta(c^2\tau^2 - R_\parallel^2 - Z^2). \tag{11.2.20}$$

Observe that the integral vanishes unless $c^2\tau^2 - R_\parallel^2 > 0$, or in other words unless $\tau > R_\parallel/c$, because otherwise the argument of the delta-function cannot be zero for any value of Z. Hence, using the standard form

$$\delta(\tau^2 - a^2) = \frac{1}{2\,|a|}\{\delta(\tau - a) + \delta(\tau + a)\},$$

we obtain

$$G_0^{(2)} = \theta(\tau)\frac{c}{2\pi}\theta(\tau - R_{\parallel}/c)$$
$$\times \int_{-\infty}^{\infty} \mathrm{d}Z \frac{\{\delta[Z - (c^2\tau^2 - R_{\parallel}^2)^{\frac{1}{2}}] + \delta[Z + (c^2\tau^2 - R_{\parallel}^2)^{\frac{1}{2}}]\}}{2(c^2\tau^2 - R_{\parallel}^2)^{\frac{1}{2}}}.$$

Provided only $\tau > R_{\parallel}/c$, both delta-functions contribute to the integral, by equal amounts; otherwise neither contributes. Thus

$$G_0^{(2)} = \theta(\tau)\theta(\tau - R_{\parallel}/c)\frac{c}{2\pi}\frac{1}{2(c^2\tau^2 - R_{\parallel}^2)^{\frac{1}{2}}}2,$$

$$G_0^{(2)}(R_{\parallel}, \tau) = \theta(\tau)\theta(\tau - R_{\parallel}/c)\frac{1}{2\pi}\frac{1}{(\tau^2 - R_{\parallel}^2/c^2)^{\frac{1}{2}}}, \tag{11.2.21}$$

which is just (11.2.9) with $R_{\parallel}$ written in place of R.

To obtain $G_0^{(1)}$ from $G_0^{(3)}$, regard $G_0^{(1)}$ as describing, in 3D, the effects of an infinitely extended plane source $\rho(\boldsymbol{R}, \tau) = \delta(\tau)\,\delta(Z)$, of unit total output per unit area in the xy-plane, flashing on and off at $\tau = 0$. By the same argument as above, one obtains

$$G_0^{(1)}(Z, \tau) = \int_{-\infty}^{\infty}\int_{-\infty}^{\infty} \mathrm{d}X\,\mathrm{d}Y\,G_0^{(3)}(X, Y, Z; \tau), \tag{11.2.22}$$

and, adopting polar coordinates again,

$$G_0^{(1)} = \theta(\tau)\frac{c}{2\pi}\int_0^{\infty} 2\pi\,\mathrm{d}R_{\parallel}\,R_{\parallel}\,\delta(c^2\tau^2 - Z^2 - R_{\parallel}^2)$$

$$= \theta(\tau)c\theta(\tau - |Z|/c)\int_0^{\infty} \tfrac{1}{2}\,\mathrm{d}R_{\parallel}^2\,\delta(c^2\tau^2 - Z^2 - R_{\parallel}^2),$$

$$G_0^{(1)}(Z, \tau) = \theta(\tau)\frac{c}{2}\theta(\tau - |Z|/c), \tag{11.2.23}$$

which is just (11.2.17b) with $|Z|$ written in place of $|X|$.

Notice that (11.2.20) represents $G_0^{(2)}$ as the *projection,* onto the equatorial plane, of $G_0^{(3)}$ regarded as a uniform distribution on the surface of a sphere of radius $c\tau$. Similarly, (11.2.22) represents $G_0^{(1)}$ as the projection of the same distribution onto its axis. Once these aspects of $G_0^{(2)}$ and $G_0^{(1)}$ are spelled out, the results (11.2.21, 23) can equally well be found geometrically rather than by analysis.

Exercise: Obtain $G_0^{(1)}$ by embedding from $G_0^{(2)}$.

11.3 A preliminary case history: point sources suddenly switched on in nD

11.3.1 Introduction

Before further comment on the crucial qualitative features of $G_0^{(n)}$ and $G_{0\tau}^{(n)}$, we interpolate a case history which may help to motivate such general observations. Readers who strongly prefer the general before the particular can reassign this illustration to Chapter 12.

Consider a system in nD, quiescent up to time $t=0$, when a point source of strength α is suddenly switched on, remaining steady ever after:

$$\rho(\boldsymbol{r}, t) = \alpha\, \delta(\boldsymbol{r})\theta(t), \tag{11.3.1}$$

$$\psi(\boldsymbol{r}, t) = 0 \quad \text{for} \quad t<0. \tag{11.3.2}$$

We determine $\psi(\boldsymbol{r}, t)$ for $t>0$, and compare it with the static solution $\psi_{\text{stat}}(\boldsymbol{r})$ of Poisson's equation, which is appropriate if one has $\rho = \alpha\, \delta(\boldsymbol{r})$ for all t. The effects of the dimensionality may then be compared with those on diffusion (Section 8.5).

From ψ one can construct the fluid velocity $\boldsymbol{v} = -\boldsymbol{\nabla}\psi$ (which is radial by symmetry); the excess pressure $\Delta p = \sigma\psi_t$; the energy density $H = \frac{1}{2}\sigma\left\{\frac{1}{c^2}\psi_t^2 + (\boldsymbol{\nabla}\psi)^2\right\}$; and the energy flux $\boldsymbol{N} = -\sigma\psi_t\boldsymbol{\nabla}\psi$. Especially interesting is the power output of the source

$$W(t) = \int \mathrm{d}V\, \sigma\rho\psi_t = \sigma\alpha\psi_t(0, t). \tag{11.3.3}$$

According to the magic rule (11.1.6), (10.4.5, 7), ψ is just the integral f_0:

$$\psi^{(n)}(\boldsymbol{r}, t) = \int_{-\infty}^{t} \mathrm{d}t' \int \mathrm{d}V'\, G_0^{(n)}(\boldsymbol{r}, t\,|\,\boldsymbol{r}', t')\alpha\delta(\boldsymbol{r}')\theta(t')$$

$$= \alpha\int_0^t \mathrm{d}t'\, G_0^{(n)}(\boldsymbol{r}, t\,|\,0, t') = \alpha\int_t^0 (-\mathrm{d}\tau) G_0^{(n)}(\boldsymbol{r}, \tau),$$

$$\psi^{(n)}(\boldsymbol{r}, t) = \alpha\int_0^t \mathrm{d}\tau\, G_0^{(n)}(\boldsymbol{r}, \tau). \tag{11.3.4}$$

In the penultimate line we have switched to the notation $G_0(\boldsymbol{R}, \tau)$, where $\boldsymbol{r}' = 0$ implies $\boldsymbol{R} = \boldsymbol{r} - 0 = \boldsymbol{r}$, and have changed the integration variable from t' to $\tau \equiv t - t'$.

11.3.2 One dimension

With $G_0^{(1)}$ from (11.2.17b), eqn (11.3.4) yields

$$\psi^{(1)}(x, t) = \alpha \frac{c}{2} \int_0^t \mathrm{d}\tau\, \theta(\tau - |x|/c).$$

The integrand survives only if $t > |x|/c$: then it is zero below $\tau = |x|/c$ and unity above. Accordingly

$$\begin{aligned} \psi^{(1)}(x, t) &= \theta(t - |x|/c) \frac{\alpha c}{2} (t - |x|/c) \\ &= \theta(t - |x|/c)\alpha(\tfrac{1}{2}ct - \tfrac{1}{2}|x|). \end{aligned} \tag{11.3.5}$$

Figure 11.1 shows this as a function of x at fixed t. For comparison,

$$\psi_{\text{stat}}^{(1)}(x) = A^{(1)} - \tfrac{1}{2}\alpha\,|x|. \tag{11.3.6}$$

Thus, in the region $|x| < ct$, the x-dependent terms of $\psi^{(1)}(x, t)$ and of $\psi_{\text{stat}}^{(1)}(x)$ agree, while the constant $A^{(1)}$ in $\psi_{\text{stat}}^{(1)}$ becomes the x-independent but t-proportional term $\alpha ct/2$.

Exercise: Accepting that $\psi^{(1)}(x, t) \equiv \theta(t - |x|/c)\{A(t) - \tfrac{1}{2}\alpha\,|x|\}$, determine the time-dependence of A by dimensional arguments, along the lines of Section 8.5.5.

With $\theta \equiv \theta(t - |x|/c)$ for short, (11.3.5) entails $\psi_t^{(1)} = \tfrac{1}{2}\alpha c\theta$, $\psi_x^{(1)} = -\tfrac{1}{2}\alpha\theta$, whence $H^{(1)} = \tfrac{1}{4}\alpha^2\sigma\theta$, and $N^{(1)} = cH^{(1)}\varepsilon(x)$. Thus, everywhere behind the expanding front at $|x| = ct$, the energy density and the energy flux are constants independent of x and t; the flux manifestly supplies energy at just the rate needed to establish H in the region swept up by

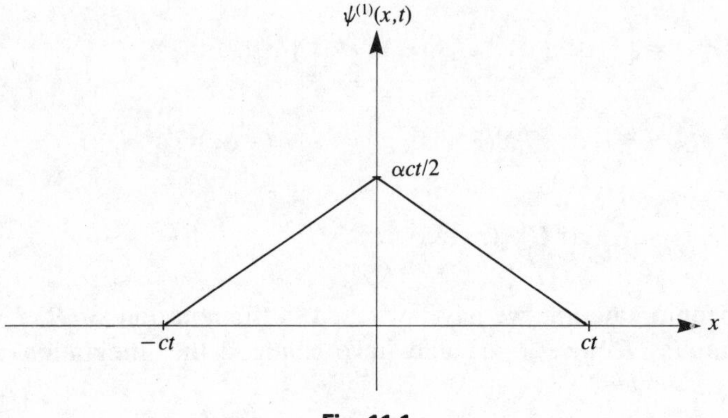

Fig. 11.1

the front. The power output (11.3.3) is

$$W^{(1)}(t) = \theta(t)\tfrac{1}{2}\alpha^2\sigma c. \tag{11.3.7}$$

11.3.3 Two dimensions

With $G_0^{(2)}$ from (11.2.12), eqn (11.3.4) yields

$$\begin{aligned}\psi^{(2)}(\mathbf{r}, t) &= \frac{\alpha}{2\pi}\int_0^t d\tau\, \theta(\tau - r/c)/(\tau^2 - r^2/c^2)^{\frac{1}{2}} \\ &= \theta(t - r/c)\frac{\alpha}{2\pi}\int_{r/c}^t d\tau/(\tau^2 - r^2/c^2)^{\frac{1}{2}},\end{aligned} \tag{11.3.8}$$

$$\psi^{(2)}(\mathbf{r}, t) = \theta(t - r/c)\frac{\alpha}{2\pi}\log\left\{\frac{ct}{r} + \left(\frac{c^2t^2}{r^2} - 1\right)^{\frac{1}{2}}\right\}. \tag{11.3.9}$$

The closed form (11.3.9) yields the derivatives, and thence H, N_r, and W. There is no need to differentiate the step-function (as we have already seen in other contexts), because the resulting factor $\delta(t - r/c)$ then multiplies the logarithm, which vanishes when $t = r/c$. One finds straightforwardly

$$\psi_t^{(2)}(\mathbf{r}, t) = \theta(t - r/c)\frac{\alpha}{2\pi}\frac{1}{(t^2 - r^2/c^2)^{\frac{1}{2}}}, \tag{11.3.10}$$

$$\psi_r^{(2)}(\mathbf{r}, t) = -\theta(r - r/c)\frac{\alpha}{2\pi}\frac{(t/r)}{(t^2 - r^2/c^2)^{\frac{1}{2}}}, \tag{11.3.11}$$

$$H^{(2)} = \theta(t - r/c)\tfrac{1}{2}\alpha^2\frac{\sigma}{(2\pi)^2}\frac{1}{c^2(t^2 - r^2/c^2)}\left\{1 + \frac{c^2t^2}{r^2}\right\}, \tag{11.3.12}$$

$$N_r^{(2)} = \theta(t - r/c)\alpha^2\frac{\sigma}{(2\pi)^2}\frac{(t/r)}{(t^2 - r^2/c^2)}; \tag{11.3.13}$$

$$W^{(2)} = \theta(t)\alpha^2\frac{\sigma}{2\pi t}. \tag{11.3.14}$$

As expected, $W^{(2)} = \lim_{r\to 0} 2\pi r N_r^{(2)}(r, t)$.

A manifest resemblance to $\psi_{\text{stat}}^{(2)}$ emerges only as $t \to \infty$ at fixed r, i.e. only if $ct \gg r$. Then (11.3.9) simplifies to

$$\psi^{(2)}(\mathbf{r}, t) \approx \frac{\alpha}{2\pi}\left\{\log\left(\frac{2ct}{r}\right) - \left(\frac{r}{2ct}\right)^2 + \cdots\right\}, \tag{11.3.15}$$

which may be compared with

$$\psi_{\text{stat}}^{(2)}(\mathbf{r}) = \alpha\left\{A^{(2)} + \frac{1}{2\pi}\log\left(\frac{1}{r}\right)\right\}. \tag{11.3.16}$$

The total energy output of the source up to any positive time (however small) is infinite, because $H^{(2)}(t) = \int_0^t \mathrm{d}t'\, W^{(2)}(t') = (\alpha^2\sigma/2\pi)\int_0^t \mathrm{d}t/t'$ diverges at its lower limit. There is another divergence as $t \to \infty$. These divergences evidently reflect the divergence of the self-energy of a static 2D point source, given formally by $\int_0^\infty 2\pi r\, \mathrm{d}r\, \frac{1}{2}\sigma(\partial\psi^{(2)}_{\text{stat}}/\partial r)^2$.

While the derivatives ψ_t and ψ_r of the closed expression (11.3.9) for ψ have been calculated unexceptionably, we should have run into the ill-defined expression $\theta(t - r/c)(\alpha/2\pi)/(\tau^2 - r^2/c^2)|_{\tau=r/c}$ if we had tried instead to obtain ψ_t say by differentiating (with respect to its upper limit) the integral representation (11.3.8) of ψ. This reflects the technical difficulty with $G^{(2)}_{0\tau}$ that we discussed *à propos* of (11.2.13) above.

11.3.4 Three dimensions

With $G_0^{(3)}$ from (11.2.6), eqn (11.3.4) yields

$$\psi^{(3)}(\mathbf{r}, t) = \alpha \int_0^t \mathrm{d}\tau\, \delta(\tau - r/c)/4\pi r = \theta(t - r/c)/4\pi r. \tag{11.3.17}$$

This, sketched in Fig. 11.2, is most remarkable. For $t > r/c$, it agrees exactly with $\psi^{(3)}_{\text{stat}} = \alpha/4\pi r$. Differentiation yields

$$\psi_t^{(3)}(\mathbf{r}, t) = \alpha\delta(t - r/c)/4\pi r, \tag{11.3.18}$$

$$\psi_r^{(3)}(\mathbf{r}, t) = -\frac{\alpha\delta(t - r/c)}{4\pi rc} - \frac{\alpha\theta(t - r/c)}{4\pi r^2}. \tag{11.3.19}$$

Thus $\psi_t^{(3)}$ vanishes except at $r = ct$. In particular, for all $t > 0$ it is zero at

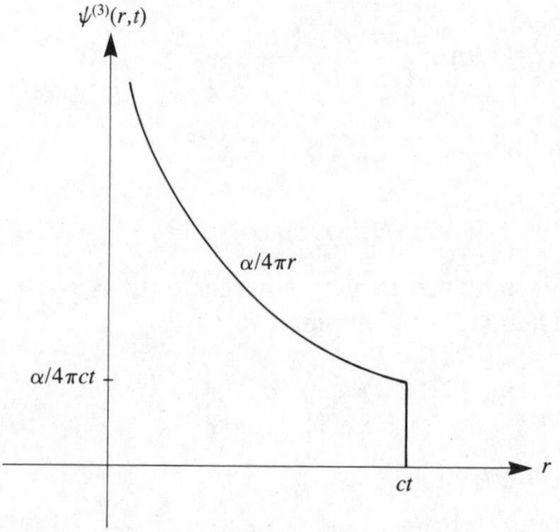

Fig. 11.2

$r = 0$, whence the power output of the source vanishes:

$$W^{(3)}(t) = 0 \quad \text{for} \quad t > 0. \tag{11.3.20}$$

Accordingly, the total energy stored in the field must have been supplied by the source instantaneously at $t = 0$, and is merely redistributed at later times. Contrast this with the finite-time power outputs $W^{(1)}(t) = \alpha^2 \frac{1}{2} \sigma c$ and $W^{(2)}(t) = \alpha^2 \sigma / 2\pi t$ as found above.

Exercise: Try to determine the time-dependence of $W^{(n)}(t)$ on dimensional grounds, bearing in mind that, from the structure of the equations, it must be proportional to $\alpha^2 \sigma$.

In 3D, H and N_r are obviously ill-defined at $r = ct$, because they involve squares of delta-functions, for which there are no rules of integration. Nevertheless, one can gain some insight into the energy balance, as follows. Everywhere strictly behind the front, i.e. at all $r < ct$, ψ_t and thereby N_r vanish, while $H = \frac{1}{2} \alpha \sigma \psi_r^2 = \alpha \sigma / 32\pi^2 r^4$ is constant in time. In other words, the total energy stored at time t_1 inside any fixed radius less than ct_1 never changes at any time later than t_1. Consequently, the energy eventually to be stored beyond $r = ct_1$, totalling

$$\int_{ct_1}^{\infty} 4\pi r^2 \, \mathrm{d}r \, H_{\text{stat}} = \alpha^2 \int_{ct_1}^{\infty} 4\pi r^2 \, \mathrm{d}r \, \sigma / 32\pi^2 r^4 = \sigma \alpha^2 / 8\pi c t_1,$$

must be precisely the energy contained in the singular distrubance exactly at the front (i.e. at $r = ct_1$), even though the latter quantity is not calculable directly. Problem 11.12 substantiates these assertions in the limit corresponding to the Gaussian representation of the factor $\delta(\boldsymbol{r})$ of ρ.

Exercise: Try to devise a similar roundabout interpretation in 2D of the energy density and of the energy flux at $r = ct$.

Notice finally that the total energy in the field at any time $t > 0$, which by the arguments above equals the total self-energy $U \equiv \int \mathrm{d}V \, H_{\text{stat}}$ of a steady point source, is formally infinite, because the integral $U = \int_0^\infty 4\pi r^2 \, \mathrm{d}r \, \alpha^2 \sigma / 32\pi^2 r^4$ diverges at its lower limit. (This does not invalidate our reasoning about the energy balance at finite times, which involved only regions finitely far from the origin.)

11.3.5 The power output determined directly from the propagator

The power output was found, above, by calculating $\psi(\boldsymbol{r}, t)$, differentiating it to obtain $\psi_t(0, t)$, and substituting this into (11.3.3).

Instead, we could have used the formula (10.5.13b), which expresses W directly in terms of the propagator; in the absence of boundaries we merely extend the volume integrals over all space. The expression (11.3.1) for ρ entails $\rho_t = \alpha\,\delta(\mathbf{r})\,\delta(t)$; we need simply substitute ρ, ρ_t, and the $G_0^{(n)}$ from Section 11.2.2, into (10.5.13b), and find

$$W^{(n)}(t) = \sigma \iint \mathrm{d}V\,\mathrm{d}V'\,\alpha\,\delta(\mathbf{r})\theta(t)\int_{-\infty}^{t}\mathrm{d}t'\,G_0^{(n)}(\mathbf{r},t\,|\,\mathbf{r}',t')\alpha\,\delta(\mathbf{r}')\,\delta(t')$$
$$= \sigma\alpha^2 G_0^{(n)}(0,t\,|\,0,0) = \sigma\alpha^2 G_0^{(n)}(R=0,\tau=t). \tag{11.3.21}$$

Accordingly, for $t>0$,

$$W^{(1)} = \sigma\alpha^2 \tfrac{1}{2}c\theta(t-R/c)\big|_{R=0} = \tfrac{1}{2}\alpha^2\sigma c,$$

$$W^{(2)} = \sigma\alpha^2 \frac{1}{2\pi}\frac{\theta(t-R/c)}{(t^2-R^2/c^2)^{\frac{1}{2}}}\bigg|_{R=0} = \frac{1}{2\pi}\frac{\alpha^2\sigma}{t},$$

$$W^{(3)} = \sigma\alpha^2\frac{\delta(t-R/c)}{4\pi R}\bigg|_{R=0} = \sigma\alpha^2\frac{\delta(t-R/c)}{4\pi ct}\bigg|_{R=0}$$
$$= 0 \quad \text{for} \quad t>0.$$

These are just the results found earlier.

11.4 Wavefronts and afterglow

The magic rule (11.1.6), (10.4.5, 7) shows that the effects of sources and of initial ψ_t are propagated by G_0, while the effects of initial ψ are propagated by $G_{0\tau}$. The $G_0^{(n)}$ and $G_{0\tau}^{(n)}$ obtained in Section 11.2 are collected in Table 11.1, and sketched in Fig. 11.3 as functions of R at fixed τ, and in Fig. 11.4 as functions of τ at fixed R.

Table 11.1 The Green's functions $G_0^{(n)}(R,\tau)$ and their time-derivatives for $\tau>0$

	$G_0^{(n)}$ (propagates sources and initial ψ_t)	$G_{0\tau}^{(n)}$ (propagates initial ψ)
3D	$\dfrac{\delta(\tau-R/c)}{4\pi R}$	$\dfrac{\delta'(\tau-R/c)}{4\pi R}$
2D	$\dfrac{1}{2\pi}\dfrac{\theta(\tau-R/c)}{[\tau^2-R^2/c^2]^{\frac{1}{2}}}$	$-\dfrac{1}{2\pi}\dfrac{\tau}{[\tau^2-R^2/c^2]^{\frac{3}{2}}}$ for $\tau>R/c$; ill-defined at $\tau=R/c$
1D	$\dfrac{c}{2}\theta(\tau-\lvert X\rvert/c)$	$\dfrac{c}{2}\delta(\tau-\lvert X\rvert/c)$

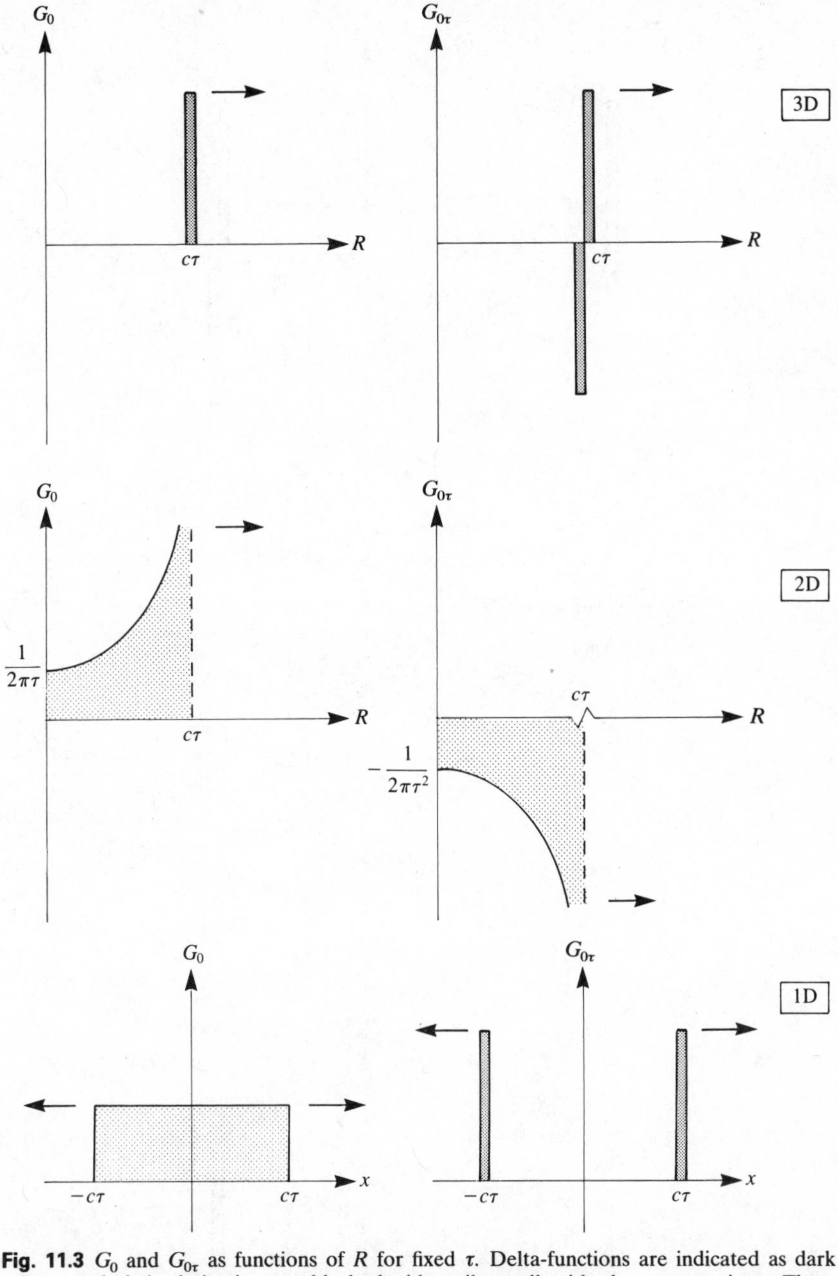

Fig. 11.3 G_0 and $G_{0\tau}$ as functions of R for fixed τ. Delta-functions are indicated as dark spikes, and their derivatives as black double spikes, all with the correct signs. These functions are in fact infinitely narrow; their strengths can be read from Table 11.1. Shading indicates ordinary functions, and is intended only to guide the eye. In 2D, the curves diverge to $\pm\infty$ as they approach the vertical asymptotes at $R = c\tau$; at $R = c\tau$, $G_{0\tau}^{(2)}$ is ill-defined. Arrows indicate the directions in which the spikes or wavefronts are moving.

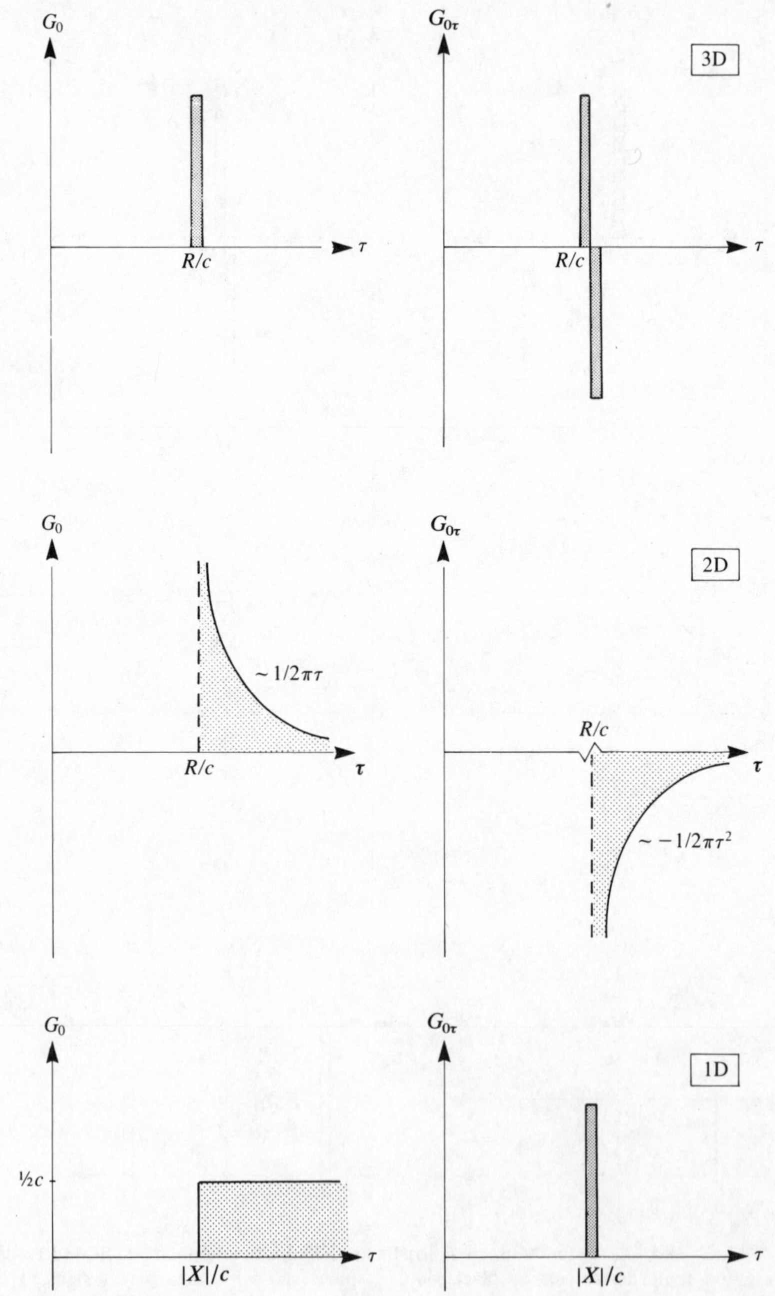

Fig. 11.4 G_0 and $G_{0\tau}$ as functions of τ for fixed R. Caption as for Fig. 11.3. In 2D, the asymptotic behaviour of the wake is indicated, appropriate when $\tau \gg R/c$.

To appreciate this information, focus on the following idealized special cases:

(a) No disturbance for $\tau < 0^-$ ($\psi = 0 = \psi_\tau$); a pulsed point source $\rho = \alpha\,\delta(\tau)\,\delta(\boldsymbol{R})$. This produces $\psi = \alpha G_0(\boldsymbol{R}, \tau)$.
(b) No sources ($\rho = 0$), but, at $\tau = 0$, initial conditions $\psi_\tau = c^2\alpha\,\delta(\boldsymbol{R})$, $\psi = 0$. This produces $\psi = \alpha G_0(\boldsymbol{R}, \tau)$. Thus, as regards their future ($\tau > 0$), cases (a) and (b) are indistinguishable.
(c) No sources, but, at $\tau = 0$, initial conditions $\psi_\tau = 0$, $\psi = c^2\beta\,\delta(\boldsymbol{R})$. This produces $\psi = \beta G_{0\tau}(\boldsymbol{R}, \tau)$.

The following points are worth stressing.

(i) *Causality.* The causes in question are the data, i.e. ρ, and the initial ψ and ψ_t; the effects are the disturbances described by the solutions ψ. The *causality property* is the feature that, irrespective of the dimensionality, there is no disturbance at R until a time lapse R/c, long enough for a signal travelling at speed c to arrive from the cause (see Appendix O).

(ii) *In 3D,* the disturbance at time τ is wholly confined to a spherical shell of radius $c\tau$. Once this shell has expanded past the observation point, no trace of its passage remains: *there is no afterglow.*

(iii) *In 2D,* an impulsive point source (cases (a) and (b)) produces at R a signal which rises from zero to infinity abruptly at $\tau = R/c$, and then diminishes gradually: *there is an indefinitely continued albeit fading afterglow.* Eventually, when $\tau \gg R/c$, the afterglow is approximated by $\psi \approx \alpha/2\pi\tau$. The remaining case (c) produces at $\tau = R/c$ the ill-natured singularity discussed *à propos* of eqn (11.2.13); this too is followed by an afterglow, fading for $\tau \gg R/c$ approximately as $\psi \sim -\beta/2\pi\tau^2$.

(iv) *In 1D,* cases (a) and (b) produce at X a signal which rises abruptly from zero to a finite value at $\tau = |X|/c$, and then remains unchanged for ever: *there is an indefinitely continued non-decaying afterglow.* For sound, such a constant value of ψ is unobservable, because it produces zero velocity $\boldsymbol{v} = -\nabla\psi$ and zero excess pressure $\Delta p = \sigma\psi_t$. But for transverse waves on a string, ψ represents the displacement, and the afterglow can be observed. By contrast, case (c) produces a delta-function signal with no afterglow.

Accordingly, in 1D as in 3D, $\boldsymbol{v}$ and Δp for sound waves are confined to the shell $R = c\tau$, but for transverse waves there is in 1D an observable afterglow in cases (a) and (b), though not in case (c).

(v) Thus, afterglows are the general rule in 2D and 1D, where only special conditions avoid them; but there is no afterglow in 3D. The 2D case is paradoxical: signals continue to arrive indefinitely from an initial disturbance that has come and gone at time zero. To see how odd this is, envisage the radio reception from a 2D sender after it has shut down. In fact 2D problems are often awkward, and we shall pay little attention to them as a rule.

(vi) *Embedding*? When 2D and 1D are embedded in 3D (Section 11.2.3 above), the afterglow can be ascribed to the signals from points on the line source or plane source physically present in 3D, and extending arbitrarily far from the observation point. But such explanations do not apply to genuinely 2D problems like waves on a stretched membrane, or to 1D waves on a stretched string, and it is truer to the physics simply to recognize that there are deep differences between waves in spaces of different dimensionalities.

(vii) In spaces with more than three dimensions, afterglow occurs for all even n. By contrast, when n is odd, the disturbance is restricted to the spherical shell $R = c\tau$ (see e.g. Courant and Hilbert (1962, Section 6.12); Garabedian (1964, Sections 7.2, 3)). In this respect therefore $n = 2$ and $n = 3$ are typical of all higher n, while $n = 1$ is quite unrepresentative.

(viii) *The non-retarded limit.* This is the formal limit $c \to \infty$ already mentioned in Section 10.1, point (vii). Here we ask specifically what happens when one tries to implement the limit in the Green's functions. In 3D, $G_0^{(3)}$ then reduces to $\delta(\tau)/4\pi R$, which is just $\delta(\tau)$ times the Green's function for Poisson's equation; the integral f_0 in the magic rule (11.1.7) consequently reduces to $\int dV' \rho(\mathbf{r}', \tau)/4\pi R$, the solution of Poisson's equation with τ treated simply as a parameter. By contrast, in 2D and 1D, $G_0(\mathbf{R}, \tau)$ lacks a sensible non-retarded limit. This mathematical fact presumably reflects the somewhat pathological features of the Poisson Green's functions, and is certainly related to the behaviour of switched-on point sources as discussed in Section 11.3 above.

(ix) *Relativistic methods.* The reader will have observed that the $G_0^{(n)}$ involve R and τ only through the Lorentz-invariant combination $\lambda^2 \equiv (c^2\tau^2 - R^2)$. Appendix O exploits this property.

11.5 Derivatives of ψ from derivatives of the data

For sound waves one is more interested in the derivatives of ψ than in ψ itself (as in the example of Section 11.3), and one might prefer to calculate them directly from the derivatives of the data. Since $(\partial/\partial t)\Box^2 = \Box^2(\partial/\partial t)$, and similarly for the Cartesian components $(\partial/\partial x, \partial/\partial y, \partial/\partial z)$ of the gradient (but only for Cartesian components), the wave equation $\Box^2\psi = \rho$ entails

$$\Box^2\psi_t = \rho_t, \qquad \Box^2\psi_x = \rho_x, \quad \text{etc.,} \tag{11.5.1a, b}$$

where $\psi_t \equiv \partial\psi/\partial t$, etc. (See the corresponding argument about ∇^2 in Section

4.3.2.) We observe that (11.5.1) relates ψ_t to ρ_t and $\nabla\psi$ to $\nabla\rho$ in exactly the same way in which the original wave equation relates ψ to ρ. The BCs (vanishing at infinity) are also the same. Therefore the magic rule (11.1.6), (10.4.5, 7) immediately yields the derivatives in terms of the corresponding data, provided of course that these data are available. Thus

$$\begin{aligned}\psi_t(\mathbf{r}, t) = &\int_{t_0}^{t} dt' \int dV'\, G_0(\mathbf{r}, t \mid \mathbf{r}', t')\rho_{t'}(\mathbf{r}', t')\\ &+ \int dV' \frac{1}{c^2} G_0(\mathbf{r}, t \mid \mathbf{r}', t_0)\psi_{t_0t_0}(\mathbf{r}', t_0)\\ &+ \frac{\partial}{\partial t}\int dV' \frac{1}{c^2} G_0(\mathbf{r}, t \mid \mathbf{r}', t_0)\psi_{t_0}(\mathbf{r}', t_0),\end{aligned} \tag{11.5.2}$$

$$\begin{aligned}\psi_x(\mathbf{r}, t) = &\int_{t_0}^{t} dt' \int dV'\, G_0(\mathbf{r}, t \mid \mathbf{r}', t')\rho_{x'}(\mathbf{r}', t')\\ &+ \int dV' \frac{1}{c^2} G_0(\mathbf{r}, t \mid \mathbf{r}', t_0)\psi_{x't_0}(\mathbf{r}', t_0)\\ &+ \frac{\partial}{\partial t}\int dV' \frac{1}{c^2} G_0(\mathbf{r}, t \mid \mathbf{r}', t_0)\psi_{x'}(\mathbf{r}', t_0), \quad \text{etc.}\end{aligned} \tag{11.5.3}$$

However, one needs to reflect a little on the status of the data that enter these integrals. There is no difficulty with the derivatives ρ_t and $\nabla\rho$, since they can be obtained by differentiating $\rho(\mathbf{r}, t)$, which is prescribed explicitly. Similarly, in (11.5.3) for $\psi_x(\mathbf{r}, t)$, differentiation of the explicitly prescribed $\psi(\mathbf{r}', t_0)$ yields $\psi_{x'}(\mathbf{r}', t_0)$, while $\psi_{t_0}(\mathbf{r}', t_0)$ yields $\psi_{x't_0} = (\partial/\partial x')\psi_{t_0}(\mathbf{r}', t_0)$. In (11.5.2), $\psi_{t_0}(\mathbf{r}', t_0)$ is likewise prescribed explicitly, but $\psi_{t_0t_0}(\mathbf{r}', t_0)$ is not. However, it can be obtained from the wave equation

$$\left(\frac{1}{c^2}\frac{\partial^2}{\partial t^2} - \nabla'^2\right)\psi(\mathbf{r}', t) = \rho(\mathbf{r}', t)$$

in the limit $t \to t_0+$:

$$\frac{1}{c^2}\psi_{t_0t_0}(\mathbf{r}', t_0) = \nabla'^2\psi(\mathbf{r}', t_0) + \rho(\mathbf{r}', t_0). \tag{11.5.4}$$

On the right, $\rho(\mathbf{r}', t_0)$ is prescribed, and $\nabla'^2\psi(\mathbf{r}', t_0)$ is obtained by differentiating $\psi(\mathbf{r}', t_0)$ as before. In particular, $\psi_{t_0t_0}$ vanishes if $\psi(\mathbf{r}, t_0)$ and ρ vanish.

Exercises: (i) Use (11.5.2) to determine $\psi_t(\mathbf{r}, t)$ and $W(t)$ in the example of Section 11.3, without first calculating $\psi(\mathbf{r}, t)$ itself. (ii) Verify (11.5.2) and (11.5.3) by differentiating both sides of the original magic rule; on the right one must then use $\partial G_0/\partial t = -\partial G_0/\partial t'$, and proceed through appropriate integrations by parts. It emerges in particular that the three terms on the right of (11.5.2) do not, individually, arise from the apparently corresponding terms of (11.1.6). (iii) Show that (11.5.2), with G_0 replaced by G_D or G_N, applies also in bounded regions under homogeneous boundary conditions. (iv) Derive the power-output formula (10.5.13b) directly from (11.5.2).

11.6 The Kirchhoff representation

Recall from Section 7.5 the Kirchhoff representation of solutions of Poisson's equation. An extension of these ideas will be indicated very briefly.

Consider the solution of $\Box^2\psi = \rho$ in some subregion V of unbounded space; V is surrounded by S, which is just a mathematically defined surface, no BCs on S being enforced by any physical means. Now we reason exactly as in deriving the magic rule in Section 10.4, but using G_0 instead of a Green's function G adapted to V, while nevertheless restricting the integrations to V and S. In this way one arrives at the first equality in (10.4.3). On the right, the second integral vanishes if the field point $\boldsymbol{r}$ is outside V; but if $\boldsymbol{r}$ is inside V, this integral yields $\psi(\boldsymbol{r}, t)$ (as in the last equality in (10.4.3)). Thus one finds

$$f_K(\boldsymbol{r}, t) + g_K(\boldsymbol{r}, t) + h_K(\boldsymbol{r}, t) = \begin{cases} \psi(\boldsymbol{r}, t) & \text{if } \boldsymbol{r} \text{ is in } V, \\ 0 & \text{if } \boldsymbol{r} \text{ is not in } V. \end{cases} \tag{11.6.1}$$

Here, f_K, g_K, h_K are defined precisely like f, g, h in the magic rule (10.4.4–7), but with G replaced by G_0.

There is no difficulty with f_K and h_K, because they are explicit constructs from the data $\rho(\boldsymbol{r}, t)$, $\psi(\boldsymbol{r}, t_0)$ and $\psi_t(\boldsymbol{r}, t_0)$. But one cannot in general construct the surface integral g_K from the data of any problem that is 'well-posed' for V, because such data prescribe *either* $\psi_S(\boldsymbol{r}, t)$ *or* $\partial_n\psi_S(\boldsymbol{r}, t)$, while g_K requires *both*, since neither G_0 nor $\partial_n G_0$ vanishes on S. In other words, ψ and $\partial_n\psi$ on S, though needed in (11.6.1), cannot be prescribed independently of each other, and in general they are not both known until the problem has been solved.

On the other hand, if an approximation to ψ and to $\partial_n\psi$ on S can be found on physical grounds, then (11.6.1) yields a corresponding approximation to ψ throughout V. This is precisely what is done in the theory of diffraction for the special case of simple-harmonic solutions (Sections 13.3–6).

Problems

11.1 The function $\psi(x, t)$ is determined by $\psi_{tt} - \psi_{xx} = -xt$ for all x, and for all $t \geqslant 0$, subject to the initial conditions $\psi(x, 0) = a \sin x$, $\psi_t(x, 0) = a \cos x$.

(i) Determine ψ by solving the equation from first principles, looking for ψ = (particular integral) + (complementary function). Guess a PI, after some trial and error, in the form constant $\times\, x^n t^m$, and write CF $= f(x+t) + g(x-t)$, fitting f and g to the initial conditions.
(ii) Determine ψ by using the magic rule. Check that the results of methods (i) and (ii) agree.

Hint: The solution is $a \sin(x+t) - xt^3/6$.

11.2 *Radiation in* 1D. An infinite string stretches along the x-axis. It is undisplaced and at rest at $t \leqslant 0$. For $t \geqslant 0$, the point at $x = 0$ is moved according to $y(0, t) = A \sin(\omega t)$, by the application of a transverse force $\alpha \sin(\omega t + \delta)$ at that point, equivalent to a source distribution (transverse force per unit length) $\rho(x, t) = \alpha\, \delta(x) \sin(\omega t + \delta)$.

(i) Use the magic rule to determine $y(x, t)$, and then adjust α and δ so that $y(0, t)$ behaves as prescribed above.
(ii) The energy flux N (energy crossing the point x per unit time from left to right) is given by (10.1.8). Determine the total power output of the source as $W(t) = N(0+, t) - N(0-, t)$, and verify explicitly that $W(t)$ = force $\times$ speed $= \alpha \sin(\omega t + \delta) A\omega \cos(\omega t)$. What is the long-time-averaged power?

11.3 *A preview of images.* At $t = 0$, an infinite string stretched along the x-axis is at rest, and its transverse displacement is $y(x, 0) = A\{\delta(x-a) - \delta(x+a)\}$. (Disregard the unphysical nature of this idealization.)

(i) Use the magic rule to calculate $y(x, t)$ at all later times. Illustrate your result by sketches of y against x at suitably chosen fixed times, distinguishing carefully between $t < a/c$ and $t > a/c$.
(ii) Note that $y(0, t) = 0$ for all t. From this, deduce the solution to the problem of a semi-infinite stretched string $x \geqslant 0$ fixed to a rigid support at $x = 0$, and satisfying the ICs $\psi(x, 0) = A\, \delta(x-a)$, $\psi_t(x, 0) = 0$.

11.4 As for Problem 11.3, but with the ICs $y(x, 0) = 0$, $y_t(x, 0) = B\{\delta(x-a) - \delta(x+a)\}$.

11.5 The function $\psi(\boldsymbol{r}, t)$ obeys $\psi(\boldsymbol{r}, 0) = 0 = \psi_t(\boldsymbol{r}, 0)$, and $\square^2\psi = \theta(t)\beta \cos(\omega t - \boldsymbol{k} \cdot \boldsymbol{r})$, where ω and $\boldsymbol{k}$ are arbitrary. Determine $\psi(\boldsymbol{r}, t)$; explain in words what physical processes your solution represents, and describe how the solution changes with ω for fixed $\boldsymbol{k}$. Be careful to cover the special case $\omega \to kc$.

Hint: You could use the magic rule, but a suitable choice of axes allows one to use d'Alembert's solution. Be prepared to apply l'Hôpital's rule.

11.6 The function $\psi(\boldsymbol{r}, t)$ obeys the homogeneous 3D wave equation, and the ICs $\psi(\boldsymbol{r}, 0) = B\cos(\boldsymbol{k} \cdot \boldsymbol{r})$, $\psi_t(\boldsymbol{r}, 0) = A\cos(\boldsymbol{k} \cdot \boldsymbol{r} + \delta)$.

(i) Determine $\psi(\boldsymbol{r}, t)$ for $t \geqslant 0$.
(ii) Describe your solution in words, and explain how A, B, and δ must be chosen to make $\psi(\boldsymbol{r}, t)$ into a single monochromatic plane wave proportional to $\cos(\boldsymbol{k} \cdot \boldsymbol{r} - \omega t)$.

11.7 *Pressure at the centre of a bursting balloon.* The velocity potential obeys the ICs $\psi(\boldsymbol{r}, 0) = 0$, $\psi_t(\boldsymbol{r}, 0) = \alpha\theta(a - r)$, which describe excess pressure proportional to α inside a balloon of radius a centred at the origin. At $t = 0$ the balloon bursts, and for $t \geqslant 0$ ψ obeys the homogeneous wave equation.

(i) Use the magic rule to determine $\psi(0, t)$ and $\psi_t(0, t)$ for $t \geqslant 0$. Sketch both as functions of time, in 3D, 2D, and 1D. (In 1D, $r \equiv |x|$.) Be careful to indicate any delta-function spikes that may be present (in the manner of Fig. 11.4).
(ii) Examine your results in the light of Sections 11.4 and 11.5.

11.8 *Pressure at the centre of an exploding shell.* The velocity potential obeys the ICs $\psi(\boldsymbol{r}, 0) = \beta\theta(a - r)$, $\psi_t(\boldsymbol{r}, 0) = 0$, and obeys the homogeneous wave equation for $t \geqslant 0$.

(i) Explain what physical situation is represented by the initial conditions (or see Section 12.2.3).
(ii) Use the magic rule to determine $\psi(0, t)$ and $\psi_t(0, t)$ for $t \geqslant 0$. Sketch both as functions of time, in 3D, 2D, and 1D, being careful to indicate any delta-function spikes.
(iii) Examine your results in the light of Section 11.4, and compare them with the results of the preceding problem.

11.9 The velocity potential in unbounded 3D space obeys the wave equation

$$\left(\frac{1}{c^2}\frac{\partial^2}{\partial t^2} - \nabla^2\right)\psi(\boldsymbol{r}, t) = A\frac{1}{T\pi^{\frac{1}{2}}}\exp(-t^2/T^2)\,\delta(\boldsymbol{r}).$$

As $t \to -\infty$, ψ and ψ_t vanish everywhere.

(i) Determine $\psi(\boldsymbol{r}, t)$ for all $\boldsymbol{r}$, t.
(ii) Hence calculate $\boldsymbol{N}(\boldsymbol{r}, t)$ for all $\boldsymbol{r}$, t.
(iii) Calculate the total energy emitted by the source, by evaluating $E = \int_{-\infty}^{\infty} dt\, 4\pi r^2 N(r, t)$. (The integral should be independent of r. Show explicitly that it is so.)
(iv) Comment briefly on the behaviour of your results as $T \to 0$.

Hint: In (iii), it may prove helpful to change the integration variable from t to $\tilde{t} = t - r/c$.

11.10 In the preceding problem, evaluate E by the method of Section 10.5.3, and check that the results agree.

11.11 Prove that $R_l(r, t)Y_{lm}(\Omega) \equiv (1/r)\phi_l(r, t)Y_{lm}(\Omega)$ satisfies the homogeneous wave equation for $r \neq 0$, provided

$$\phi_l(r, t) = r^{l+1}\left(\frac{1}{r}\frac{\partial}{\partial r}\right)^l \left\{\frac{f(r - ct) + g(r + ct)}{r}\right\},$$

where the functions f and g are arbitrary. In other words, for $r \neq 0$, ϕ_l satisfies

$$\left\{\frac{1}{c^2}\frac{\partial^2}{\partial t^2} - \frac{\partial^2}{\partial r^2} + \frac{l(l+1)}{r^2}\right\}\phi_l(r, t) = 0.$$

(This result is the basis for determining the $l > 0$ components in the polar decomposition of the propagator $K_0^{(3)}$, i.e. for generalizing to $l > 0$ the method used in Appendix N.3 to determine the isotropic ($l = 0$) component.)

11.12 *An extended steady source switched on in* 3D. (Section 11.3.4 considered the power $W(t)$ emitted by the 3D point source $\rho(\boldsymbol{r}, t) = \alpha\, \delta(\boldsymbol{r})\theta(t)$, with the startling result that an infinite amount of energy is emitted instantaneously at $t = 0$, and none at any later time. Here we reinforce these conclusions by regarding the point source as the limit of an indefinitely shrinking but finitely extended distribution, namely as $\delta(\boldsymbol{r}) = \lim_{a \to 0} (1/\pi^{\frac{3}{2}}a^3) \exp(-r^2/a^2)$.)

For $t \leq 0$, ψ is zero. The source distribution is $\rho(\boldsymbol{r}, t) = (\alpha/\pi^{\frac{3}{2}}a^3) \exp(-r^2/a^2)\theta(t)$.

(i) Use eqns (10.5.23) to show that the energy emitted up to time t is

$$H(t) = \frac{\alpha^2}{a}\frac{\sigma}{(2\pi)^{\frac{3}{2}}}\{1 - \exp(-c^2t^2/2a^2)\}\theta(t), \tag{1}$$

whence

$$W(t) \equiv \frac{dH}{dt} = \frac{\alpha^2}{a} \frac{\sigma}{(2\pi)^{\frac{3}{2}}} \frac{c^2 t}{a^2} \exp(-c^2t^2/2a^2). \tag{2}$$

(ii) Prove that $\lim_{a\to 0} (c^2t/a^2) \exp(-c^2t^2/2a^2) = \delta(t)$

(in the strong sense that $\int_0^\infty dt\, \delta(t) = 1$). This shows that

$$\lim_{a\to 0} W(t) = \lim_{a\to 0} \frac{\alpha^2}{a} \frac{\sigma}{(2\pi)^{\frac{3}{2}}} \delta(t), \tag{3}$$

confirming the conclusions of Section 11.3.4 that W is a delta-function times a divergent coefficient.

11.13 *Generalization of a theorem of Gauss* (see Problem 5.11). A source distribution $\rho(\boldsymbol{r}, t)$ vanishes outside a region V surrounded by the closed surface S; $\boldsymbol{r}_0$ is any point outside V, and $\boldsymbol{r}_i$ is any point inside; G_0 is the usual unbounded-space Green's function, and G_D the interior Dirichlet Green's function for V.

(i) Prove that

$$G_0(\boldsymbol{r}_0, t_0 \mid \boldsymbol{r}_i, t_i) = -\int_{t_i}^{t_0} dt' \int_S dS'\, G_0(\boldsymbol{r}_0, t_0 \mid \boldsymbol{r}', t')\partial_n' G_D(\boldsymbol{r}', t' \mid \boldsymbol{r}_i, t_i). \tag{1}$$

(In this problem, t_0 refers to the observation time at $\boldsymbol{r}_0$, and not, as elsewhere in the book, to an initial time.)

(ii) By acting on both sides of (1) with $\int_{t_1}^{t_0} dt_i \int_V dV_i\, \rho(\boldsymbol{r}_i, t_i)$, where t_1 is chosen so that $t_1 < t_0$, show that

$$\int_{t_1}^{t_0} dt_i \int_V dV_i G_0(\boldsymbol{r}_0, t_0 \mid \boldsymbol{r}_i, t_i)\rho(\boldsymbol{r}_i, t_i)$$
$$= \int_{t_1}^{t_0} dt' \int_S dS'\, G_0(\boldsymbol{r}_0, t_0 \mid \boldsymbol{r}', t')\sigma(\boldsymbol{r}', t'), \tag{2}$$

$$\sigma(\boldsymbol{r}', t') \equiv -\int_{t_1}^{t'} dt_i \int_V dV_i\, \partial_n' G_D(\boldsymbol{r}', t' \mid \boldsymbol{r}_i, t_i)\rho(r_i, t_i). \tag{3}$$

(iii) Explain how (2, 3) allow one to choose a surface-source distribution on S which at all points outside V, and at all times, gives rise to the same signal as does the actual source distribution $\rho(\boldsymbol{r}, t)$ inside V.

Hints: In (i), start from the integral

$$\int_{t_1}^{t_2} \mathrm{d}t' \int_V \mathrm{d}V' \,\{G_0(\boldsymbol{r}_0, t_0 \mid \boldsymbol{r}', t')\nabla'^2 G_D(\boldsymbol{r}', t' \mid \boldsymbol{r}_i, t_i)$$
$$- (\nabla'^2 G_0(\boldsymbol{r}_0, t_0 \mid \boldsymbol{r}', t'))G_D(\boldsymbol{r}', t' \mid \boldsymbol{r}_i, t_i)\}.$$

Evaluate this using Green's theorem; then again by using the wave equation for G_0 and G_D to re-express $\nabla'^2 G_0$ and $\nabla'^2 G_D$; and equate the results. In (ii), repeated use can be made of the fact that $G_0(|)$ and $G_D(|)$ both vanish unless the time to the left of the upright is later than the time to its right.

11.14 Establish the analogous result using the Neumann Green's function for V.

12 The wave equation: III. Examples

12.1 The initial-value problem in unbounded 1D space

In unbounded space free of sources ($\Box^2\psi = 0$), the initial values

$$\psi_t(\mathbf{r}, t_0) \equiv \alpha(\mathbf{r}), \qquad \psi(\mathbf{r}, t_0) = \beta(\mathbf{r}) \tag{12.1.1}$$

evolve according to the magic rule (11.1.6) into

$$\psi(\mathbf{r}, t) = h_{10}(\mathbf{r}, t) + h_{20}(\mathbf{r}, t)$$
$$= \int dV' \frac{1}{c^2} G_0(R, \tau)\alpha(\mathbf{r}') + \frac{\partial}{\partial t}\int dV' \frac{1}{c^2} G_0(R, \tau)\beta(\mathbf{r}'). \tag{12.1.2}$$

Here $\mathbf{R} = \mathbf{r} - \mathbf{r}'$, and $\tau \equiv t - t_0$. Evidently, h_{20} is obtained from h_{10} by replacing $\alpha \to \beta$ and acting on the result with $\partial/\partial t = \partial/\partial\tau$. In 1D we have x, x' instead of $\mathbf{r}, \mathbf{r}'$ and $G_0^{(1)}(X, \tau) = \frac{1}{2}c\theta(\tau - |X|/c)$. Thus

$$h_{10}(x, t) = \frac{1}{2c}\int_{-\infty}^{\infty} dx'\, \theta(\tau - |x - x'|/c)\alpha(x')$$
$$= \frac{1}{2c}\int_{x-c\tau}^{x+c\tau} dx'\, \alpha(x'), \tag{12.1.3}$$

$$h_{20}(x, t) = \frac{\partial}{\partial\tau}\frac{1}{2c}\int_{x-c\tau}^{x+c\tau} dx'\, \beta(x')$$
$$= \frac{1}{2c}\{c\beta(x + c\tau) - (-c)\beta(x - c\tau)\},$$

$$h_{20}(x, t) = \tfrac{1}{2}\{\beta(x + c\tau) + \beta(x - c\tau)\}. \tag{12.1.4}$$

Substituting from (12.1.3, 4) into (12.1.2) we obtain the end-result

$$\psi(x, t) = \frac{1}{2c}\int_{x-c(t-t_0)}^{x+c(t-t_0)} dx'\, \alpha(x') + \tfrac{1}{2}\{\beta(x + c(t - t_0)) + \beta(x - c(t - t_0))\}. \tag{12.1.5}$$

This is known as *d'Alembert's solution.* Appendix N.1 derives it by a different method.

(i) Note how $\psi(x, t)$ depends on all the initial $\psi_t \equiv \alpha$ *within* a distance $c\tau$ of x, but depends on the initial $\psi \equiv \beta$ only at the two points *precisely* at that distance. Relativistically (Appendix O) one says that ψ depends on α *within*, but on β only *on* the past (backward) light cone.

(ii) The behaviour of h_{20} is particularly simple: the initial shape β splits into two identical shapes $\frac{1}{2}\beta$; both shapes move bodily with speed c, $\frac{1}{2}\beta(x-c(t-t_0))$ rightwards, and $\frac{1}{2}\beta(x+c(t-t_0))$ leftwards. The relation of h_{10} to the initial values α is less immediate, though the comments on $G_0^{(1)}$ in Section 11.4 afford some insight.

(iii) To illustrate the contrast between h_{10} and h_{20}, we choose similarly-shaped initial distributions for ψ and ψ_t,

$$\alpha(x) = A/\cosh^2(x/a), \qquad \beta(x) = B/\cosh^2(x/b).$$

These are symmetric peaks, with maxima at $x=0$. Then (12.1.3) yields

$$\begin{aligned} h_{10}(x,t) &= \frac{A}{2c}\int_{x-ct}^{x+ct} \mathrm{d}x' \, \frac{1}{\cosh^2(x'/a)} \\ &= \frac{Aa}{2c}\int_{(x-ct)/a}^{(x+ct)/a} \frac{\mathrm{d}\xi}{\cosh^2 \xi} \\ &= \frac{Aa}{2c}\left\{\tanh\left(\frac{x+ct}{a}\right) - \tanh\left(\frac{x-ct}{a}\right)\right\}. \end{aligned}$$

After some manipulation this reduces to

$$h_{10}(x,t) = \frac{2Aa}{c}\,\frac{\sinh(2ct/a)}{\{\cosh(2x/a)+\cosh(2ct/a)\}}. \tag{12.1.7}$$

Evidently h_{10} as a function of x peaks at $x=0$ for all t, but the width of the peak, at half-maximum say, grows steadily.

Exercises: (i) Verify that $h_{10t}(x,0) = \alpha(x)$. (ii) Find the leading approximations to (12.1.7) in the regimes $ct/a \to \infty$, x/a fixed; (iii) ct/a fixed, $x/a \to \infty$; (iv) $ct/a \to \infty$, $x/a \to \infty$, ct/x fixed >1; (v) $ct/a \to \infty$, $x/a \to \infty$, ct/x fixed <1.

By contrast, h_{20} is given immediately by

$$h_{20}(x,t) = \frac{1}{2}\left\{\frac{1}{\cosh^2\left(\dfrac{x-ct}{b}\right)} + \frac{1}{\cosh^2\left(\dfrac{x-ct}{b}\right)}\right\}, \tag{12.1.8}$$

having a trough at $x=0$, and two peaks receding left and right with speeds that approach c as $ct/b \to \infty$.

12.2 The initial-value problem in unbounded 3D space

12.2.1 Poisson's solution

There are several ways of expressing h_{10} and h_{20} in 3D; unusually, the more explicit the notation the harder it seems to be to absorb its message. Equations (12.1.1, 2) with $G_0^{(3)}(R, \tau) = c\,\delta(c\tau - R)/4\pi R$, entail

$$h_{10}(\boldsymbol{r}, t) = \frac{1}{c}\int \mathrm{d}V'\,\frac{\delta(c\tau - R)}{4\pi R}\,\alpha(\boldsymbol{r}'). \tag{12.2.1}$$

By virtue of the delta-function, h_{10} and h_{20} can be expressed in terms of the averages defined by (4.3.12): for any function $F(\boldsymbol{r}')$ of position $\boldsymbol{r}'$,

$$M_{\boldsymbol{r},c\tau}\{F\} \equiv \left\{\begin{matrix}\text{average of } F(\boldsymbol{r}') \text{ over the surface of the}\\ \text{sphere } |\boldsymbol{r}' - \boldsymbol{r}| = c\tau \text{ of radius } c\tau \text{ and centre } \boldsymbol{r}\end{matrix}\right\}. \tag{12.2.2}$$

To implement this idea explicitly, we first define $\boldsymbol{R}' \equiv \boldsymbol{r}' - \boldsymbol{r} \equiv -\boldsymbol{R}$, so that $\boldsymbol{r}' = \boldsymbol{r} + \boldsymbol{R}'$, and the sphere in question is $R' = R = c\tau$. We call it the averaging or the information-gathering sphere. Now change integration variables in (12.2.1) from $\boldsymbol{r}'$ to $\boldsymbol{R}' \equiv (R', \Omega')$:

$$\begin{aligned} h_{10}(\boldsymbol{r}, t) &= \frac{1}{c}\int \mathrm{d}\Omega' \int_0^\infty \mathrm{d}R'\,R'^2\,\frac{\delta(c\tau - R')}{4\pi R'}\,\alpha(\boldsymbol{r}') \\ &= \frac{1}{c}\int \mathrm{d}\Omega'\,\frac{c\tau}{4\pi}\,\alpha(\boldsymbol{r}') = \tau \int \frac{\mathrm{d}\Omega'}{4\pi}\,\alpha(\boldsymbol{r}'), \end{aligned} \tag{12.2.3a}$$

$$\boldsymbol{r}' = \boldsymbol{r} + \boldsymbol{R}' = \boldsymbol{r} + (c\tau, \Omega'), \qquad \tau = (t - t_0). \tag{12.2.3b}$$

In other words

$$h_{10}(\boldsymbol{r}, t) = \tau M_{\boldsymbol{r},c\tau}\{\alpha\}. \tag{12.2.4}$$

By the remark just below (12.1.2), this entails

$$h_{20}(\boldsymbol{r}, t) = \frac{\partial}{\partial \tau}[\tau M_{\boldsymbol{r},c\tau}\{\beta\}]. \tag{12.2.5}$$

Combining (12.2.4, 5), we obtain *Poisson's solution*:

$$\begin{aligned} \psi(\boldsymbol{r}, t) &= h_{10}(\boldsymbol{r}, t) + h_{20}(\boldsymbol{r}, t) \\ &= (t - t_0)M_{\boldsymbol{r},c(t-t_0)}\{\psi_t(\boldsymbol{r}', t_0)\} \\ &\quad + \frac{\partial}{\partial t}[(t - t_0)M_{\boldsymbol{r},c(t-t_0)}\{\psi(\boldsymbol{r}', t_0)\}]. \end{aligned} \tag{12.2.6}$$

12.2.2 The bursting balloon: h_{10}

Excess pressure Q inside a sphere of radius a and centre 0 is suddently released at $t = 0$. We seek the excess pressure and the flow velocity at a point $r > a$.

The ICs are $\psi(\boldsymbol{r}, 0) \equiv \beta(\boldsymbol{r}) = 0$, and $\Delta p(\boldsymbol{r}, 0) = \sigma\psi_t(\boldsymbol{r}, 0) = \sigma\alpha(\boldsymbol{r}) = Q\theta(a-r)$. Then $h_{20} = 0$, and, by (12.2.6) with $t = 0$,

$$\psi(\boldsymbol{r}, t) = h_{10}(\boldsymbol{r}, t) = tM_{\boldsymbol{r},ct}\left\{\frac{Q}{\sigma}\,\theta(a-r)\right\}. \tag{12.2.7}$$

Thus M is the average of $(Q/\sigma)\theta(a-r)$ over the surface of the steadily-expanding information-gathering sphere of radius ct shown in Fig. 12.1. Unless this sphere intersects the initial boundary sphere of radius a, M vanishes. Thus $\psi = 0$ unless $(r-a) \leqslant ct \leqslant (r+a)$, conditions equivalent to $-a \leqslant (ct-r) \leqslant a$, i.e. to $a^2 > (ct-r)^2$. When ct does satisfy this condition, the requisite average of $\theta(a-r)$ is $\Omega/4\pi$, where Ω is just the solid angle in the cone whose half-opening angle is χ. This is obvious, because on the portion of the information-gathering sphere within the cone, $\theta(a-r)$ has the value 1, while outside the cone it has the value 0. But

$$\Omega = \int_0^{\chi} 2\pi \sin\chi \, d\chi = 2\pi(1-\cos\chi). \tag{12.2.8}$$

By the cosine rule, $\cos\chi = [r^2 + c^2t^2 - a^2]/2ctr$,

$$(1-\cos\chi) = [2ctr - r^2 - c^2t^2 + a^2]/2ctr = [a^2 - (ct-r)^2]/2ctr;$$

accordingly

$$M = \frac{Q}{\sigma}\frac{1}{4\pi}\frac{2\pi[a^2-(ct-r)^2]}{2ctr} = \frac{Q}{\sigma}\frac{[a^2-(ct-r)^2]}{4ctr}. \tag{12.2.9}$$

We substitute this into (12.2.7), insert a step-function to embody the

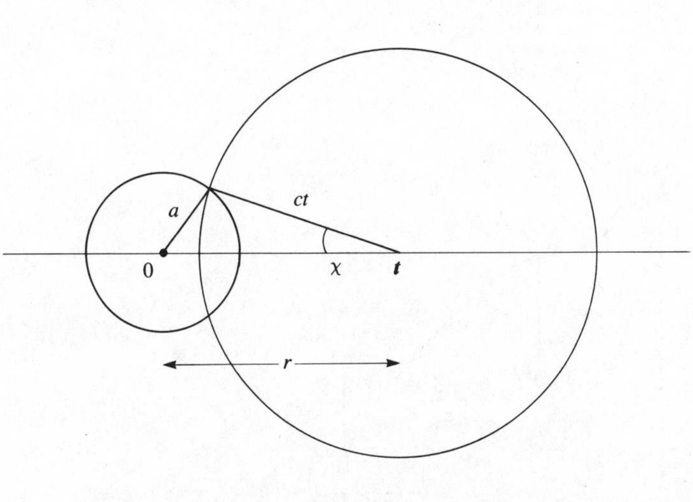

Fig. 12.1

restriction on ct, and find

$$\psi(\boldsymbol{r}, t) = \theta(a^2 - (ct - r)^2)\,\frac{Q}{\sigma}\,\frac{[a^2 - (ct - r)^2]}{4cr}, \qquad (r > a). \tag{12.2.10}$$

When differentiating this pattern $\theta(f(x))f(x)g(x)$, we need not differentiate the step-function: in other words

$$\frac{\mathrm{d}[\theta(f)fg]}{\mathrm{d}x} = \left[\delta(f)f'fg + \theta(f)\,\frac{\mathrm{d}(fg)}{\mathrm{d}x}\right] = \theta\,\frac{\mathrm{d}(fg)}{\mathrm{d}x},$$

because $\delta(f)f = 0$. Accordingly, with $\theta \equiv \theta(a^2 - (ct - r)^2) = \theta(ct - a + r)\theta(r + a - ct)$ for short,

$$\Delta p(\boldsymbol{r}, t) = \sigma\,\frac{\partial\psi(\boldsymbol{r}, t)}{\partial t} = \theta \cdot Q \cdot \frac{r - ct}{r}, \qquad (r > a). \tag{12.2.11}$$

This is sketched in Fig. 12.2a. A different derivation is given at the end of Appendix M. Note that the time-integral of Δp vanishes:

$$\int_0^\infty \mathrm{d}t\,\Delta p(\boldsymbol{r}, t) = 0. \tag{12.2.12}$$

Exercise: Explain why.

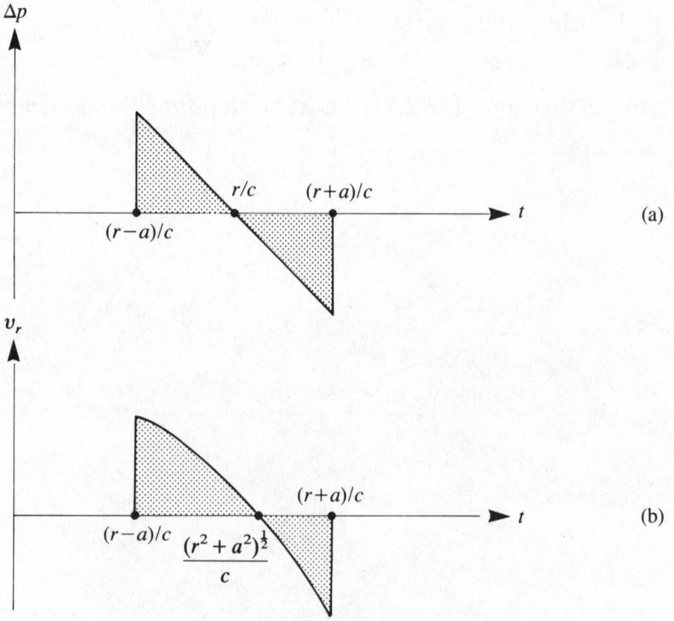

Fig. 12.2

Next,

$$v_r(r, t) = -\frac{\partial \psi}{\partial r} = \theta \frac{Q}{4\sigma c} \frac{(a^2 + r^2 - c^2 t^2)}{r^2}, \qquad (r > a), \tag{12.2.13}$$

which is sketched in Fig. 12.2b.

Exercise: Determine Δp and v_r at an interior point $r < a$.

The solutions (12.2.11, 13) are constrained by energy conservation; the time-integrated outflow of energy through any sphere of radius $r > a$ must equal the total energy $E = \int \mathrm{d}V\, H(\mathbf{r}, 0)$ initially stored in the balloon. The energy density H is given by (10.1.3) (see also Appendix K); in our case $E = (4\pi a^3/3)(Q^2/2c^2\sigma)$. Accordingly we expect that

$$\frac{4\pi a^3}{3} \frac{Q^2}{2c^2\sigma} = \int_0^\infty \mathrm{d}t\, 4\pi r^2 v_r\, \Delta p. \tag{12.2.14}$$

Exercise: Verify by explicit integration that (12.2.11, 13) satisfy (12.2.14).

12.2.3 The exploding spherical shell: h_{20}

Consider now the example complementary to the last, namely the ICs $\alpha = 0$, $\beta = B\theta(a - r)$, so that $h_{10} = 0$ and $\psi = h_{20}$. First we ask what physics these conditions represent. Clearly $\Delta p = \sigma\psi_t$ is initially zero everywhere, while $v_r = -\partial\psi/\partial r = B\,\delta(r - a)$. Thus at $r = a$ the fluid is initially moving outwards, but everywhere else it is at rest. The radial momentum that initially resides on a surface element $\delta \mathbf{S} = \hat{\mathbf{r}} a^2\, \delta\Omega$ of this exploding shell is given by the volume integral of $\sigma \mathbf{v}$ over the cone defined by the solid angle $\delta\Omega$; it equals

$$\hat{\mathbf{r}} \int_0^\infty \delta\Omega \mathrm{d}r\, r^2 \sigma v_r = \hat{\mathbf{r}}\, \delta\Omega\sigma \int \mathrm{d}r\, r^2 B\, \delta(r - a)$$

$$= \hat{\mathbf{r}}\, \delta\Omega a^2 \sigma B = \hat{\mathbf{r}}\, \delta S \sigma B, \tag{12.2.15}$$

i.e. it has a finite value per unit area of the shell. (Nevertheless the situation is highly idealized, in that the initial energy $\int \mathrm{d}V \tfrac{1}{2}\sigma v_r^2$ is infinite.)

For a change we now consider an interior point $r < a$. Figure 12.3a shows that for $ct < (a - r)$ the information-gathering sphere lies wholly inside the initial sphere. Hence

$$M\{\beta\} = B, \qquad h_{20} = \frac{\partial}{\partial t}[tB] = B,$$

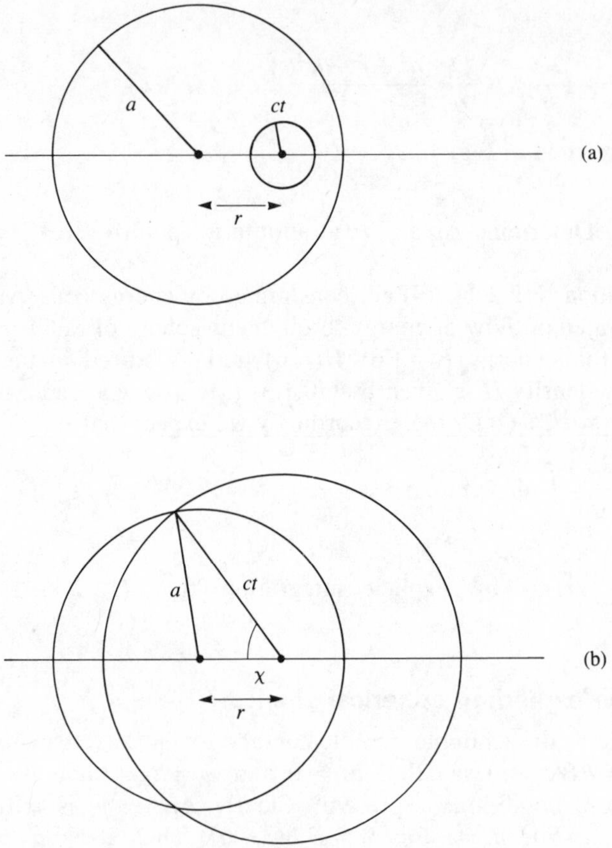

Fig. 12.3

and nothing happens because ψ_t and $\nabla\psi$ both vanish. Similarly, when $ct > (a + r)$, the information-gathering sphere lies wholly outside the initial sphere, $M = 0$, and again nothing happens. It remains to consider the period $(a - r) < ct < (a + r)$, as shown in Fig. 12.3b. Reasoning as for the bursting balloon one finds

$$M = B\Omega/4\pi = B\tfrac{1}{2}(1 - \cos\chi), \qquad \cos\chi = (c^2t^2 + r^2 - a^2)/2ctr,$$

$$M\{\beta\} = B\frac{[a^2 - (r - ct)^2]}{4ctr}, \tag{12.2.16}$$

$$\psi = \frac{\partial}{\partial t}[tM] = B(r - ct)/2r,$$

$$\psi = \tfrac{1}{2}B(1 - ct/r), \qquad (a - r) < ct < (a + v). \tag{12.2.17}$$

Collecting the results, and writing the step-functions conveniently for differentiation, we have

$$\psi(\boldsymbol{r}, t) = \theta(a - r - ct)B + \theta(ct - a + r)\theta(a + r - ct)\tfrac{1}{2}B(1 - ct/r). \tag{12.2.18}$$

Though M, naturally, is continuous at $ct = a \pm r$, one notes that the solution ψ itself is not. Straightforward differentiation yields

$$\begin{aligned}\Delta p(\boldsymbol{r}, t) = \sigma\psi_t = \tfrac{1}{2}c\sigma B\{&-\delta(ct - a + r)a/r \\ &- \theta(ct - a + r)\theta(a + r - ct)(1/r) + \delta(a + r - ct)a/r\},\end{aligned} \tag{12.2.19}$$

$$\begin{aligned}v_r(\boldsymbol{r}, t) = -\partial\psi/\partial r = \tfrac{1}{2}B\{&\delta(ct - a + r)a/r \\ &- \theta(ct - a + r)\theta(a + r - ct)ct/r^2 + \delta(a + r - ct)a/r\}.\end{aligned} \tag{12.2.20}$$

ψ, Δp and v_r are sketched in Fig. 12.4; the heavy spikes represent delta-functions, whose coefficients can be identified from (12.2.19, 20).

Exercises: (i) Verify (12.2.19, 20). (ii) Verify that (12.2.12) now fails, and explain why. (iii) Evaluate $\int_0^\infty dt\, v_r(r, t)$. (iv) Determine and sketch $\Delta p(0, t)$. (This is probably easier from (12.2.6) than from (12.2.19).)

12.2.4 Comments

(i) One must remember that Poisson's solution applies only in 3D. In 1D it is replaced by d'Alembert's solution. In 2D there is no comparably simple analogue, and we skip this case.

(ii) Propagation does not smooth the singularities of the initial data: witness the Green's functions themselves (Section 11.4), and the examples in Sections 12.2.2, 3 above (see e.g. Figs 12.2, 4). This lack of smoothing in the (hyperbolic) wave equation contrasts with Poisson's equation (elliptic), and with the diffusion equation (parabolic): see Sections 5.2, point (vi), and 8.3.1, point (i).

(iii) Note how un-obvious it is from mere inspection that Poisson's solution (12.2.6) indeed satisfies the homogeneous wave equation.

(iv) It is instructive to verify explicitly that (12.2.6) satisfies the two initial conditions (12.1.1). For the present consider only cases where $\psi(\boldsymbol{r}, t_0) \equiv \beta(\boldsymbol{r})$ and $\psi_t(\boldsymbol{r}, t_0) \equiv \alpha(\boldsymbol{r})$ are continuous, and $\nabla^2\beta$ is bounded.

As $t \to t_0$, h_{10} vanishes because of its factor $\tau \equiv t - t_0$. In h_{20}, the part where $\partial/\partial t$ acts on M vanishes for the same reason. Thus only the other component $M_{\boldsymbol{r},c(t-t_0)}\{\beta\}$ of h_{20} survives in ψ. In the limit it becomes the average of β over a sphere of zero radius centred on $\boldsymbol{r}$, and for continuous functions this is just $\beta(\boldsymbol{r})$ as required.

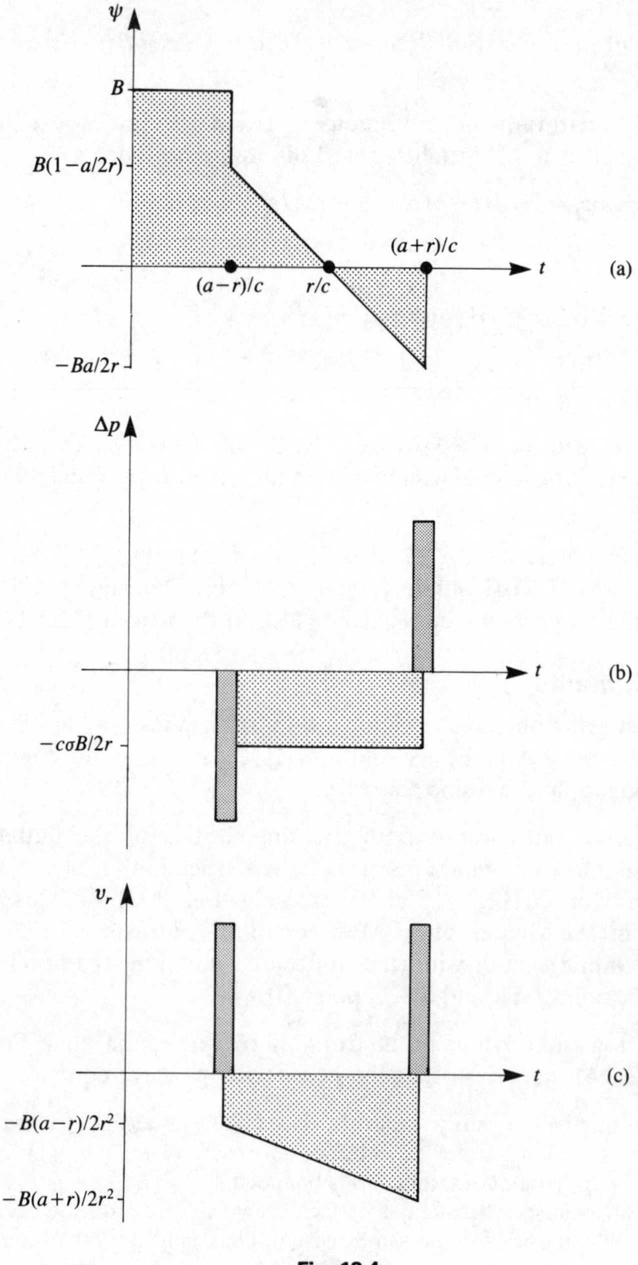

Fig. 12.4

To determine $\lim_{t\to t_0} \psi_t$, we need

$$\lim_{\tau\to 0}\frac{\partial}{\partial\tau}\left\{[\tau M_{r,c\tau}\{\alpha\}]+\frac{\partial}{\partial\tau}[\tau M_{r,c\tau}\{\beta\}]\right\}.$$

Carry out $\partial/\partial\tau$: any term still prefaced by τ then vanishes in the limit, and is dropped. The remainder is

$$M_{r,0}\{\alpha\}+\lim_{\tau\to 0}\frac{\partial^2}{\partial\tau^2}[\tau M_{r,c\tau}\{\beta\}]=M_{r,0}\{\alpha\}+2\lim_{\tau\to 0}\frac{\partial}{\partial\tau}M_{r,c\tau}\{\beta\}. \tag{12.2.21}$$

The first term is just $\alpha(\boldsymbol{r})$, i.e. the expression we want. It remains to show that the second term vanishes.

The crucial observation is that $M_{r,c(\tau+\delta\tau)}\{\beta\}$ is the average of β over a sphere whose radius is bigger by $c\,\delta\tau$ than the radius of the averaging sphere in $M_{r,c\tau}\{\beta\}$. On this slightly larger sphere, the function to be averaged is $\beta(\boldsymbol{r}')+c\,\delta\tau\dfrac{\partial\beta}{\partial R'}+\cdots$, where $\boldsymbol{R}'\equiv\boldsymbol{r}'-\boldsymbol{r}$ as in Section 12.2.1. Accordingly

$$\begin{aligned}M_{r,c(\tau+\delta\tau)}\{\beta(\boldsymbol{r}')\}&=M_{r,c\tau}\{\beta(\boldsymbol{r}')\}+\delta\tau\frac{\partial}{\partial\tau}M_{r,c\tau}\{\beta(\boldsymbol{r}')\}+\cdots,\\&=M_{r,c\tau}\{\beta(\boldsymbol{r}')+c\,\delta\tau\,\partial\beta(\boldsymbol{r}')/\partial R'+\cdots\}\\&=M_{r,c\tau}\{\beta(\boldsymbol{r}')\}+c\,\delta\tau M_{r,c\tau}\{\partial\beta(\boldsymbol{r}')/\partial R'\}+\cdots,\end{aligned}$$

$$\frac{\partial}{\partial\tau}M_{r,c\tau}\{\beta(\boldsymbol{r}')\}=cM_{r,c\tau}\left\{\frac{\partial\beta(\boldsymbol{r}')}{\partial R'}\right\}. \tag{12.2.22}$$

The penultimate step relies simply on the linearity of the average, $M(A+B)=M(A)+M(B)$.

Now let S denote the surface of the averaging sphere $R'=c\tau$, and V its interior. By Gauss's theorem, the expression on the right of (12.2.22) becomes

$$\frac{c}{4\pi R'^2}\int_S \mathrm{d}S'\,\frac{\partial\beta(\boldsymbol{r}')}{\partial R'}=\frac{c}{4\pi R'^2}\int_V \mathrm{d}V'\,\nabla'^2\beta(\boldsymbol{r}'). \tag{12.2.23}$$

Since $\nabla^2\beta$ is bounded by assumption, we have

$$\lim_{\tau\to 0}(\text{RHS of (12.2.23)})<\lim_{R'=c\tau\to 0}\frac{c}{4\pi R'^2}\frac{4\pi R'^3}{3}\text{constant}=0. \quad\blacksquare \tag{12.2.24}$$

(v) One can now appreciate (once again) the enormous generalizing power of explicit constructions like Poisson's solution, especially if the initial data are discontinuous. The value of $\psi(\boldsymbol{r},t_0)=\beta(\boldsymbol{r})$ precisely at a point of discontinuity $\boldsymbol{r}_1$ is, from a physical point of view, irrelevant; correspondingly, the self-consistency check just carried out becomes ambiguous, because the value of $\beta(\boldsymbol{r}_1)$ could be changed by a physically empty mathematical redefinition, while the value of $\lim_{t\to t_0}\psi(\boldsymbol{r},t)$ found above, namely $\lim_{\tau\to 0}M_{r,c\tau}\{\beta\}$, is fully determinate. (Similar ideas occurred in Section 7.2.) In such cases physical interest usually attaches only to the function explicitly given by Poisson's solution.

Similar remarks apply to d'Alembert's solution in 1D.

12.3 Radiation sources in unbounded space

We consider a system where sources $\rho(\mathbf{r}, t)$ have acted after initial quiescence at $t \to -\infty$. Thus $\Box^2\psi = \rho$, and the magic rule yields the solution $\psi = f_0$:

$$\psi(\mathbf{r}, t) = \int_{-\infty}^{t} \mathrm{d}t' \int \mathrm{d}V'\, G_0(R, \tau)\rho(\mathbf{r}', t'), \tag{12.3.1a}$$

$$R \equiv \mathbf{r} - \mathbf{r}', \qquad \tau \equiv t - t'. \tag{12.3.1b}$$

In 3D,

$$\psi^{(3)}(\mathbf{r}, t) = \int_{-\infty}^{t} \mathrm{d}t' \int \mathrm{d}V'\, \frac{\delta(t - t' - R/c)}{4\pi R}\rho(\mathbf{r}', t'). \tag{12.3.2}$$

Integrate over t' first: at any fixed $\mathbf{r}'$, the delta-function picks out the value of ρ at a time

$$t' = t - R/c \equiv t_{\text{ret}}(\mathbf{r}, t) \equiv \tilde{t}(\mathbf{r}, t), \tag{12.3.3}$$

earlier than t by as long as it takes a signal with speed c to travel from the source point $\mathbf{r}'$ to the field point $\mathbf{r}$. It is absolutely essential to remember, always, that $\tilde{t}$ is a function of $\mathbf{r}$, $\mathbf{r}'$, and t. Possible notations include

$$\rho(\mathbf{r}', t - R/c) \equiv [\rho(\mathbf{r}', t)]_{\text{ret}} \equiv \rho(\mathbf{r}', \tilde{t}). \tag{12.3.4}$$

Thus

$$\begin{aligned}\psi^{(3)}(\mathbf{r}, t) &= \int \mathrm{d}V'\, [\rho(\mathbf{r}', t)]_{\text{ret}}/4\pi R \\ &= \int \mathrm{d}V'\, \rho(\mathbf{r}', t - R/c)/4\pi R = \int \mathrm{d}V'\, \rho(\mathbf{r}', \tilde{t})/4\pi R. \end{aligned}\tag{12.3.5}$$

From each point $\mathbf{r}'$, $\psi^{(3)}(\mathbf{r}, t)$ draws a contribution emitted at the unique time $\tilde{t}$.

In 2D and 1D things are quite different. In 2D, (12.3.1) yields

$$\begin{aligned}\psi^{(2)}(\mathbf{r}, t) &= \int_{-\infty}^{t} \mathrm{d}t' \int \mathrm{d}V' \frac{1}{2\pi} \frac{\theta(t - t' - R/c)}{[(t - t')^2 - R^2/c^2]^{\frac{1}{2}}} \rho(\mathbf{r}', t') \\ &= \int \mathrm{d}V' \int_{-\infty}^{\tilde{t}} \mathrm{d}t' \frac{1}{2\pi} \frac{\rho(\mathbf{r}', t')}{[(t - t')^2 - R^2/c^2]^{\frac{1}{2}}}. \end{aligned}\tag{12.3.6}$$

Thus, from each point $\mathbf{r}'$, $\psi^{(2)}(\mathbf{r}, t)$ draws contributions emitted at all times t' from $-\infty$ up to $\tilde{t}$.

In 1D,

$$\psi^{(1)}(x, t) = \int_{-\infty}^{t} \mathrm{d}t' \int_{-\infty}^{\infty} \mathrm{d}x' \frac{c}{2} \theta(t - t' - |X|/c)\rho(x', t')$$

$$= \int_{-\infty}^{\infty} \mathrm{d}x' \int_{-\infty}^{\tilde{t}} \mathrm{d}t' \frac{c}{2} \rho(x', t'). \qquad (12.3.7)$$

As in 2D, each source point $\boldsymbol{r}'$ contributes at all times up to $\tilde{t}$.

The special cases $\rho = \delta(\boldsymbol{r})\theta(t)$ of (12.3–7) have already occurred in Section 11.3.

Throughout the rest of the book we shall consider 3D almost exclusively. The reader may however enjoy adapting the main results to 2D and 1D.

12.4 Harmonic sources

Though they are covered by the Helmholtz equation, it is perhaps more instructive to explore harmonic point sources

$$\rho(\boldsymbol{r}, t) = Q\, \delta(\boldsymbol{r}) \cos[\omega t] \qquad (12.4.1)$$

directly from (12.3.5).

Of course (12.4.1) is an idealization: in some cases it might be more realistic to replace $Q\,\delta(\boldsymbol{r})$ by a continuous peak. Or, as suggested at the end of Appendix K, (12.4.1) might be an approximation to a small sphere centred on the origin and having variable radius

$$\xi = a + h \sin[\omega t], \qquad (12.4.2)$$

where $h \ll a$, and (by hindsight) $a \ll c/\omega \equiv \lambda/2\pi$. From distances $r \gg a$ the sphere then appears as a volume-source of negligible extent, having total strength $\int \mathrm{d}V\, \rho(\boldsymbol{r}, t) = \mathrm{d(volume)}/\mathrm{d}t = 4\pi a^2 h\omega \cos[\omega t]$, so that, by comparison,

$$Q = 4\pi a^2 h \omega. \qquad (12.4.3)$$

Strictly harmonic time-dependence (i.e. for all t) is another idealization, as discussed at the end of this section.

Substituting (12.4.1) into (12.3.5) we find

$$[\rho(\boldsymbol{r}', \tilde{t})] = \rho(\boldsymbol{r}', t - R/c) = Q\, \delta(\boldsymbol{r}') \cos[\omega(t - R/c)], \qquad (12.4.4)$$

$$\psi(\boldsymbol{r}, t) = \int \mathrm{d}V'\, Q\, \delta(\boldsymbol{r}') \cos[\omega(t - |\boldsymbol{r} - \boldsymbol{r}'|/c)]/4\pi\, |\boldsymbol{r} - \boldsymbol{r}'|,$$

$$\psi(\boldsymbol{r}, t) = \frac{Q}{4\pi r} \cos[\omega(t - r/c)]. \qquad (12.4.5)$$

Differentiation yields

$$\Delta p = \sigma \frac{\partial \psi}{\partial t} = -\sigma \frac{Q}{4\pi} \frac{\omega}{r} \sin[\omega(t - r/c)], \tag{12.4.6}$$

$$\boldsymbol{v} = -\nabla\psi = \hat{\boldsymbol{r}} \frac{Q}{4\pi} \left\{ \frac{1}{r^2} \cos[\omega(t - r/c)] - \frac{\omega}{cr} \sin[\omega(t - r/c)] \right\}$$

$$\equiv \hat{\boldsymbol{r}} v_r, \tag{12.4.7}$$

$$\boldsymbol{N} = \boldsymbol{v}\,\Delta p \equiv \hat{\boldsymbol{r}} N_r,$$

$$N_r = \sigma \left(\frac{Q}{4\pi}\right)^2 \left\{ -\frac{\omega}{r^3} \sin[\;]\cos[\;] + \frac{\omega^2}{cr^2} \sin^2[\;] \right\}$$

$$= \sigma \left(\frac{Q}{4\pi}\right)^2 \frac{1}{2} \left\{ -\frac{\omega}{r^3} \sin[2\omega(t - r/c)] + \frac{\omega^2}{cr^2} (1 - \cos[2\omega(t - r/c)]) \right\}. \tag{12.4.8}$$

Exercise: Check dimensions throughout (12.4.6–8).

The region $r \ll c/\omega$ is called the *near-field*. Here v_r is dominated by the first term of (12.4.7); if the phase lag $\omega r/c$ in the argument of the cosine is negligible, then this term reduces to $Q \cos[\omega t]/4\pi r^2$, which is just the flow from a *static* point source having the strength $Q \cos[\omega t]$ that the actual source has at time t.

Exercise: Expand v_r in powers of $\omega r/c$, and show that the leading correction to $v_r \approx Q \cos[\omega t]/4\pi r^2$ is only of relative order $(\omega r/c)^2$, and not $(\omega r/c)$ as (12.4.7) might suggest at first sight.

In the *far-field* region $r \gg c/\omega$, the second term of (12.4.7) dominates; as $r \to \infty$ this falls only like $1/r$.

As regards the energy flow, observe first that the time-average $\bar{N}_r(r)$ of $N_r(r, t)$ and the *time-averaged power outflow* $4\pi r^2 \bar{N}_r(r) \equiv \bar{W}_{\text{rad}}$ through a sphere of radius r centred on the source are given by

$$\bar{N}_r(r) = \sigma \left(\frac{Q}{4\pi}\right)^2 \frac{1}{2} \frac{\omega^2}{cr^2}, \qquad \bar{W}_{\text{rad}} = \frac{\sigma Q^2 \omega^2}{8\pi c}. \tag{12.4.9a,b}$$

(Naturally, $\bar{W}_{\text{rad}}$ is independent of r.) For a pulsating sphere, (12.4.3) yields

$$\bar{W}_{\text{rad}} = 2\pi\sigma a^4 h^2 \omega^4/c. \tag{12.4.10}$$

The difference between near-field and far-field reappears in the time-dependence (12.4.8) of $N_r(r, t)$ and of $W(r, t) \equiv 4\pi r^2 N_r(r, t)$. As $r \to 0$, the instantaneous values of $W(r, t)$ evidently diverge like

$-(\sigma Q^2\omega/8\pi r)\sin[2\omega t]$. Thus the actual energy flow in and out of the source is far more violent than one can see merely from $\bar{W}_{\text{rad}}$. In the far-field, an oscillatory energy flow is added to the time average, giving $W(r, t) \sim (\sigma Q^2\omega^2/8\pi c)(1 - \cos[2\omega(t - r/c)])$.

By contrast to the pulsating sphere, a small body oscillating in position or orientation but without change of volume would have to be mimicked by a source of zero total strength, in the sense that $\int dV\,\rho(\boldsymbol{r}, t) = 0$. Mathematically the simplest example is the point dipole

$$\rho(\boldsymbol{r}, t) = -\boldsymbol{p}\cdot\boldsymbol{\nabla}\,\delta(\boldsymbol{r})\cos[\omega t], \tag{12.4.11}$$

which one can think of as two equal-strength point sources in antiphase:

$$\rho = Q\cos[\omega t]\{\delta(\boldsymbol{r} - \boldsymbol{a}/2) - \delta(\boldsymbol{r} + \boldsymbol{a}/2)\},$$

in the limit where their separation $\boldsymbol{a} \to 0$, with $Q\boldsymbol{a} = \boldsymbol{p}$ fixed. (In practice, a point dipole might be approximated for instance by a body rotating rigidly around any axis not passing through the centre of mass of the displaced fluid.) By the light of Sections 4.4.4 and 11.5, ψ, $\partial\psi/\partial t$, and $\boldsymbol{\nabla}\psi$ are obtained immediately from the corresponding solutions (12.4.5) for a point source, by setting $Q = 1$ and then acting with $-\boldsymbol{p}\cdot\boldsymbol{\nabla}$. This gives

$$\psi(\boldsymbol{r}, t) = (\boldsymbol{p}\cdot\hat{\boldsymbol{r}})\left\{\frac{\omega}{4\pi c r}\sin[\omega(t - r/c)] + \frac{1}{4\pi r^2}\cos[\omega(t - r/c)]\right\}. \tag{12.4.12}$$

In the far-field, and to leading order in $1/r$, derivatives need act only on the trigonometric functions:

$$\Delta p = (\boldsymbol{p}\cdot\hat{\boldsymbol{r}})\frac{\sigma\omega^2}{4\pi c r}\cos[\omega/t - r/c)] + O(1/r^2), \tag{12.4.13}$$

$$\boldsymbol{v} = \hat{\boldsymbol{r}}(\boldsymbol{p}\cdot\hat{\boldsymbol{r}})\frac{\omega^2}{4\pi c^2 r}\cos[\omega(t - r/c)] + O(1/r^2), \tag{12.4.14}$$

$$\boldsymbol{N} = \Delta p\,\boldsymbol{v} = \hat{\boldsymbol{r}}(\boldsymbol{p}\cdot\hat{\boldsymbol{r}})^2\frac{\sigma\omega^4}{(4\pi)^2c^3r^2}\cos^2[\omega(t - r/c)] + O(1/r^3). \tag{12.4.15}$$

The factor $(\boldsymbol{p}\cdot\hat{\boldsymbol{r}})^2$ makes the asymptotic energy flux anisotropic, and proportional to $\cos^2\theta$ if we choose the polar axis along $\boldsymbol{p}$. The time-averaged radiated power is

$$\bar{W}_{\text{rad}} = \lim_{r\to\infty}\int d\Omega\, r^2 N_r = \frac{\sigma\omega^4}{(4\pi)^2c^3}\frac{1}{2}\int d\Omega\,(\boldsymbol{p}\cdot\hat{\boldsymbol{r}})^2,$$

$$\bar{W}_{\text{rad}} = \sigma\omega^4p^2/24\pi c^3. \tag{12.4.16}$$

Alluding to the model just described for the dipole, we set $p = Qa$, and compare (12.4.16) with the power output (12.4.9b) of an isotropic point source of strength Q:

$$\frac{\bar{W}\text{ (dipole)}}{\bar{W}\text{ (point source)}}=\frac{\sigma\omega^4Q^2a^2/24\pi c^3}{\sigma\omega^2Q^2/8\pi c}=\frac{\omega^2a^2}{3c^2}=\frac{1}{3}\left(\frac{2\pi a}{\lambda}\right)^2. \tag{12.4.17}$$

Reverting to isotropic sources, instead of a point source as above one could consider a pulsating sphere of finite radius (12.4.2), still taking $h \ll a$ but not now neglecting $a\omega/c$. From the point of view of the magic rule this constitutes a Neumann BVP, prescribing $v_r = -\partial\psi/\partial r$ at $r = a$; the solution is an integral g rather than an integral f. But in order to evaluate g one needs the (isotropic part of the) Neumann Green's function for the exterior of the sphere, which is not considered in this book. Instead, it is much simpler to determine directly the appropriate isotropic solution of the Helmholtz equation (Section 13.1.5). It turns out that the expression (12.4.10) for $\bar{W}_{\text{rad}}$ is then merely divided by $[1+(a\omega/c)^2]$.

Finally, realizable sources cannot of course vary like $\cos[\omega t]$ for all t; it would be better to consider say $\rho = Q\,\delta(\boldsymbol{r})\theta(t-t_0)\cos[\omega t]$, taking the limit $(t-t_0)\to\infty$ if appropriate. The expressions obtained above then apply in the region $r < c(t-t_0)$, i.e. inside the leading wavefront that was emitted at the start-up time t_0. By confining attention to this region we avoid the singularities associated with the start-up. It is one of the perks in 3D that these singularities can be sidestepped so easily: Section 11.3 shows that in 2D and 1D even a source $Q\,\delta(\boldsymbol{r})\theta(t-t_0)$ would produce a time-varying ψ and a non-zero power-flow at all $r < c(t-t_0)$. In other words N and W would not then vanish when the frequency vanishes, as they do in 3D.

Exercise: Find $\psi(\boldsymbol{r}, t)$ for $\rho = Q\,\delta(\boldsymbol{r})\theta(t-t_0)\cos[\omega t]$, and verify the claims just made.

12.5 The Liénard–Wiechert potential

Acoustics interests us mainly as a fair yet far simpler analogue to electromagnetism. There, charge conservation forbids an isolated charge from changing its magnitude, so that the two simplest types of radiation source are either a time-varying point dipole of fixed position, or an accelerating point charge of fixed magnitude. Nevertheless, for simplicity we shall discuss isotropic time-varying acoustic point sources rather than point dipoles. Moreover, only subsonic source-speeds will be admitted.

Consider a point source of total strength $Q(t)$, situated at the point $\boldsymbol{s}(t)$; the source trajectory $\boldsymbol{s}(t)$ is an assigned function of time. Then the source density is

$$\rho(\boldsymbol{r}, t) = Q(t)\,\delta(\boldsymbol{r}-\boldsymbol{s}(t)). \tag{12.5.1}$$

Rather than use (12.3.5), we revert to the basic integral f_0 of the magic rule; with $G_0 = \delta(t - t' - R/c)/4\pi R$, this yields

$$\psi(\boldsymbol{r}, t) = \int_{-\infty}^{t} \mathrm{d}t' \int \mathrm{d}V' \frac{\delta(t - t' - R/c)}{4\pi R} Q(t')\, \delta(\boldsymbol{r}' - \boldsymbol{s}(t')). \tag{12.5.2}$$

Now we perform the volume rather than the time integral first, and find

$$\psi(\boldsymbol{r}, t) = \int_{-\infty}^{t} \mathrm{d}t' \frac{\delta(t - t' - |\boldsymbol{r} - \boldsymbol{s}(t')|/c)}{4\pi\, |\boldsymbol{r} - \boldsymbol{s}(t')|} Q(t'). \tag{12.5.3}$$

Clearly, an important role will again be played by that value $\tilde{t}$ of t' which makes the argument of the delta-function vanish:

$$t - \tilde{t} - |\boldsymbol{r} - \boldsymbol{s}(\tilde{t})|/c = 0. \tag{12.5.4}$$

(It will be shown later that for subsonic source speeds the solution $\tilde{t}$ of this equation is unique.) One needs

$$\boldsymbol{s}(\tilde{t}) \equiv \tilde{\boldsymbol{s}}; \tag{12.5.5a}$$

$$\boldsymbol{r} - \tilde{\boldsymbol{s}} = \tilde{\boldsymbol{R}}, \qquad |\tilde{\boldsymbol{R}}| \equiv \tilde{R}; \tag{12.5.5b}$$

$$\mathrm{d}\boldsymbol{s}/\mathrm{d}t \equiv \boldsymbol{u}(t), \qquad \boldsymbol{u}(\tilde{t}) \equiv \tilde{\boldsymbol{u}}; \tag{12.5.5c}$$

$$\mathrm{d}\boldsymbol{u}/\mathrm{d}t = \mathrm{d}^2\boldsymbol{s}/\mathrm{d}t^2 \equiv \boldsymbol{a}(t), \qquad \boldsymbol{a}(\tilde{t}) \equiv \tilde{\boldsymbol{a}}; \tag{12.5.5d}$$

$$Q(\tilde{t}) \equiv \tilde{Q}; \tag{12.5.5e}$$

and also

$$\tilde{R} - \tilde{\boldsymbol{u}} \cdot \tilde{\boldsymbol{R}}/c \equiv \tilde{P}. \tag{12.5.6}$$

As already stated in Section 12.3, it is absolutely essential to bear in mind, always, that the retarded time $\tilde{t}$ is a function of the independent variables $\boldsymbol{r}$ and t; so therefore is any function of $\tilde{t}$, and so in particular are $\tilde{\boldsymbol{s}}$, $\tilde{\boldsymbol{R}}$, $\tilde{R}$, $\tilde{\boldsymbol{u}}$, $\tilde{\boldsymbol{a}}$, and $\tilde{P}$. Evidently (12.5.4) entails

$$t - \tilde{t} = \tilde{R}/c. \tag{12.5.7}$$

Unfortunately, the function $\tilde{t}(\boldsymbol{r}, t)$ is not given directly, but must be determined by solving (12.5.4), and we shall see that this ultimately essential task can be quite wearisome. However, even in advance of assigning $\boldsymbol{s}(t)$ and solving for $\tilde{t}$, one can deduce useful relations in terms of the retarded variables. Meanwhile, $\tilde{t}$ is easy to find only in the very special case where $\boldsymbol{s}$ is fixed ($\tilde{\boldsymbol{s}} = \boldsymbol{s} =$ constant), as for instance in Section 12.4. Of course the fictitious non-retarded limit $c \to \infty$ entails $\tilde{t} = t$ regardless of $\boldsymbol{s}(t)$.

Reverting to (12.5.3), we now express the integral *in terms of retarded variables* by appeal to the standard result

$$\psi(\boldsymbol{r}, t) = \int \mathrm{d}t'\, \delta(f(t'))\, g(t') = \left. \frac{g(t')}{|\mathrm{d}f/\mathrm{d}t'|} \right|_{\text{at } f(t')=0}. \tag{12.5.8}$$

For given $\boldsymbol{r}$ and t, we identify

$$f(t') = t - t' - |\boldsymbol{r} - \boldsymbol{s}(t')|/c, \tag{12.5.9a}$$

$$g(t') = Q(t')/4\pi\,|\boldsymbol{r} - \boldsymbol{s}(t')|. \tag{12.5.9b}$$

In view of (12.5.4), the prescription $f(t') = 0$ implies $t' = \tilde{t}$, whence in (12.5.8) the numerator is $g(\tilde{t}) = \tilde{Q}/4\pi\tilde{R}$. In the denominator, for arbitrary t' we have

$$\begin{aligned}\frac{\mathrm{d}f}{\mathrm{d}t'} &= -1 - \frac{1}{c}\frac{\mathrm{d}}{\mathrm{d}t'}[(\boldsymbol{r} - \boldsymbol{s}(t'))^2]^{\frac{1}{2}}\\ &= -1 + \frac{(\boldsymbol{r} - \boldsymbol{s}(t'))\cdot \mathrm{d}\boldsymbol{s}(t')/\mathrm{d}t'}{c\,|\boldsymbol{r} - \boldsymbol{s}(t')|}\\ &= -1 + \frac{(\boldsymbol{r} - \boldsymbol{s}(t'))\cdot \boldsymbol{u}(t')}{c\,|\boldsymbol{r} - \boldsymbol{s}(t')|}.\end{aligned}$$

When one sets $t' = \tilde{t}$ and takes the modulus as required in (12.5.8), this yields

$$\left|\frac{\mathrm{d}f}{\mathrm{d}t'}\right|_{t'=\tilde{t}} = 1 - \frac{\tilde{\boldsymbol{R}}\cdot\tilde{\boldsymbol{u}}}{c\tilde{R}} = \frac{1}{\tilde{R}}(\tilde{R} - \tilde{\boldsymbol{R}}\cdot\boldsymbol{u}/c) = \frac{\tilde{P}}{\tilde{R}}.$$

Substituting back into (12.5.3) we obtain the *Liénard–Wiechert potential* $\psi = (\tilde{Q}/4\pi\tilde{R})/(\tilde{P}/\tilde{R})$, i.e.

$$\psi(\boldsymbol{r}, t) = \frac{\tilde{Q}}{4\pi\tilde{P}} = \frac{Q(\tilde{t})}{4\pi(\tilde{R} - \tilde{\boldsymbol{R}}\cdot\tilde{\boldsymbol{u}}/c)}, \tag{12.5.10a}$$

$$\psi(\boldsymbol{r}, t) = \left[\frac{Q}{4\pi(R - \boldsymbol{R}\cdot\boldsymbol{u}/c)}\right]_{\text{ret}}. \tag{12.5.10b}$$

The derivatives of ψ, needed for the energy flux, are easy to calculate only if the source is stationary. Then we choose it as the origin, which reduces (12.5.10) to

$$\psi(\boldsymbol{r}, t) = \frac{1}{4\pi r}Q(t - r/c) = \frac{[Q]_{\text{ret}}}{4\pi r} = \frac{\tilde{Q}}{4\pi r}. \tag{12.5.11}$$

With $\dot{Q}$ denoting the derivative of Q with respect to its argument, one finds

$$\frac{\partial\psi}{\partial t} = \frac{\dot{Q}(t - r/c)}{4\pi r} \equiv \frac{\dot{\tilde{Q}}}{4\pi r}, \tag{12.5.12}$$

$$\boldsymbol{v} = -\nabla\psi = -\hat{\boldsymbol{r}}\left\{-\frac{1}{4\pi r^2}Q(t - r/c) + \frac{1}{4\pi r^2}\dot{Q}(t - r/c)\left(-\frac{1}{c}\right)\right\},$$

$$\boldsymbol{v} = \hat{\boldsymbol{r}}\left\{\frac{\tilde{Q}}{4\pi r^2} + \frac{\dot{\tilde{Q}}}{4\pi rc}\right\}, \tag{12.5.13}$$

$$\boldsymbol{N}(\boldsymbol{r}, t) = \sigma \frac{\partial \psi}{\partial t} \boldsymbol{v} = \hat{\boldsymbol{r}} \frac{\sigma}{(4\pi)^2} \left\{ \frac{\tilde{Q}\dot{\tilde{Q}}}{r^3} + \frac{\dot{\tilde{Q}}^2}{cr^2} \right\} \tag{12.5.14a}$$

$$= \hat{\boldsymbol{r}} \frac{\sigma}{(4\pi)^2} \left\{ \frac{1}{r^3} \frac{1}{2} \frac{\mathrm{d}}{\mathrm{d}t} \tilde{Q}^2 + \frac{\dot{\tilde{Q}}^2}{cr^2} \right\} \equiv \boldsymbol{N}_1 + \boldsymbol{N}_2, \tag{12.5.14b,c}$$

where $\tilde{Q}(r, t) = Q(t - r/c)$ as before. The first term is a total time-derivative, a fact with remarkable consequences both at small and at large r.

Consider the net energy outflow through a sphere S of radius r, from time t_1 to t_2. Since the integrand is isotropic, one has $\int_S \mathrm{d}\boldsymbol{S} \cdot \boldsymbol{N} = \int \mathrm{d}\Omega\, r^2 N_r = 4\pi r^2 N_r$, whence the net outflow is

$$\int_{t_1}^{t_2} \mathrm{d}t \int \mathrm{d}\boldsymbol{S} \cdot \boldsymbol{N}(\boldsymbol{r}, t) = \int_{t_1}^{t_2} \mathrm{d}t \frac{\sigma}{4\pi} \left\{ \frac{1}{r} \frac{1}{2} \frac{\mathrm{d}}{\mathrm{d}t} Q^2(t - r/c) + \frac{1}{c} \dot{Q}^2(t - r/c) \right\} \tag{12.5.15a}$$

$$= \frac{\sigma}{4\pi} \frac{1}{2} \left[Q^2\left(t_2 - \frac{r}{c}\right) - Q^2\left(t_1 - \frac{r}{c}\right) \right] + \int_{t_1}^{t_2} \mathrm{d}t \frac{\sigma}{4\pi c} \dot{Q}^2(t - r/c). \tag{12.5.15b}$$

The net outflow E from $t_1 = -\infty$ to $t_2 = +\infty$ is

$$E(r) = \frac{1}{r} \frac{\sigma}{4\pi} \tfrac{1}{2}[Q^2(\infty) - Q^2(-\infty)] + \int_{-\infty}^{\infty} \mathrm{d}t \frac{\sigma}{4\pi} \dot{Q}^2(t - r/c) \tag{12.5.16a}$$

$$\equiv E_1 + E_2. \tag{12.5.16b}$$

The term E_1 is just the difference between the total energies residing beyond S in the final and in the initial static fields, given by

$$\int_r^{\infty} 4\pi r^2 \,\mathrm{d}r\, H = \int_r^{\infty} 4\pi r^2 \,\mathrm{d}r\, \tfrac{1}{2}\sigma(\boldsymbol{\nabla}\psi)^2$$

$$= \int_r^{\infty} 4\pi r^2 \,\mathrm{d}r\, \tfrac{1}{2}\sigma \left(\frac{Q}{4\pi r^2} \right)^2 = \frac{1}{r} \frac{\sigma}{4\pi} \tfrac{1}{2} Q^2. \tag{12.5.17}$$

We stress that E_1 (like the first term in (12.5.15b)) depends only on the initial and final values of Q, and not on its intervening history. Notice also that E_1 diverges as $r \to 0$, unless $Q^2(+\infty) = Q^2(-\infty)$.

By sharp contrast to E_1, the component E_2 (i) does depend on the detailed history of $Q(t)$, and not only on its initial and final values; (ii) is non-negative (vanishing only if Q never changes at all); and (iii) is independent of r. In particular, only E_2 (i.e. only the contribution from $\boldsymbol{N}_2$ in (12.5.14)) survives in the limit $r \to \infty$. Accordingly, it proves useful to *define* the '*radiated power*' $W_{\mathrm{rad}}(t)$ in a somewhat roundabout fashion,

by

$$W_{\text{rad}}(t) \equiv \lim_{r\to\infty} \int_S \mathrm{d}\boldsymbol{S} \cdot \boldsymbol{N}(\boldsymbol{r}, t + r/c) \tag{12.5.18a}$$

$$= \lim_{r\to\infty} \int \mathrm{d}\Omega\, r^2 N_r(t, t + r/c), \tag{12.5.18b}$$

$$W_{\text{rad}}(t) = \int_S \mathrm{d}\boldsymbol{S} \cdot \boldsymbol{N}_2(\boldsymbol{r}, t + r/c), \tag{12.5.18c}$$

$$W_{\text{rad}}(t) = \sigma \dot{Q}^2(t)/4\pi c. \tag{12.5.18d}$$

Notice that the first integral, whose $r \to \infty$ limit defines $W_{\text{rad}}(t)$, is evaluated at a later time $t + r/c$. In this sense $W_{\text{rad}}(t)$ is defined prospectively, with a view to the fact that

$$\tilde{Q}(t + r/c) = Q(t + r/c - r/c) = Q(t). \tag{12.5.19}$$

Notice also that implementing the limit $r \to \infty$ in (12.5.18a,b) has the same effect as replacing $\boldsymbol{N}$ by $\boldsymbol{N}_2$; and that $\boldsymbol{N} \to \boldsymbol{N}_2$ makes the integral independent of r.

It should be stressed that $W_{\text{rad}}(t)$ differs radically from the instantaneous power output $W(t)$ of the source at time t. Indeed, as a formal definition of $W(t)$ one might reasonably adopt

$$W(t) \equiv \lim_{r\to 0} \int_S \mathrm{d}\boldsymbol{S} \cdot \boldsymbol{N}(\boldsymbol{r}, t + r/c) \tag{12.5.20a}$$

$$= \lim_{r\to 0} \frac{1}{r} \frac{\sigma}{4\pi} \frac{\mathrm{d}}{\mathrm{d}t} \tfrac{1}{2} \dot{Q}^2(t) + \frac{\sigma}{4\pi c} \dot{Q}^2(t). \tag{12.5.20b}$$

This should be compared with W_{rad} in (12.5.18). They differ by the first term of (12.5.20b), which, though formally divergent, is a total time-derivative (essentially because $\boldsymbol{N}_1$ in (12.5.14) is such a derivative). Crucially, this entails that for harmonic sources the difference contributes nothing to the time-averages over any whole number of periods T:

$$T^{-1} \int_t^{t+T} \mathrm{d}t'\, W(t') = T^{-1} \int_t^{t+T} \mathrm{d}t'\, W_{\text{rad}}(t').$$

Nor, as we have seen, does the difference contribute to the net energy outflow

$$\int_{-\infty}^{\infty} \mathrm{d}t'\, W(t') = \int_{-\infty}^{\infty} \mathrm{d}t'\, W_{\text{rad}}(t')$$

for arbitrarily time-varying sources, provided only that $Q^2(-\infty) = Q^2(+\infty)$. In these special but important cases, divergences can be excluded from our expressions altogether if the time integrals are evaluated before taking the limit $r \to 0$.

Instead of (12.5.20a), we could just as plausibly adopt the alternative definition

$$W(t) = \lim_{r \to 0} \int_S \mathrm{d}\boldsymbol{S} \cdot \boldsymbol{N}(\boldsymbol{r}, t). \tag{12.5.21a}$$

By expanding $Q(t - r/c) = Q(t) - (r/c)\dot{Q}(t) + \cdots$ (and similarly for $\dot{Q}$), and discarding terms that vanish as $r \to 0$, one can show that this entails

$$W(t) = \lim_{r \to 0} \frac{1}{r} \frac{\sigma}{4\pi} \frac{\mathrm{d}}{\mathrm{d}t} \tfrac{1}{2} Q^2(t) - \frac{\sigma}{4\pi c} \frac{\mathrm{d}^2}{\mathrm{d}t^2} \tfrac{1}{2} Q^2(t) + \frac{\sigma}{4\pi c} \dot{Q}^2(t), \tag{12.5.21b}$$

differing from (12.5.20b) merely by the middle term, another (but non-divergent) time-derivative.

Exercise: Verify (12.5.21b).

The ambiguity merely reflects the fact that a point source is an idealization: different physical approximations to such a source can have different instantaneous power outputs, albeit we have just seen that in many important cases the difference between (12.5.20) and (12.5.21) is irrelevant.

Summarizing, we see that any $\dot{Q} \neq 0$ radiates energy. Two special cases have been discussed already, $Q \propto \cos[\omega(t)]$ in Section 12.4, and $Q \propto \theta(t)$ in Section 11.3.4.

Exercise: Re-derive the results in those sections from (12.5.10–20).

By contrast, radiation induced by acceleration of constant-strength sources will be discussed in Section 12.7.

Some comments are now in order.

(i) Although the physical significance of the prescription $[\cdots]_{\mathrm{ret}}$ in (12.5.10) is clear, the evaluation of such an expression for a moving source is very far from trivial, and one needs to explain to oneself what it involves. The contents of $[\cdots]_{\mathrm{ret}}$ must be evaluated at some earlier time $\tilde{t}$ and for the corresponding point $\boldsymbol{s}(\tilde{t}) = \tilde{\boldsymbol{s}}$ on the trajectory of the source; and $\tilde{t}$ itself must be determined by solving (12.5.4). In other words, $\tilde{\boldsymbol{s}}$ and $\tilde{t}$ are the place and time where and when the source must have emitted a signal which, having travelled at speed c, is received at the field point $\boldsymbol{r}$ at time t.

(ii) As promised earlier, we prove that the solution $\tilde{t}$ of (12.5.4) is unique if u is always less than c.

Shift the origin of space and time to the field point, i.e. take $\boldsymbol{r}=0=t$. Then we are concerned with the roots of

$$c\tilde{t}=|\boldsymbol{s}(\tilde{t})|=s(\tilde{t}). \tag{12.5.22}$$

We proceed by *reductio ad absurdum*. Suppose there are (at least) two distinct roots $\tilde{t}_1$ and $\tilde{t}_2$: then, by subtraction,

$$c\,|\tilde{t}_1-\tilde{t}_2|=|s(\tilde{t}_1)-s(\tilde{t}_2)|. \tag{12.5.23}$$

On the left, $c\,|\tilde{t}_1-\tilde{t}_2|$ is the distance travelled by sound in time $|\tilde{t}_1-\tilde{t}_2|$. On the right, $|s(\tilde{t}_1)-s(\tilde{t}_2)|$ is the change in the distance from source to origin during the same time interval. This change cannot exceed $|\tilde{t}_1-\tilde{t}_2|\,|u_{\max}|$, where $|u_{\max}|$ is the maximum speed of the source. Thus for subsonic sources, where $|u_{\max}|<c$, the equality (12.5.18) cannot hold, and (12.5.17) cannot have had two distinct roots after all. ■

(iii) The Liénard–Wiechert potential can be written as $(Q(\tilde{t})/4\pi\tilde{R})(1-\hat{\tilde{\boldsymbol{R}}}\cdot\tilde{\boldsymbol{u}}/c)^{-1}$. Mathematically, the factor $(1-\hat{\tilde{\boldsymbol{R}}}\cdot\tilde{\boldsymbol{u}}/c)^{-1}$ is the Jacobian of the transformation from t' in (12.5.3) to the new integration variable $f(t')$ defined in (12.5.9a). A vivid physical interpretation of this factor is given by Feynman *et al.* (1964, Section 21.5).

(iv) One could now proceed directly to the general problem of radiation (Section 12.7), which is essentially the problem of calculating the derivatives of ψ for an arbitrarily assigned $\boldsymbol{s}(t)$. Or one could consider first the special case of a source with constant velocity $\boldsymbol{u}$ (Section 12.6), which does not radiate when Q is constant, but where all the field variables can be obtained in explicitly in terms of $\boldsymbol{r}$ and t. The reader should order the next two sections as she prefers.

12.6 Point sources in uniform motion

To evaluate the Liénard–Wiechert potential $\psi=\tilde{Q}/4\pi\tilde{P}$ for a uniformly moving point source amounts to determining $\tilde{t}(\boldsymbol{r},t)$ and thence the denominator $\tilde{P}$. We take the source to travel along the x-axis with constant speed u, passing the origin at $t=0$:

$$\boldsymbol{s}(t)=\hat{\boldsymbol{x}}ut=(ut,0,0). \tag{12.6.1}$$

Once $\psi(\boldsymbol{r},t)$ is known explicitly, its derivatives (and thence $\boldsymbol{N}$) can be found without the detour through $\partial\tilde{t}/\partial t$ and $\boldsymbol{\nabla}\tilde{t}$ that is needed in Section 12.7 for accelerated sources. The main point of the exercise is to illustrate the complexities masked by the elegance of (12.5.10) even in this simplest non-trivial example. Incidentally we can explain the Doppler effect for a moving source, and verify that an unaccelerated source cannot radiate if $\dot{Q}=0$.

The retarded time $\tilde{t}$ is found by substituting the trajectory (12.6.1) into (12.5.4):

$$c(t-\tilde{t}) = |\boldsymbol{r}-\boldsymbol{s}(\tilde{t})| = |\boldsymbol{r}-\hat{\boldsymbol{x}}u\tilde{t}| = [(x-u\tilde{t})^2+y^2+z^2]^{\frac{1}{2}}. \tag{12.6.2}$$

We square and rearrange this into a quadratic equation for $\tilde{t}$:

$$(1-u^2/c^2)\tilde{t}^2 - 2(t-xu/c^2)\tilde{t} + (t^2-r^2/c^2) = 0,$$

$$\tilde{t} = \frac{1}{(1-u^2/c^2)}\{(t-xu/c^2) \pm [(t-xu/c)^2-(1-u^2/c^2)(t^2-r^2/c^2)]^{\frac{1}{2}}\}. \tag{12.6.3}$$

The lower root is appropriate, because $u/c \to 0$ must entail $\tilde{t} = -r/c$ (as for a fixed source). The radicand simplifies somewhat, and eventually one finds

$$\tilde{t}(\boldsymbol{r}, t) = \frac{1}{(1-u^2/c^2)}\left\{(t-xu/c^2) - \frac{1}{c}[(x-ut)^2+(1-u^2/c^2)(y^2+z^2)]^{\frac{1}{2}}\right\}, \tag{12.6.4}$$

which is the desired explicit expression for $\tilde{t}$.

If the field point $\boldsymbol{r}$ lies on the trajectory, so that $y=0=z$, then $\tilde{t}$ simplifies much further. The square root (always positive) becomes $|x-ut| = \pm(x-ut)$; the upper (lower) sign applies when the source at ut is approaching (receding from) the observer at x. Thus

$$\tilde{t}(x,0,0,t) = \frac{1}{(1-u^2/c^2)}\{t-xu/c^2 \mp (x-ut)/c\}$$
$$= \frac{(t \mp x/c)(1 \pm u/c)}{(1-u^2/c^2)},$$

$$\tilde{t}(x,0,0,t) = \frac{(t \mp x/c)}{(1 \mp u/c)}, \qquad (x \gtrless ut\text{: approach (recession)}). \tag{12.6.5}$$

Exercise: Re-derive (12.6.5) by setting $y=0=z$ directly in (12.6.2).

The denominator $\tilde{P} \equiv \tilde{R} - \tilde{\boldsymbol{u}}\cdot\tilde{\boldsymbol{R}}/c$ of ψ follows straightforwardly. By (12.5.7), $\tilde{R} = c(t-\tilde{t})$. Because $\tilde{\boldsymbol{R}} = \boldsymbol{r}-\tilde{\boldsymbol{s}} = \boldsymbol{r}-\hat{\boldsymbol{x}}u\tilde{t}$, while $\boldsymbol{u} = u\hat{\boldsymbol{x}}$, we have $\tilde{\boldsymbol{u}}\cdot\tilde{\boldsymbol{R}}/c = u\hat{\boldsymbol{x}}\cdot(\boldsymbol{r}-\hat{\boldsymbol{x}}u\tilde{t})/c = u(x-u\tilde{t})/c$, whence

$$\tilde{P} = c(t-\tilde{t}) - u(x-u\tilde{t})/c = c(t-xu/c^2) - c\tilde{t}(1-u^2/c^2).$$

On substitution for $c\tilde{t}(1-u^2/c^2)$ from (12.6.3), all terms but the square root cancel, yielding

$$\tilde{P} = [(x-ut)^2+(1-u^2/c^2)(y^2+z^2)]^{\frac{1}{2}}. \tag{12.6.6}$$

Accordingly, (12.5.10) becomes†

$$\psi(\boldsymbol{r}, t) = \frac{\tilde{Q}}{4\pi\tilde{P}} = \frac{Q(\tilde{t})}{4\pi} \frac{1}{[(x-ut)^2 + (1-u^2/c^2)(y^2+z^2)]^{\frac{1}{2}}} \tag{12.6.7a}$$

$$= \frac{Q(\tilde{t})}{4\pi} \frac{1}{\sqrt{1-u^2/c^2}} \frac{1}{\left[\left(\dfrac{x-ut}{\sqrt{1-u^2/c^2}}\right)^2 + y^2 + z^2\right]^{\frac{1}{2}}}. \tag{12.6.7b}$$

This embodies the acoustic Doppler frequency shift for source moving and observer at rest with respect to the medium. Consider a harmonic source $Q(t) = Q \cos[\omega t]$, and for simplicity only head-on encounters, where $\boldsymbol{r} = (x, 0, 0)$, and $\tilde{t}$ is given by (12.6.5). The signal as received, namely (12.6.7), has a slowly-varying amplitude governed by the denominator, while its oscillations are described by the numerator

$$Q(\tilde{t}) = Q \cos\left[\frac{\omega}{(1 \mp u/c)}(t \mp x/c)\right].$$

Thus the observed frequency for approach (recession) is $\omega/(1 \mp u/c)$ instead of the source frequency ω.

For the rest of this section we consider only $\tilde{Q} = Q =$ constant. Then explicit differentiation of (12.6.7a) say yields

$$\Delta p = \sigma \frac{\partial \psi}{\partial t} = \sigma \frac{Q}{4\pi} \frac{(x-ut)u}{\tilde{P}^3}, \tag{12.6.8}$$

$$\boldsymbol{v} = -\left(\frac{\partial \psi}{\partial x}, \frac{\partial \psi}{\partial y}, \frac{\partial \psi}{\partial z}\right) = \frac{Q}{4\pi} \frac{1}{\tilde{P}^3} \{(x-ut), (1-u^2/c^2)y, (1-u^2/c^2)z\}, \tag{12.6.9}$$

$$H = \frac{1}{2}\left\{\frac{(\Delta p)^2}{\sigma c^2} + \sigma \boldsymbol{v}^2\right\}$$

$$= \tfrac{1}{2}\sigma \left(\frac{Q}{4\pi}\right)^2 \frac{1}{\tilde{P}^6} \left\{\left(1 + \frac{u^2}{c^2}\right)(x-ut)^2 + \left(1 - \frac{u^2}{c^2}\right)^2 (y^2+z^2)\right\}, \tag{12.6.10}$$

† The relativistic equation governing a Lorentz-scalar field ψ generated *in vacuo* by a constant-magnitude point source $\rho(\boldsymbol{r}, t) = Q\,\delta(\boldsymbol{r} - \boldsymbol{s}(t))$ is $\Box^2\psi = Q(1 - \dot{s}^2/c^2)^{\frac{1}{2}}\,\delta(\boldsymbol{r} - \boldsymbol{s}(t))$, where c now is the speed of light. If $\dot{s}^2/c^2 \ll 1$, this reduces to (12.5.1) as it should. Accordingly, for a uniformly moving source (12.6.1), ψ is obtained simply by multiplying (12.6.7) by $(1-u^2/c^2)^{\frac{1}{2}}$:

$$\psi(\boldsymbol{r}, t) = (Q/4\pi)\left[\left(\frac{x-ut}{\sqrt{1-u^2/c^2}}\right)^2 + y^2 + z^2\right]^{-\frac{1}{2}}.$$

The same expression follows from an appropriate Lorentz transformation of the stationary-source solution $\psi = Q/4\pi r$; Feynman *et al.* (1964, Section 25.5) give the analogous argument for the electrostatic potential (which is the time-component of a four-vector field).

$$\boldsymbol{N} = \Delta p \boldsymbol{v} = \sigma\left(\frac{Q}{4\pi}\right)^2 \frac{(x-ut)u}{\tilde{P}^6}\{(x-ut), (1-u^2/c^2)y, (1-u^2/c^2)z\}. \tag{12.6.11}$$

The implications are best explored through simple special cases. For instance, at fixed $\boldsymbol{r}$, Δp varies with time as

$$\Delta p = -\frac{\sigma Q u}{4\pi}\frac{\xi}{(\xi^2+\gamma^2)^{\frac{3}{2}}}, \tag{12.6.12a}$$

$$\xi \equiv (ut - x), \qquad \gamma \equiv (1-u^2/c^2)(y^2+z^2). \tag{12.6.12b}$$

This is very like the dispersive ('real') component of a resonance response curve: Δp has a zero at $\xi = 0$, and a maximum (minimum) at $\xi = \mp\gamma/2^{\frac{1}{2}}$, whose magnitude is proportional to $1/\gamma^2$. Thus at high but still subsonic speeds, i.e. as $u \to c-$, $\gamma \to 0$, Δp contracts into two very sharp and very large pulses antisymmetrically disposed about the time when the source is closest to the observation point.

Exercises: (i) Show that the flow $\boldsymbol{v}(\boldsymbol{r}, t)$ observed at any fixed $\boldsymbol{r}$ appears to emanate from a point on the trajectory that lags behind the source by a distance $(x-ut)u^2/(c^2-u^2)$. (ii) Sketch the magnitude of $\boldsymbol{v}(\boldsymbol{r}, t)$ as a function of t.

The energy density (12.6.10) depends only on $(x-ut)$, y, and z, i.e. only on the coordinate differences between source and observer. Hence, as the source moves, H moves with it, and (12.6.11) is just the flux responsible for this rearrangement. But there is no radiation in the sense of an energy outflow say through a cylinder with its axis along the trajectory and with arbitrarily large radius. The machinery for calculating radiative power flows will be set up in the next section. For the present we need note only that the total area of a surface receding to infinity increases proportionally to r^2; hence, in the limit $r \to \infty$, the net power outflow $\int_S \mathrm{d}\boldsymbol{S}\cdot\boldsymbol{N}$ vanishes if N falls faster than $1/r^2$. But (12.6.6) shows that $r \to \infty$ entails $\tilde{P} \sim O(r)$; therefore, in (12.6.11), the denominator behaves like $\tilde{P}^6 \sim r^6$, while the numerator behaves like r^2. Thus $N \sim r^{-4}$, and $\int_S \mathrm{d}\boldsymbol{S}\cdot\boldsymbol{N} \sim r^{-2}$ indeed vanishes as claimed.

12.7 Radiation from accelerated sources: Larmor's formula

12.7.1 The secret of radiation

The Liénard–Wiechert potential of a constant-strength source is

$$\psi(\boldsymbol{r}, t) = \frac{Q}{4\pi\tilde{P}} = \frac{Q}{4\pi(\tilde{R} - \tilde{\boldsymbol{u}}\cdot\tilde{\boldsymbol{R}}/c)}; \tag{12.7.1}$$

the symbols are defined in (12.5.5–7). If the source accelerates ($\tilde{\boldsymbol{a}} \equiv \mathrm{d}\tilde{\boldsymbol{u}}/\mathrm{d}\tilde{t} \neq 0$), it turns out that, like ψ itself, the far-fields $\partial\psi/\partial t$ and $\boldsymbol{\nabla}\psi$ fall only as $1/\tilde{R}$. This allows finite amounts of energy to detach themselves from the source: they do not follow its subsequent motion, and escape to infinity as radiation. We proceed to calculate the radiated power. The secret of the $O(1/\tilde{R})$ fields is that such components arise (only) from the action of the derivatives on the velocity $\tilde{\boldsymbol{u}}$ which features in the denominator of the potential. This is the crucial point that needs watching.

12.7.2 The fields

Evidently

$$\frac{\partial\psi}{\partial t} = -\frac{Q}{4\pi\tilde{P}^2}\frac{\partial}{\partial t}(\tilde{R} - \tilde{\boldsymbol{u}}\cdot\tilde{\boldsymbol{R}}/c), \tag{12.7.2a}$$

$$-\boldsymbol{\nabla}\psi = \frac{Q}{4\pi\tilde{P}^2}\boldsymbol{\nabla}(\tilde{R} - \tilde{\boldsymbol{u}}\cdot\tilde{\boldsymbol{R}}/c). \tag{12.7.2b}$$

To evaluate the derivatives in the numerators, we first need some other more basic derivatives. We shall progress through $\partial\tilde{t}/\partial t$, $\boldsymbol{\nabla}\tilde{t}$, $\partial\tilde{R}/\partial t$, etc. to $\partial\tilde{\boldsymbol{s}}/\partial t$, $\partial\tilde{\boldsymbol{u}}/\partial t$, etc. Although all this is needed for the full results (12.7.16, 17) below, the terms responsible for radiation stem exclusively from the derivatives of $\tilde{\boldsymbol{u}}$, as already stated. They all feature the acceleration $\tilde{\boldsymbol{a}}$ explicitly. The reader willing to take their dominance over all other terms on trust needs the detail only as far as (12.7.9) below. In (12.7.2) she can then use (12.7.7–9) to write

$$\frac{\partial(\tilde{\boldsymbol{u}}\cdot\tilde{\boldsymbol{R}})}{\partial t} \approx \tilde{\boldsymbol{R}}\cdot\frac{\partial\tilde{\boldsymbol{u}}}{\partial t} = (\tilde{\boldsymbol{R}}\cdot\tilde{\boldsymbol{a}})\frac{\partial\tilde{t}}{\partial t} = (\tilde{\boldsymbol{R}}\cdot\tilde{\boldsymbol{a}})\tilde{R}/\tilde{P}, \tag{12.7.2c}$$

$$\boldsymbol{\nabla}(\tilde{\boldsymbol{u}}\cdot\tilde{\boldsymbol{R}}) \approx (\tilde{\boldsymbol{R}}\cdot\tilde{\boldsymbol{a}})\boldsymbol{\nabla}\tilde{t} = -(\tilde{\boldsymbol{R}}\cdot\tilde{\boldsymbol{a}})\tilde{\boldsymbol{R}}/c\tilde{P}, \tag{12.7.2d}$$

drop $\partial\tilde{R}/\partial t$ and $\boldsymbol{\nabla}\hat{R}$, and thus arrive at the radiated fields (12.7.18, 19) directly.

For convenience we assemble the basic relations

$$\tilde{R} = [\tilde{\boldsymbol{R}}^2]^{\frac{1}{2}} = [(\boldsymbol{r} - \tilde{\boldsymbol{s}})^2]^{\frac{1}{2}} \equiv [(\boldsymbol{r} - \boldsymbol{s}(\tilde{t}))^2]^{\frac{1}{2}} = c(t - \tilde{t}), \tag{12.7.3}$$

$$\tilde{R} = c(t - \tilde{t}), \qquad \tilde{P} = \tilde{R} - \tilde{\boldsymbol{u}}\cdot\tilde{\boldsymbol{R}}/c. \tag{12.7.4, 5}$$

Since the source trajectory $\boldsymbol{s}$ is an assigned function of time,

$$\frac{\partial\tilde{\boldsymbol{s}}}{\partial t} \equiv \frac{\partial\boldsymbol{s}(\tilde{t})}{\partial t} = \frac{\mathrm{d}\boldsymbol{s}(\tilde{t})}{\mathrm{d}\tilde{t}}\frac{\partial\tilde{t}}{\partial t} = \boldsymbol{u}(\tilde{t})\frac{\partial\tilde{t}}{\partial t} = \tilde{\boldsymbol{u}}\frac{\partial\tilde{t}}{\partial t}, \tag{12.7.6}$$

and by the same token

$$\frac{\partial \tilde{\boldsymbol{s}}}{\partial r_i} = \tilde{\boldsymbol{u}} \frac{\partial \tilde{t}}{\partial r_i}, \qquad \frac{\partial \tilde{\boldsymbol{u}}}{\partial t} = \tilde{\boldsymbol{a}} \frac{\partial \tilde{t}}{\partial t}, \qquad \frac{\partial \tilde{u}_j}{\partial r_i} = \tilde{a}_j \frac{\partial \tilde{t}}{\partial r_i}, \tag{12.7.7a,b,c}$$

which explains the key role destined for $\partial \tilde{t}/\partial t$ and $\boldsymbol{\nabla} \tilde{t}$.

In order to find $\partial \tilde{t}/\partial t$, we act with $\partial/\partial t$ on (12.7.3) and then use (12.7.6):

$$c\left(1 - \frac{\partial \tilde{t}}{\partial t}\right) = \frac{1}{2} \frac{1}{[(\boldsymbol{r} - \tilde{\boldsymbol{s}})^2]^{\frac{1}{2}}} 2(\boldsymbol{r} - \tilde{\boldsymbol{s}}) \cdot \left(-\frac{\partial \tilde{\boldsymbol{s}}}{\partial t}\right) = -\frac{1}{\tilde{R}} \tilde{\boldsymbol{R}} \cdot \frac{\partial \tilde{\boldsymbol{s}}}{\partial t} = -\frac{1}{\tilde{R}} \tilde{\boldsymbol{R}} \cdot \tilde{\boldsymbol{u}} \frac{\partial \tilde{t}}{\partial t};$$

solving this for $\partial \tilde{t}/\partial t$, one obtains

$$\frac{\partial \tilde{t}}{\partial t} = \frac{\tilde{R}}{\tilde{R} - \tilde{\boldsymbol{u}} \cdot \tilde{\boldsymbol{R}}/c} = \frac{\tilde{R}}{\tilde{P}}. \tag{12.7.8}$$

Similarly, to find $\boldsymbol{\nabla} \tilde{t}$, we act on (12.7.3) with $\partial/\partial r_i$, and use $\partial r_j/\partial r_i = \delta_{ij}$, and (12.7.7a):

$$-c \frac{\partial \tilde{t}}{\partial r_i} = \frac{1}{\tilde{R}} \sum_{j=1}^{3} \tilde{R}_j \frac{\partial}{\partial r_i}(r_j - \tilde{s}_j) = \frac{1}{\tilde{R}} \sum_{j=1}^{3} \tilde{R}_j \left(\delta_{ij} - \tilde{u}_j \frac{\partial \tilde{t}}{\partial r_j}\right)$$

$$-c \frac{\partial \tilde{t}}{\partial r_i} = \frac{1}{\tilde{R}} \left(\tilde{R}_i - (\tilde{\boldsymbol{u}} \cdot \tilde{\boldsymbol{R}}) \frac{\partial \tilde{t}}{\partial r_i}\right),$$

$$\boldsymbol{\nabla} \tilde{t} = -\frac{(\tilde{\boldsymbol{R}}/c)}{\tilde{R} - \tilde{\boldsymbol{u}} \cdot \tilde{\boldsymbol{R}}/c} = -\tilde{\boldsymbol{R}}/c\tilde{P}. \tag{12.7.9}$$

Next, the derivatives of $\tilde{R}$ are found in terms of those of $\tilde{t}$ through (12.7.4):

$$\frac{\partial \tilde{R}}{\partial t} = c\left(1 - \frac{\partial \tilde{t}}{\partial t}\right) = c\left(1 - \frac{\tilde{R}}{\tilde{R} - \tilde{\boldsymbol{u}} \cdot \tilde{\boldsymbol{R}}/c}\right) = -(\tilde{\boldsymbol{u}} \cdot \tilde{\boldsymbol{R}})/\tilde{P}, \tag{12.7.10}$$

where the second step quotes (12.7.8). From (12.7.4, 9) (seeing that $\boldsymbol{\nabla} t = 0$ because $\boldsymbol{r}$ and t are independent) we obtain

$$\boldsymbol{\nabla} \tilde{R} = -c \boldsymbol{\nabla} \tilde{t} = \tilde{\boldsymbol{R}}/\tilde{P}. \tag{12.7.11}$$

Reasoning along the same lines and quoting judiciously from (12.7.7) yields the other relations we need:

$$\partial \tilde{\boldsymbol{s}}/\partial t = \tilde{\boldsymbol{u}} \tilde{R}/\tilde{P}, \qquad \partial \tilde{s}_j/\partial r_i = -\tilde{u}_j \tilde{R}_i/c\tilde{P}, \tag{12.7.12a,b}$$

$$\partial \tilde{\boldsymbol{u}}/\partial t = \tilde{\boldsymbol{a}} \tilde{R}/\tilde{P}, \qquad \partial \tilde{u}_j/\partial r_i = -\tilde{a}_j \tilde{R}_i/c\tilde{P}, \tag{12.7.13a,b}$$

$$\partial(\tilde{\boldsymbol{u}} \cdot \tilde{\boldsymbol{R}})/\partial t = (\tilde{\boldsymbol{a}} \cdot \tilde{\boldsymbol{R}} - \tilde{u}^2)\tilde{R}/\tilde{P}, \tag{12.7.14}$$

$$\boldsymbol{\nabla}(\tilde{\boldsymbol{u}} \cdot \tilde{\boldsymbol{R}}) = \{\tilde{\boldsymbol{R}}(-\tilde{\boldsymbol{a}} \cdot \tilde{\boldsymbol{R}} + \tilde{u}^2)/c\tilde{P} + \tilde{\boldsymbol{u}}\}. \tag{12.7.15}$$

Exercise: Derive these.

Substituting into (12.7.2a, b) we can now write down the retarded fields:

$$\frac{\partial\psi}{\partial t}=\frac{Q}{4\pi\tilde{P}^3}\left\{[(\tilde{\boldsymbol{a}}\cdot\tilde{\boldsymbol{R}})-\tilde{u}^2]\frac{\tilde{R}}{c}+\tilde{\boldsymbol{u}}\cdot\tilde{\boldsymbol{R}}\right\}, \tag{12.7.16}$$

$$-\nabla\psi=\frac{Q}{4\pi\tilde{P}^3}\{\tilde{\boldsymbol{R}}[(\tilde{\boldsymbol{a}}\cdot\tilde{\boldsymbol{R}})/c^2+(1-\tilde{u}^2/c^2)]-\tilde{\boldsymbol{u}}\tilde{P}/c\}. \tag{12.7.17}$$

For uniform motion ($\tilde{\boldsymbol{u}}=\boldsymbol{u}=u\hat{\boldsymbol{x}}$, $\tilde{\boldsymbol{a}}=0$) they reduce as they must to the expressions in Section 12.6.

Bearing in mind that $\tilde{R}\to\infty$ entails $\tilde{P}\sim O(\tilde{R})\to\infty$, it is now clear that only the $(\tilde{\boldsymbol{a}}\cdot\tilde{\boldsymbol{R}})$-proportional components contribute to the leading order $1/\tilde{R}$. Discarding terms of higher order in $1/\tilde{R}$ we find

$$\frac{\partial\psi}{\partial t}\sim\frac{Q\tilde{R}}{4\pi c\tilde{P}^3}(\tilde{\boldsymbol{a}}\cdot\tilde{\boldsymbol{R}})=\frac{Q}{4\pi c\tilde{R}}\frac{(\tilde{\boldsymbol{a}}\cdot\hat{\boldsymbol{R}})}{(1-\tilde{\boldsymbol{u}}\cdot\hat{\boldsymbol{R}}/c)^3}, \tag{12.7.18}$$

$$-\nabla\psi\sim\frac{Q}{4\pi c^2\tilde{P}^3}(\tilde{\boldsymbol{a}}\cdot\tilde{\boldsymbol{R}})\tilde{\boldsymbol{R}}=\hat{\boldsymbol{R}}\frac{Q}{4\pi c^2\tilde{R}}\frac{(\tilde{\boldsymbol{a}}\cdot\hat{\boldsymbol{R}})}{(1-\tilde{\boldsymbol{u}}\cdot\hat{\boldsymbol{R}}/c)^3}, \tag{12.7.19}$$

$$\begin{aligned}H&=\tfrac{1}{2}\sigma\left\{\frac{1}{c^2}\left(\frac{\partial\psi}{\partial t}\right)^2+(\nabla\psi)^2\right\}\\&\sim\frac{\sigma}{c^4}\left(\frac{Q}{4\pi}\right)^2\frac{\tilde{R}^2}{\tilde{P}^6}(\tilde{\boldsymbol{a}}\cdot\tilde{\boldsymbol{R}})^2=\frac{\sigma}{c^4}\left(\frac{Q}{4\pi}\right)^2\frac{1}{\tilde{R}^2}\frac{(\tilde{\boldsymbol{a}}\cdot\hat{\boldsymbol{R}})^2}{(1-\tilde{\boldsymbol{u}}\cdot\hat{\boldsymbol{R}}/c)^6},\end{aligned} \tag{12.7.20}$$

$$\boldsymbol{N}=-\sigma\frac{\partial\psi}{\partial t}\nabla\psi\sim\hat{\boldsymbol{R}}\frac{\sigma}{c^3}\left(\frac{Q}{4\pi}\right)^2\frac{(\tilde{\boldsymbol{a}}\cdot\hat{\boldsymbol{R}})^2}{\tilde{R}^2(1-\tilde{\boldsymbol{u}}\cdot\hat{\boldsymbol{R}}/c)^6}\sim\hat{\boldsymbol{R}}cH. \tag{12.7.21}$$

If $\tilde{u}/c\ll 1$, then $\tilde{P}/\tilde{R}=(1-\tilde{\boldsymbol{u}}\cdot\hat{\boldsymbol{R}}/c)$ can be replaced by unity:

$$\boldsymbol{N}\approx\hat{\boldsymbol{R}}\frac{\sigma}{c^3}\left(\frac{Q}{4\pi}\right)^2\frac{(\tilde{\boldsymbol{a}}\cdot\hat{\boldsymbol{R}})^2}{\tilde{R}^2},\qquad(\tilde{R}\to\infty,\ \tilde{u}/c\ll 1). \tag{12.7.22}$$

12.7.3 The radiated power

To calculate the power radiated by an arbitrarily (though still subsonically) moving as opposed to a stationary source, two new ideas are needed.

First we must determine the rate at which energy crosses a surface element $\mathrm{d}\boldsymbol{S}$ that moves with velocity $\boldsymbol{u}$. This rate $\mathrm{d}W$ is simply the sum of the rate appropriate when $\boldsymbol{u}=0$ (the only case we have considered hitherto), namely $\mathrm{d}\boldsymbol{S}\cdot\boldsymbol{N}$, and of the rate appropriate when $\boldsymbol{N}=0$, namely $-\mathrm{d}\boldsymbol{S}\cdot\boldsymbol{u}H$. Accordingly,

$$\mathrm{d}W=\mathrm{d}\boldsymbol{S}\cdot\{\boldsymbol{N}-\boldsymbol{u}H\}. \tag{17.7.23}$$

In particular, the power outflow at time t through a closed surface S

moving rigidly with velocity $\boldsymbol{u}$ is

$$\int_S \mathrm{d}\boldsymbol{S} \cdot \{\boldsymbol{N}(\boldsymbol{r}, t) - \boldsymbol{u}H(\boldsymbol{r}, t)\}. \tag{17.7.24}$$

Here, in keeping with our standard notation, $\boldsymbol{r}$ in the arguments of $\boldsymbol{N}$ and H is the position that the surface element $\mathrm{d}\boldsymbol{S}$ occupies at time t.

To explain the second idea, we recall that hitherto we have viewed the field point $\boldsymbol{r}$ and the observation time t as the independent variables, and the retarded variables $\tilde{t}$, $\tilde{\boldsymbol{s}} \equiv \boldsymbol{s}(\tilde{t})$, $\tilde{\boldsymbol{u}}$, $\tilde{\boldsymbol{a}}$, and $\tilde{\boldsymbol{R}} \equiv \boldsymbol{r} - \tilde{\boldsymbol{s}}$ as functions (which can be quite awkward to determine) of $\boldsymbol{r}$ and t. This is how we found the central relations (12.7.16–21), which we shall continue to exploit, albeit with a somewhat differently slanted notation. The crucial step is to reverse the view just described, and to adapt the notation accordingly: now we start by choosing a time t and the corresponding point $\boldsymbol{s}(t)$ on the trajectory of the source; specify certain field points by $\boldsymbol{r} = \boldsymbol{s}(t) + \boldsymbol{R}$; consider the energy flow at $\boldsymbol{r}$ only at the time $t + R/c$; and treat t and $\boldsymbol{R}$ as our independent variables. For our present purposes, the tildes previously reserved for retarded variables are unnecessary, and we now drop them. For instance, we now write $\boldsymbol{u} = \dot{\boldsymbol{s}}(t)$, $\boldsymbol{a} = \ddot{\boldsymbol{s}}(t)$, etc. In this new notation the fields at a point $\boldsymbol{r}$ and at time $t + R/c$ are given simply by (12.7.16–20) with the tildes removed.

We are now in a position to define the total power radiated by an arbitrarily moving point source,† through an appropriate generalization of (12.5.18a). To this end, consider a sphere S of time-independent radius R, which at time $t + R/c$ is centred on $\boldsymbol{s}(t)$, so that (at time $t + R/c$) every point of S moves with velocity $\boldsymbol{u}(t)$. Then we *define* $W_{\mathrm{rad}}(t)$ as the total power outflow through S at time $t + R/c$, in the limit $R \to \infty$. In view of (12.7.24), this reads

$$W_{\mathrm{rad}}(t) \equiv \lim_{R\to\infty} \int_S \mathrm{d}\boldsymbol{S} \cdot \{\boldsymbol{N}(\boldsymbol{r}, t + R/c) - \boldsymbol{u}(t)H(\boldsymbol{r}, t + R/c)\}, \tag{12.7.25}$$

$$\boldsymbol{r} \equiv \boldsymbol{s}(t) + \boldsymbol{R}. \tag{12.7.26}$$

By the light of the preceding remarks about notation, the integrand can be found immediately from (12.7.21). Writing $\boldsymbol{R} \equiv (R, \Omega)$ and

$$\boldsymbol{\beta} \equiv \boldsymbol{u}/c, \qquad \beta^2 = u^2/c^2,$$

† For moving sources, there is a difference between power emitted by the source and power received by stationary detectors, a difference which sometimes goes unrecognized. Here we are concerned only with the former. A clear discussion (for electromagnetic radiation *in vacuo*) is given by G. B. Rybicki and A. P. Lightman (1979), *Radiative processes in astrophysics*, Section 4.8; Wiley, New York 1979.

we have

$$\mathrm{d}\boldsymbol{S}\cdot\{\boldsymbol{N}-\boldsymbol{u}H\}\sim \mathrm{d}\Omega\, R^2\hat{\boldsymbol{R}}\cdot\{\boldsymbol{N}-\boldsymbol{u}H\}=\mathrm{d}\Omega\, R^2(1-\hat{\boldsymbol{R}}\cdot\boldsymbol{\beta})\hat{\boldsymbol{R}}\cdot\boldsymbol{N}, \tag{12.7.27}$$

whence‡

$$W_{\mathrm{rad}}(t)=\frac{\sigma}{c^3}\left(\frac{Q}{4\pi}\right)^2\int \mathrm{d}\Omega\,(\boldsymbol{a}\cdot\hat{\boldsymbol{R}})^2/(1-\hat{\boldsymbol{R}}\cdot\boldsymbol{\beta})^5. \tag{12.7.28}$$

The integral is straightforward but tedious. But in the regime $u/c\ll 1$ we can replace the denominator by unity; since $\int \mathrm{d}\Omega(\boldsymbol{a}\cdot\hat{\boldsymbol{R}})^2=4\pi a^2/3$, one then has (the acoustic analogue of) *Larmor's formula*

$$W_{\mathrm{rad}}\approx \sigma Q^2a^2/12\pi c^3, \qquad ((u/c)^2\ll 1). \tag{12.7.29}$$

Exercises: (i) By expanding the integrand of (12.7.28) in powers of $\hat{\boldsymbol{R}}\cdot\boldsymbol{\beta}$, show that the leading correction to Larmor's formula is only of relative order $(u/c)^2$ rather than (u/c). (ii) Verify that W_{rad} in (12.7.28) is invariant under time-reversal, in the sense that it remains unchanged on reversing the motion, i.e. on replacing $\boldsymbol{s}(t)$ by $\boldsymbol{s}(-t)$.

It is interesting to compare Larmor's formula for an accelerating constant-magnitude source with (12.5.16) for a stationary but time-varying source. Their radiated power outputs are equal if $(\dot{Q}/Q)^2=\frac{1}{3}(a/c)^2$.

In terms of $\hat{\boldsymbol{u}}\cdot\hat{\boldsymbol{a}}\equiv\cos\alpha$, eqn (12.7.28) eventually yields

$$W_{\mathrm{rad}}=\frac{\sigma Q^2a^2}{12\pi c^3},\frac{1}{(1-\beta^2)^4}\{1-\beta^2(1-6\cos^2\alpha)\}. \tag{12.7.30}$$

Finally, we determine the instantaneous power output $W(t)$, defining it by analogy to (12.5.20a) and (12.7.26) as

$$W(t)\equiv\lim_{R\to 0} W(R,t), \tag{12.7.31}$$

$$W(R,t)=\int_S \mathrm{d}\boldsymbol{S}\cdot\{\boldsymbol{N}(\boldsymbol{r},t+R/c)-\boldsymbol{u}(t)H(\boldsymbol{r},t+R/c)\}$$
$$=\int \mathrm{d}\Omega\, R\boldsymbol{R}\cdot\{\boldsymbol{N}-\boldsymbol{u}H\}. \tag{12.7.32}$$

Tedious substitution via $\boldsymbol{N}$ and H from the exact expressions (12.7.16, 17) for the

‡ Notice that the total power outflow (also at time $t+R/c$) through a *stationary* sphere instantaneously coincident with S is $(\sigma/c^3)(Q/4\pi)^2\int \mathrm{d}\Omega\,(\boldsymbol{a}\cdot\hat{\boldsymbol{R}})^2/(1-\hat{\boldsymbol{R}}\cdot\boldsymbol{\beta})^6$.

fields eventually yields

$$W(R, t) = \int \mathrm{d}\Omega\, Rc\sigma\left(\frac{Q}{4\pi P^3}\right)^2 \left\{\left(\frac{\boldsymbol{a}\cdot\boldsymbol{R}}{c^2}\right)^2 R^2 P + \left(\frac{\boldsymbol{a}\cdot\boldsymbol{R}}{c^2}\right)\left[R^2(1-2\beta^2) + (\boldsymbol{R}\cdot\boldsymbol{\beta})^2\right]P\right.$$

$$\left. + R^3(1-\beta^2)\left[-\beta^2 + \tfrac{1}{2}(\hat{\boldsymbol{R}}\cdot\boldsymbol{\beta})(1+2\beta^2) - \tfrac{1}{2}(\hat{\boldsymbol{R}}\cdot\boldsymbol{\beta})^3\right]\right\}$$

$$\equiv W_1(R, t) + W_2(R, t) + W_3(R, t), \tag{12.7.33}$$

where $P = R(1 - \hat{\boldsymbol{R}}\cdot\boldsymbol{\beta})$, and where W_1, W_2, W_3 are, respectively, the components quadratic in $\boldsymbol{a}$, linear in $\boldsymbol{a}$, and independent of $\boldsymbol{a}$.

The component W_1 is identically the same as W_{rad}, eqn (12.7.28); this integral is independent of R, making the limit $R \to 0$ redundant. The component $W_3(R, t)$ vanishes. Though this is by no means obvious from mere inspection, it is expected from the fact that the expression for W_3 (being independent of $\boldsymbol{a}$) formally coincides with the total power outflow through the sphere S associated with a *uniformly* moving source, as described in problem 12.11.

Exercise: Verify by explicit integration that $W_3 = 0$.

Finally, it is not too difficult to show that

$$W_2(R, t) = \frac{\mathrm{d}}{\mathrm{d}t}\left\{\frac{\sigma Q^2}{8\pi R}\cdot\frac{(1-\beta^2/3)}{(1-\beta^2)^2}\right\}$$

$$= \mathrm{d}E(u, R)/\mathrm{d}t, \tag{12.7.32}$$

where $E(u, R)$ is simply the total energy outside the sphere S in the special case of a source having (at all times) the constant velocity $\boldsymbol{u} = c\boldsymbol{\beta}$ (see Problem 12.11).

Accordingly, the total instantaneous power output of an arbitrarily (but subsonically) moving constant-strength source, as defined by (12.7.31), is

$$W(t) = W_{\text{rad}}(t) + \lim_{R\to 0} \mathrm{d}E\,(u, R)/\mathrm{d}t. \tag{12.7.33}$$

As for a stationary but variable-strength source in Section 12.5, we see that $W(t)$ differs from $W_{\text{rad}}(t)$ only by a total time-derivative (albeit a formally divergent one). The consequences, commented on following eqns (12.5.20, 21) are also the same.

12.8 Boundaries and reflections

12.8.1 Images

The Green's function G_0 for unbounded space proves useful also in regions bounded by planes. This is especially so in a halfspace, say $z \geqslant 0$ in 3D. There, under Dirichlet or Neumann boundary conditions (upper and lower signs respectively),

$$G(\boldsymbol{r}, t \mid \boldsymbol{r}', t') = G_0(\boldsymbol{r}, t \mid \boldsymbol{r}', t') \mp G_0(\boldsymbol{r}, t \mid \tilde{\boldsymbol{r}}', t'), \tag{12.8.1a}$$

$$\tilde{\boldsymbol{r}}' \equiv (x', y', -z'). \tag{12.8.1b}$$

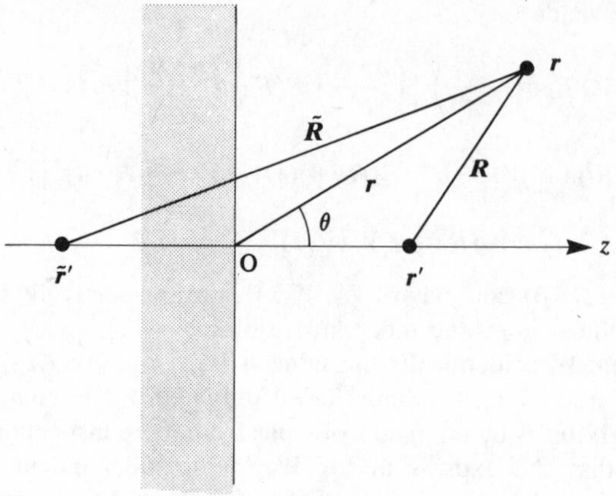

Fig. 12.5

(In this section the tilde refers to the image position as in Sections 5.3.3 and 9.4.4, and not as elsewhere in this chapter to retardation.) It is easy to verify that these expressions satisfy all the defining relations of G, just as they did for diffusion. Source, image, and field points $\boldsymbol{r}'$, $\tilde{\boldsymbol{r}}'$, $\boldsymbol{r}$, and the vectors $\boldsymbol{R} = \boldsymbol{r} - \boldsymbol{r}'$, $\tilde{\boldsymbol{R}} = \boldsymbol{r} - \tilde{\boldsymbol{r}}'$ are drawn in Fig. 12.5. As always for plane boundaries, the image source is equal and opposite (equal) to the true source under DBCs (NBCs).

As an example, consider radiation from a source $Q(t)$ stationary at $\boldsymbol{r}'$:

$$\rho(\boldsymbol{r}, t) = Q(t)\,\delta(\boldsymbol{r} - \boldsymbol{r}'). \tag{12.8.2}$$

Then, with $\psi = f(\boldsymbol{r}, t) = \int dt'' \int dV''\, G(\boldsymbol{r}, t \,|\, \boldsymbol{r}'', t'')\rho(\boldsymbol{r}'', t'')$, the magic rule yields

$$\psi(\boldsymbol{r}, t) = Q(t - R/c)/4\pi R \mp Q(t - \tilde{R}/c)/4\pi\tilde{R}. \tag{12.8.3}$$

In the physical halfspace $z \geqslant 0$ one observes interference between the appropriately retarded signals from source and image. For simplicity, specialize to a harmonic source,

$$Q(t) = Q\cos[\omega t], \qquad k \equiv 2\pi/\lambda \equiv \omega/c; \tag{12.8.4}$$

$$\psi(\boldsymbol{r}, t) = \frac{Q}{4\pi}\{\cos[\omega t - kR]/R \mp \cos[\omega t - k\tilde{R}]/\tilde{R}\}. \tag{12.8.5}$$

In order to determine the radiated power, we first evaluate ψ to $O(1/r)$ as $r \to \infty$ at fixed θ. Then

$$R = (r^2 + z'^2 - 2rz'\cos\theta) = r(1 - 2z'\cos\theta/r + z^2/r^2)^{\frac{1}{2}},$$

$$R = r - z'\cos\theta + O(1/r), \qquad 1/R = 1/r + O(1/r^2), \tag{12.8.6a}$$

$$\tilde{R} = r + z'\cos\theta + O(1/r), \qquad 1/\tilde{R} = 1/r + O(1/r^2), \tag{12.8.6b}$$

whence

$$\psi(r, \theta, t) \approx \frac{Q}{4\pi r}\{\cos[\omega t - kr + kz'\cos\theta] \mp \cos[\omega t - kr - kz'\cos\theta]\}$$

$$= \begin{cases} -\dfrac{Q}{4\pi r} 2\sin[\omega t - kr]\sin[kz'\cos\theta], & \text{(DBCs);} \quad (12.8.7a) \\ \dfrac{Q}{4\pi r} 2\cos[\omega t - kr]\cos[kz'\cos\theta], & \text{(NBCs).} \quad (12.8.7b) \end{cases}$$

As the source approaches the boundary ($z' \to 0$), the Neumann potential approaches twice the far-field potential for the isolated source. The Dirichlet potential approaches $-(2z'Q/4\pi r)\cos\theta\,\sin[\omega t - kr]$, which is the far-field potential for an isolated dipole $\boldsymbol{p} = \hat{\boldsymbol{z}}2z' \cdot Q$, eqn (12.4.12). This might have been expected from the fact that in the limit, under these BCs, source and image jointly constitute just these systems.

To order $1/r^2$ in the energy flux $\boldsymbol{N} = -\sigma(\partial\psi/\partial t)\boldsymbol{\nabla}\psi$, the derivatives act only on the trigonometric factors (see Sections 12.4, 7), and for the time-average one finds

$$\bar{\boldsymbol{N}}(\boldsymbol{r}) \approx \hat{\boldsymbol{r}}\frac{\sigma\omega^2 Q^2}{16\pi^2 c^2}\frac{1}{r^2}\{1 \mp \cos[2kz'\cos\theta]\}. \quad (12.8.8a)$$

Finally, the time-averaged radiated power $\bar{W}_{\text{rad}}$ is the integral $\int d\Omega\, r^2 N_r$ over the right-hand hemisphere $\theta \geqslant 0$:

$$\bar{W}_{\text{rad}} = \int_0^1 2\pi r^2\, d\cos\theta \cdot N_r,$$

$$\bar{W}_{\text{rad}} = \frac{\sigma\omega^2 Q^2}{8\pi c}\left\{1 \mp \frac{\sin[2kz']}{2kz'}\right\}. \quad (12.8.8b)$$

Compare this with the power $\bar{W}^{(0)}_{\text{rad}} \equiv \sigma\omega^2 Q^2/8\pi c$, (12.4.9b), radiated by the same source in unbounded space, and observe what happens as $z' \to 0$. Under DBCs, W vanishes: obviously, because $\psi(\boldsymbol{r}, t)$ itself then vanishes always and everywhere, since $z' \to 0$ entails $\tilde{\boldsymbol{R}} \to \boldsymbol{R}$. In other words, the effective dipole strength $p = 2z'Q$ mentioned earlier vanishes with z'. Under NBCs, $\bar{W}_{\text{rad}} \to 2\bar{W}^{(0)}_{\text{rad}}$; this comes about because the effective source strength is doubled, whence $\boldsymbol{N}(\boldsymbol{r}, t)$ is quadrupled, but acts only over a hemisphere instead of a full sphere. (Analogous effects operate in electromagnetism, and persist in quantum electrodynamics, where the spontaneous decay rate of an excited atom is altered by proximity to a mirror.)

The method of images extends to the slab $0 \leqslant z \leqslant L$ exactly as it did for Poisson's and for the diffusion equation (see (5.3.19) and (9.4.16)). For this region one has the multiple-image representation

$$G_{D,N}(\boldsymbol{r}, t \mid \boldsymbol{r}', t') = \sum_{N=-\infty}^{\infty} \{G_0(x, y, z, t \mid x', y', 2NL + z', t') \mp G_0(x, y, z, t \mid x', y', 2NL - z', t')\}, \tag{12.8.9}$$

where the minus (plus) sign applies to G_D (G_N).

Exercises: (i) Verify by substitution that (12.8.9) satisfies the defining differential equation and the BCs. (ii) For Poisson's equation under NBCs, it was pointed out just below eqn (6.4.7) that the would-be multiple-image series analogous to (12.8.9) diverges. Show that for the wave equation the series (12.8.9) does converge, and explain what properties of the respective free-space Green's functions G_0 are responsible for this difference. (iii) Verify that (12.8.9) agrees with the normal-mode expansion constructed along the lines of Section 10.5.1.

12.8.2 Causality with boundaries

For any bounded region one may choose to write

$$G(\boldsymbol{r}, t \mid \boldsymbol{r}', t') = G_0(\boldsymbol{r}, t \mid \boldsymbol{r}', t') + \chi(\boldsymbol{r}, t \mid \boldsymbol{r}', t'). \tag{12.8.10}$$

Since G and G_0 obey the same differential equation (with the same inhomogeneous term) and the same initial conditions, we have

$$\Box^2 \chi = 0, \tag{12.8.11}$$

$$\chi(\boldsymbol{r}, t \mid \boldsymbol{r}', t')\big|_{t=t'} = 0 = \chi_t(\boldsymbol{r}, t \mid \boldsymbol{r}', t')\big|_{t=t'}; \tag{12.8.12}$$

and χ obviously shares the symmetries common to G and G_0, e.g. time-translation invariance and the relation $\chi(\boldsymbol{r}, t \mid \boldsymbol{r}', t') = \chi(\boldsymbol{r}', t \mid \boldsymbol{r}, t')$. Finally, the homogeneous BCs on G imply for χ the BCs

$$\left.\begin{array}{ll} \text{DBCs:} & \chi(\boldsymbol{r}, t \mid \boldsymbol{r}', t') = -G_0(\boldsymbol{r}, t \mid \boldsymbol{r}', t') \\ \text{NBCs:} & \partial_n \chi(\boldsymbol{r}, t \mid \boldsymbol{r}', t') = -\partial_n G_0(\boldsymbol{r}, t \mid \boldsymbol{r}', t') \end{array}\right\} \quad \begin{array}{l} \boldsymbol{r} \text{ on } S, \\ t > t'. \end{array} \tag{12.8.13}$$

The image term in (12.8.1a) is a prime example of such an addend χ.

The integral equation we shall derive for χ proves convenient both in approximations and in general arguments. Here we shall use it to establish a causality property evident to intuition, namely that $G(\boldsymbol{r}, t \mid \boldsymbol{r}', t') = G_0(\boldsymbol{r}, t \mid \boldsymbol{r}', t')$, i.e. that $\chi = 0$, until t is large enough for a signal travelling at speed c to have reached $\boldsymbol{r}$ via the boundary after emission from $\boldsymbol{r}'$ at time t'.

For simplicity we consider only NBCs. Regarding χ as a function of $\boldsymbol{r}$ and t defined by (12.8.11–13), we can express it as the appropriate surface integral g in the magic rule, namely as the first integral in (10.4.6), constructed with the appropriate Green's function G (of which χ itself forms part). Because the single-primed variables have been preempted, the integration variables in g now carry double primes:

$$\chi(\boldsymbol{r}, t \mid \boldsymbol{r}', t') = \int_{t'}^{t} \mathrm{d}t'' \int_S \mathrm{d}S''\, G(\boldsymbol{r}, t \mid \boldsymbol{r}'', t'')\, \partial_n'' \chi_s(\boldsymbol{r}'', t'' \mid \boldsymbol{r}', t'). \tag{12.8.14}$$

Substituting $G = G_0 + \chi$, and for $\partial''_n \chi_s$ from (12.8.13), one finds

$$\chi(\boldsymbol{r}, t \mid \boldsymbol{r}', t') = -\int_{t'}^{t} dt'' \int_S dS'' \{G_0(\boldsymbol{r}, t \mid \boldsymbol{r}'', t'') + \chi(\boldsymbol{r}, t \mid \boldsymbol{r}'', t'')\}\, \partial''_n G_0(\boldsymbol{r}'', t'' \mid \boldsymbol{r}', t')$$
$$\equiv \chi_1 + \chi_2. \tag{12.8.15}$$

This is an integral equation for χ, inhomogeneous on account of the first term χ_1, whose integrand is known because it involves only the two G_0s. It proves crucial that the function χ defined by (12.8.11–13), and therefore the solution of (12.8.15), is unique (Section 10.2).

Theorem

$$\chi(\boldsymbol{r}, t \mid \boldsymbol{r}', t') = 0 \quad \text{unless} \quad t > t' + \min_{(\boldsymbol{r}'')} \{|\boldsymbol{r} - \boldsymbol{r}''|/c + |\boldsymbol{r}'' - \boldsymbol{r}'|/c\}, \tag{12.8.16}$$

where $\min_{(\boldsymbol{r}'')}\{\cdots\}$ denotes the minimum value of $\{\cdots\}$ as the point $\boldsymbol{r}''$ varies over the boundary S.

Proof: (i) By the causality property of G_0 (Section 11.4), the inhomogeneous term χ_1 vanishes unless, for some point $\boldsymbol{r}''$ on S, $t > t'' + |\boldsymbol{r} - \boldsymbol{r}''|/c$ (because otherwise the first G_0 in the integrand vanishes), and unless, for the same point $\boldsymbol{r}''$, $t'' > t' + |\boldsymbol{r}'' - \boldsymbol{r}'|/c$ (because otherwise $\partial''_n G_0$ vanishes). Hence $\chi_1 = 0$ unless $t > t' + \min_{(\boldsymbol{r}'')}\{|\boldsymbol{r} - \boldsymbol{r}''|/c + |\boldsymbol{r}'' - \boldsymbol{r}'|/c\}$. Thus χ_1 does have the property (12.8.16).

(ii) If $\chi_1 = 0$, then the integral equation is satisfied by $\chi = 0$ (i.e. χ_2 zero as well as χ_1). Because the solution is unique, $\chi = 0$ if $\chi_1 = 0$. ■

Problems

12.1 The function $\psi(x, t)$ obeys the homogeneous 1D wave equation. At $t=0$, $\psi=\beta(x)$ and $\psi_t=\alpha(x)$, where $\alpha(x)$ and $\beta(x)$ are strictly zero when $|x|>a$. What further conditions, if any, must be imposed on α and β in order to guarantee that both $\psi(x, t)$ and $\psi_t(x, t)$ vanish for $|x| \leqslant a$ as $t \to \infty$?

12.2 The function $\psi(\boldsymbol{r}, t)$ obeys the homogeneous 3D wave equation. At $t=0$, $\psi=\beta(\boldsymbol{r})$ and $\psi_t=\alpha(\boldsymbol{r})$, where $\alpha(\boldsymbol{r})$ and $\beta(\boldsymbol{r})$ are strictly zero when $|\boldsymbol{r}|>R$. What further conditions, if any, must be imposed on α and β in order to guarantee that both $\psi(\boldsymbol{r}, t)$ and $\psi_t(\boldsymbol{r}, t)$ vanish for $|\boldsymbol{r}| \leqslant R$ as $t \to \infty$?

12.3 The function $\psi(\boldsymbol{r}, t)$ obeys the homogeneous wave equation; $\psi(\boldsymbol{r}, 0)=\beta(\boldsymbol{r})$ and $\psi_t(\boldsymbol{r}, 0)=\alpha(\boldsymbol{r})$ entail $\psi(\boldsymbol{r}, T)=B(\boldsymbol{r})$ and $\psi_t(\boldsymbol{r}, T)=A(\boldsymbol{r})$.

(i) Show that $\psi(\boldsymbol{r}, T)=B(\boldsymbol{r})$ and $\psi_t(\boldsymbol{r}, T)=-A(\boldsymbol{r})$ entail $\psi(\boldsymbol{r}, 2T)=\beta(\boldsymbol{r})$ and $\psi_t(\boldsymbol{r}, 2T)=-\alpha(\boldsymbol{r})$.
(ii) In 1D, $\psi(x, 0)=1/(x^2+1)$, $\psi_t(x, 0)=0$. Determine $\psi(x, t)$ and $\psi_t(x, t)$ for all $t \gtrless 0$.
(iii) How must $\psi(x, 0)$ and $\psi_t(x, 0)$ be chosen to ensure that $\psi(x, T)=1/(x^2+1)$ and $\psi_t(x, T)=0$? Verify your answer explicitly through d'Alembert's solution.

Hint: Recall Sections 10.3.1, 10.6, 9.2.2, and 9.5.

12.4 Gas at excess pressure λ is contained between two infinitely extended parallel walls, one at $z=0$, the other at $z=-A$. Initially, the gas is at rest everywhere. At time $t=0$ both walls disintegrate suddenly, ceasing to have any effect.

(i) Write down initial values of ψ and ψ_t that can represent these conditions.
(ii) Use Poisson's solution to determine the velocity potential, excess pressure, and gas velocity in the region $z>0$ for all $t>0$. Sketch all these as functions of time for given $z>0$.
(iii) Solve the same problem by d'Alembert's method, and check that the results agree.

12.5 For the bursting balloon (Section 12.2.2), use Poisson's solution to determine and sketch the excess pressure and the flow velocity as functions of time at a point $r<a$. The origin needs special attention: verify that your result there agrees with that of Problem 11.7.

12.6 At $t = 0$, velocity and excess pressure are everywhere zero, except that in an infinitely extended parallel-sided and very thin slab of gas the velocity normal to the slab is suddenly raised to a very high value.

(i) Before doing any calculations, guess what the velocity distribution will be at later times.
(ii) Use d'Alembert's solution to determine $\psi(z, t)$ and $-\psi_z(z, t)$, given $\psi(z, 0) = u\theta(-z)$, $\psi_t(z, 0) = 0$.
(iii) Was your guess in (i) correct? If not, why not?

12.7 The ICs in Problem 12.6(ii) imply that $v_3(z, 0) = u\,\delta(z)$, $v_{3t}(z, 0) = 0$. Since $v_3(z, t)$ satisfies the homogeneous wave equation, determine it directly (i.e. without first calculating ψ) from Poisson's solution (see Section 11.5), and verify that your results agree.

12.8 (i) For the exploding shell (Section 12.2.3), use Poisson's solution to find and sketch the velocity potential, excess pressure, and velocity, as functions of time for a point $r > a$.
(ii) When $(r - a) \ll a$, and for small enough times, one would on physical grounds expect the same results as in Problem 12.6. Is this expectation correct?

12.9 Solve Problems 12.5 and 12.8 by using the magic rule with the polar decomposition of the propagator (Appendices M and N). Is this quicker or slower than evaluating Poisson's solution?

12.10 *Quadrupole radiation.* Two equal and opposite simple-harmonic point dipoles $\pm\boldsymbol{p}\cos(\omega t)$ are situated at $\boldsymbol{r} = \boldsymbol{s}$ and $\boldsymbol{r} = 0$ respectively:

$$\rho(\boldsymbol{r}, t) = \{-\boldsymbol{p}\cdot\nabla\,\delta(\boldsymbol{r} - \boldsymbol{s}) + \boldsymbol{p}\cdot\nabla\,\delta(\boldsymbol{r})\}\cos(\omega t).$$

Write $\boldsymbol{p} = p\hat{\boldsymbol{n}}$, $\boldsymbol{s} = s\hat{\boldsymbol{m}}$, where $\hat{\boldsymbol{n}}$, $\hat{\boldsymbol{m}}$ are unit vectors, and consider the limit $p \to \infty$, $s \to 0$, with fixed $sp = Q$, $\hat{\boldsymbol{n}}$, $\hat{\boldsymbol{m}}$.

(i) Determine by dimensional arguments how the power output W depends on σ, c, and ω, assuming it is proportional to Q^2.
(ii) Calculate the time-averaged energy flux $\bar{\boldsymbol{N}}$ as $r \to \infty$.
(iii) Verify that, for given Q, $\bar{W} = \lim_{r\to\infty} \int d\Omega\, r^2 \bar{N}_r$ is maximum (minimum) when $\hat{\boldsymbol{n}}$ and $\hat{\boldsymbol{m}}$ are parallel (orthogonal), and that its maximum value is three times its minimum value.
(iv) Sketch the angular distribution of $\bar{\boldsymbol{N}}$ in these two cases.

12.11 *Self-energies.* A point source of constant strength Q moves at constant velocity $\dot{\mathbf{s}}(t) = \mathbf{u}$, where $u < c$.

(i) Write down the integral $E(u, R) \equiv \int_V \mathrm{d}V\, H(\mathbf{r}, t)$, where V is the region outside a (moving) sphere S having fixed radius R, and centred at $\mathbf{s}(t - R/c)$. (In other words, S is the locus of points reached at time t by the disturbance emitted from the source at time $t - R/c$.)
(ii) Convince yourself that E is independent of time
(iii) By an appropriate change of integration variables, show that

$$E(u, R) = \frac{\sigma Q^2}{8\pi R} \cdot \frac{(1 + u^2/3c^2)}{(1 - u^2/c^2)^2}.$$

12.12 An infinitely long line source of constant strength $\boldsymbol{A}$ per unit length stretches along the x-axis, and moves along the x-axis at constant speed $u < c$. Show that at a fixed observation point (x, y, z), the excess pressure and the x-component of the flow velocity vanish, and that the y- and z-components of the flow velocity have the same values as they would have for $u = 0$.

12.13 *Rotating point source.* A point source of constant strength Q circulates steadily round a circle:

$$\rho(\mathbf{r}, t) = Q\, \delta(\mathbf{r} - \mathbf{s}(t)),$$
$$\mathbf{s}(t) \equiv (b \cos(\omega t), b \sin(\omega t), 0),$$

where $\omega b/c \ll 1$.

(i) Determine the total radiated power $\bar{W}$ from Larmor's formula.
(ii) Approximate $Q\, \delta(\mathbf{r} - \mathbf{s}) \approx Q\{\delta(\mathbf{r}) - \mathbf{s} \cdot \nabla\, \delta(\mathbf{r})\}$; calculate ψ, Δp, $\boldsymbol{v}$, and $\boldsymbol{N}$ to leading order in s and $1/r$; thence determine the angular distribution of the radiation. Verify that $\lim_{r\to\infty} \int \mathrm{d}\Omega\, r^2 \bar{N}_r$ reproduces $\bar{W}_{\mathrm{rad}}$ found in (i).

Hint: With the approximate form of ρ you will find that in the far-field the effects of $Q\, \delta(\mathbf{r})$ can be ignored; there are effectively two dipoles oscillating at right angles to each other, and out of phase by $\pi/2$. (How would the radiation be affected if these two dipoles were to oscillate in phase?)

12.14 *Are two rotating sources better than one?* Two point sources, both of constant magnitude Q, circulate steadily round a circle at opposite ends of a diameter:

$$\rho(\mathbf{r}, t) = Q\{\delta(\mathbf{r} - \mathbf{s}(t)) + \delta(\mathbf{r} + \mathbf{s}(t))\},$$

with $s(t)$ as in the preceding problem.

(i) Before doing any calculations, would you expect the presence of the second source to increase or to decrease $\bar{W}_{rad}$?
(ii) Using the approximation

$$\delta(\boldsymbol{r} \mp \boldsymbol{s}) \approx \left\{1 \mp (\boldsymbol{s} \cdot \boldsymbol{\nabla}) + \frac{1}{2!}(\boldsymbol{s} \cdot \boldsymbol{\nabla})^2\right\} \delta(\boldsymbol{r}),$$

calculate $\bar{W}_{rad}$ to leading order in s and in $1/r$. If your guess in (i) was incorrect, explain why.

Hint: The solution of Problem 12.10 may help.

12.15 *Pulsating jet.* In the halfspace $z \geqslant 0$, the velocity potential ψ obeys the homogeneous wave equation; ψ and ψ_t vanish at $t = -\infty$; and

$$\psi_z(x, y, 0, t) = \alpha \operatorname{Re} \theta(a - |x|)\theta(a - |y|) \exp(-i\omega t).$$

(i) Taking $r \gg a$, but making no assumptions about $\omega a/c$, write down ψ, and then $\boldsymbol{N}(\boldsymbol{r}, t)$, accurately to order $1/r$.
(ii) Describe the angular distribution of the radiation when $\omega a/c \ll 1$, and when $\omega a/c \gg 1$.

12.16 *Causality in a* U-*tube.* Explain why the theorem (12.8.16) is (as expected) uninformative if the straight-line path from $\boldsymbol{r}'$ to $\boldsymbol{r}$ passes outside the region V (as for instance between two points near opposite ends of a U-tube).

12.17 Derive the theorem (12.8.16) under DBCs.

12.18 A point source of constant strength Q must be moved from rest at $\boldsymbol{r} = 0$ at time $t = 0$ to rest at $\boldsymbol{r} = (0, 0, L)$ at time $t = T$. Show that the energy radiated is at least $\sigma Q^2 L^2/\pi c^3 T^3$.

V The Helmholtz equation and diffraction

Summary

The Helmholtz equation: $-(\nabla^2 + k^2)\psi(\boldsymbol{r}) = \rho(\boldsymbol{r})$. Can arise from the wave equation

$$\left(\frac{1}{c^2}\frac{\partial^2}{\partial t^2} - \nabla^2\right)\Psi = P$$

if $P(\boldsymbol{r}, t) = \rho(\boldsymbol{r})\exp(-i\omega t)$, if all prescribed boundary values are $\propto \exp(-i\omega t)$, and if $\Psi(\boldsymbol{r}, t) = \psi(\boldsymbol{r})\exp(-i\omega t)$. Then $k = \omega/c$.

BCs for finite regions: as for Poisson's equation. Then the homogeneous problem has only discrete eigenvalues k_n^2.

On boundaries at infinity: one generally requires 'outgoing-wave' conditions: in 3D, $\psi(\boldsymbol{r} \to \infty) \sim \exp(ikr)/r$.

Green's function: in unbounded space with outgoing-wave BCs:

- $G_0^{(3)} = \exp(ikR)/4\pi R$;
- $G_0^{(2)} = \dfrac{i}{4}H_0^{(1)}(kR)$: $\begin{cases} H_0^{(1)}(z \to \infty) \sim \sqrt{2/\pi z}\exp i(z - \pi/4) \\ H_0^{(1)}(z \to 0) \sim (2i/\pi)\log z \end{cases}$
- $G_0^{(1)} = \dfrac{i}{2k}\exp(ik|X|)$, where $\boldsymbol{R} \equiv \boldsymbol{r} - \boldsymbol{r}'$, $\quad X \equiv |x - x'|$.

Kirchhoff representation in 3D: V bounded by S; no sources in V; outgoing-wave BCs on any infinitely remote parts of S.

- If $\boldsymbol{r}$ is in V:

$$\psi(\boldsymbol{r}) = \int_S \mathrm{d}S' \left\{\frac{\exp(ikR)}{4\pi R}\,\partial_n'\psi(\boldsymbol{r}') - \left(\partial_n' \frac{\exp(ikR)}{4\pi R}\right)\psi(\boldsymbol{r}')\right\}.$$

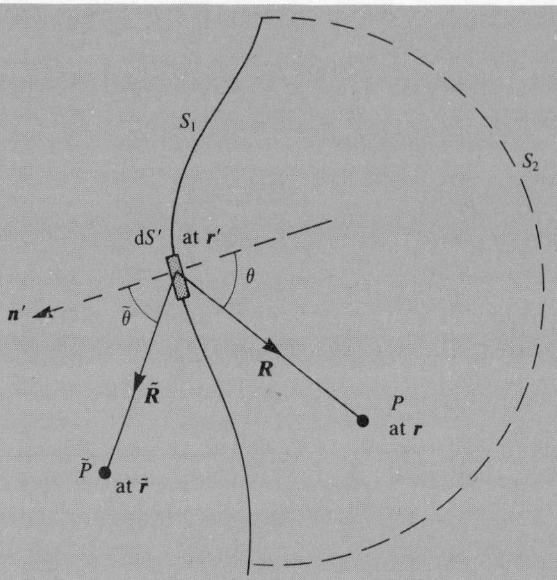

Kirchhoff's form of the Huygens–Fresnel principle:

- Wave emitted from $\tilde{P}$: $\alpha \exp(ik\tilde{R})/4\pi\tilde{R}$.
- Outgoing-Wave BCs on S_2.
- Unobstructed parts of S_1 are $\equiv S_3$.
- Approximations: $k\tilde{R}$, $kR \gg 1$ plus St Venant's hypothesis. Then

$$\psi(\boldsymbol{r}) \approx -\frac{\alpha}{(4\pi)^2} ik \int_{S_3} \mathrm{d}S' \frac{\exp(ik\tilde{R})}{\tilde{R}} \cdot \frac{\exp(ikR)}{R} (\cos\theta + \cos\tilde{\theta}).$$

- For the source replaced by an incident plane wave $A \exp(i\boldsymbol{k}\cdot\boldsymbol{r})$:

$$\psi(\boldsymbol{r}) \approx -A \frac{ik}{4\pi} \int_{S_3} \mathrm{d}S' \exp(i\boldsymbol{k}\cdot\boldsymbol{r}') \frac{\exp(ikR)}{R} (\cos\theta + \cos\tilde{\theta}).$$

13 The Helmholtz equation

The Helmholtz equation determines the $\boldsymbol{r}$-dependence of those solutions of the wave equation that are simple-harmonic in time. We have already covered its essentials for finite regions, where it governs normal modes (recall Sections 3.2 (point v), 4.5, 8.2, 10.3.2, and 10.5.2). Therefore, from Section 13.1.3 onwards, we shall consider only infinite regions, where the equation governs the emission, scattering, and diffraction of monochromatic waves. Emission has been considered in Section 12.4. Scattering is a vast subject not addressed in this book. Instead, we shall conclude with some simple examples of the diffraction of scalar waves: diffraction phenomena were crucial in the history of the wave theory of light, and they remain the best illustration of some of the more peculiar ways in which waves can propagate.

For definiteness, we shall concentrate on the Helmholtz equation as a special case just of the wave equation. Frequency and wavenumber are then connected by $\omega = ck$, and we shall use the labels ω and k interchangeably, depending simply on whether we wish to stress the underlying variation in time (ω) or the variation in space (k).

13.1 Introduction

13.1.1 Provenance

Consider the wave equation $\Box^2\Psi(\boldsymbol{r}, t) = \rho(\boldsymbol{r}, t)$ in cases where all the data are simple-harmonic (or zero): for instance, we might have

$$\rho_\omega(\boldsymbol{r}, t) = \rho_\omega(\boldsymbol{r}) \exp(-i\omega t), \tag{13.1.1}$$

and similar expressions for any prescribed boundary values. We can then look for simple-harmonic solutions in the form

$$\Psi_\omega(\boldsymbol{r}, t) = \psi_\omega(\boldsymbol{r}) \exp(-i\omega t). \tag{13.1.2}$$

On substituting (13.1.1, 2) into the wave equation, the factors $\exp(-i\omega t)$ cancel, and one obtains the Helmholtz equation

$$(\nabla^2 + \omega^2/c^2)\psi_\omega(\boldsymbol{r}) \equiv (\nabla^2 + k^2)\psi_k(\boldsymbol{r}) = -\rho_\omega(\boldsymbol{r}). \tag{13.1.3}$$

As a rule, the requisite physical quantity would be the real (or perhaps the imaginary) part of the complex function $\Psi_\omega(\boldsymbol{r}, t)$, and similarly for the source term $\rho_\omega(\boldsymbol{r}, t)$. Without loss of generality, the frequency ω in (13.1.1, 2) is taken as non-negative (except for a brief excursion at the start of Section 13.2.1, which has no bearing on any expressions elsewhere in this chapter). Under these conventions, $\psi_\omega(\boldsymbol{r})$ and $\rho_\omega(\boldsymbol{r})$ can be complex, and we shall see in Section 13.1.4 how this allows one to accommodate running as well as standing waves. The suffix ω (or k) will sometimes be dropped if this can cause no confusion.

13.1.2 Finite regions

The Helmholtz equation is elliptic, like Poisson's. It is therefore well-posed under the same conditions, and in a finite region V, bounded by a closed surface S, it is subject in the same way to the Fredholm alternative, as described in Section 3.5.6 which should be re-read at this point (see also Section 4.5, Problem 5.12, and Section 10.5.2).

The Green's function is defined as usual by

$$(\nabla^2 + k^2)G_k(\boldsymbol{r} \mid \boldsymbol{r}') = -\delta(\boldsymbol{r} - \boldsymbol{r}'), \tag{13.1.4}$$

plus the homogeneous version of the BCs that apply to the solution of (13.1.3). In terms of the eigenfunctions of $-\nabla^2$ subject to the same homogeneous BCs ($-\nabla^2\phi_p = k_p^2\phi_p$), one has

$$G_k(\boldsymbol{r} \mid \boldsymbol{r}') = \sum_p \frac{\phi_p^*(\boldsymbol{r}')\phi_p(\boldsymbol{r})}{k^2 - k_p^2}, \tag{13.1.5}$$

which fails to exist (only) in the 'special case' where $k^2 = k_p^2$. If $k^2 \neq k_p^2$, then the magic rule is derived by exactly the same steps as for Poisson's equation in Sections 5.2 or 6.3, but free now of the special complications of the Poisson–Neumann case that were troublesome in Chapter 6. In the derivation, the terms with the k^2 from the Helmholtz equations cancel at the outset (e.g. on the left of (5.2.1)); in terms of the appropriate Green's function the end-result takes identically the same form as for Poisson's equation:

$$\psi_{\mathrm{D,N}}(\boldsymbol{r}) = f_{\mathrm{D,N}}(\boldsymbol{r}) + g_{\mathrm{D,N}}(\boldsymbol{r}), \tag{13.1.6a}$$

$$f_{\mathrm{D,N}}(\boldsymbol{r}) = \int_V \mathrm{d}V'\, G_{\mathrm{D,N}}(\boldsymbol{r} \mid \boldsymbol{r}')\rho(\boldsymbol{r}'), \tag{13.1.6b}$$

$$g_{\mathrm{D}}(\boldsymbol{r}) = -\int_S \mathrm{d}S'\, \psi_S(\boldsymbol{r}')\, \partial_n' G_{\mathrm{D}}(\boldsymbol{r} \mid \boldsymbol{r}'), \tag{13.1.6c}$$

$$g_{\mathrm{N}}(\boldsymbol{r}) = \int_S \mathrm{d}S'\, (\partial_n' \psi_S(\boldsymbol{r}'))G_{\mathrm{N}}(\boldsymbol{r} \mid \boldsymbol{r}'). \tag{13.1.6d}$$

13.1.3 Unbounded regions

Infinite systems, unlike finite ones, can oscillate harmonically at any frequency (have a continuous spectrum). Hence any positive k^2 is equal to an eigenvalue of $-\nabla^2$, and, formally, all inhomogeneous problems present the special case; but this causes no serious difficulty. The physical reason is that the energy supplied by the sources escapes to infinity, and does not entail amplitudes increasing with time at any fixed point. Thus, under physically appropriate BCs at infinity, we shall find that the equation always has a solution, and that, correspondingly, the Green's function always exists.

The homogeneous equation reads

$$(\nabla^2 + k^2)\psi = 0. \tag{13.1.7}$$

In most systems, the source density $\rho(\boldsymbol{r})$ in (13.1.3) vanishes beyond some limited region, so that ψ obeys (13.1.7) asymptotically as $r \to \infty$ in a fixed direction Ω. Thus, writing

$$\nabla^2 = \nabla_r^2 + \frac{1}{r^2}\nabla_\Omega^2,$$

and setting $\psi(\boldsymbol{r}) \equiv \phi(\boldsymbol{r})/r$ in the familiar way, one finds

$$\left(\frac{\partial^2}{\partial r^2} + \frac{1}{r^2}\nabla_\Omega^2 + k^2\right)\phi(r, \Omega) \underset{(r\to\infty)}{\simeq} \left(\frac{\partial^2}{\partial r^2} + k^2\right)\phi = 0. \tag{13.1.8}$$

Therefore

$$\phi(r \to \infty, \Omega) \sim \exp(\pm ikr) \times \text{constant}, \tag{13.1.9}$$

and ψ itself must behave according to

$$\psi(r \to \infty, \Omega) \sim \frac{1}{r}\{A(\Omega)\exp(ikr) + B(\Omega)\exp(-ikr)\}, \tag{13.1.10}$$

where A, B are functions at most of Ω but not of r. Thus, when we come to impose 'boundary' or rather asymptotic conditions at infinity, this can amount only to imposing constraints on A and/or B. (We concern ourselves now only with $k^2 > 0$; but see Section 8.2 and Problem 13.5 for physical examples where $k^2 < 0$.)

Certain simple solutions of the homogeneous equation (13.1.7) recur frequently; when $k^2 \to 0$, they generally reduce to one of the harmonic (in space) functions of Section 4.3. In Cartesians one has $\psi_{\boldsymbol{K}} = \exp(i\boldsymbol{K}\cdot\boldsymbol{r})$ with $K^2 = k^2$, where $\boldsymbol{K}$ may be complex. In spherical polars, and in terms of the spherical Bessel functions j_l and n_l (see Appendix A.5), one has

$$\psi_{lm} = \{a_{lm}n_l(kr) + b_{lm}j_l(kr)\}Y_{lm}(\Omega), \tag{13.1.11}$$

with the important isotropic special case

$$\psi_{00} = \left\{a_{00}\frac{\cos(kr)}{kr} + b_{00}\frac{\sin(kr)}{kr}\right\}Y_{00}. \tag{13.1.12}$$

The functions $j_l(kr)Y_{lm}(\Omega)$ obey (13.1.7) even at $\boldsymbol{r}=0$. But the n_l are singular there, and in any region that includes $\boldsymbol{r}=0$ they are Green's functions rather than solutions of the homogeneous equation. This is readily illustrated with j_0 and n_0. For $\boldsymbol{r}\neq 0$, both are easily seen to obey $(\nabla^2+k^2)\psi=0$, because $\exp(ikr)/(kr)$ obeys it. As $r\to 0$,

$$j_0(kr)=\frac{\sin(kr)}{kr}\sim\left(1-\frac{1}{3!}k^2r^2+\cdots\right),$$

whence $(\nabla^2+k^2)j_0(kr)=0$ even at the origin. By contrast,

$$n_0(kr)=\frac{\cos(kr)}{kr}\sim(1/kr-kr/2!+\cdots),$$

whence $(\nabla^2+k^2)n_0(kr)\underset{r\to 0}{\simeq}(\nabla^2+k^2)(1/kr)$; the Laplacean dominates, so that

$$(\nabla^2+k^2)n_0(kr)\sim\nabla^2(1/kr)=-(4\pi/k)\,\delta(\boldsymbol{r}).$$

Therefore $n_0(kr)$ is a solution not of (13.1.7) but (up to a constant factor) of (13.1.4).

As $kr\to\infty$, eqns (A.5.26, 27) entail

$$\psi_{lm}(kr\to\infty)\sim(kr)^{-1}\{a_{lm}\sin[kr-\tfrac{1}{2}\pi(l+1)] + b_{lm}\cos[kr-\tfrac{1}{2}\pi(l+1)]\}\,Y_{lm}(\Omega), \tag{13.1.13}$$

which illustrates the asymptotics prescribed by (1.1.10).

13.1.4 Time-dependence by convention: boundary conditions and the intensity

The Helmholtz equation itself makes no mention of time. But if a definite time-dependent factor is adopted by convention, then one can distinguish standing from travelling waves. From now on, unless otherwise stated, we adopt the prefactor $\exp(-i\omega t)$ as in (13.1.2); then $\psi(\boldsymbol{r})=\exp(i\boldsymbol{k}\cdot\boldsymbol{r})$ implies $\Psi_\omega(\boldsymbol{r},t)=\exp(-i\omega t+i\boldsymbol{k}\cdot\boldsymbol{r})$, with $\operatorname{Re}\Psi_\omega=\cos(\omega t-\boldsymbol{k}\cdot\boldsymbol{r})$, which is a plane wave travelling in the direction $\hat{\boldsymbol{k}}$ with phase velocity $\omega/k=ck/k=c$. By contrast, $\psi(\boldsymbol{r})=\cos(\boldsymbol{k}\cdot\boldsymbol{r}+\delta)$ entails $\operatorname{Re}\Psi_\omega=\cos(\omega t)\cos(\boldsymbol{k}\cdot\boldsymbol{r}+\delta)$, which is a standing wave. A little reflection shows that running waves can be described only by complex ψ. Thus, in 3D (spherical) polars, $\sin(kr)/r$ and $\cos(kr)/r$ represent (isotropic) standing waves, while $\exp(\pm ikr)/r$ are travelling waves: $\exp(ikr)/r$ moves outwards, and is called an outgoing or diverging wave, while $\exp(-ikr)/r$ moves inwards and is called an incoming or converging wave. The preceding subsection showed that, taken separately, outgoing and incoming waves are singular at the origin; $\sin(kr)/r=(\exp(ikr)-\exp(-ikr))/2ir$ is the unique linear combination regular even at $\boldsymbol{r}=0$.

In the 3D asymptotic form (13.1.10), the appropriate combination of $\exp(\pm ikr)/r$ depends on the physics. For instance, physically realizable sources produce only outgoing waves; therefore, in problems featuring such sources, one imposes the so-called '*outgoing-wave*' or '*Sommerfeld radiation condition*', namely

$$\psi(r \to \infty, \Omega) \sim A(\Omega) \exp(ikr)/r. \tag{13.1.14}$$

In other words one requires that $B = 0$ in (13.1.10).

A corresponding condition must be imposed on the 3D Green's function: we require that solution of (13.1.4) which satisfies

$$G_\omega^{(3)}(\boldsymbol{r} \mid \boldsymbol{r}') \underset{|\boldsymbol{r}| \to \infty}{\simeq} \text{constant} \cdot \exp(ikr)/r. \tag{13.1.15}$$

Then the magic rule

$$\psi_\omega(\boldsymbol{r}) = \int \mathrm{d}V' \, G_\omega(\boldsymbol{r} \mid \boldsymbol{r}') \rho_\omega(\boldsymbol{r}') \tag{13.1.16}$$

yields a solution of (13.1.3) guaranteed to obey the BCs (13.1.14). (Notice that, in order to avoid confusion with $\omega = 0$, we do not now employ the suffix 0 on G to indicate free-space functions.)

Because strictly monochromatic motion is unrealizable in practice, one might well regard the reasoning just presented in favour of outgoing-wave BCs as merely plausible rather than compelling. Therefore Section 13.2 will indicate how the same BCs follow from far less restrictive arguments, by Fourier transformation from the underlying wave equation, subject only to the more clearly-motivated causal BCs on waves (described in Section 11.1). This approach has other advantages too: in any number of dimensions, it delivers the Helmholtz Green's function $G_\omega^{(n)}(\boldsymbol{r} \mid \boldsymbol{r}')$ simply as the Fourier transform (with respect to time) of the wave Green's function $G^{(n)}(R, \tau)$. Thus we need not determine, in advance, the 2D and 1D versions of the asymptotics (13.1.10) or (13.1.15): since the $G_\omega^{(n)}(\boldsymbol{r} \mid \boldsymbol{r}') \equiv G_\omega^{(n)}(R)$ will be obtained explicitly, their asymptotics as $r \to \infty$ (i.e. $R \to \infty$) is revealed by inspection, and the magic rule (13.1.16) automatically transfers this information to the solution of the general inhomogeneous Helmholtz equation. (Alternatively, the $G_\omega^{(n)}(R)$ are obtained in Appendix F.3 through contour integration; the method starts from the version of (13.1.5) appropriate to unbounded space according to the rules of Section 4.5.2, namely from $G_\omega^{(n)}(R) = (2\pi)^{-n} \int \mathrm{d}^n l \exp(i\boldsymbol{l} \cdot \boldsymbol{R})/(l^2 - k^2)$, and uses the outgoing-wave BC to resolve the ambiguity arising from the vanishing denominator.)

Finally, we need an expression for the *intensity* $I(\boldsymbol{r}, t)$ of a monochromatic wave field, in terms of our (generally complex) solutions of the Helmholtz equation. Here one must be especially careful with the conventions regarding time-dependence: in terms of Ψ_ω from (13.1.2), the physical solution of the wave equation is (with $\psi_{1,2}$ real),

$$\operatorname{Re}\Psi(\boldsymbol{r}, t) = \operatorname{Re}\psi(\boldsymbol{r})\exp(-i\omega t) = \operatorname{Re}\{\psi_1(\boldsymbol{r}) + i\psi_2(\boldsymbol{r})\}\exp(-i\omega t). \tag{13.1.17}$$

We consider only time-averaged quantities, and in particular the time-averages of the obvious candidates

$$H = \tfrac{1}{2}\sigma\left\{\frac{1}{c^2}(\operatorname{Re}\partial\Psi/\partial t)^2 + (\operatorname{Re}\nabla\Psi)^2\right\}, \quad \boldsymbol{N} = -\sigma(\operatorname{Re}\partial\Psi/\partial t)(\operatorname{Re}\nabla\Psi),$$

and of the squared pressure $(\Delta p)^2 = \sigma^2(\operatorname{Re}\partial\Psi/\partial t)^2$. One has

$$\bar{H} = \tfrac{1}{2}\cdot\tfrac{1}{2}\sigma\{k^2|\psi|^2 + |\nabla\psi|^2\}, \tag{13.1.18}$$

$$\bar{\boldsymbol{N}} = \tfrac{1}{2}\sigma kc\{-\psi_2\nabla\psi_1 + \psi_1\nabla\psi_2\} = \tfrac{1}{2}\sigma kc\tfrac{1}{2}\{\psi^*\nabla\psi - (\nabla\psi^*)\psi\}, \tag{13.1.19}$$

$$\overline{(\Delta p)^2} = \tfrac{1}{2}\sigma^2(kc)^2|\psi|^2. \tag{13.1.20}$$

Exercise: Verify this.

The fact that $\bar{\boldsymbol{N}}$ vanishes when ψ is essentially real reminds one that real ψ cannot describe travelling waves. We follow common practice by *defining* intensity as the mean-squared pressure $\overline{(\Delta p)^2}$, proportional to $|\psi|^2$. (The electromagnetic analogue would be the mean-squared electric or magnetic field.) For a plane wave all three candidates are proportional to $|\psi|^2$, and therefore to each other, and this proportionality persists as a reasonable approximation provided the field consists of plane waves all propagating in a narrow cone of directions (the so-called paraxial approximation). For a wider spread of directions, the most convenient definition of intensity depends on the physical nature of the detectors, and $\overline{(\Delta p)^2}$ often fits the bill.

13.1.5 Example: Radiation from a pulsating sphere

We determine the time-averaged power W radiated by the pulsating sphere of radius $\xi = a + h\sin[kct]$ already considered in Section 12.4, still with $h \ll a$, but placing no restriction now on ka. Setting $\Psi = \operatorname{Re}\{\exp(-ikct)\psi(\boldsymbol{r})\}$, we need for ψ a solution of the homogeneous Helmholtz equation that is isotropic (by symmetry), and satisfies at $r = a$ the BC

$$v_r \equiv -\partial\Psi/\partial r = \mathrm{d}\xi/\mathrm{d}t = hkc\cos(kct) = \operatorname{Re}\{hkc\exp(-ikct)\},$$

whence $-\partial\psi/\partial r = hkc$. As $r \to \infty$, we require only outgoing waves. Therefore $\psi = A\exp(ikr)/r$; the constant A is determined by the inner

BC:

$$A \exp(ika)(-ik/a + 1/a^2) = hkc,$$

whence $A = -hkca^2 \exp(-ika)/(1 - ika)$, and

$$\psi(r) = -\frac{hcka^2 \exp[ik(r-a)]}{r(1-ika)}. \tag{13.1.21}$$

$\bar{N}_r$ to order $1/r^2$ is obtained from (13.1.19) by letting the derivatives act only on the factor $\exp(ikr)$ of ψ, and we find

$$\bar{N}_r(r \to \infty) \sim \frac{1}{2}\frac{\sigma kc}{r^2}\frac{(hcka^2)^2}{|1-ika|^2}k = \frac{\sigma a^4 h^2 k^4 c^3}{2r^2[1+(ka)^2]}. \tag{13.1.22}$$

Thus

$$\bar{W} = \lim_{r\to\infty} 4\pi r^2 N_r = 2\pi\sigma a^4 h^2 k^4 c^3/[1+(ka)^2];$$

this is just the point-source ($a \to 0$) result (12.4.10) divided by $[1+(ka)^2]$, as anticipated at the end of Section 12.4. When $ka \gg 1$, we have $\bar{W} \approx 2\pi\sigma a^2 h^2 k^2 c^3$; unsurprisingly, $\bar{W}$ has become proportional to the surface area of the sphere.

Exercise: Inside this vibrating sphere, the solution must take the form $\psi = A \sin(kr)/r$. Determine A; show that it diverges when $\tan(ka) = ka$, and explain why.

13.2 The Green's functions in unbounded space: Fourier transformation of the wave equation

13.2.1 Fourier integrals and the boundary conditions

Though from the point of view of the wave equation simple-harmonic time-dependence is very special, yet *any* function of time can be represented mathematically as a Fourier integral, and by so representing Ψ and ρ in the wave equation, one obtains valuable insight into the Helmholtz equation. In particular, this method yields, without effort, the outgoing-wave BCs, and the Helmholtz Green's function, as anticipated above.

In the wave equation we now take $\Psi(\mathbf{r}, t)$ and $\rho(\mathbf{r}, t)$ as real (unlike Ψ_ω and ρ_ω in (13.1.2, 1)), and represent them as Fourier integrals:

$$\Psi(\mathbf{r}, t) = \int_{-\infty}^{\infty} d\omega \exp(-i\omega t)\, \psi_\omega(\mathbf{r}), \tag{13.2.1a}$$

$$\psi_\omega(\mathbf{r}) = \frac{1}{2\pi}\int_{-\infty}^{\infty} dt \exp(i\omega t)\, \Psi(\mathbf{r}, t), \tag{13.2.1b}$$

$$\rho(\boldsymbol{r}, t) = \int_{-\infty}^{\infty} \mathrm{d}\omega \exp(-i\omega t)\rho_\omega(\boldsymbol{r}), \tag{13.2.2a}$$

$$\rho_\omega(\boldsymbol{r}) = \frac{1}{2\pi}\int_{-\infty}^{\infty} \mathrm{d}t \exp(i\omega t)\rho(\boldsymbol{r}, t). \tag{13.2.2b}$$

Here (in contrast to Section 13.1) the frequency variable ω assumes negative as well as positive values. However, the function $\psi_{-|\omega|}$ is determined automatically by $\psi_{|\omega|}$, in virtue of the fact that (by assumption) Ψ is real. Complex-conjugating (13.2.1a) we find

$$\Psi^*(\boldsymbol{r}, t) = \int_{-\infty}^{\infty} \mathrm{d}\omega \exp(i\omega t)\psi_\omega^*(\boldsymbol{r}) = \int_{-\infty}^{\infty} \mathrm{d}\omega \exp(-i\omega t)\psi_{-\omega}^*(\boldsymbol{r}), \tag{13.2.3}$$

where the last step follows by changing the integration variable from ω to $-\omega$. Since $\Psi^* = \Psi$, the RHSs of (13.2.1a) and (13.2.3) coincide; on equating their integrands we obtain the *reality condition* for ψ, and similarly for ρ:

$$\psi_{-\omega}^*(\boldsymbol{r}) = \psi_\omega(\boldsymbol{r}), \qquad \rho_{-\omega}^*(\boldsymbol{r}) = \rho_\omega(\boldsymbol{r}). \tag{13.2.4}$$

Thus the knowledge of ψ_ω for positive ω determines ψ_ω for all ω, and thereby suffices to determine Ψ.

The reality condition allows (13.2.1a) to be re-expressed as

$$\Psi(\boldsymbol{r}, t) = \int_0^{\infty} \mathrm{d}\omega \{\exp(-i\omega t)\psi_\omega(\boldsymbol{r}) + \exp(i\omega t)\psi_\omega^*(\boldsymbol{r})\}$$

$$= \int_0^{\infty} \mathrm{d}\omega \exp(-i\omega t)\psi_\omega(\boldsymbol{r}) + \text{c.c.}, \tag{13.2.5}$$

where c.c. stands for complex conjugate.

Exercise: Verify this.

If we substitute (13.2.1a, 2a) into the wave equation $\Box^2\Psi = \rho$, take $\Box^2$ under the integral $\int \mathrm{d}\omega$, and finally equate Fourier components, we naturally recover the Helmholtz equation (13.1.3).

Exercise: Verify this in detail.

In 3D, the Helmholtz boundary conditions at infinity are obtained from (13.2.5) by substituting for $\psi_\omega(\boldsymbol{r} \to \infty)$ from (13.1.10):

$$\Psi(r \to \infty, \Omega, t) \sim \frac{1}{r}\int_0^{\infty} \mathrm{d}\omega \exp(-i\omega t)$$

$$\times \{A_\omega(\Omega)\exp(ikr) + B_\omega(\Omega)\exp(-ikr)\}$$

$$+ \text{c.c.}, \qquad (\omega = ck). \tag{13.2.6}$$

The causality condition, as explained in Section 11.1, demands that solutions Ψ of the wave equation have no incoming components at large distances and at large positive times (far in the future). But the component

$$\frac{1}{r}\int_0^\infty \mathrm{d}\omega\, B_\omega(\Omega)\exp\{-ik(ct+r)\}$$

is precisely such an incoming wave group; therefore, from the causality condition on waves, we deduce, for solutions of the Helmholtz equation, precisely the outgoing-waves-only BC $B=0$, as already anticipated in Section 13.1.4.

Exercise: Devise analogous arguments for 2D and 1D.

13.2.2 The Helmholtz Green's functions in *n*D

The quickest way to the Helmholtz Green's functions in any number of dimensions bypasses the argument just given (albeit only in 3D) for the outgoing-wave BCs, and proceeds directly from the Fourier integrals (13.2.1, 2) and from the magic rules. On the one hand, we have the magic rule for the wave equation

$$\Psi(\boldsymbol{r},t)=\int_{-\infty}^{\infty}\mathrm{d}t'\int \mathrm{d}V'\, G(R,\tau)\rho(\boldsymbol{r}',t'). \tag{13.2.7}$$

(Recall $\boldsymbol{R}\equiv\boldsymbol{r}-\boldsymbol{r}'$, $\tau\equiv t-t'$, and that we are continuing to omit the suffix 0 identifying unbounded space.) On the other hand, appealing successively to (13.2.1a), to the Helmholtz magic rule (13.1.6b), and to the inverse Fourier transform (13.2.2b), we have

$$\begin{aligned}\Psi(\boldsymbol{r},t)&=\int_{-\infty}^{\infty}\mathrm{d}\omega\exp(-i\omega t)\psi_\omega(\boldsymbol{r})\\ &=\int_{-\infty}^{\infty}\mathrm{d}\omega\exp(-i\omega t)\int\mathrm{d}V'\,G_\omega(\boldsymbol{r}\mid\boldsymbol{r}')\rho_\omega(\boldsymbol{r}')\\ &=\int_{-\infty}^{\infty}\mathrm{d}\omega\exp(-i\omega t)\int\mathrm{d}V'\,G_\omega(\boldsymbol{r}\mid\boldsymbol{r}')\frac{1}{2\pi}\int_{-\infty}^{\infty}\mathrm{d}t'\exp(i\omega t')\rho(\boldsymbol{r}',t')\\ &=\int_{-\infty}^{\infty}\mathrm{d}t'\int\mathrm{d}V'\left\{\frac{1}{2\pi}\int_{-\infty}^{\infty}\mathrm{d}\omega\exp(-i\omega(t-t'))G_\omega(\boldsymbol{r}\mid\boldsymbol{r}')\right\}\rho(\boldsymbol{r}',t').\end{aligned} \tag{13.2.8}$$

Comparison with (13.2.7) shows that the expression in curly brackets is just $G(R, \tau)$:

$$G(R, \tau) = \frac{1}{2\pi}\int_{-\infty}^{\infty} d\omega \exp(-i\omega(t-t'))G_\omega(\mathbf{r} \mid \mathbf{r}')$$
$$= \frac{1}{2\pi}\int_{-\infty}^{\infty} d\omega \exp(-i\omega\tau)G_\omega(\mathbf{r} \mid \mathbf{r}'). \tag{13.2.9}$$

Fourier inversion yields the end-result

$$G_\omega(\mathbf{r} \mid \mathbf{r}') = \int_{-\infty}^{\infty} d\tau \exp(i\omega\tau)G(R, \tau)$$
$$\equiv G_\omega(R) \equiv G_k(R). \tag{13.2.10}$$

The $G^{(n)}(R, \tau)$ from Section 11.2.2 now deliver the $G_k^{(n)}(R)$ simply by integration.

In 3D, the calculation is elementary:

$$G_k^{(3)}(R) = \int_{-\infty}^{\infty} d\tau \exp(ikc\tau)\frac{\delta(\tau - R/c)}{4\pi R} = \frac{\exp(ikR)}{4\pi R}. \tag{13.2.11}$$

In 2D,

$$G_k^{(2)}(R) = \int_{-\infty}^{\infty} d\tau \exp(ikc\tau)\frac{\theta(\tau)\theta(\tau - R/c)}{2\pi[\tau^2 - R^2/c^2]^{\frac{1}{2}}}.$$

Let $\tau \equiv Rx/c$:

$$G_k^{(2)}(R) = \frac{1}{2\pi}\int_1^{\infty} dx \frac{\exp(ixkR)}{(x^2-1)^{\frac{1}{2}}} = \frac{i}{4}H_0^{(1)}(kR). \tag{13.2.12}$$

The x-integral is one of the standard representations of the Bessel function $H_0^{(1)}$, which is itself a standard shorthand: $H_0^{(1)}(z) = J_0(z) + iY_0(z)$. Many applications need only the asymptotics

$$H_0^{(1)}(z \to 0) \sim (2i/\pi)\log z, \tag{13.2.13a}$$
$$H_0^{(1)}(z \to \infty) \sim (2/\pi z)^{\frac{1}{2}} \exp[i(z - \pi/4)]. \tag{13.2.13b}$$

In 1D, the integral (13.2.10) with $G_0^{(1)} = \frac{1}{2}\theta(\tau)\theta(\tau - |X|/c)$ fails to converge at its upper limit. We interpret it by introducing a cutoff factor $\exp(-\eta\tau)$ under the integral, and then taking the limit $\eta \to 0+$ at the end of the calculation; under this limit $\exp(-\eta\tau) \to 1$ for any fixed τ.

(The same device is used in Appendix O.) Thus

$$G_k^{(1)}(X) = \lim_{\eta\to 0}\frac{1}{2}c\int_{|X|/c}^{\infty} d\tau \exp(ikc\tau - \eta\tau)$$

$$= \lim_{\eta\to 0}\frac{1}{2}c\frac{\exp(ik|X| - \eta|X|/c)}{(\eta - ikc)} = \frac{i}{2k}\exp(ik|X|). \quad (13.2.14)$$

Inspection reveals that, as expected, the $G_k^{(n)}$ obey outgoing-wave BCs.

Exercise: Verify by substitution that $G_k^{(3)}$ and $G_k^{(1)}$ from (13.2.11, 14) obey the defining differential equation (13.1.3).

In the limit $k\to 0$ at fixed R, $G_k^{(3)}$ evidently reduces to the Poisson Green's function $\frac{1}{4}\pi R$. Equations (13.2.12, 13a) show that $G_k^{(2)}$ also reduces to the 2D Poisson Green's function, up to a constant addend.

Exercise: Elucidate the relation between the 1D Poisson Green's function (4.4.9c) and the behaviour of $G_k^{(1)}$ as $k\to 0$.

For a half-space, the *method of images* yields the Green's function exactly as before. Under DBCs and NBCs respectively, one has in 3D (see Figs 5.1, 12.5)

$$G_{D,N}(\mathbf{r}\,|\,\mathbf{r}') = \exp(ikR)/4\pi R \mp \exp(ik\tilde{R})/4\pi\tilde{R}. \quad (13.2.15)$$

Exercise: For a monochromatic point source near a reflecting plane, re-derive (12.8.5–9) directly from the Helmholtz instead of the wave equation.

13.2.3 Example: The form factor of a radiation source

We determine the far-field and the time-averaged power for a monochromatic but otherwise arbitrary localized source distribution $\rho(\mathbf{r})$ in 3D. The results are used to compare the power radiated by a point source $\rho_1 \equiv Q\,\delta(\mathbf{r})$ and by the hollow-shell distribution $\rho_2 = Q\,\delta(r - a)/4\pi a^2$ with the same total strength $\int dV\rho = Q$.

One proceeds by showing that

$$\psi(r\to\infty, \Omega) \sim \frac{\exp(ikr)}{4\pi r}F(\Omega), \quad (13.2.16)$$

where $F(\Omega)$ is called the *form factor* of the source. A unit point-source, $\rho = \delta(\mathbf{r})$, gives the isotropic field $\psi = \exp(ikr)/4\pi r$, and (at all r)

$\bar{W} = \sigma k^2 c/8\pi$ (see (12.4.1, 9)); hence the power it radiates into solid angle $\delta\Omega$ is $\bar{W}\,\delta\Omega/4\pi = \sigma k^2 c\,\delta\Omega/32\pi^2$. By comparison, the power per unit solid angle radiated according to (13.2.16), and the total power, are

$$\frac{\mathrm{d}W}{\mathrm{d}\Omega} = \frac{\sigma k^2 c}{32\pi^2}|F(\Omega)|^2, \qquad W = \int \frac{\mathrm{d}W}{\mathrm{d}\Omega}\,\mathrm{d}\Omega. \tag{13.2.17}$$

It remains to establish (3.2.16). The magic rule gives

$$\psi(\boldsymbol{r}) = \int \mathrm{d}V'\,\rho(\boldsymbol{r}')\exp(ikR)/4\pi R, \tag{13.2.18}$$

which we approximate to $O(1/r)$. Thus (see (12.8.6))

$$R \equiv |\boldsymbol{r}-\boldsymbol{r}'| = [r^2 + r'^2 - 2r\hat{\boldsymbol{r}}\cdot\boldsymbol{r}']^{\frac{1}{2}} \sim r - \hat{\boldsymbol{r}}\cdot\boldsymbol{r}' + O(r'^2/r), \tag{13.2.19a}$$

$$1/R \sim 1/r + O(1/r^2), \tag{13.2.19b}$$

$$\exp(ikR) \sim \exp(ikr)\exp(-ik\hat{\boldsymbol{r}}\cdot\boldsymbol{r}') + O(kr'^2/r). \tag{13.2.19c}$$

Substitution into (13.2.18) and comparison with (13.2.16) yield

$$\psi(r,\Omega) \sim \int \mathrm{d}V'\,\rho(\boldsymbol{r}')\frac{\exp(ikr)\exp(-ik\hat{\boldsymbol{r}}\cdot\boldsymbol{r}')}{4\pi r}$$
$$= \frac{\exp(ikr)}{4\pi r}\int \mathrm{d}V'\,\rho(\boldsymbol{r}')\exp(-ik\hat{\boldsymbol{r}}\cdot\boldsymbol{r}'),$$

$$F(\Omega) = \int \mathrm{d}V'\,\rho(\boldsymbol{r}')\exp(-ik\hat{\boldsymbol{r}}\cdot\boldsymbol{r}'). \tag{13.2.20}$$

Thus $(2\pi)^{-3}F(\Omega)$ is just the 3D Fourier transform of the source distribution, corresponding to $\boldsymbol{k} = k\hat{\boldsymbol{r}}$, namely to a wavenumber of magnitude k and directed towards the observation point $\boldsymbol{r}$. This approach is complementary to a multipole expansion of ρ, and is usually labour-saving when the wavelength is comparable to or smaller than the dimensions of the source.

For the hollow shell ρ_2, we evaluate $F(\Omega)$ by choosing the polar axis along $\hat{\boldsymbol{r}}$; then the exponent $-ik\hat{\boldsymbol{r}}\cdot\boldsymbol{r}' = -ikr'\cos\theta'$, and

$$F(\Omega) = \int_0^\infty \mathrm{d}r'\,r'^2 \int \mathrm{d}\Omega'\,\frac{Q\,\delta(r'-a)}{4\pi a^2}\exp(-ikr'\cos\theta')$$
$$= \frac{Q}{4\pi}\int_0^{2\pi}\mathrm{d}\phi'\int_{-1}^{1}\mathrm{d}\cos\theta'\exp(-ika\cos\theta') = Q\frac{\sin(ka)}{ka}.$$

Finally, from (13.2.17),

$$\bar{W} = \frac{\sigma k^2 c}{32\pi^2}\int \mathrm{d}\Omega\left(\frac{Q\sin(ka)}{ka}\right)^2 = \frac{\sigma k^2 c}{8\pi}Q^2\left(\frac{\sin(ka)}{ka}\right)^2. \tag{13.2.21}$$

(In this special (because isotropic) case, the integral F is of the standard type considered in Appendix F.1. A multipole expansion would have been equally convenient, because only the monopole ($l = 0$) is present.)

As $ka \to \infty$, $\bar{W}$ vanishes. Surprisingly, it vanishes also when $ka = n\pi$, $n = 1, 2, 3, \ldots$, i.e. at $2a = 2n\pi/k = n\lambda$, when the diameter equals a whole number of wavelengths.

Exercise: Explain why this conclusion does not conflict with the results for the pulsating sphere in Section 13.1.5, which exhibit no such zeros.

13.3 The Huygens–Fresnel model of diffraction

(i) *Diffraction* is the word applied, loosely, to the propagation of waves almost but not quite along straight lines, i.e. almost but not quite according to the rules of geometrical optics (which include Fermat's principle, and the laws of reflection and refraction). Our view of this field is still conditioned by the observations that were important in developing the wave theory of light, not because of historical piety, but because they happen to be quite startling in themselves, gratifyingly inaccessible to untutored (mathematically undeveloped) intuition, and because their practical relevance remains undiminished. Section 13.5 tackles some such problems through the Helmholtz equation; but first we set the stage by describing, briefly, Fresnel's elaboration of Huygens' principle. This amounts to a quite specific mathematical *model* for describing the phenomena; the enormous success of the model then challenges the theorist to anchor it to first principles, which is the task attempted by Kirchhoff.

We consider only scalar waves, thus necessarily ignoring any role played by polarization; and we remain in 3D. Excellent detailed accounts are given in several books on optics, e.g. those by Ditchburn (1952); Hecht and Zajac (1974); Born and Wolf (1975). For Huygens' principle in general see Baker and Copson (1950).

(ii) *Huygens' principle* asserts the following.† Every point on a wavefront emits so-called secondary waves. At any later time, the wavefront is the envelope of the secondary waves in the forward direction. This accounts for geometrical optics. Note that the principle refers to a time-dependent process: the wavefront is envisaged as a one-off disturbance moving through space, and each successive wavefront takes care of itself, so to speak.

† While Fresnel's and Kirchhoff's work is easily understood by to-day's reader, Huygens' is not, and the present writer has not studied it seriously. The following account derives from secondary sources, scrutinized for sense and consistency, but not for historical accuracy.

This formulation leaves open many obvious questions, to some of which later theories supply answers. (a) Why are the secondary waves ineffective except where they touch the envelope? (It turns out that they do in fact have effects elsewhere, and it is precisely these effects that are responsible for the diffractive corrections to geometrical optics, when the corrections are weak. When they are not weak, this whole approach becomes inapplicable.) (b) Why are no secondary waves emitted backward? (There are, except in the exactly reverse direction.) (c) What determines the intensities and phases of the secondary waves?

(iii) *Fresnel's construction* is an elaboration (over a hundred years later) of Huygens' principle, adapted to monochromatic waves, and for our purposes time-independent; it aims to describe the diffractive corrections to the geometrical-optics limiting case which was Huygens' chief concern. We describe it briefly.

Suppose the region V containing the observation (field) point P is separated from all sources by a surface S_1, as in Fig. 13.1. A typical element of S_1 is shown as dS', situated at $\boldsymbol{r}'$. The broken line is the surface normal; by our usual convention the vector $d\boldsymbol{S}' = \boldsymbol{n}'\, dS'$ points along the outward unit normal $\boldsymbol{n}'$, i.e. to the left. (Fresnel's model considers only the (open) surface S_1. In the next section, Kirchhoff's reasoning requires the closed boundary S of V, which we write as $S = S_1 + S_2$, completing S_1, for definiteness, by the appropriate portion

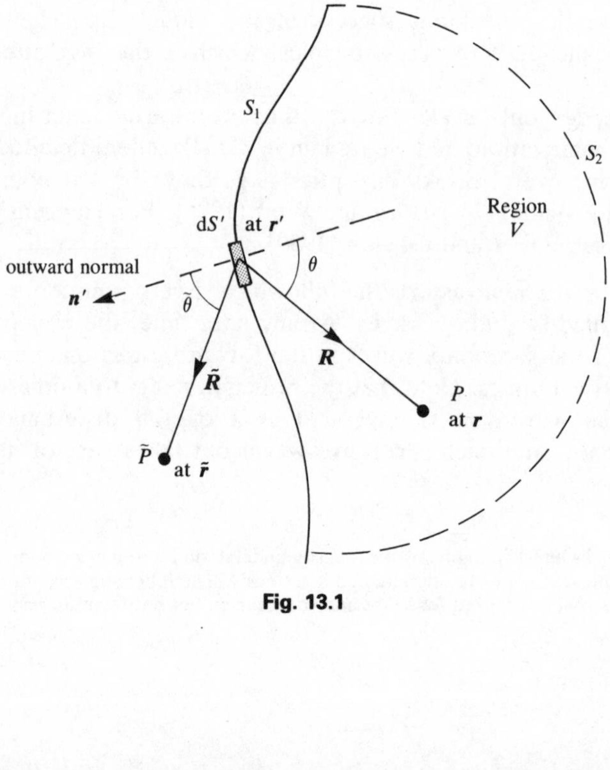

Fig. 13.1

S_2 of a sphere at infinity to the right. For instance, if S_1 is flat, then S_2 is a hemisphere. But S_2 plays no part in the present section.)

For simplicity we consider only two types of source. The first is a single point source $\tilde{P}$, of strength α, at $\tilde{\boldsymbol{r}}$. (In this section and the next, the tilde refers to real sources, and not to images; and we define $\boldsymbol{R} = \boldsymbol{r} - \boldsymbol{r}'$, $\tilde{\boldsymbol{R}} = \tilde{\boldsymbol{r}} - \boldsymbol{r}'$.) Then the signal incident on S_1 is

$$\psi_0(\boldsymbol{r}') = \frac{\alpha}{4\pi\tilde{R}} \exp(ik\tilde{R}). \tag{13.3.1a}$$

The second type of source produces an incident plane wave

$$\psi_0(\boldsymbol{r}') = A \exp(i\boldsymbol{k} \cdot \boldsymbol{r}'); \tag{13.3.1b}$$

this may be regarded as a limiting case of (13.3.1a), where $\tilde{R} \to \infty$ with $\alpha/4\pi\tilde{R} \equiv A$ fixed. What matters is that everywhere on S_1 we can identify a unique incident direction, and assign a definite value to the angle $\tilde{\theta}$. For the plane wave, $\tilde{\theta}$ is the angle between $d\boldsymbol{S}$ and $-\boldsymbol{k}$. Notice that θ and $\tilde{\theta}$ are measured from opposite directions of the surface normal: straight-through transmission from $\tilde{P}$ to P along a normal would correspond to $\theta = 0 = \tilde{\theta}$.

Fresnel explicates Huygens' idea by *assuming* that the secondary wave at P, due to the secondary source on dS', is

$$d\psi(P) = dS'\, \beta\psi_0(\boldsymbol{r}') f(\tilde{\theta}, \theta) \frac{\exp(ikR)}{R}. \tag{13.3.2}$$

Here ψ_0 is the incident wave at $\boldsymbol{r}'$; β is a proportionality constant, to be determined later from a self-consistency condition. The function $f(\tilde{\theta}, \theta)$ is called the *inclination factor*; its absolute normalization is absorbed into β, and we choose $f(0, 0) = 1$. Fresnel assumed that f diminishes gradually with increasing θ and/or $\tilde{\theta}$. More specifically he assumed that $f(0, \pi/2) = 0$, but in fact it is $f(0, \pi)$ that vanishes. The factor $\exp(ikR)$ accommodates the phase change over the optical path from $\boldsymbol{r}'$ to P (just as $\exp(ik\tilde{R})$ in (13.3.1a) accommodates the phase change from the source to $\boldsymbol{r}'$); and the factor $1/R$ accommodates the inverse square law for intensities. The total signal received at P is

$$\psi(P) = \int d\psi(P) = \beta \int_{S_1} dS'\, \psi_0(\boldsymbol{r}') f \exp(ikR)/R. \tag{13.3.3}$$

As it stands, (13.3.3) applies only to an unobstructed surface. This is not yet very interesting, because, if the model makes any sense at all, the result must be simply $\psi(P) = \psi_0(\boldsymbol{r})$; for instance, the point source producing (13.3.1a) at $\boldsymbol{r}'$ must also produce

$$\psi(P) = \alpha \exp(ik\,|\boldsymbol{r} - \tilde{\boldsymbol{r}}|)/4\pi\,|\boldsymbol{r} - \tilde{\boldsymbol{r}}|$$

at $\boldsymbol{r}$. To make (13.3.3) useful it must be adapted to partially obstructed surfaces. Here one adopts *St Venant's hypothesis* and its obvious generalizations. Suppose parts of S_1, say S_4, are totally opaque, i.e. 'blacked out' either by truly black (absorbing) or by perfectly reflecting screens, while the other parts $S_3 = S_1 - S_4$ are clear (unobstructed). Then, by St Venant's hypothesis one replaces ψ_0 in (13.3.3) by zero on the opaque parts S_4, and leaves it unchanged (equal to the incident signal) on the clear parts S_3. (If parts of S_1 transmit with reduced amplitude or with a change in phase, one makes the obvious changes. Thus a change of phase $\phi(\boldsymbol{r}', \tilde{\theta}, \theta)$ at $\boldsymbol{r}'$ simply inserts a factor $\exp(i\phi)$ into $d\psi$.) For instance, for the point source shown in Fig. 13.1, the Huygens–Fresnel model asserts

$$\psi(P) = \frac{\alpha\beta}{4\pi}\int_{S_3} dS' \frac{\exp(ik\tilde{R})}{\tilde{R}} f(\tilde{\theta}, \theta)\frac{\exp(ikR)}{R}. \tag{13.3.4}$$

Once β and f are specified, this constitutes an explicit prediction waiting merely to be evaluated.

Most of the assumptions of the Fresnel model are *ad hoc*. It is all the more remarkable that such integrals can account, quantitatively, for most of optical diffraction and interference in the near-geometric short-wave regime, where the wavelength $\lambda = 2\pi/k$ is well below the least linear dimension of the apertures and obstacles, and also well below R, $\tilde{R}$, so that

$$kR \gg 1, \qquad k\tilde{R} \gg 1. \tag{13.3.5}$$

(In quantum mechanics, this short-wave regime is called the WKB or semi-classical regime: the Schroedinger waves then propagate approximately along the rays prescribed by geometrical optics, which in turn correspond to the particle trajectories prescribed by classical mechanics.) The empirical evidence that lends force to assertion (13.3.4) is discussed in texts on optics. Some such examples will be given in Section 13.5, after the status of the Fresnel model has been clarified *vis-à-vis* the Helmholtz equation. Often the variation of $1/R$ and $1/\tilde{R}$ under the integral is unimportant, and the variation of f is either unimportant as well, or else only its gradual decline is relevant. What matters crucially is the variation of the phase factor $\exp(ik(R + \tilde{R}))$; it is the different approximations to this phase factor that distinguish between the so-called Fraunhofer and Fresnel regimes which will be discussed later.

(iv) The proportionality constant β is determined by the self-consistency condition that an unobstructed plane wave $A\exp(ikz)$ continue to propagate as such, i.e. with constant amplitude and appropriately increasing phase. If in (13.3.3) one sets $\psi_0(\boldsymbol{r}') = A\exp(ikz')|_{z'=0} = A$,

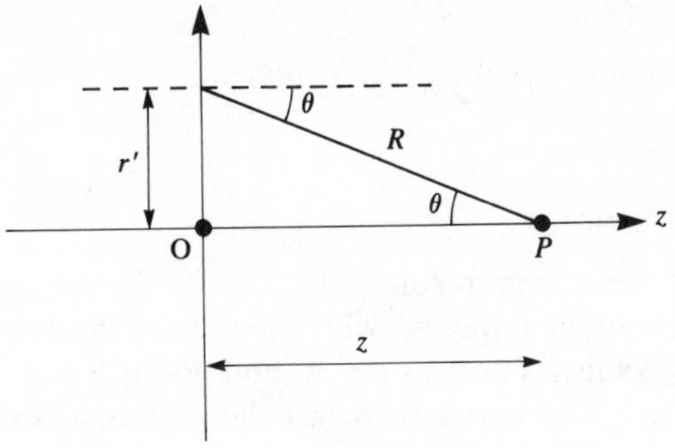

Fig. 13.2

and if one extends the integral over the entire xy-plane, then it should reproduce $\psi(P) = \psi(z) = A \exp(ikz)$.

To implement this condition, we choose cylindrical-polar coordinates as shown in Fig. 13.2. Then $R = (z^2 + r'^2)^{\frac{1}{2}}$, so that

$$\int dS' \cdots = \int_0^{2\pi} d\phi' \int_0^{\infty} dr'\, r' \cdots = 2\pi \int_0^{\infty} dr'\, r' \cdots = 2\pi \int_z^{\infty} dR\, R \cdots,$$

where the second step follows from the azimuthal symmetry and the third on changing the integration variable from r' to R. One has $\tilde{\theta} = 0$, and the inclination factor f becomes a function only of θ, where $\cos\theta = z/R$. With z fixed, θ is zero at $r' = 0$ (when $R = z$), and $\theta = \pi/2$ as r', $R \to \infty$. Then (13.3.3) becomes

$$\psi(z) = \beta 2\pi A \int_z^{\infty} dR\, R \frac{\exp(ikR)}{R} f = \beta 2\pi A \int_z^{\infty} dR \exp(ikR) f. \quad (13.3.6)$$

Here one meets a snag, because with $f \to$ constant the integral fails to converge at its upper limit (The traditional argument through Fresnel zones merely obscures the difficulty without solving it. Of course this does not detract from the merits of the zone construction in other problems.) If, with Fresnel, we assumed that f vanishes at $\theta = \pi/2$, i.e. as $R \to \infty$, then without any significant loss of generality we could choose $f = \exp(-\eta R)$, and take the limit $\eta \to 0$ at the end of the calculation. The integral would become

$$\lim_{\eta \to 0} \int_z^{\infty} dR \exp(ikR - \eta R) = \lim_{\eta \to 0} -\frac{\exp(ikz - \eta z)}{ik - \eta} = \frac{i}{k} \exp(ikz); \quad (13.3.7)$$

substituting back into (13.3.6) and setting $\psi(z) = A \exp(ikz)$ on the left, one finds

$$A \exp(ikz) = \beta(2\pi Ai/k) \exp(ikz).$$

Therefore

$$\beta = -ik/2\pi = -i/\lambda, \tag{13.3.8}$$

which is in fact the correct result.

However, Kirchhoff's theory will show that in this case $f = \frac{1}{2}(1 + \cos\theta)$, so that f vanishes only at $\theta = \pi$, and not at $\theta = \pi/2$ as assumed by Fresnel. Therefore it is better to recast the argument so that it becomes less dependent on the convergence of the integral at its upper limit. To this end we focus on $\partial\psi/\partial z$ instead of ψ, and demand only that $\partial\psi/\partial z$ tend to $(\partial/\partial z)A \exp(ikz) = ikA \exp(ikz)$ in the limit $kz \to \infty$ (which brings us under the umbrella of the underlying assumption $kR \gg 1$).

The integral (13.3.6) depends on z in two ways: through its lower limit, and through f, which we think of as a function of $\cos\theta = z/R$. Then

$$\partial f(\cos\theta)/\partial z = f'(\cos\theta)\, \partial \cos\theta/\partial z = f'/R,$$

where f', like f, is by assumption a gradually-varying function. Differentiating with respect to z we require

$$ikA \exp(ikz) = \lim_{kz\to\infty} \frac{\partial\psi}{\partial z}$$

$$= \lim_{kz\to\infty} \beta 2\pi A\left\{-\exp(ikz) f(\theta = 0) + \int_z^\infty dR \exp(ikR)\frac{f'}{R}\right\}. \tag{13.3.9}$$

On changing the integration variable to $KR \equiv \xi$, the remaining integral becomes $\int_{kz}^{\infty} d\xi \exp(i\xi) f'/\xi$. This converges at its upper limit even if $f' \to$ constant, and therefore vanishes unambiguously as $kz \to \infty$. Thus only the first term survives on the right of (13.3.9), correctly reproducing Fresnel's result (13.3.8) for β.

Remarkably, the factor $-i = \exp(-2\pi i/4)$ of β implies that, on their emission, the phase of the secondary waves $d\psi$, eqn (13.3.2), is ahead of the incident wave ψ_0 by one quarter-period. The self-consistency argument shows that this must be so, but has not explained why it is so.

13.4 Kirchhoff's diffraction theory

The Huygens–Fresnel formulae (13.3.3, 4) are enormously successful in describing and predicting the phenomena, but at this stage they are not rooted in any coherent theory, and therefore one cannot tell what their success really signifies. The theorist's task is to derive the formulae from the wave equation, which for monochromatic waves means the Helmholtz equation. The basic tool is the Kirchhoff representation; here we shall need it only for solutions of the homogeneous equation (source-free in our chosen region V). It is derived by precisely the same steps as it was for Poisson's equation (Section 7.5), and reads

$$\psi(\boldsymbol{r}) = \int_S \mathrm{d}S' \, \{G_0(R)\, \partial'_n \psi_S(\boldsymbol{r}') - (\partial'_n G_0(R))\psi_S(\boldsymbol{r}')\}, \qquad (\boldsymbol{r} \text{ in } V), \tag{13.4.1}$$

where $G_0(R) = \exp(ikR)/4\pi R$, S is the closed surface surrounding V, and ψ_S, $\partial_n\psi_S$ are the exact (but initially unknown) solution and its normal derivative on S.

Recall that if $\boldsymbol{r}$ lies outside V, then the RHS vanishes; it does not, then, represent the solution at all. In particular, the RHS is discontinuous as $\boldsymbol{r}$ crosses S; $\psi_S(\boldsymbol{r}')$ and $\partial'_n\psi_S(\boldsymbol{r}')$ in (13.4.1) are the limiting values as $\boldsymbol{r}'$ tends to S from within V.

Since $\boldsymbol{R} = \boldsymbol{r} - \boldsymbol{r}'$, we can write

$$\begin{aligned}\boldsymbol{\nabla} G_0(R) &= -\boldsymbol{\nabla}' G_0(R) = (\boldsymbol{\nabla} R)\frac{\mathrm{d}G_0}{\mathrm{d}R} = \hat{\boldsymbol{R}}\frac{\mathrm{d}G_0}{\mathrm{d}R} \\ &= \hat{\boldsymbol{R}}\left(\frac{ik}{4\pi R} - \frac{1}{4\pi R^2}\right)\exp(ikR) = \hat{\boldsymbol{R}}\frac{ik}{4\pi R}\exp(ikR)(1 - 1/ikR).\end{aligned} \tag{13.4.2}$$

For short wavelengths, when $kR \gg 1$, the second term is only a small correction (one recognizes the far-field approximation familiar from Chapters 11 and 12).

For definiteness, we envisage the situation shown in Fig. 13.1, namely a single point-source at $\tilde{P}$, giving $\psi_0(\boldsymbol{r}') = \alpha \exp(ik\tilde{R})/4\pi\tilde{R}$. According to Fresnel, this yields (13.3.4). (Afterwards it will be easy to make the changes appropriate to an incident plane wave.) That Fresnel's formula should be traceable to the Kirchhoff representation (13.4.1) is surprising, because the latter features both ψ_S and $\partial_n\psi_S$, while the former features only the incident wave but not its gradient. As the argument proceeds, one should keep track, or one should review afterwards, at what stage the various approximations are introduced, i.e. which are needed for each successive conclusion.

We start with the Kirchhoff representation applied over the surface $S = S_1 + S_2$ shown in Fig. 13.1. At this stage, ψ is the true but unknown solution of the Helmholtz equation, elicited by the given source, and subject to whatever boundary conditions are enforced by the screens. The surface S_1 is chosen so that the screens form part of S_1.

One important fact is that $\int_{S_2} \mathrm{d}S' \cdots$ vanishes as the radius, say R_2, of the sphere S_2 tends to infinity (the origin being chosen at any convenient point near P). We can make this plausible as follows (a proof is given by

Stratton, 1941, Section 8.13). In an obvious sense, any signal reaching S_2 has come from the source at $\tilde{P}$ through S_1. Hence, to leading order, $\psi(\text{on } S_2) \sim \text{constant} \cdot \exp(ikR_2)/R_2$, where the constant depends on the pattern of obstructions on S_1. Ultimately, such behaviour is ensured by the outgoing-wave BCs built into G_0. Because G_0, $\partial_n G_0$, ψ_S, and $\partial_n \psi_S$ all vanish like $1/R_2$, the integrand of (13.4.1) evaluated on S_2 vanishes like $1/R_2^2$. The snag is that this is not enough to make $\int_{S_2} dS' \cdots$ vanish, because $dS' = d\Omega' R_2^2$, or in other words because the area of S_2 increases like R_2^2; hence the two integrals in (13.4.1) taken separately do not decrease at all as $R_2 \to \infty$. The day is saved by the fact that each integrand behaves like $\text{constant} \cdot \exp(2ikR_2)/R_2^2$, with the same constant in both, so that from the difference between them the terms of order $1/R_2^2$ cancel. This leaves terms of order $1/R_2^3$ at most, which do vanish in the limit even when combined with $dS' = d\Omega\, R_2^2$. The conclusion is that, in (13.4.1), we can and now do replace $\int_S$ by $\int_{S_1}$. No approximations have been made so far. Under outgoing-wave BCs the absence of any contribution from S_2 is physically almost obvious: no influences on $\psi(\boldsymbol{r})$ are reflected from infinitely far downstream when there is nothing there.

The next step is the crucial one. In the short-wavelength regime (described just below (13.3.4)) it might seem physically reasonable to approximate Kirchhoff's integrand in (13.4.1) by adopting St Venant's hypothesis on S_1, as spelled out in Section 13.3. Accordingly, on the unobstructed parts S_3 of S_1, we set

$$\psi_S(\boldsymbol{r}') \approx \psi_0(\boldsymbol{r}') = \frac{\alpha}{4\pi\tilde{R}} \exp(ik\tilde{R}), \tag{13.4.3a}$$

$$\boldsymbol{\nabla}' \psi_S(\boldsymbol{r}') \approx \boldsymbol{\nabla}' \psi_0(\boldsymbol{r}') = -\hat{\tilde{\boldsymbol{R}}} \frac{ik}{4\pi\tilde{R}} \exp(ik\tilde{R})(1 - 1/ik\tilde{R}). \tag{13.4.3b}$$

On the (inward-facing side of the) opaque parts S_4 we set both ψ_S and $\partial_n \psi_S$ equal to zero. Thus the Kirchhoff representation plus the St Venant's hypothesis yield

$$\psi(\boldsymbol{r}) \approx \frac{\alpha}{(4\pi)^2} \int_{S_3} dS' \left[\frac{\exp(ikR)}{R} \left(\partial_n' \frac{\exp(ik\tilde{R})}{\tilde{R}} \right) - \left(\partial_n' \frac{\exp(ikR)}{R} \right) \frac{\exp(ik\tilde{R})}{\tilde{R}} \right]. \tag{13.4.4}$$

Finally, for short wavelengths the gradients $\boldsymbol{\nabla} G_0$ and $\boldsymbol{\nabla} \psi_0$ are approximated by dropping $1/ikR$, $1/ik\tilde{R}$ relative to unity. This turns $[\cdots]$

in (13.4.4) into

$$[\cdots] = ik\boldsymbol{n}' \cdot (-\hat{\tilde{\boldsymbol{R}}} + \hat{\boldsymbol{R}}) \frac{\exp(ik\tilde{R})}{\tilde{R}} \frac{\exp(ikR)}{R}$$

$$= -ik(\cos\tilde{\theta} + \cos\theta) \frac{\exp(ik\tilde{R})}{\tilde{R}} \frac{\exp(ikR)}{R}, \qquad (13.4.5)$$

where some care is necessary with the signs (see Fig. 13.1 for $\boldsymbol{R}$, $\tilde{\boldsymbol{R}}$, $\boldsymbol{n}'$, θ, $\tilde{\theta}$). Thus one obtains *Kirchhoff's diffraction formula* for a point source:

$$\psi(\boldsymbol{r}) \approx -\frac{\alpha}{(4\pi)^2} 2ik \int_{S_3} dS' \frac{\exp(ik\tilde{R})}{\tilde{R}} \frac{1}{2}(\cos\tilde{\theta} + \cos\theta) \frac{\exp(ikR)}{R}. \qquad (13.4.6)$$

For an incident plane wave we merely replace $\alpha \exp(ik\tilde{R})/4\pi\tilde{R} \to A \exp(i\boldsymbol{k} \cdot \boldsymbol{r}')$, and obtain

$$\psi(\boldsymbol{r}) \approx -A \frac{ik}{2\pi} \int_{S_3} dS' \exp(i\boldsymbol{k} \cdot \boldsymbol{r}') \tfrac{1}{2}(\cos\tilde{\theta} + \cos\theta) \frac{\exp(ikR)}{R}. \qquad (13.4.7)$$

Kirchhoff's formula (13.4.6) should be compared with Fresnel's formula (13.3.4). They agree, provided we identify

$$f(\tilde{\theta}, \theta) = \tfrac{1}{2}(\cos\tilde{\theta} + \cos\theta) \qquad (13.4.8)$$

(recall that $f(0, 0) = 1$), and provided $\alpha\beta/4\pi = -\alpha(2ik)/(4\pi)^2$, i.e. $\beta = -ik/2\pi$, which is just the Fresnel self-consistency condition found earlier.

Some preliminary comments are now in order. More searching criticism is reserved to Section 13.6.

(i) The manifest symmetry of the integrand of (13.4.6) between source point and field point merely reflects the reciprocity property of the exact Green's function; in other words our approximations have respected this symmetry.

(ii) The inclination factor f is now fully specified. It vanishes only if $\theta = \pi - \tilde{\theta}$. Of course this cannot happen when the surface S excludes the sources, as in Fig. 13.1. But the zero of f does become relevant if source and field point are on the same side of the screens (books on optics discuss such configurations).

(iii) From (13.4.4) onwards, the short-wavelength approximations

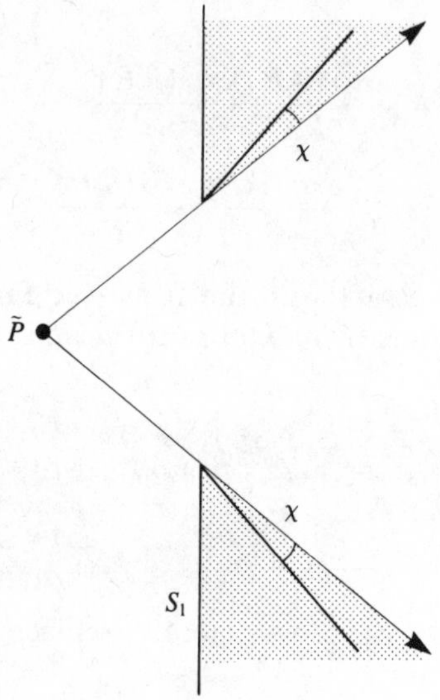

Fig. 13.3

kR, $k\tilde{R} \gg 1$ are merely convenient but not necessary, and could be redressed simply by multiplying $\cos\theta$ and $\cos\tilde{\theta}$ in the integrand of (13.4.6) by $(1 - 1/ikR)$ and $(1 - 1/ik\tilde{R})$ respectively. But for wavelengths that are not short, the underlying St Venant's hypothesis ceases to be plausible. Suppose for instance that a typical slot has width or radius a. The examples in the next section show that appreciable deviations from ray propagation, i.e. diffraction angles χ as sketched in Fig. 13.3, are of order $\chi \sim \lambda/a$. (The figure also indicates the geometrical shadow region.) If $\lambda/a \ll 1$, then the actual intensity on the downstream face of the obstructions is small (at least over most of it), and St Venant's hypothesis cannot be very bad. But it is bad if $\lambda/a \gtrsim 1$.

(iv) The crucial step is the approximation $\psi_S \to \psi_0$, $\partial_n\psi_S \to \partial_n\psi_0$ on S_3, and $\psi_S \to 0$, $\partial_n\psi_S \to 0$ on S_4, i.e. the adoption of St Venant's hypothesis. Though this may be plausible when $\lambda/a \ll 1$, it cannot in any convenient way be improved step by step.

(v) The functions $\psi(\mathbf{r})$ given by (13.4.4) and, approximately, by (13.4.6, 7), are certainly solutions of the Helmholtz equation throughout V (see Section 7.5). But they do not reproduce the St Venant's hypothesis values on S_1, neither on S_3, nor on the opaque parts S_4. (Indeed, it is impossible for both ψ and $\partial_n\psi$ to vanish on the latter: one can prove that if a solution and its normal derivative both vanish over any finite part of S, then the solution vanishes identically everywhere.)

Such inconsistency is not necessarily objectionable: after all, one is aiming to approximate the true solution as well as possible in V, rather than to justify St Venant's hypothesis on S. This point is discussed further in Section 13.6.

13.5 Applications of Kirchhoff's theory

13.5.1 Babinet's principle

St Venant's hypothesis as incorporated either into (13.4.4) or into the short-wavelength forms (13.4.6, 7) has an immediate consequence for *complementary screens*. These are two geometrically identical surfaces ($S_1' = S_1'' = S_1$), but such that the clear parts of S_1' are opaque on S_1'', and vice versa. Thus $S_3' = S_4''$ and $S_4' = S_3''$. The signals $\psi'(P)$ and $\psi''(P)$ are given by Kirchhoff's integral extended over S_3' and $S_3'' = S_4'$ respectively. Therefore they sum (mathematically) to an integral extended over $S_3' + S_3'' = S_3' + S_4' = S_1'$, i.e. over the entire surface S_1. But by the Fresnel consistency condition this integral reproduces the unobstructed incident signal $\psi_0(P)$, so that one has Babinet's principle:

$$\psi'(P) + \psi''(P) = \psi_0(P). \tag{13.5.1}$$

(Admittedly 'principle' is an overstatement for a mere consequence of St Venant's hypothesis, which is an approximation that can fail to different degrees on the two screens.)

The full force of Babinet's principle becomes apparent when the source is imaged onto a receiving surface, say onto a photographic film, by means of a system of lenses. On the receiving surface, $\psi_0(P)$ is zero except at the image point (provided we can neglect geometrical-optics aberration, and diffraction at the lens edges and other stops). On inserting either one of a pair of complementary screens, diffraction will produce faint illumination at points P other than the geometrical image. At all such points, $\psi'(P) = -\psi''(P)$, whence $I'(P) \equiv |\psi(P')|^2 = I''(P)$. In other words the intensity (at points other than the geometrical image) remains unaffected when the screens are interchanged, even if the clear spaces on them have very different areas.

We shall meet another remarkable consequence of the principle in Section 13.5.4.

13.5.2 Diffraction by a plane screen

For simplicity we consider only a plane wave $A \exp(ikz)$ incident normally onto a flat screen occupying the xy-plane; until further notice we take the screen to be opaque except for a localized slit system having linear dimensions of order a. As shown in Fig. 13.4 we choose the origin

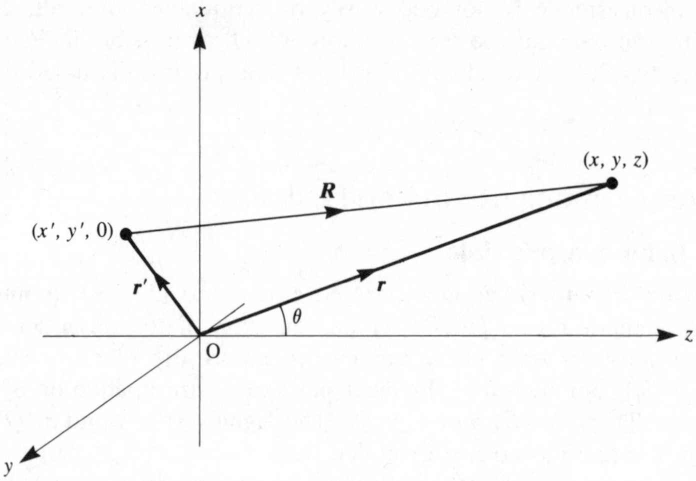

Fig. 13.4

on the screen near the slits; the observation point is $\boldsymbol{r} = (x, y, z)$, and a typical point on S is $\boldsymbol{r}' = (x', y', 0)$, with $x', y' < O(a)$. The Kirchhoff formula (13.4.7) reads

$$\psi(\boldsymbol{r}) = -\frac{Aik}{2\pi}\int_{S_3} \mathrm{d}S' \, \tfrac{1}{2}(1+\cos\theta)\frac{\exp(ikR)}{R}, \tag{13.5.2}$$

where

$$\int \mathrm{d}S' = \iint \mathrm{d}x' \, \mathrm{d}y'; \qquad \cos\theta = z/R;$$

and

$$R = [(x-x')^2 + (y-y')^2 + z^2]^{\frac{1}{2}}. \tag{13.5.3}$$

We simplify even further, to cases where $\cos\theta$ remains close to unity, and the variation of $\cos\theta$ and of $1/R$ under the integral is unimportant. These restrictions are obeyed in many practically important cases. Then $\frac{1}{2}(1+\cos\theta) \approx 1$, and $1/R \approx 1/r$, so that

$$\psi(\boldsymbol{r}) \approx -\frac{Aik}{2\pi r}\int_{S_3} \mathrm{d}S' \exp(ikR). \tag{13.5.4}$$

Much more care is needed to capture the variation of the phase kR, i.e. of the exponent of the phase factor $\exp(ikR)$. While the denominator $1/R$ can tolerate an error δR provided only the *relative* error $\delta R/R$ is small ($\delta R/R \ll 1$), the phase kR must be evaluated accurately enough to ensure, everywhere on S_3, that the *absolute* error $k\,\delta R$ is much less than unity (say $k\,\delta R \ll \pi$); obviously so, because $\exp(ikR)$ changes sign when kR changes by π.

With an eye on R as in (13.5.3), and on the assumed restriction $x', y' \lesssim a$, one commonly distinguishes two regimes. The Fraunhofer regime, whose approximations are the simpler, applies as $kr \to \infty$ with $\boldsymbol{r}$ in a fixed direction; then the direction cosines of $\boldsymbol{r} \equiv (r, \theta, \phi)$, namely

$$\frac{x}{r} \equiv \cos\alpha = \sin\theta\cos\phi, \qquad \frac{y}{r} \equiv \cos\beta = \sin\theta\sin\phi, \qquad \frac{z}{r} = \cos\theta, \tag{13.5.5}$$

remain fixed, Since $\cos^2\alpha + \cos^2\beta + \cos^2\theta = 1$, the underlying assumption that $\cos\theta \approx 1$ implies that $\cos\alpha$ and $\cos\beta$ are numerically small, but in the Fraunhofer regime they remain fixed, i.e. they do not decrease as kr rises. This regime covers all sufficiently distant observation points; it applies to instruments (e.g. interferometers) where source and screen are viewed from the observation point through a telescope, which makes them optically remote.

By contrast, the Fresnel regime, whose approximations are harder to handle, applies if the observation point is limited laterally to distances comparable to the slit dimensions, so that $x, y \lesssim O(a)$, while $kz \to \infty$. In this case $\cos\alpha$ and $\cos\beta$ obviously diminish as kz rises. We shall quantify the formal requirements '$kr \to \infty$' and '$kz \to \infty$' presently.

13.5.3 Fraunhofer diffraction

As $r \to \infty$ with fixed α, β, the phase kR is expanded in powers of $1/r$, on principles that are familiar by now (see (12.8.6) and (13.2.19)). From (13.5.3, 5) we have

$$\begin{aligned} kR &= kr\left[1 - \frac{2(x'\cos\alpha + y'\cos\beta)}{r} + \left(\frac{r'}{r}\right)^2\right]^{\frac{1}{2}} \\ &= kr\left[1 - \frac{(x'\cos\alpha + y'\cos\beta)}{r} + O(r'^2/r^2)\right]. \end{aligned} \tag{13.5.6a}$$

Since $r'^2 \lesssim a^2$, the third term is negligible provided

$$kr(a^2/r^2) \ll 1 \Rightarrow r \gg ka^2 \Rightarrow kr \gg (ka)^2, \tag{13.5.6b}$$

which we now assume. It proves convenient to define the 'diffracted wave-vector' $\boldsymbol{k}_d$ and its component $\boldsymbol{\kappa}$ parallel to the screen by

$$\boldsymbol{k}_d \equiv k(\cos\alpha, \cos\beta, \cos\theta) \equiv \boldsymbol{\kappa} + \hat{\boldsymbol{z}}k\cos\theta, \tag{13.5.7a}$$

$$\boldsymbol{\kappa} \equiv (k\cos\alpha, k\cos\beta, 0) = k\sin\theta(\cos\phi, \sin\phi, 0). \tag{13.5.7b}$$

Evidently $\boldsymbol{k}_d$ is a vector of magnitude k, directed from the (localized) slit system towards the distant observation point $\boldsymbol{r}$; and

$$\kappa^2 = k^2 \sin^2\theta \approx k^2\theta^2, \tag{13.5.7c}$$

so that $\kappa^2 \ll k^2$ because in the Fraunhofer regime θ is small.

Exercise: To first order in θ, show that $\boldsymbol{\kappa} \approx \boldsymbol{k}_d - k\hat{\boldsymbol{z}}$, or in other words that $\boldsymbol{\kappa}$ is the change in wave-vector of an incident wave redirected by diffraction in the direction of observation.

Equations (13.5.6, 7b) together with $\boldsymbol{r}' = (x', y', 0)$ allow us to write

$$kR \approx kr - k(\cos\alpha \cdot x' + \cos\beta \cdot y') = kr - \boldsymbol{\kappa}\cdot\boldsymbol{r}'. \tag{13.5.8}$$

Substitution into (13.5.4) yields the Fraunhofer end-result

$$\psi(\boldsymbol{r}) \approx -\frac{Aik}{2\pi r}\exp(ikr)\iint_{S_3} \mathrm{d}x'\,\mathrm{d}y'\exp(-i\boldsymbol{\kappa}\cdot\boldsymbol{r}'). \tag{13.5.9}$$

The angular distribution and the intensity of the diffracted waves is governed by the final integral, which is simply the 2D Fourier transform of the slit pattern. (One is reminded of two other Fourier transforms, the form factor (13.2.20) governing emission from an extended source, and the first Born approximation to the scattering amplitude in quantum mechanics.) The integral depends, through $\boldsymbol{\kappa}$, only on the direction cosines $\cos\alpha$ and $\cos\beta$ of $\boldsymbol{r}$. In the forward direction ($\theta = 0$, $\cos\theta = 1$, $\cos\alpha = 0 = \cos\beta$) one has $\boldsymbol{\kappa} = 0$, the phase factor $\exp(i\boldsymbol{\kappa}\cdot\boldsymbol{r}')$ is unity, and

$$\psi(0, 0, z) \approx -(Aik \cdot S_3/2\pi)(\exp(ikr)/r),$$

where S_3 now stands for the total area of the slits. Perhaps unexpectedly, the forward intensity $|\psi|^2$ is therefore proportional to S_3^2 rather than to S_3.

As an example, consider a rectangular slit defined by $-a \leqslant x' \leqslant a$, $-b \leqslant y' \leqslant b$. Then the double integral factorizes:

$$\left\{\int_{-a}^{a} \mathrm{d}x'\exp(i\kappa_1 x')\right\}\left\{\int_{-b}^{b} \mathrm{d}y'\exp(i\kappa_2 y')\right\}$$
$$= \left\{2a\frac{\sin(a\kappa_1)}{a\kappa_1}\right\}\left\{2b\frac{\sin(b\kappa_2)}{b\kappa_2}\right\},$$

$$\psi(\boldsymbol{r}) = -\frac{Aik}{2\pi r}\exp(ikr)(4ab)\frac{\sin(a\kappa_1)}{a\kappa_1}\cdot\frac{\sin(b\kappa_2)}{b\kappa_2}. \tag{13.5.10}$$

The two final factors produce the familiar diffraction pattern, a bright central maximum at $\kappa_1 = 0 = \kappa_2$, flanked by much fainter intensity maxima separated by zeros at

$$a\kappa_1 = n_1\pi, \qquad b\kappa_2 = n_2\pi, \qquad n_{1,2} = \pm 1, \pm 2, \ldots .$$

Mathematically speaking, the slit pattern S_3 can be determined from its Fourier transform, i.e. from the angular distribution of the diffracted waves, but only if the latter is known for all values of $\boldsymbol{\kappa}$. Of course, in practice $\kappa_{1,2}$ can vary only over a range limited by the assumption that θ is small, and by the magnitude of the incident wavenumber k, and k itself cannot be increased indefinitely. Moreover, often one measures only the intensity $|\psi|^2$ rather than ψ itself. Under such conditions diffraction alone cannot reveal the slit pattern from scratch. However, if some information is available independently, then diffraction can sometimes supply the rest. For instance, if we know from the outset that S_3 is a rectangle, then $|\psi|^2$ measured as a function of α and β for given k suffices to determine the values of the slit widths a and b.

13.5.4 Fresnel diffraction

For large z with fixed $x, y, x', y' \lesssim O(a)$, the phase kR is expanded in powers of $1/z$ (rather than of $1/r$ as for Fraunhofer diffraction):

$$kR = kz\left[1 + \frac{(x-x')^2 + (y-y')^2}{z^2}\right]^{\frac{1}{2}}$$

$$= kz\left[1 + \frac{1}{2}\frac{(x-x')^2 + (y-y')^2}{z^2} + O(a/z)^4\right]. \qquad (13.5.11)$$

The third term is negligible if

$$kz(a^4/z^4) \ll 1 \;\Rightarrow\; (kz)^3 \gg (ka)^4 \;\Rightarrow\; kz \gg (ka)^{\frac{4}{3}}. \qquad (13.5.12)$$

In these circumstances $z \sim r$, so that the Fresnel criterion could also be written as

$$kr \gg (ka)^{\frac{4}{3}}. \qquad (13.5.13)$$

But $ka \gg 1$ (by the basic short-wavelength assumption); therefore $(ka)^2 \gg (ka)^{\frac{4}{3}}$, so that the Fraunhofer criterion (13.5.6b) is more restrictive than the Fresnel criterion (13.5.13). Thus, provided $\cos\theta \approx 1$, Fresnel validates Fraunhofer but not vice versa. (When $\cos\theta$ is unrestricted, the relation between the two regimes is more subtle. A clear account is given by Rees (1987).) Under the Fresnel regime, substitution from (13.5.11) into (13.5.4) yields

$$\psi(\boldsymbol{r}) \approx -\frac{Aik}{2\pi z}\exp(ikz)\iint_{S_3} \mathrm{d}x'\,\mathrm{d}y'\,\exp\left\{\frac{ik([x-x')^2 + (y-y')^2}{2z}\right\}. \qquad (13.5.14)$$

As a rule such Gaussian integrals are harder to evaluate than the Fourier transforms in the Fraunhofer formula (13.5.9). Hence one resorts to the Fresnel approximation only when the Fraunhofer criterion (13.5.7) fails. Moreover, in some Fresnel problems one must respect the variation of the factors $\frac{1}{2}(1+\cos\theta)/R$ under the integral (13.5.2): e.g. in diffraction from a straight edge, where S_3 extends to infinity.

The only very simple example, but a very spectacular one, is the circular slit (or concentric ring system) and an observation point on the axis $(x=0=y)$. Consider a circular aperture of radius a, using plane polar coordinates $\mathbf{r}=(r', \phi')$. Then the integral in (13.5.14) yields

$$2\pi\int_0^a dr'\, r' \exp(ikr'^2/2z) = \frac{2\pi z}{ik}\{\exp(ika^2/2z)-1\},$$

$$\psi(0,0,z) \approx A\exp(ikz)\{1-\exp(ika^2/2z)\}. \qquad (13.5.15)$$

Therefore the intensity is given by

$$I/I_0 = |\psi|^2/|A|^2 = |1-\exp(ika^2/2z)|^2 = 4\sin^2(ka^2/4z). \qquad (13.5.16)$$

This is an oscillatory function of z (albeit the zeros are far from equally spaced), with maxima four times the intensity of the wholly unobstructed beam. Thus there are points on the axis where the intensity is increased fourfold when most of the incident beam is blacked out.

Exercises: (i) Show that, as $z\to\infty$, (13.5.15) reduces to the Fraunhofer expression in the forward direction, as given following (13.5.9) above. (ii) Where on the axis is the intensity zero furthest from the screen?

Although the formulae (13.5.14–16) were deduced only for slits of limited extent, Babinet's principle (13.5.1) determines the amplitude on the axis behind the complementary screen, namely behind a circular obstacle of radius a, the rest of the xy-plane being clear. Here $\psi_0(P)=A\exp(ikz)$, $\psi'(P)$ is given by (3.5.15), whence Babinet's principle yields $\psi=\psi''(P)=\psi_0(P)-\psi'(P)$, namely

$$\psi(0,0,z) = A\exp(ikz) - A\exp(ikz)(1-\exp(ika^2/2z)),$$

$$\psi(0,0,z) = A\exp(ikz+ika^2/2z), \qquad (13.5.17)$$

$$I/I_0 = |\psi|^2/|A|^2 = 1. \qquad (13.5.18)$$

In other words, once z is large enough to satisfy the Fresnel criterion (13.5.12), the intensity on the axis behind a circular obstacle, i.e. at the centre of the geometrical shadow, is exactly the same as it would have been with no obstacle at all.

13.6 A critique of Kirchhoff's theory

Surprisingly perhaps in view of its empirical success, Kirchhoff's theory rewards its critic at least twice. First, it is instructive to explore its connection with the magic rule for 'well-posed' Helmholtz BVPs. Second, one learns that the reasons why a theory succeeds need not be the reasons why it was thought plausible in the first place. By 'Kirchhoff's theory' we mean ψ written as the Kirchhoff representation, with the boundary values dictated by the St Venant's hypothesis (SVH).

To be specific, we consider only a plane wave $\psi_0 = A\exp(ikz)$ normally incident on the xy-plane ($\equiv S_1$), which is opaque except for a circular slot ($\equiv S_3$) of radius a centred on the origin.

As already pointed out, Kirchhoff's expressions (13.4.4), and, to a good approximation, (13.4.7), call them ψ_K, certainly solve the Helmholtz equation for $z \geqslant 0$. If Kirchhoff's theory were at least approximately self-consistent (i.e. if ψ_K once written down conformed to the SVH), then its success would imply that the SVH indeed furnishes an adequate approximation to the true boundary values; namely to those one would find by solving the Helmholtz equation everywhere ($z<0$ as well as $z>0$), subject to proper boundary conditions on both faces of the screen, and with ψ_0 prescribed merely as that component of ψ identifiable as incident from the far left. But we shall see presently that ψ_K is not at all self-consistent in this sense, and shall then consider the implications.

Suppose one knows the true ψ_S *or* the true $\partial_n\psi_S$ on S_1, and adopts Sommerfeld BCs on the far-right hemisphere S_2. We assert (without proof, but claiming it as obvious on physical grounds) that in the magic rule one can then use the exact Dirichlet or Neumann Green's functions (13.2.15) given for our flat S_1 by the image method; and that there is no contribution from S_2. It is easy to verify that on S_1, i.e. at $z'=0$, $\partial G_D(\mathbf{r}\,|\,\mathbf{r}')/\partial z' = 2\,\partial G_0(\mathbf{r}\,|\,\mathbf{r}')/\partial z'$, and $G_N(\mathbf{r}\,|\,\mathbf{r}') = 2G_0(\mathbf{r}\,|\,\mathbf{r}')$. Consequently the magic rule yields (with $f_{D,N}=0$ in our source-free region $z \geqslant 0$)

$$\psi_D(\mathbf{r}) = -2\int_{S_1} dS'\;\psi_S(\mathbf{r}')\,\partial'_n G_0(\mathbf{r}\,|\,\mathbf{r}'), \tag{13.6.1a}$$

$$\psi_N(\mathbf{r}) = 2\int_{S_1} dS'\;(\partial'_n\psi_S(\mathbf{r}'))G_0(\mathbf{r}\,|\,\mathbf{r}'). \tag{13.6.1b}$$

Comparison with the Kirchhoff representation (13.4.1) reveals that

$$\psi_K(\mathbf{r}) = \tfrac{1}{2}\{\psi_D(\mathbf{r}) + \psi_N(\mathbf{r})\}. \tag{13.6.2}$$

It is important to appreciate clearly what this means. If the two alternative input functions ψ_S and $\partial_n\psi_S$ in (13.6.1) were the true surface values, then ψ_D and ψ_N would coincide for all $z \geqslant 0$, and would reproduce both the inputs on S_1. But Kirchhoff's theory adopts instead the inputs dictated by the SVH, and in that case ψ_D and ψ_N as defined by (13.6.1) differ.

To test self-consistency, we use (13.6.1) plus the SVH to evaluate ψ_D and ψ_N

on the axis. In this evaluation we do not now make any mathematical approximations at all (unlike the Fraunhofer and Fresnel procedures in Sections 13.5.3, 4). In ψ_D, one notes that

$$\partial_n' G_0(\boldsymbol{r}\mid\boldsymbol{r}') = -(\partial R/\partial z')(\mathrm{d}/\mathrm{d}R)(\exp(ikR)/4\pi R),$$

and that, at $z'=0$, $-\partial R/\partial z' = z/R$. Accordingly

$$\psi_D(0,0,z) = -2\int_0^a 2\pi r'\,\mathrm{d}r'\,A\left(\frac{z}{R}\right)\frac{\mathrm{d}}{\mathrm{d}R}\left(\frac{\exp(ikR)}{4\pi R}\right)$$

$$= -2A\int_z^{(z^2+a^2)^{\frac{1}{2}}} 2\pi\,\mathrm{d}R\,R\left(\frac{z}{R}\right)\frac{\mathrm{d}}{\mathrm{d}R}\left(\frac{\exp(ikR)}{4\pi R}\right),$$

$$\psi_D(0,0,z) = A\left\{\exp(ikz) - \frac{z}{(z^2+a^2)^{\frac{1}{2}}}\exp(ik(z^2+a^2)^{\frac{1}{2}})\right\}. \tag{13.6.3}$$

Naturally $\psi_D(0) = \psi_0(0) = A$, and we see that

$$\left.\frac{\partial\psi_D}{\partial z}\right|_{z=0} = Aik\left(1 - \frac{\exp(ika)}{ika}\right). \tag{13.6.4}$$

Reasoning similarly, one obtains

$$\psi_N(0,0,z) = A\{\exp(ikz) - \exp(ik(z^2+a^2)^{\frac{1}{2}})\}. \tag{13.6.5}$$

Naturally $\partial\psi_N/\partial z = \partial\psi_0/\partial z = Aik$ at $z=0$, and we see that

$$\psi_N(0) = A(1-\exp(ika)). \tag{13.6.6}$$

These results combine according to (13.6.2) to yield

$$\psi_K(0,0,z) = A\left\{\exp(ikz) - \frac{1}{2}\left(1 + \frac{z}{(z^2+a^2)^{\frac{1}{2}}}\right)\exp(ik(z^2+a^2)^{\frac{1}{2}})\right\}, \tag{13.6.7}$$

$$\psi_K(0) = A\{1 - \tfrac{1}{2}\exp(ika)\}, \tag{13.6.8a}$$

$$\left.\frac{\partial\psi_K}{\partial z}\right|_{z=0} = Aik\left\{1 - \frac{1}{2}\frac{\exp(ika)}{ika}\right\}. \tag{13.6.8b}$$

Thus ψ_K and $\partial\psi_K/\partial z$ at the centre of the slot differ quite appreciably from the SVH $\psi_0 = A$, $\partial\psi_0/\partial z = Aik$. For instance, if $ka = 2n\pi$, then $\psi_D(0)/\psi_0(0) = 1$, $\psi_N(0)/\psi_0(0) = 0$, while $\psi_K(0)/\psi_0(0) = \frac{1}{2}$. In this special case, eqn (13.6.7) for ψ_K along the axis has been confirmed experimentally, including the value $\psi_K(0) = \frac{1}{2}\psi_0(0)$. The measurements are quoted by Marchand and Wolf (1966) and by Stamnes (1986). (In particular, their close agreement with $\psi_K = \frac{1}{2}\psi_0$ at the centre of the slot shows at once that, on the surface, ψ_D alone, or ψ_N alone, would be altogether inadequate.)

Thus the Kirchhoff theory is certainly far from self-consistent. On the other hand, it is close to the experimental truth. Therefore one suspects that the surface values of ψ_K and of $\partial_n\psi_K$, whatever they may be, are good approximations to the true surface values arising from the incident wave as modified by the screen. From this point of view the lack of self-consistency is good rather than bad: the true surface values are different from the SVH, and, possibly by luck, once ψ_K is written down for whatever reason, it approximates to the true rather than to the SVH values.

Before accepting such a conclusion, one must at least verify that the surface distribution of ψ_K is not intrinsically unreasonable on physical grounds. Though the details are beyond our scope, this has been done, with reasonably encouraging results, by Marchand and Wolf (1966) (see also Stamnes 1986).

Exercise: Under the approximations of the Fraunhofer and Fresnel formulae (13.5.9, 14), show that $\psi_D \approx \psi_N \approx \psi_K$ (in sharp contrast to the behaviour near the screen as described by (13.6.3–8)).

Problems

13.1 *Resonators.* Determine the ratio between the lowest eigenfrequencies of a spherical and a cubic resonator having equal volumes, and both subject to DBCs.

Hint: The numerical value of the ratio sphere/cube is approximately 0.931. You may rely on the fact that in a spherically symmetric system the lowest normal mode is isotropic. For spherical Bessel functions, see Appendix A.

13.2 Determine the ratio between the lowest *non-zero* eigenfrequencies of the resonators in Problem 13.1, but subject to NBCs.

Hint: The ratio is approximately 1.068; you will have to solve an appropriate eigenvalue equation numerically. You may rely on the fact that in a spherically symmetric system the lowest-but-one normal mode has either $l = 0$ or $l = 1$.

13.3 The same as Problem 13.1, but comparing circle and square.

Hint: See Appendix A for lowest root of $J_0(z) = 0$.

13.4 *Reflection and transmission.* Two semi-infinite rods along the x-axis are firmly joined at the origin. The mass per unit length, and the speed of longitudinal sound waves, are (σ_1, c_1) and (σ_2, c_2) for $x < 0$ and $x > 0$ respectively. The energy flux is as usual given by $N_{1,2} = -\sigma_{1,2}\psi_t\psi_x$. The continuity condition $\psi_x(0-, t) = \psi_x(0+, t)$ is evidently satisfied at all times.

(i) Determine the second requisite matching condition, from the fact that energy cannot accumulate at the join, whence $N(0-, t) = N(0+, t)$ for all t.
(ii) Monochromatic waves of frequency ω arrive from the left. From the matching conditions, determine the ratios of reflected and transmitted to incident amplitudes.
(iii) Determine the ratios of reflected and transmitted to incident energies, and check that these ratios add to unity.
(iv) Do the ratios in (ii) and (iii) depend on the frequency?
(v) Would the ratios in (ii) and (iii) be different for waves arriving from the right?

13.5 *Klein–Gordon equation and Yukawa potential.* The Klein–Gordon relativistic wave equation for a real scalar meson field in the presence of sources reads

$$\left(\frac{1}{c^2}\frac{\partial^2}{\partial t^2} - \nabla^2 - \left(\frac{mc}{\hbar}\right)^2\right)\psi(\boldsymbol{r}, t) = \rho(\boldsymbol{r}, t).$$

Determine the static (time-independent) solution, vanishing at infinity, in the presence of a static point source (notice the sign) $\rho = -g\,\delta(\boldsymbol{r} - \boldsymbol{r}_1)$. Calling this solution $gG(\boldsymbol{r} \mid \boldsymbol{r}_1)$, the mutual potential energy of two such sources situated at $\boldsymbol{r}_1$ and $\boldsymbol{r}_2$ is $g^2G(\boldsymbol{r}_2 \mid \boldsymbol{r}_1)$. Given that nuclear forces have a strength of roughly 40 MeV, and a range of roughly 10^{-15} m, and assuming that they can be roughly represented by such a potential, estimate the parameters g and m, expressing your estimates in units that measure energy in MeV, and distance in Fermis (1 Fermi $= 10^{-15}$ m).

13.6 (i) Determine the Green's function for the Helmholtz equation inside the unit square ($0 \leqslant x, y \leqslant 1$), and subject to DBCs, in the form of a double Fourier series in x and y.
(ii) Determine the same Green's function in the form

$$G(\boldsymbol{r} \mid \boldsymbol{r}') = \sum_{n=1}^{\infty} \sin(n\pi x)\sin(n\pi x')f_n(y \mid y').$$

(iii) Using both forms in turn, write down the solution of the Helmholtz equation $-(\nabla^2 + k^2)\psi = 0$ in the unit square, given that ψ vanishes along the left-hand, top, and right-hand edges, and that along the bottom edge $\psi(x, 0) = \delta(x - \frac{1}{2})$.

Hint: In determining f_n in (ii), be careful to distinguish between the two cases $n\pi \lesseqgtr k$.

13.7 *Directional antenna.* A monochromatic line source extends along the z-axis from $z = -L$ to $z = +L$. It has strength α per unit length, but the phase can be controlled through the function $\chi(z)$:

$$\rho(\boldsymbol{r}, t) = \alpha\,\delta(x)\,\delta(y)\theta(L - |z|)\cos(kct + \chi(z)).$$

The object is to maximize the power radiated per unit solid angle in the direction (θ, ϕ).

(i) How should $\chi(z)$ be chosen to achieve this?
(ii) Show that, when the optimum is achieved, power emission in the opposite direction is weaker by a factor $(\sin(2kL\cos\theta)/2kL\cos\theta)^2$.

Hint: Conformably with (13.1.1), the complex source function in the underlying wave equation may be taken as

$$\rho_\omega(\boldsymbol{r}, t) = \alpha\,\delta(x)\,\delta(y)\theta(L - |z|)\exp[-i(kct + \chi(z)],$$

whose real part is the actual source density given above. Then the complex source function entering the Helmholtz equation is

$$\rho_\omega(\boldsymbol{r}) = \alpha\,\delta(x)\,\delta(y)\theta(L - |z|)\exp[-i\chi(z)].$$

13.8 Consider the intensity ratio $f \equiv |\psi(r, \theta, \phi)/\psi(r, 0, 0)|^2$ for Fraunhofer diffraction from a square slit, as described by eqn (13.5.10) with $a = b$. Assuming $ka \gg 1$, and $\theta \ll 1$, determine the smallest value θ_1 of θ where f vanishes:

(i) When $\phi = 0$ (diffraction in a plane parallel to an edge).
(ii) When $\phi = \pi/4$ (diffraction in a plane parallel to a diagonal).
(iii) Determine θ_1 as a function of ϕ; draw a polar diagram of the innermost dark band of the diffraction pattern, i.e. a (ρ, ϕ) diagram with $\rho(\phi)$ proportional to $\theta_1(\phi)$.

13.9 *Fraunhofer diffraction from a circular slot.* For a normally incident plane wave and a circular slot of radius a ($ka \gg 1$), and in the usual small-θ regime, evaluate the intensity ratio $f \equiv |\psi(r, \theta, \phi)/\psi(r, 0, 0)|^2$ as a function of θ from (13.5.9). If $ka = 10$, what are the two smallest values of θ where $f = 0$?

Hint: $\int_0^{z_1} dz z J_0(z) = z_1 J_1(z_1)$, and $J_1(z) = 0$ at $z = 0$, 3.83, 7.02, ...

13.10 *Fresnel diffraction.* A flat screen is pierced by a circular slot of radius a. It is illuminated by a monochromatic point source of strength α situated on the axis, a distance $\tilde{z}$ from the screen.

(i) Calculate the illumination on the axis downstream as a function of z, using approximations corresponding to those in Section 13.5.4, but with $\psi_0 = \alpha \exp(ik\tilde{R})/4\pi\hat{R}$.
(ii) Verify that in the limit $\tilde{z} \to \infty$ with $\alpha/\tilde{z}$ fixed, your result reduces to (13.5.16).
(iii) In the same approximation, calculate the intensity on the axis behind the complementary screen.

13.11 *A burning glass.* A circular slot of radius a is illuminated by a monochromatic plane wave of unit intensity, incident normally. According to eqns (13.5.14–16), the intensity $I(z)$ along the axis has maxima at

$$z = z_n = ka^2/2\pi(2n + 1) = a^2/\lambda(2n + 1), \qquad n = 0, 1, 2, \ldots,$$

with $I(z_n) = 4$. The slot is now covered with a transparent film of variable optical thickness, which shifts the phase of the transmitted light be a factor $\exp(i\phi(r')) = \exp(i(a^2 - r'^2)/b^2)$. Here r' is radial distance measured in the plane of the screen, and b is a design parameter.

(i) Determine the new intensity $J(z)$ along the axis, and compare it with $I(z)$.
(ii) Verify that, for fixed z, a suitable choice of b can achieve $(J(z))_{\max} = (ka^2/2z)^2$, so that, in particular, $(J(z_n))_{\max} = \pi^2(2n+1)^2$. (Note that this is an enhancement by a factor $[\pi(n+\frac{1}{2})]^2$.)
(iii) Show that no phase-shift function $\phi(r')$ whatever can increase $J(z)$ above $(ka^2/2z)^2$.

Appendices

A Notation and formulary

This appendix collects notation and some particularly useful standard results, mainly for convenient reference.

A.1 Implications and proofs

The open arrow $\Rightarrow$ represents 'it follows that'. For instance, $(a = b) \Rightarrow (c = d)$ means 'if $a = b$, then $c = d$'. The inverse may or may not hold. If it does, i.e. one also has $(c = d) \Rightarrow (a = b)$, and if one wishes to stress this, one writes $(a = b) \rightleftharpoons (c = d)$.

The symbol ■ marks the end of a proof, usually of an explicitly stated theorem or lemma, but sometimes just of an assertion in the text.

A.2 Varieties of equality, and orders of magnitude

(i) *Equality by definition* is written $a \equiv b$, and can serve to define either a or b. For instance, $\operatorname{erf}(x) \equiv (2/\pi^{\frac{1}{2}}) \int_0^x dt \exp(-t^2)$ simply assigns the more compact symbol $\operatorname{erf}(x)$ to represent the integral written out explicitly on the right. The symbol $\equiv$ reminds the reader not to spend time looking for the reason for the equality, except perhaps for the hindsight that makes such a definition convenient.

(ii) *Approximate equality* is written $\approx$, usually stressing its numerical aspects: e.g. $\pi \approx 22/7$. What percentage or absolute error can be tolerated by such a relation always depends on the context, and is not a matter for general definition.

(iii) *Asymptotic equality.* The relation $f(x) \sim g(x)$ indicates that in some limit (stated or clear from the context),

$$\lim f(x)/g(x) = 1. \tag{A.2.1}$$

We read 'f is asymptotically equal to g'. For instance, as $x \to \infty$, $(x^2 + 1)^{\frac{1}{2}} \sim x$. Similarly, $(1 + 3x)/(2 + x^2) \sim 3/x$, because $(1 + 3x) \sim 3x$, $(2 + x^2) \sim x^2$, whence $(1 + 3x)/((2 + x^2) \sim 3x/x^2 = 3/x$. Notice that $f \sim g$ does not imply that the difference between f and g vanishes in the limit; in other words $(f \sim g) \not\Rightarrow f - g \to 0$. For instance, as $x \to \infty$, $[(x^2 + 1)^{\frac{1}{2}} + x^{\frac{1}{2}}] \sim x$ even though $\{[(x^2 + 1)^{\frac{1}{2}} + x^{\frac{1}{2}}] - x\} \sim x^{\frac{1}{2}} \to \infty$.

(iv) In principle the '*much less than*' symbol $\ll$ should be defined asymptotically with reference to some specific limit:

$$f(x) \ll g(x) \rightleftharpoons \lim \frac{|f(x)|}{|g(x)|} = 0. \tag{A.2.2}$$

The symbol $\gg$ is just the converse: $f \ll g \rightleftharpoons g \gg f$. In these relations (by contrast to < and >) signs are disregarded, even though this admits the seemingly paradoxical assertion that, as $x \to \infty$, $x \ll -x^2$. However, the same symbol is often used in a supposedly obvious but not uniquely defined numerical sense, as in $10^{-6} \ll 1$. The numerical disproportion between a and b required to validate $a \ll b$ then depends on the context; in this sense $\ll$ and $\gg$, jointly, are opposites to $\sim$.

(v) The *order of magnitude* symbol $O(\ )$ is, roughly speaking, a weaker version of $\sim$. In some limit,

$$f(x) = O(g(x)) \;\Rightarrow\; \lim \frac{|f(x)|}{|g(x)|} = \text{constant} \neq 0. \tag{A.2.3}$$

For instance, $\phi(x) \equiv (ax^2 + bx + c)^{\frac{1}{2}} = O(x)$ as $x \to \infty$, but $\phi(x) \sim x$ only if $a = 1$. On occasion we may write $f(x) \sim O(g(x))$, meaning exactly the same as (A.2.3). To indicate that f is of lower order than g, we write

$$g(x) < O(g(x)) \;\Rightarrow\; \lim \frac{|f|}{|g|} = 0. \tag{A.2.4}$$

To leave open the possibility that the constant in (A.2.3) is zero, we would write

$$f(x) \lessapprox O(g(x)) \;\Rightarrow\; \lim \frac{|f|}{|g|} = \text{constant} < \infty. \tag{A.2.5}$$

(Some texts use $f = O(g)$ in the sense of our (A.2.5).)

A very common use of $O(\ )$ is illustrated by

$$(x^2 + 1)^{\frac{1}{2}} = x(1 + 1/x^2)^{\frac{1}{2}} \underset{x\to\infty}{=} x(1 + 1/2x^2 + O(x^{-4}))$$

$$= x + 1/2x + O(x^{-3}), \tag{A.2.6}$$

which is a more informative version of $(x^2 + 1)^{\frac{1}{2}} \sim x$. (Contrary to our definition of $\sim$, the last form of (A.2.6) is sometimes written as $(x^2 + 1)^{\frac{1}{2}} \sim x + 1/2x$, meaning simply that any terms that have been dropped are of lower order than the last term retained. One can legitimize this usage through a formal definition, but we shall not need to.)

A.3 Miscellanea

A.3.1 The Kronecker delta

The Kronecker delta symbol $\delta_{n,m}$ is defined only when its indices n, m are integers; the comma between them is sometimes omitted:

$$\delta_{n,m} \equiv \delta_{nm} \equiv \begin{cases} 1 & \text{if} \quad n = m, \\ 0 & \text{if} \quad n \neq m. \end{cases} \tag{A.3.1}$$

For instance, $\sum_{n=15}^{\infty} n^2\,\delta_{n,13}=0$, while $\sum_{n=0}^{\infty} n^2\,\delta_{n,13}=169$. Somewhat more generally,

$$\sum_{n=n_1}^{n_2} \delta_{nm} f_m = \begin{cases} f_m & \text{if} \quad n_1 \leqslant m \leqslant n_2, \\ 0 & \text{otherwise.} \end{cases} \tag{A.3.2}$$

The delta-function is a generalization of the Kronecker delta to continuously variable indices.

A.3.2 Greater and lesser

Given two numbers, or two functions, x_1 and x_2,

$$x_< \equiv \min(x_1, x_2) \qquad x_> \equiv \max(x_1, x_2). \tag{A.3.3}$$

In other words $x_<$ ($x_>$) is the lesser (greater) of x_1 and x_2. For instance, if $x_1 = y$ and $x_2 = y^2$, then $x_< = x_1$ and $x_> = x_2$ if $y<0$ or $y>1$, while $x_< = x_2$ and $x_> = x_1$ if $0<y<1$.

A.3.3 One-sided limits

The limits $\lim_{x\to a+}$ and $\lim_{x\to a-}$ mean, respectively, that x tends to a from the right or the left. Evidently, $\lim_{x\to a+} f(x) \neq \lim_{x\to a-} f(x)$ means that $f(x)$ is discontinuous at $x=a$. For instance, $\lim_{x\to 0\pm} \tan(\pi/2+x) = \mp\infty$, while $f(x) \equiv \int_0^x dy\,\delta(y-a)$ entails $\lim_{x\to a+} f(x)=1$, $\lim_{x\to a-} f(x)=0$. For brevity, one writes

$$x\pm \equiv x \pm 0 \equiv \lim_{\varepsilon\to 0\pm} (x+\varepsilon). \tag{A.3.4}$$

A.3.4 l'Hôpital's theorem and 0/0

If $f(a)=0$ and $g(a)=0$, then $f(a)/g(a)$ as it stands is meaningless (not defined). But if $f(x)$ and $g(x)$ can be expanded in Taylor series around $x=a$, then

$$\lim_{x\to a} \frac{f(x)}{g(x)} = \lim_{x\to a} \frac{(x-a)f'(a) + \frac{1}{2!}(x-a)^2 f''(a) + \cdots}{(x-a)g'(a) + \frac{1}{2!}(x-a)^2 g''(a) + \cdots}. \tag{A.3.5}$$

In most applications, $f'(a)$ and $g'(a)$ do not both vanish, whence

$$\lim_{x\to a} \frac{f(a)}{g(a)} = f'(a)/g'(a); \tag{A.3.6}$$

this is called l'Hôpital's theorem (or l'Hôpital's rule). For instance,

$$\lim_{x\to 0}\frac{\sin x}{x}=\lim_{x\to 0}\frac{\cos x}{1}=1;$$

$$\lim_{x\to 0}\frac{\sin(x^2)}{x}=0;\qquad \lim_{x\to 0}\frac{\sin x}{x^2}=\infty.$$

Of course, it can happen that $f'(a)$ and $g'(a)$ do both vanish, while $f''(a)$ and $g''(a)$ do not; then $\lim_{x\to a} f(x)/g(x)=f''(a)/g''(a)$. For instance, $\lim_{x\to 0}\sin(x^2)/x^2=1$.

A.4 Some useful integrals

Most of the integrals given below, with only minimal comment, are standard, and can be evaluated with the help of any good text on the calculus (e.g. Courant 1937). But as regards professional routine this remark is misleading: normally one saves time by consulting tables of integrals as a first rather than as a last resort. Good collections include Dwight (1961), Groebner and Hofreiter (1957, 1958), Gradshteyn and Ryzhik (1980). Abramowitz and Stegun (1968) summarize the properties of the error function and of the exponential integral.

$$\int\frac{dx}{(x^2+a^2)}=\frac{1}{a}\tan^{-1}\left(\frac{x}{a}\right); \tag{A.4.1a}$$

$$\int_0^\infty dx/(x^2+a^2)=\pi/2a,\qquad \int_0^\infty dx/(x^2+a^2)^2=\pi/4a^3, \tag{A.4.1b,c}$$

where (c) follows from (b) by differentiation with respect to a or a^2.

$$\int_0^\infty dx\exp(-\alpha x)=1/\alpha,\qquad (\mathrm{Re}\,\alpha>0). \tag{A.4.2}$$

On setting $\alpha=a+ib$, the real and imaginary parts yield

$$\int_0^\infty dx\exp(-ax)\cos(bx)=\frac{a}{a^2+b^2}, \tag{A.4.3a}$$

$$\int_0^\infty dx\exp(-ax)\sin(bx)=\frac{b}{(a^2+b^2)}. \tag{A.4.3b}$$

Repeated differentiation of (A.4.2) with respect to α leads to

$$\int_0^\infty dx\,x^n\exp(-\alpha x)=n!/\alpha^{n+1}. \tag{A.4.4}$$

The Dirichlet integral $\int_0^\infty dx\sin(ax)/x=\pi\varepsilon(a)/2$ is evaluated in Appendix B.

The *exponential integral* cannot be expressed in terms of simpler functions:

$$E_1(\alpha) \equiv \int_\alpha^\infty \frac{\mathrm{d}x}{x} \exp(-x) = -\gamma - \log \alpha - \sum_{n=1}^{\infty} \frac{(-\alpha)^n}{n!\,n}, \tag{A.4.5a}$$

where $\gamma \approx 0.577$ is Euler's constant. Asymptotically (see also Section 8.5.3),

$$E_1(\alpha \to \infty) \sim \frac{\exp(-\alpha)}{\alpha}\{1 - 1!/\alpha + 2!/\alpha^2 - 3!/\alpha^3 + \cdots\}. \tag{A.4.5b}$$

Cauchy's theorem yields

$$\int_{-\infty}^{\infty} \mathrm{d}x \cos(x/b)/(x^2 + a^2) = (\pi/a)\exp(-a/b). \tag{A.4.6}$$

The *Gaussian integral* is evaluated by a trick:

$$J_0(\alpha) \equiv \int_0^\infty \mathrm{d}x \exp(-\alpha^2 x^2),$$

$$\begin{aligned} J_0^2(\alpha) &= \int_0^\infty \mathrm{d}x \exp(-\alpha^2 x^2) \int_0^\alpha \mathrm{d}y \exp(-\alpha^2 y^2) \\ &= \int_0^\infty \int_0^\infty \mathrm{d}x\,\mathrm{d}y \exp(-\alpha^2(x^2 + y^2)) \\ &= \int_0^{\pi/2} \mathrm{d}\phi \int_0^\infty \mathrm{d}r\, r \exp(-\alpha^2 r^2) \\ &= \frac{\pi}{2} \int_0^\infty \tfrac{1}{2}\,\mathrm{d}r^2 \exp(-\alpha^2 r^2) = \pi/4\alpha^2; \end{aligned} \tag{A.4.7}$$

$$J_0(\alpha) \equiv \int_0^\infty \mathrm{d}x \exp(-\alpha^2 x^2) = \pi^{\frac{1}{2}}/2\alpha, \tag{A.4.8a}$$

$$\int_{-\infty}^{\infty} \mathrm{d}x \exp(-\alpha^2 x^2) = \pi^{\frac{1}{2}}/\alpha. \tag{A.4.8b}$$

By making α^2 complex, one finds, eventually,

$$\begin{aligned} \int_{-\infty}^{\infty} \mathrm{d}x \exp(-a^2 x^2 + ikx) &= \int_{-\infty}^{\infty} \mathrm{d}x \exp(-a^2 x^2) \cos(kx) \\ &= \frac{\pi^{\frac{1}{2}}}{a} \exp(-k^2/4a^2). \end{aligned} \tag{A.4.9}$$

Other useful integrals follow by differentiation with respect to k or to a^2.

Repeated differentiation of $J_0(\alpha)$ with respect to α^2 yields

$$J_{2n}(\alpha) \equiv \int_0^\infty \mathrm{d}x\, x^{2n} \exp(-\alpha^2 x^2)$$
$$= \frac{1\cdot 3\cdots(2n-1)}{2^{n+1}}\cdot\frac{\pi^{\frac{1}{2}}}{\alpha^{2n+1}}. \tag{A.4.10}$$

By contrast, a change of integration variable from x to x^2 and appeal to (A.4.4) yield

$$J_{2m+1}(\alpha) \equiv \int_0^\infty \mathrm{d}x\, x^{2m+1} \exp(-\alpha^2 x^2) = \tfrac{1}{2} m!/\alpha^{2m+2}. \tag{A.4.11}$$

(In fact both (A.4.10, 11) are special cases of

$$\int_0^\infty \mathrm{d}x\, x^s \exp(-\alpha^2 x^2) = \Gamma(\tfrac{1}{2} + s/2)/2\alpha^{s+1},$$

were s need not be an integer.)

A finite instead of the infinite upper limit in (A.4.8) leads to the *error function*, which cannot be expressed in more elementary terms, except asymptotically (see also Sections 8.5.2, 4)):

$$\mathrm{erf}(x) \equiv (2/\pi^{\frac{1}{2}}) \int_0^x \mathrm{d}y \exp(-y^2), \tag{A.4.12a}$$

$$\mathrm{erf}(x) = \frac{2x}{\pi^{\frac{1}{2}}}\left\{1 - \frac{x^2}{1!\,3} + \frac{x^4}{2!\,5} - \frac{x^6}{3!\,7} + \cdots\right\}, \tag{A.4.12b}$$

$$\mathrm{erf}(x\to\infty) \sim 1 - \frac{\exp(-x^2)}{x\pi^{\frac{1}{2}}}\left\{1 - \frac{1}{2x^2} + \frac{1\cdot 3}{(2x^2)^2} - \frac{1\cdot 3\cdot 5}{(2x^2)^3} + \cdots\right\}. \tag{A.4.12c}$$

The error function also occurs as

$$\int_0^\infty \mathrm{d}x \exp(-\alpha^2 x^2)\frac{\sin(kx)}{x} = \tfrac{1}{2}\pi\, \mathrm{erf}(k/2\alpha). \tag{A.4.13}$$

A.5 Some useful Fourier series

For $-\pi < x < \pi$,

$$x = 2\left\{\frac{\sin x}{1} - \frac{\sin 2x}{2} + \frac{\sin 3x}{3} - \cdots\right\} \equiv F_1(x), \tag{A.5.1}$$

where $F_1(x)$ is defined to equal the series for all x, so that $F(x + 2\pi) = F(x)$.

$$x^2 = \tfrac{1}{3}\pi^2 - 4\left\{\frac{\cos x}{1^2} - \frac{\cos 2x}{2^2} + \frac{\cos 3x}{3^2} - \cdots\right\}. \tag{A.5.2}$$

The second series follows by integrating the first. Further,

$$\varepsilon(x) = \frac{4}{\pi}\left\{\frac{\sin x}{1} + \frac{\sin 3x}{3} + \frac{\sin 5x}{5} + \cdots\right\}, \tag{A.5.3}$$

$$|x| = \tfrac{1}{2}\pi - \frac{4}{\pi}\left\{\frac{\cos x}{1^2} + \frac{\cos 3x}{3^2} + \frac{\cos 5x}{5^2} + \cdots\right\}. \tag{A.5.4}$$

Again the second series follows from the first by integration.

It is instructive to compare (A.5.1, 2) with the following.

For $0 < x < 2\pi$,

$$x = \pi - 2\left\{\frac{\sin x}{1} + \frac{\sin 2x}{2} + \frac{\sin 3x}{3} + \cdots\right\} \equiv F_2(x), \tag{A.5.5a}$$

$$(x - \pi) = -2\left\{\frac{\sin x}{1} + \frac{\sin 2x}{2} + \frac{\sin 3x}{3} + \cdots\right\} \equiv F_3(x), \tag{A.5.5b}$$

$$(x - \pi)^2 = \tfrac{1}{3}\pi^2 + 4\left\{\frac{\cos x}{1^2} + \frac{\cos 2x}{2^2} + \frac{\cos 3x}{3^2} + \cdots\right\}. \tag{A.5.6}$$

The functions $F_1(x)$ and $F_2(x)$ are drawn in Fig. A.1. Evidently they coincide over $0 < x < \pi$, but differ over the other half-range $\pi < x < 2\pi$.

Exercises: (i) Verify (trigonometrically) that $F_3(x) = F_1(x \pm \pi)$ for all x. (ii) Draw the graphs corresponding to Fig. A.1 for (A.5.2) and (A.5.6). (iii) Differentiate (A.5.3), and examine the result in the light of Section 1.3.

A.6 Vector analysis

We shall need the vector operators grad ($\text{grad}\, a \equiv \nabla a$), div ($\text{div}\, \boldsymbol{A} \equiv \nabla \cdot \boldsymbol{A}$), and the Laplace operator div grad $\equiv \nabla^2$. We shall not need curl, because we consider only irrotational flow and only scalar waves.

Gauss's theorem reads

$$\int_V \mathrm{d}V \,\text{div}\, \boldsymbol{A} = \int_S \mathrm{d}\boldsymbol{S} \cdot \boldsymbol{A} = \int_S \mathrm{d}S\, A_n, \tag{A.6.1}$$

where S is the closed surface bounding the region V, and the vector surface element $\mathrm{d}\boldsymbol{S} = \boldsymbol{n}\,\mathrm{d}S$ points along the outward unit normal $\boldsymbol{n}$ (see Section 3.1 and Appendix G).

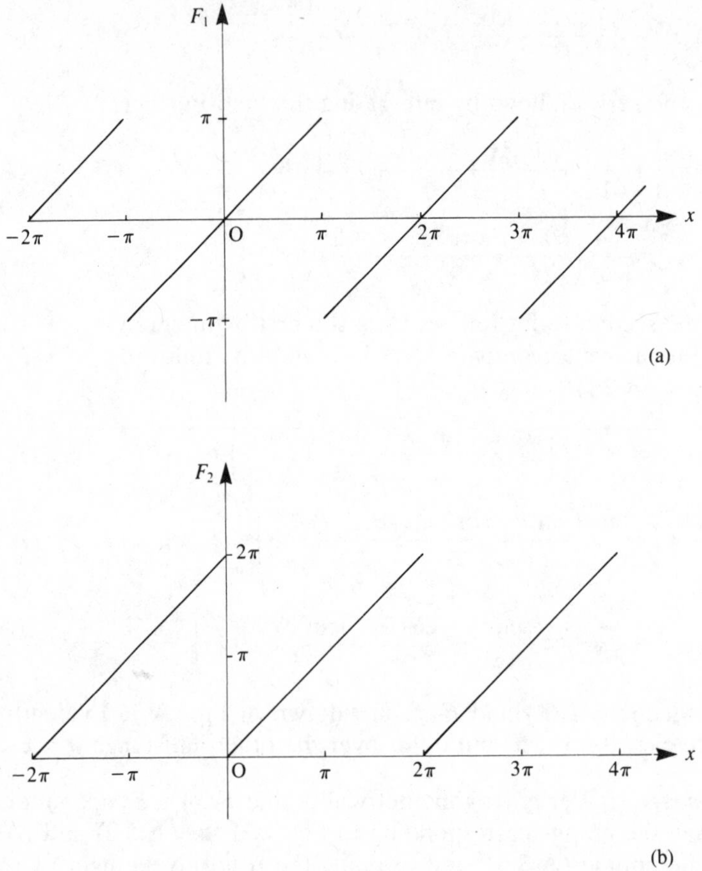

Fig. A.1

The identity

$$\operatorname{div}(a \operatorname{grad} b) = a\nabla^2 b + \operatorname{grad} a \cdot \operatorname{grad} b \tag{A.6.2}$$

applied to the difference $\operatorname{div}(a \operatorname{grad} b - b \operatorname{grad} a)$ then leads to *Green's theorem*:

$$\int_V \mathrm{d}V \{a\nabla^2 b - b\nabla^2 a\} = \int_S \mathrm{d}\mathbf{S} \cdot \{a\boldsymbol{\nabla} b - (\boldsymbol{\nabla} a)b\}. \tag{A.6.3}$$

A.7 The Laplace operator and the Helmholtz equation in spherical polar coordinates

This and the next section quote standard expressions without proof or comment. The material is covered in all good books on mathematical physics: see e.g. Arfken (1973), Butkov (1968), or Riley (1974). Abramowitz and Stegun (1968) provide exceptionally convenient summaries of the properties of the so-called special functions (including, in particular, Bessel functions).

A.7.1 The Laplace operator in 3D

Using the notation $\boldsymbol{r} \equiv (r, \theta, \phi) \equiv (r, \Omega)$ explained in Section 1.4.1,

$$\nabla^2 \equiv \nabla_r^2 + \frac{1}{r^2} \nabla_\Omega^2, \tag{A.7.1}$$

$$\nabla_r^2 \equiv \frac{1}{r^2} \frac{\partial}{\partial r}\left(r^2 \frac{\partial}{\partial r}\right) = \frac{\partial^2}{\partial r^2} + \frac{2}{r} \frac{\partial}{\partial r}, \tag{A.7.2}$$

$$\nabla_\Omega^2 \equiv \frac{1}{\sin\theta} \frac{\partial}{\partial \theta}\left(\sin\theta \frac{\partial}{\partial \theta}\right) + \frac{1}{\sin^2\theta} \frac{\partial^2}{\partial \phi^2}. \tag{A.7.3}$$

In quantum mechanics, the operator $-\nabla_\Omega^2$ is familiar as the squared orbital angular momentum:

$$\boldsymbol{L} \equiv \boldsymbol{r} \wedge \boldsymbol{p} \equiv -i\boldsymbol{r} \wedge \boldsymbol{\nabla}, \qquad L_z = -i\frac{\partial}{\partial \phi}, \qquad \boldsymbol{L}^2 = -\nabla_\Omega^2. \tag{A.7.4}$$

A.7.2 Spherical harmonics and the Legendre polynomials

Consider the simultaneous eigenfunctions of $-\nabla_\Omega^2$ and of $-i\,\partial/\partial\phi$ under the boundary conditions that they be single-valued and non-singular. These eigenfunctions are the *spherical harmonics,* taken as normalized:

$$-\nabla_\Omega^2 Y_{lm} = l(l+1) Y_{lm}, \qquad -i\frac{\partial}{\partial \phi} Y_{lm} = m Y_{lm}, \tag{A.7.5a,b}$$

$$l = 0, 1, 2, \ldots, \qquad m = 0, \pm 1, \pm 2, \ldots, \pm l. \tag{A.7.6}$$

$$\int \mathrm{d}\Omega\, Y^*_{l'm'} Y_{lm} = \delta_{ll'}\, \delta_{mm'}. \tag{A.7.7}$$

The orthogonality (A.7.7) follows from the fact that $-\nabla_\Omega^2$ and $-i\,\partial/\partial\phi$ are Hermitean operators under the stated boundary conditions.

The first few harmonics are

$$Y_{00} = (4\pi)^{-\frac{1}{2}}, \tag{A.7.8}$$

$$Y_{10} = (3/4\pi)^{\frac{1}{2}} \cos\theta, \qquad Y_{1,\pm 1} = \mp(3/8\pi)^{\frac{1}{2}} \sin\theta \exp(\pm i\phi), \tag{A.7.9}$$

$$\begin{aligned} Y_{20} &= (5/16\pi)^{\frac{1}{2}}(3\cos^2\theta - 1), \\ Y_{2,\pm 1} &= \mp(15/8\pi)^{\frac{1}{2}} \cos\theta \sin\theta \exp(\pm i\phi), \\ Y_{2,\pm 2} &= (15/32\pi)^{\frac{1}{2}} \sin^2\theta \exp(\pm 2i\phi). \end{aligned} \tag{A.7.10}$$

Appendix E rewrites them in Cartesians.

The general expression for Y_{lm} is

$$Y_{lm} = N_{lm}(-1)^m P_l^m(\cos\theta)\exp(im\phi). \tag{A.7.11}$$

The P_l^m are the associated Legendre polynomials, defined for $m \geqslant 0$ by $P_l^m(z) = (1-z^2)^{m/2}(d/dz)^m P_l(z)$, and for $m<0$ by $P_l^{-m}(z) = (-1)^m[(l-m)!/(l+m)!]P_l^m(z)$. It follows that $Y_{l,-m} = (-1)^m Y_{l,m}^*$. The phase factors $(-1)^m$ are conventional (but must be observed religiously in any calculation using tables of vector additions coefficients). The norming constant is

$$N_{lm} = \left[\frac{(2l+1)}{4\pi}\frac{(l-m)!}{(l+m)!}\right]^{\frac{1}{2}}. \tag{A.7.12}$$

The generating function for the P_l and the important addition theorem for the Y_{lm} are quoted at the start of Appendix H.3 (see also Section 1.4.9).

A.7.3 The radial Helmholtz equation: spherical Bessel functions

The Helmholtz equation $-(\nabla^2 + k^2)\psi(\mathbf{r}) = 0$ has separable solutions $\psi = R_l(k, r)Y_{lm}(\Omega)$, whose radial factors satisfy

$$\left(-\nabla_r^2 + \frac{l(l+1)}{r^2} - k^2\right)R_l(k, r) = 0. \tag{A.7.13}$$

A convenient pair of linearly independent solutions (for $r \neq 0$) are the *spherical Bessel functions* $j_l(kr) = (\pi/2kr)^{\frac{1}{2}}J_{l+\frac{1}{2}}(kr)$, and $n_l(kr) = (\pi/2kr)^{\frac{1}{2}}N_{l+\frac{1}{2}}(kr)$; but only j_l satisfies the homogeneous equation (A.7.13) at $r=0$. (Some texts denote our n_l by y_l.) They are generated by

$$j_l(z) = z^l\left(-\frac{1}{z}\frac{d}{dz}\right)^l \frac{\sin z}{z}, \qquad n_l(z) = -z^l\left(-\frac{1}{z}\frac{d}{dz}\right)^l \frac{\cos z}{z}. \tag{A.7.14,15}$$

For instance,

$$j_0(z) = \frac{\sin z}{z}; \qquad j_1 = \frac{\sin z}{z^2} - \frac{\cos z}{z}; \qquad j_2 = \left(\frac{3}{z^3} - \frac{1}{z}\right)\sin z - \frac{3}{z^2}\cos z; \tag{A.7.16}$$

$$n_0(z) = -\frac{\cos z}{z}; \qquad n_1 = -\frac{\cos z}{z^2} - \frac{\sin z}{z};$$

$$n_2 = \left(-\frac{3}{z^3} + \frac{1}{z}\right)\cos z - \frac{3}{z^2}\sin z. \tag{A.7.17}$$

The series expansions in powers of z, and the asymptotics as $z \to \infty$ at fixed l, are

$$j_l(z) = \frac{z^l}{1\cdot 3\cdots(2l+1)}\left\{1 - \frac{z^2}{2(2l+3)} + \cdots\right\} \tag{A.7.18}$$

$$n_l(z) = -\frac{1\cdot 3\cdots(2l-1)}{z^{l+1}}\left\{1 + \frac{z^2}{2(2l-1)} + \cdots\right\} \tag{A.7.19}$$

$$j_l(z\to\infty) \sim \frac{1}{z}\cos\left[z - \tfrac{1}{2}\pi(l+1)\right], \tag{A.7.20}$$

$$n_l(z\to\infty) \sim \frac{1}{z}\sin\left[z - \tfrac{1}{2}\pi(l+1)\right]. \tag{A.7.21}$$

Solutions of the Laplace equation emerge in the limit $k^2 \to 0$, i.e. $z \equiv kr \to 0$; then the leading terms of (A.7.18, 19) yield $R_l \sim \text{constant}\cdot r^l$ and $R_l \sim \text{constant}/r^{l+1}$ respectively, which are the functions discussed in Section 4.3.

A.8 The Laplace operator and the Helmholtz equation in plane polar coordinates: Bessel functions

In 2D one has

$$\nabla^2 \equiv \nabla_r^2 + \frac{1}{r^2}\frac{\partial^2}{\partial\phi^2}, \qquad \nabla_r^2 = \frac{1}{r}\frac{\partial}{\partial r}\left(r\frac{\partial}{\partial r}\right) = \frac{\partial^2}{\partial r^2} + \frac{1}{r}\frac{\partial}{\partial r}. \tag{A.8.1, 2}$$

The single-valued separated solutions of the Helmholtz equation may be written as $\psi(\mathbf{r}) = R_{|m|}(k, r)\exp(im\phi)$, $m = 0, \pm 1, \pm 2, \ldots$; the radial factors satisfy

$$\left(-\nabla_r^2 + \frac{m^2}{r^2} - k^2\right)R_{|m|}(k, r) = 0, \qquad |m| = 0, 1, 2, \ldots. \tag{A.8.3}$$

To save writing, we now drop the modulus signs and take m as positive. A convenient pair of linearly independent solutions (for $r \neq 0$) are the Bessel functions $J_m(kr)$ and $N_m(kr)$; but only J_m satisfies the homogeneous equation (A.8.3) at $r = 0$. (Some texts denote N_m by Y_m.) The power series and the asymptotic forms (as $z \to \infty$ for fixed m) are

$$J_m(z) = \left(\frac{z}{2}\right)^m\left\{1 - \frac{(z/2)^2}{1!\,m!} + \frac{(z/2)^4}{2!\,(m+1)!}\cdots\right\}, \tag{A.8.4}$$

$$N_0(z\to 0) \sim \frac{2}{\pi}(\log z + \gamma - \log 2), \tag{A.8.5a}$$

$$N_m(z\to 0) \sim -\frac{(m-1)!}{\pi}\left(\frac{2}{z}\right)^m, \qquad m \geqslant 1 \tag{A.8.5b}$$

$$J_m(z \to \infty) \sim (2/\pi z)^{\frac{1}{2}} \cos\,[z - \tfrac{1}{2}\pi(m + \tfrac{1}{2})], \tag{A.8.6}$$

$$N_m(z \to \infty) \sim (2/\pi z)^{\frac{1}{2}} \sin\,[z - \tfrac{1}{2}\pi(m + \tfrac{1}{2})]. \tag{A.8.7}$$

Solutions of the Laplace equation follow in the limit $k \to 0$, as in 3D.

Eigenvalue problems in cylindrical enclosures require the zeros of J_m or of J'_m. Define $J_m(a_m^{(n)}) = 0$, so that $a_m^{(n)}$ is the nth root; for $m \geqslant 1$ we do not count the zero at the origin. One has

$$a_0^{(1)} = 2.405, \qquad a_0^{(2)} = 5.520, \qquad a_0^{(3)} = 8.654; \tag{A.8.8}$$

$$a_1^{(1)} = 3.832, \qquad a_1^{(2)} = 7.016, \qquad a_1^{(3)} = 10.173. \tag{A.8.9}$$

Similarly, define $J'_m(b_m^{(n)}) = 0$, not counting the zero of J'_0 at the origin:

$$b_0^{(1)} = 3.832, \qquad b_0^{(2)} = 7.016, \qquad b_0^{(3)} = 10.173, \tag{A.8.10}$$

$$b_1^{(1)} = 1.841, \qquad b_1^{(2)} = 5.331, \qquad b_1^{(3)} = 8.536. \tag{A.8.11}$$

Exercise: The roots (A.8.8–11) are very close to the approximate values derived from the asymptotic form (A.8.6) for J_m. Verify this statement.

A.9 Abbreviations

BC	boundary condition
BVP	boundary-value problem
D	Dirichlet
DBC	Dirichlet boundary condition
IC	initial condition
IVP	initial-value problem
LHS	left-hand side
N	Neumann
NBC	Neumann boundary condition
ODE	ordinary differential equation
PDE	partial differential equation
RHS	right-hand side
SVH	St Venant's hypothesis

Table A.1 The Greek alphabet

α	A	alpha	ν	N	nu
β	B	beta	ξ	Ξ	xsi
γ	Γ	gamma	o	O	omicron
δ, ∂	Δ	delta	π	Π	pi
ε	E	epsilon	ρ	P	rho
ζ	Z	zeta	σ	Σ	sigma
η	H	eta	τ	T	tau
θ	Θ	theta	υ	Y	upsilon
ι	I	iota	ϕ	Φ	phi
κ	K	kappa	χ	X	chi
λ	Λ	lambda	ψ	Ψ	psi
μ	M	mu	ω	Ω	omega

B The Dirichlet integral

The (conditionally convergent) integral

$$J(a) \equiv \int_{-\infty}^{\infty} \mathrm{d}\xi \sin(a\xi)/\xi = 2\int_{0}^{\infty} \mathrm{d}\xi \sin(a\xi)/\xi \tag{B.1}$$

appeared in Section 1.2. Evidently it vanishes when $a = 0$, while $J(-a) = -J(a)$ because $\sin(-a\xi) = -\sin(a\xi)$. Thus we need evaluate $J(a)$ only for strictly positive a. Remarkably, it is then independent of the magnitude of a, since changing the integration variable to $x = a\xi$ gives (for $a > 0$):

$$\int_{-\infty}^{\infty} \mathrm{d}x\, a \sin(x)/ax = \int_{-\infty}^{\infty} \mathrm{d}x \sin(x)/x = J(1). \tag{B.2}$$

Hence we need evaluate only $J(1)$, and do so by two methods. The first is elementary but relies on a trick. (Rigorous yet elementary methods are apt to be somewhat laborious: see e.g. Courant (1937).) The second method is standard but needs Cauchy's theorem about functions of a complex variable.

B.1 Elementary method

Interpret $J(1)$ as $\lim_{\varepsilon \to 0} 2\int_0^\infty \mathrm{d}x \exp(-\varepsilon x) \sin(x)/x$. The trick is to write $1/x = \int_0^\infty \mathrm{d}\lambda \exp(-\lambda x)$ under the integral, and then to reverse the order of the x and λ integrations. The cutoff factor $\exp(-\varepsilon x)$ serves to legitimize this interchange. One finds

$$\begin{aligned} J(1) &= \lim_{\varepsilon \to 0} 2\int_0^\infty \mathrm{d}\lambda \int_0^\infty \mathrm{d}x \exp(-(\lambda+\varepsilon)x) \sin x \\ &= 2\int_0^\infty \mathrm{d}\lambda \int_0^\infty \mathrm{d}x \exp(-\lambda x) \sin x. \end{aligned} \tag{B.3}$$

But (cf. Appendix A), $\int_0^\infty \mathrm{d}x \exp(-\lambda x) \sin x = 1/(\lambda^2+1)$. Substitute this into (B.3):

$$J(1) = 2\int_0^\infty \mathrm{d}\lambda/(\lambda^2+1) = 2(\pi/2) = \pi. \tag{B.4}$$

Thus the end-result reads

$$J(0) = 0, \tag{B.5a}$$

$$J(a) \equiv \int_{-\infty}^{\infty} \mathrm{d}\xi \sin(a\xi)/\xi = \varepsilon(a)\pi = \pm\pi \quad \text{for} \quad a \gtrless 0. \tag{B.5b}$$

B.2 Complex-variable method

Consider the contour integral

$$K \equiv \int_C \mathrm{d}z \exp(iz)/z \tag{B.6}$$

where the contour C runs as shown in Fig. B.1, from $-\infty$ to $-\eta$ and from $+\eta$ to $+\infty$ along the real axis, avoiding the origin (where the integrand has a pole) by a small semicircle of radius η in the upper halfplane, centred on the origin. (A semicircle in the lower halfplane would do, but makes the calculations longer.) We shall eventually take the limit $\eta \to 0$.

Thus $K = K_1 + K_2$, with K_1 from the straight parts,

$$K_1 = \left\{ \int_{-\infty}^{-\eta} + \int_{\eta}^{\infty} \right\} \mathrm{d}z \frac{(\cos z + i \sin z)}{z}.$$

The $\cos(z)/z$ component is odd, and cancels from the integral by symmetry (positive and negative contributions cancel each other identically). Hence

$$K_1 = i\left\{ \int_{-\infty}^{-\eta} + \int_{\eta}^{\infty} \right\} \mathrm{d}z \frac{\sin z}{z} \xrightarrow[(\eta \to 0)]{} i \int_{-\infty}^{\infty} \mathrm{d}z \frac{\sin z}{z} = iJ(1). \tag{B.7}$$

In the contribution K_2 from the semicircle, write $z = \eta \exp(i\theta)$, and change the integration variable to θ:

$$K_2 = \int_{\pi}^{0} \frac{\exp(i\theta)(i\,\mathrm{d}\theta)\exp(i\eta\exp(i\theta))}{(\exp(i\theta))} \xrightarrow[(\eta \to 0)]{} \int_{\pi}^{0} i\,\mathrm{d}\theta = -i\pi. \tag{B.8}$$

Hence $K \equiv K_1 + K_2 = i(J(1) - \pi)$.

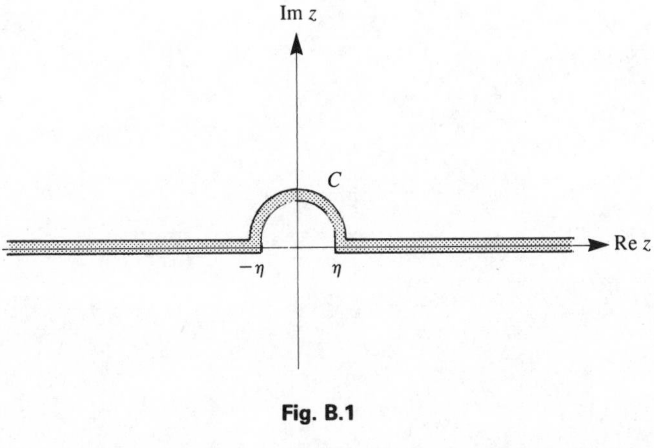

Fig. B.1

Now evaluate K another way, by closing the contour with a semicircle at infinity in the upper halfplane. This large semicircle contributes nothing, because the integrand vanishes as $\text{Im}\, z \to +\infty$, on account of the factor $\exp(iz)$. Hence, by Cauchy's theorem, the integral is zero (because the integrand is analytic everywhere on and inside the closed contour: the pole at $z = 0$ is excluded by the small semicircle). Therefore $K = 0$, whence $J(1) = \pi$. ■

C Fourier analysis in the light of the delta-function

We apply the representation established in Section 1.2,

$$\delta(x-y) = (2\pi)^{-1}\int_{-\infty}^{\infty} dk \exp[ik(x-y)], \tag{C.1}$$

in order to derive some of the crucial formulae in Fourier analysis. The point is that delta-functions make the provenance and the structure of these relations much more transparent. Other comments will follow later.

(i) *The Fourier inversion theorem:*

$$F(x) = \int_{-\infty}^{\infty} dk \exp(ikx)\tilde{F}(k) \rightleftharpoons \tilde{F}(k) = \frac{1}{2\pi}\int_{-\infty}^{\infty} dy \exp(-iky)F(y). \tag{C.2}$$

Proof: Start from the identity $F(x) = \int_{-\infty}^{\infty} dy\, F(y)\,\delta(x-y)$. Substitute (C.1) on the right, and interchange the orders of the y and k integrations:

$$F(x) = \int_{-\infty}^{\infty} dy\, F(y)\frac{1}{2\pi}\int_{-\infty}^{\infty} dk \exp(ik(x-y))$$

$$= \int_{-\infty}^{\infty} dk \exp(ikx)\left\{\frac{1}{2\pi}\int_{-\infty}^{\infty} dy \exp(-iky)F(y)\right\}. \tag{C.3}$$

This identifies $\tilde{F}(k)$ by comparison with (C.2). ■

(ii) *Parseval's theorem:*

$$\int_{-\infty}^{\infty} dx\, |F(x)|^2 = 2\pi\int_{-\infty}^{\infty} dk\, |\tilde{F}(k)|^2. \tag{C.4}$$

Proof: Substitute (C.2) for $F(x)$ and its complex conjugate for $F^*(x)$

$$\int dx\, F^*(x)F(x)$$

$$= \int dx \int dk\, \tilde{F}^*(k)\exp(-ikx)\int dk'\, \tilde{F}(k')\exp(ik'x)$$

$$= \iint dk\, dk'\, \tilde{F}^*(k)\tilde{F}(k')\int dx \exp(i(k-k')x)$$

$$= \iint dk\, dk'\, \tilde{F}^*(k)\tilde{F}(k')2\pi\,\delta(k-k')$$

$$= 2\pi\int dk\, |\tilde{F}(k)|^2.$$ ■

(ii) *Convolutions.* If $c(x) = \int_{-\infty}^{\infty} \mathrm{d}y\, a(x-y)b(y) = \int_{-\infty}^{\infty} \mathrm{d}y\, a(y)b(x-y)$, then the Fourier transforms $\tilde{c}$, $\tilde{a}$, $\tilde{b}$ are related by

$$\tilde{c}(k) = 2\pi\tilde{a}(k)\tilde{b}(k). \tag{C.5}$$

Proof:

$$\begin{aligned}
2\pi\tilde{c}(K) &= \int \mathrm{d}x\, c(x) \exp(-iKx) \\
&= \int \mathrm{d}x \exp(-iKx) \int \mathrm{d}y\, a(x-y)b(y) \\
&= \iint \mathrm{d}x\, \mathrm{d}y \exp(-iKx) \\
&\quad \times \iint \mathrm{d}k\, \mathrm{d}k'\, \tilde{a}(k) \exp(ik(x-y))\tilde{b}(k') \exp(ik'y) \\
&= \iint \mathrm{d}k\, \mathrm{d}k'\, \tilde{a}(k)\tilde{b}(k') \int \mathrm{d}x \exp(i(k-K)x) \int \mathrm{d}y \exp(i(k'-K)y) \\
&= \iint \mathrm{d}k\, \mathrm{d}k'\, \tilde{a}(k)\tilde{b}(k') 2\pi\, \delta(k-K) 2\pi\, \delta(k'-K) \\
&= (2\pi)^2 \tilde{a}(K)\tilde{b}(K).
\end{aligned}$$

■

(iv) *The inversion theorem for series.* For functions $F(x)$ defined over $-\pi \leqslant x \leqslant \pi$, the analogue of (C.2) reads

$$F(x) = \sum_{n=-\infty}^{\infty} c_n \exp(inx) \rightleftharpoons c_n = \frac{1}{2\pi} \int_{-\pi}^{\pi} \mathrm{d}y \exp(-iny) F(y). \tag{C.6}$$

Proof: Start from the representation (1.3.12), namely from $\delta(x-y) = (2\pi)^{-1} \sum_{n=-\infty}^{\infty} \exp(in(x-y))$. The steps are the same as in the proof of (C.2); briefly,

$$\begin{aligned}
F(x) &= \int_{-\pi}^{\pi} \mathrm{d}y\, F(y)\, \delta(x-y) \\
&= \int_{-\pi}^{\pi} \mathrm{d}y\, F(y)(2\pi)^{-1} \sum_n \exp(in(x-y)) \\
&= \sum_{n=-\infty}^{\infty} \exp(inx) \left\{ \frac{1}{2\pi} \int_{-\pi}^{\pi} \mathrm{d}y\, F(y) \exp(-iny) \right\}.
\end{aligned}$$

■

The fact underlying this proof is that

$$\frac{1}{2\pi} \lim_{N\to\infty} \sum_{n=-N}^{N} \exp(inx) = \delta(x), \qquad (|x| \leqslant \pi), \tag{C.7}$$

(see eqn (1.3.12)). Some further insight is obtained by evaluating this geometric sum explicitly. The problem is familiar from the theory of diffraction gratings. Since the first term is $\exp(-iNx)$ and there are $(2N+1)$ terms, one has

$$\sum_{-N}^{N} \exp(inx) = \exp(-iNx)\frac{[1-\exp(ix(2N+1))}{[1-\exp(ix)]}$$
$$= \frac{\exp(i\tfrac{1}{2}x)[\exp(-i(N+\tfrac{1}{2})x) - \exp(i(N+\tfrac{1}{2})x)]}{\exp(i\tfrac{1}{2}x)[\exp(-i\tfrac{1}{2}x) - \exp(i\tfrac{1}{2}x)]}$$
$$= \frac{\sin[(N+\tfrac{1}{2})x]}{\sin[\tfrac{1}{2}x]}. \tag{C.8}$$

Thus, the LHS of (C.7) becomes

$$\frac{1}{2\pi}\lim_{N\to\infty}\frac{\sin[(N+\tfrac{1}{2})x]}{\sin[\tfrac{1}{2}x]} = \lim_{N\to\infty}\left\{\frac{\sin[(N+\tfrac{1}{2})x]}{\pi x}\right\}\left\{\frac{x}{2\sin[\tfrac{1}{2}x]}\right\}. \tag{C.9}$$

But the limit of the first curly brackets is $\delta(x)$, by virtue of the Dirichlet representation (1.2.6): identify $1/\varepsilon \to \infty$ there with $(N+\tfrac{1}{2}) \to \infty$ here. Accordingly,

$$\lim_{N\to\infty}\sum_{-N}^{N}\exp(inx) = \delta(x)\{x/2\sin(x/2)\} = \delta(x),$$

because, by virtue of (1.1.6b), $x/2\sin(x/2)$ when multiplied by $\delta(x)$ can be replaced by its limit as $x \to 0$, which is unity. ■

A little thought by the light of Section 1.3 shows that in their essentials the arguments for the inversion and Parseval theorems apply equally to any complete orthonormal set of functions, since they hinge simply on orthonormality and on the closure property.

(v) *Poisson's summation formula.* Mathematically speaking, Poisson's formula is a remarkable device for converting certain infinite series which converge rather slowly into other series that may converge much faster. As regards Green's functions, the formula can link normal-mode expansions to the multiple-image series (see e.g. Section 9.4.4).

Theorem: Given a function $F(\xi)$,

$$S \equiv \sum_{N=-\infty}^{\infty} F(2\pi N) = \sum_{N=-\infty}^{\infty}\frac{1}{2\pi}\int_{-\infty}^{\infty} d\xi\, F(\xi)\exp(iN\xi)$$
$$= \sum_{N=-\infty}^{\infty}\tilde{F}(N). \tag{C.10}$$

Proof:

$$S \equiv \sum_{N=-\infty}^{\infty} F(2\pi N) = \sum_{N=-\infty}^{\infty} \int_{-\infty}^{\infty} \mathrm{d}\xi\, \delta(\xi - 2\pi N) F(\xi)$$

$$= \int_{-\infty}^{\infty} \mathrm{d}\xi \left\{ \sum_{N=-\infty}^{\infty} \delta(\xi - 2\pi N) \right\} F(\xi).$$

But, by (1.3.13),

$$\left\{ \sum_{N=-\infty}^{\infty} \delta(\xi - 2\pi N) \right\} = (2\pi)^{-1} \sum_{N=-\infty}^{\infty} \exp(iN\xi);$$

hence

$$S = \int_{-\infty}^{\infty} \mathrm{d}\xi \left\{ \frac{1}{2\pi} \sum_{N=-\infty}^{\infty} \exp(iN\xi) \right\} F(\xi)$$

$$= \sum_{N=-\infty}^{\infty} \frac{1}{2\pi} \int_{-\infty}^{\infty} \mathrm{d}\xi\, F(\xi) \exp(iN\xi). \qquad \blacksquare$$

Example:

$$F(\xi) = \exp(-\alpha^2\xi^2),$$

$$\int_{-\infty}^{\infty} \mathrm{d}\xi \exp(-\alpha^2\xi^2) \exp(iN\xi) = \frac{\pi^{\frac{1}{2}}}{2\alpha} \exp(-N^2/4\alpha^2):$$

$$\sum_{N=-\infty}^{\infty} \exp(-\alpha^2(2\pi N)^2) = 1 + 2 \sum_{N=1}^{\infty} \exp(-\alpha^2(2\pi N)^2)$$

$$= \sum_{N=-\infty}^{\infty} \frac{1}{2\pi} \frac{\pi^{\frac{1}{2}}}{2\alpha} \exp(-N^2/4\alpha^2)$$

$$= \frac{1}{2(2\pi)^{\frac{1}{2}}} \frac{1}{\alpha} \left\{ 1 + 2 \sum_{N=1}^{\infty} \exp(-N^2/4\alpha^2) \right\}. \qquad \text{(C.11)}$$

To appreciate this relation, envisage the problem of evaluating the leftmost series to high accuracy when α^2 is very small. Obviously one would need to sum a great many terms. By contrast, the rightmost series then converges very rapidly.

Exercise: Determine how many terms must be summed on the left, and on the right, to guarantee a relative accuracy of $1:10^4$, when $\alpha^2 = 10^{-6}$.

The theorem is readily adapted to an arbitrary repetition distance λ

instead of 2π:

$$\sum_{N=-\infty}^{\infty} F(\lambda N) = \sum_{N=-\infty}^{\infty} \frac{1}{\lambda} \int_{-\infty}^{\infty} d\xi \, F(\xi) \exp(2\pi i N\xi/\lambda). \tag{C.12}$$

Exercise: Prove this.

There are, naturally, quite other mathematical problems associated with Fourier analysis. Basically, they concern the sense in which the series or integral represents the function $F(x)$, e.g. the sense in which the finite sum $(2\pi)^{-1} \sum_{-N}^{N} c_n \exp(inx)$ converges to $F(x)$ as $N \to \infty$, while the c_n are prescribed by (C.6). (Recall for instance the remarks, in small print near the beginning of Section 1.3, on non-uniform convergence, especially about the case where $F(x)$ is singular or, unlike the functions $\exp(inx)$, is not itself periodic.) From the point of view of mathematical rigour these problems should be resolved before one can legitimize in advance the reordering of operations (sums, integrals, and limits) in the formal proofs offered above. However, they are far more subtle than the plain inversion theorems, and we do not pursue them. Indeed, while some introductions to Fourier analysis tend to obscure the essence of the inversion theorems (C.2) and (C.6) because they are afraid to discuss them in terms of delta-functions, others obscure it by focusing prematurely on the far deeper convergence problems. Good yet elementary mathematical treatments are given e.g. by Courant (1937), Lighthill (1958), and Schwartz (1966). Lighthill (1987) gives a vivid account of the contrasts between Dirac's and the orthodox mathematical approaches.

In this book we adopt the procedure that generally proves most efficient in theoretical physics: at first, operations are performed in whatever order is convenient, but the end-results are checked for consistency and for physical sense. Results that appear paradoxical are, as a rule, easily traced to an unwarranted reordering of limit, sum, and/or integral, and the mistake can then be rectified. Interesting cases where such subtleties matter occur, for instance, in Appendix I and in Section 9.4.4.

A fascinatingly well-motivated and accessible introduction to the rigorous results on Fourier analysis is given by Champeney (1987).

D Brownian motion

We sketch the Einstein–Smoluchowski–Langevin account of Brownian motion, mainly as a spectacular case history illustrating the solution of IVPs by the Green's function technique of Section 2.2. A natural extension of the argument then illuminates the provenance of the diffusion equation (Chapters 8 and 9) and of its characteristic irreversibility. An excellent detailed discussion is given by Reif (1965), Sections 15.5–10.

Small but macroscopic particles of mass m suspended in a fluid at temperature T undergo random *Brownian motion*. For simplicity we consider only 1D motion without any applied forces: e.g. vertical motion under gravity and charged particles in an electric field are excluded. The different Brownian particles move independently of each other; statistical averages over many such particles under macroscopically identical conditions ('ensemble averages') are denoted by $\langle \cdots \rangle$. Let $x(t)$ be the position of a particle at time t; the equation of motion is taken to be

$$\ddot{x}(t) + \gamma \dot{x}(t) = F(t)/m. \tag{D.1}$$

(By contrast to Section 2.2, the dependent and independent variables are now called x and t instead of ψ and x.) Both the familiar resistive (energy-dissipative) force $-\eta\dot{x}$ $(\gamma \equiv \eta/m)$ and the fluctuating residual force $F(t)$ are due to the bombardment of the particle by the thermally agitated molecules of the fluid. The basic idea is that, once the total bombardment force is subdivided in this way, it should be a good approximation to treat $F(t)$ (unlike $-\eta\dot{x}$) as lacking any preferred direction. More specifically, we take $F(t)$ as fluctuating at random, on a microscopically short time-scale τ comparable to the mean time between successive collisions of fluid molecules with the particle. Hence we assume not only that $F(t)$ averages to zero:

$$\langle F(t) \rangle = 0, \tag{D.2}$$

but also that there is effectively no correlation at all between the values of F at macroscopically distinguishable times. This is quantified by the Langevin relation

$$\langle F(t)F(t') \rangle = C\,\delta(t - t'). \tag{D.3}$$

Though the RHS should really be a peak with width of order τ, it is approximated by a delta-function for convenience. This leaves γ^{-1} as the only time-scale in the problem.

Studying the motion of particles governed by (D.1), we shall determine, first, the constant C in (D.3), by determining the mean-square velocity $\langle \dot{x}^2(t)\rangle$ and appealing to the equipartition theorem as $t\to\infty$ (i.e. $t\gamma \gg 1$); and second, determine the mean-square displacement $\langle x^2(t)\rangle$, which is the quantity directly measured in Brownian motion, and which also serves as a bridge to the diffusion equation.

Since $F(t)$ is not known, there is obviously no hope at all of any explicit solution of (D.1); the Green's function technique is now essential, because it alone can deal conveniently with arbitrary $F(t)$.

Without significant loss of generality (cf. the exercise below) we can focus on particles that at time $t=0$ start from rest at the origin: $x(0)=0=\dot{x}(0)$. Then in the solution (2.2.14), $x=x_c+x_p$, the complementary function x_c is zero. The appropriate Green's function is defined by

$$\{\partial^2/\partial t^2+\gamma\,\partial/\partial t\}G(t\,|\,t')=-\delta(t-t'), \tag{D.4}$$

subject to the IC $G=0$ for $t\leqslant t'$. It is easily verified that

$$G(t\,|\,t')=-\theta(t-t')\frac{1}{\gamma}\{1-\exp(-\gamma(t-t'))\}, \tag{D.5}$$

whence the motion of the particle is given by

$$x(t)=x_p(t)=-\int_0^t \mathrm{d}t'\; G(t\,|\,t')F(t')/m, \tag{D.6a}$$

$$x(t)=\int_0^t \mathrm{d}t'\;\{1-\exp(-\gamma(t-t'))\}F(t')/\gamma m. \tag{D.6b}$$

(The minus signs in (D.5) and (D.6a) stem from our conventional minus signs in (2.1.1, 2), which entail $f=-F/m$, as in (2.1.6).)

Differentiation of (D.6b) yields the velocity

$$\dot{x}(t)=\int_0^t \mathrm{d}t'\exp(-\gamma(t-t'))F(t')/m. \tag{D.7}$$

There is no contribution from differentiating with respect to the upper limit in (D.6b), because the integrand vanishes there. Thus $\mathrm{d}/\mathrm{d}t$ applied to the right of (D.6b) is effective only when it acts on the integrand.

The mean-square velocity is evaluated straightforwardly:

$$\begin{aligned}\langle \dot{x}^2(t)\rangle &= \left\langle \int_0^t \mathrm{d}t'\exp(-\gamma(t-t'))\frac{F(t')}{m}\int_0^t \mathrm{d}t''\exp(-\gamma(t-t''))\frac{F(t'')}{m}\right\rangle\\ &= \int_0^t \mathrm{d}t'\exp(-\gamma(t-t'))\int_0^t \mathrm{d}t''\exp(-\gamma(t-t''))\langle F(t')F(t'')\rangle/m^2\\ &= \int_0^t \mathrm{d}t'\exp(-\gamma(t-t'))\int_0^t \mathrm{d}t''\exp(-\gamma(t-t''))C\,\delta(t'-t'')/m^2\\ &= \frac{C}{m^2}\int_0^t \mathrm{d}t'\exp(-2\gamma(t-t'))=\frac{C}{2m^2\gamma}(1-\exp(-2\gamma t)).\end{aligned} \tag{D.8}$$

In the first step we have used (D.7), twice; in the second step, we have isolated the two random factors which need averaging; in the third step we have appealed to (D.3).

When $t \gg 1/\gamma$ (i.e. at times that are large enough on a macroscopic scale, and not merely compared to the microscopic correlation time τ), (D.8) implies

$$\langle \dot{x}^2 \rangle \sim C/2m^2\gamma. \tag{D.9}$$

This is the mean-square velocity after a long enough wait, i.e. in thermodynamic equilibrium. But the equipartition theorem (derived from the Boltzmann distribution) asserts that at equilibrium the average 1D kinetic energy is $\frac{1}{2}kT$, where k is Boltzmann's constant. Hence, at large times as in (D.9), one has

$$\tfrac{1}{2}m\langle \dot{x}^2 \rangle = \tfrac{1}{2}m\frac{C}{2m^2\gamma} = \frac{C}{4m\gamma} = \tfrac{1}{2}kT,$$

$$C = 2m\gamma kT = 2\eta kT. \tag{D.10}$$

Thus Langevin's constant C, which measures the mean-square random microscopically fluctuating force, is fully determined by kT and by the macroscopic resistive (dissipative) coefficient η. This is the archetypal example of the 'fluctuation-dissipation theorem' (see e.g. Reif 1965).

For the mean-square displacement, we use (D.6b) twice, then (D.3), and finally (D.10):

$$\begin{aligned}\langle x^2(t)\rangle &= \int_0^t \mathrm{d}t'\,(1-\exp[-\gamma(t-t')])\\ &\quad\times \int_0^t \mathrm{d}t''\,(1-\exp[-\gamma(t-t'')])\,\frac{\langle F(t')F(t'')\rangle}{\gamma^2 m^2}\\ &= \frac{C}{\gamma^2 m^2}\int_0^t \mathrm{d}t'\,(1-\exp[-\gamma(t-t')])\\ &\quad\times \int_0^t \mathrm{d}t''\,(1-\exp[-\gamma(t-t'')])\,\delta(t'-t'')\\ &= \frac{2kT}{\eta}\int_0^t \mathrm{d}t\,(1-\exp[-\gamma(t-t')])^2,\end{aligned}$$

$$\langle x^2(t)\rangle = \frac{2kT}{\eta}\left\{t - \frac{2}{\gamma}(1-\exp[-\gamma t]) + \frac{1}{2\gamma}(1-\exp[-2\gamma t])\right\}. \tag{D.11}$$

At large times ($t\gamma \gg 1$) this yields Einstein's epoch-making result

$$\langle x^2(t)\rangle \sim \left(\frac{2kT}{\eta}\right)t. \tag{D.12}$$

Thus the mean-*square* deviation from the initial position $x(0) = 0$ eventually increases *linearly* with the time t, a relation strongly reminiscent of a random walk (Reif 1965; Feller 1970). The rate of increase is governed wholly by T and by η.

Exercise: Show that the crucial results (D.9, 10, 12) ensue even for an ensemble of particles not all starting from rest at the origin, provided that the initial values $x(0) = a$ and $\dot{x}(0) = u$ have finite averages $\langle a \rangle$ and $\langle u \rangle$, and are uncorrelated with F in the sense that $\langle aF(t) \rangle = \langle a \rangle \langle F(t) \rangle = 0$, $\langle uF(t) \rangle = \langle u \rangle \langle F(t) \rangle = 0$.

The ensemble of Brownian particles that set out from the origin at $t = 0$ will, at time t, be distributed so that $K(x; t)\,dx$ is the fraction of the ensemble in the range dx around x. (We ignore the initial velocity distribution because at times $t \gg 1/\gamma$ it has become irrelevant, as implied by the exercise above.) If $K(x; t)$ were known and if it could be identified as a solution of some simple (linear) and sensibly interpretable partial differential equation, satisfying the IC $K(x; 0) = \delta(x)$ and the conservation law $\int_{-\infty}^{\infty} dx\, K(x; t) = 1$, then one would expect the same equation to govern the time evolution $\psi(x, t)$ of other initial distributions $\psi(x, 0)$ as well. (In the language of Section 8.3, K would be the *propagator* of the equation.)

Obviously we cannot determine $K(x; t)$ uniquely without more physics, because (D.12) supplies only the mean-square value

$$\langle x^2(t) \rangle \equiv \int_{-\infty}^{\infty} dx\, K(x; t) x^2 = (2kT/\eta)t, \tag{D.13}$$

which is compatible with many different distributions K. But, fortunately, a physically plausible assumption that does deliver a unique $K(x; t)$ is suggested by the random-walk property of $\langle x^2(t) \rangle$ already pointed out *à propos* of (D.12). The Langevin relation (D.3) does not exclude the possibility that higher moments of the force are correlated at different times (e.g. $\langle F^2(t) F^2(t') \rangle$ could be non-zero even when $t \neq t'$); but it seems natural to extend (D.3) by assuming that $F(t)$ is totally uncorrelated with $F(t')$ whenever $t \neq t'$, and to guarantee this mathematically in the simplest and most familiar way, by positing that $F(t)$ is a Gaussian random process. (For such a process, the second-order correlation determines all the higher-order ones: see e.g. Feller 1970.)

Granted this assumption, the displacement of any one particle (at times $t \gg 1/\gamma$) is the resultant of very many small and effectively uncorrelated random displacements (we do not spell out the details: imagine the time t subdivided into very many intervals $t/n > 1/\gamma$). Then the law of large numbers (the 'central limit theorem') entails that the distribution $K(x; t)$ is a Gaussian, i.e. of the form $N \exp(-x^2/a^2)$, where N and a are constants. But a Gaussian with zero mean (as here) is uniquely specified by its mean-square deviation $\langle x^2 \rangle = a^2/2 = 2kTt/\eta$, given in our case by (D.12); hence we can write

$$K(x; t) = [4\pi kT/\eta]^{-\frac{1}{2}} \exp\{-x^2/[4kTt/\eta]\}. \tag{D.14}$$

(The representation (1.2.3) shows that $K(x; t \to 0) = \delta(x)$, as required.)

The linear partial differential equation governing K is easily recognized as the 1D diffusion equation from Chapter 8, namely

$$\{D\,\partial/\partial t - \partial^2/\partial x^2\}K(x;t) = 0, \tag{D.15}$$

with its propagator

$$K(x;t) = [4\pi Dt]^{-\frac{1}{2}} \exp\{-x^2/4Dt\}. \tag{D.16}$$

Comparison between (D.14) and (D.16) relates the diffusion constant D to kT and to η, and to Langevin's constant:

$$D = kT/\eta = C/2\eta^2. \tag{D.17}$$

Evidently, the diffusion equation (at least for Brownian particles) can be regarded as a consequence of whatever physics underlies the Einstein–Smoluchowski–Langevin theory, supplemented by the assumption that F is wholly uncorrelated at different times. One of the most notorious features of the diffusion equation is that it is irreversible, i.e. not invariant under time-reversal, as discussed in Section 9.5; in this crucial respect it is unlike the basic laws of molecular dynamics and of electromagnetism, which are reversible. Hence it is clear that the diffusion equation can be related to these more fundamental truths only through some further physical assumption of a statistical nature. A little reflection shows that the present argument makes this assumption at the very beginning, by introducing into the equation of motion (D.1) the force $-m\gamma\dot{x}$ which is explicitly irreversible (it changes sign under time reversal i.e. under $t \to -t$ which entails $\mathrm{d}/\mathrm{d}t \to -\mathrm{d}/\mathrm{d}t$). Indeed, one of the merits of this approach is that it allows subsequent discussions of the very deep problem of irreversibility to focus on the formally very simple equation (D.1).

E Degeneracy and reality properties of complete orthonormal sets

In discussing Green's functions we have repeatedly appealed to the fact that for the operator $-\nabla^2$ one can choose a complete orthonormal set of eigenfunctions all of which are essentially real (see e.g. Sections 1.3, 2.4, 5.1). As we shall see, this is a special case of a general result for Hermitean operators L that are also real in the sense that $L^* = L$, where the asterisk denotes complex conjugation.

Recall first that a function $f(\boldsymbol{r})$ is said to be *essentially real* if one can find a constant C (independent of $\boldsymbol{r}$) such that $f^*(\boldsymbol{r}) = Cf(\boldsymbol{r})$. Evidently C has modulus 1, and we could write $C = \exp(-2i\,\delta)$ with constant δ, i.e. $f^*(\boldsymbol{r}) = \exp(-2i\,\delta)f(\boldsymbol{r})$, implying $f(\boldsymbol{r}) = |f(\boldsymbol{r})| \exp(i\,\delta)$, $f^*(\boldsymbol{r}) = |f(\boldsymbol{r})| \exp(-i\,\delta)$. Since members of a complete orthonormal set are arbitrary up to complex factors of modulus 1, essential reality rather than reality is the important property to look for, being unaffected by such factors. (However, for brevity one often speaks of 'reality' when 'essential reality' is all that is needed and implied.)

If L is Hermitean, then its eigenvalues are real, and its eigenfunctions belonging to different eigenvalues are orthogonal. Hence we write

$$L\phi_n = \lambda_n \phi_n, \qquad \lambda_n = \lambda_n^*, \tag{E.1a,b}$$

$$\int dV\, \phi_{n'}^* \phi_n = \delta_{nn'} \quad \text{if} \quad \lambda_n \neq \lambda_{n'}. \tag{E.2}$$

If $\lambda_n = \lambda_{n'}$ while ϕ_n and $\phi_{n'}$ are linearly independent (i.e. $\phi_n \neq C\phi_{n'}$), we say that the eigenvalue λ_n is degenerate, and that ϕ_n, $\phi_{n'}$ are mutually degenerate eigenfunctions. For short, we also say under such conditions that the operator L itself is degenerate.

From now on we consider only real $L = L^*$. Then complex conjugation of (E.1) yields

$$L\phi_n^* = \lambda_n \phi_n^*. \tag{E.3}$$

Thus, ϕ_n and ϕ_n^* belong to the same eigenvalue. There are two possibilities.

Either $\phi_n = C_n \phi_n^*$ for all n, i.e. every ϕ_n is essentially real, and our goal is achieved automatically. This is necessarily the case if L is non-degenerate. (It is shown at the end of this appendix that L is automatically non-degenerate in the important special case where it is one-dimensional, non-singular, and subject to homogeneous (rather than, say, periodic) boundary conditions.

Or, for some n, $\phi_n \neq C_n\phi_n^*$. Then, taking the sum and the difference of (E.1a) and (E.3), we note that

$$L(\phi_n + \phi_n^*)/2 = L \operatorname{Re} \phi_n = \lambda_n \operatorname{Re} \phi_n, \tag{E.4}$$

$$L(\phi_n - \phi_n^*)/2i = L \operatorname{Im} \phi_n = \lambda_n \operatorname{Im} \phi_n. \tag{E.5}$$

For simplicity we pursue only the case (by far the most common) where ϕ_n and ϕ_n^* are mutually orthonormal in the sense of (E.2):

$$1 = \int dV\, \phi_n^* \phi_n, \tag{E.6}$$

$$0 = \int dV\, (\phi_n^*)^* \phi_n = \int dV\, \phi_n^2, \qquad 0 = \int dV\, \phi_n^{*}, \tag{E.7a,b}$$

where (E.7b) follows from (E.7a) by complex conjugation. In turn, (E.6, 7) entail

$$\int dV\, (\sqrt{2} \operatorname{Re} \phi_n)^2 = 1 = \int dV\, (\sqrt{2} \operatorname{Im} \phi_n)^2, \tag{E.8}$$

$$\int dV\, (\operatorname{Re} \phi_n)(\operatorname{Im} \phi_n) = 0. \tag{E.9}$$

Exercise: Prove (E.8, 9).

But (E.4, 5, 8, 9) simply state that $\sqrt{2} \operatorname{Re} \phi_n$ and $\sqrt{2} \operatorname{Im} \phi_n$ are eigenfunctions of L, both belonging to the eigenvalue λ_n, and that they are mutually orthonormal. Moreover, ϕ_n and ϕ_n^* are of course linear combinations of $\operatorname{Re} \phi_n$ and $\operatorname{Im} \phi_n$. Hence, in the complete orthonormal set of eigenfunctions of L, we simply replace ϕ_n and ϕ_n^* by $\sqrt{2} \operatorname{Re} \phi_n$ and $\sqrt{2} \operatorname{Im} \phi_n$, and modify the labelling of the new set appropriately. In this way one secures an essentially real complete orthonormal set of eigenfunctions, as desired.

The most important examples of such rearrangements are furnished by the harmonic functions on the unit circle and the unit sphere. On the circle, these are the single-valued eigenfunctions of $(\boldsymbol{r} \wedge (-i\nabla))_z^2 = -\partial^2/\partial\phi^2$, belonging to the eigenvalue m^2, namely $\{(2\pi)^{-\frac{1}{2}} \exp(im\phi)\}$, $m = 0, \pm 1, \pm 2, \ldots$. This complex set can be rearranged into $\{(2\pi)^{-\frac{1}{2}}, \pi^{-\frac{1}{2}} \cos(m\phi), \pi^{-\frac{1}{2}} \sin(m\phi)\}$, $m = 1, 2, 3, \ldots$. Notice the different labelling of the two sets.

On the sphere, the harmonics are the single-valued eigenfunctions of

$$(\boldsymbol{r} \wedge (-i\nabla))^2 = -\left\{\frac{1}{\sin\theta}\frac{\partial}{\partial\theta}\left(\sin\theta\frac{\partial}{\partial\theta}\right) + \frac{1}{\sin^2\theta}\frac{\partial^2}{\partial\phi^2}\right\},$$

belonging to the eigenvalues $l(l+1)$, namely the $\{Y_{lm}(\Omega)\}$, $l =$

0, 1, 2; $m = 0, \pm 1, \ldots, \pm l$ (cf. Appendix A). This set can be rearranged into $\{Y_{00}, Y_{l0}, \sqrt{2}\,\mathrm{Re}\,Y_{lm}, \sqrt{2}\,\mathrm{Im}\,Y_{lm}\}$, $l = 0, 1, 2, \ldots; m = 1, 2, \ldots, l$.

Whether this particular choice of the members of the real set is the most convenient depends on the circumstances. For instance, for $l = 1$ one has

$$\{Y_{10}, \sqrt{2}\,\mathrm{Re}\,Y_{11}, \sqrt{2}\,\mathrm{Im}\,Y_{11}\} = \left(\frac{3}{4\pi}\right)^{\frac{1}{2}}\{z/r, -x/r, -y/r\},$$

which does look natural from a Cartesian viewpoint. By contrast, for $l = 2$, the corresponding sequence reads

$$\{(5/16\pi)^{\frac{1}{2}}(3z^2/r^2 - 1), -(15/4\pi)^{\frac{1}{2}}zx/r^2, -15/4\pi)^{\frac{1}{2}}zy/r^2,$$
$$(15/16\pi)^{\frac{1}{2}}(x^2 - y^2)/r^2, (15/4\pi)^{\frac{1}{2}}xy/r^2\},$$

whereas from a Cartesian point of view one might prefer other linear combinations proportional to the obvious elements of the symmetric traceless tensor, namely to $(3z^2/r^2 - 1)$, $(3x^2/r^2 - 1)$, $(3y^2/r^2 - 1)$, xy/r^2, yz/r^2, zx/r^2.

The degeneracies in these examples obviously stem from the invariance of the problem under rotations. While it can happen that particular eigenfunctions are degenerate even when the problem has no physically significant invariance properties, such 'accidental degeneracies' affect only a finite subset of $\{\phi_n\}$. By contrast, in homogeneous BVPs with one-dimensional and non-singular operators L of the type considered in Chapter 2, there can be no degeneracy at all.

Proof: Let $L\phi_1 = \lambda_1\phi_1$ and $L\phi_2 = \lambda_1\phi_2$. Then, by the theorems proved in Section 2.1, the Wronskian $W\{\phi_1, \phi_2\} \equiv (\phi_1\phi_2' - \phi_1'\phi_2)$ is non-zero for all x if ϕ_1 and ϕ_2 are linearly independent, and zero for all x if ϕ_1 and ϕ_2 are linearly dependent (i.e. if $\phi_2 = C\phi_1$). But, by virtue of the homogeneous BCs, W certainly vanishes at both boundary points, where either ϕ_1 and ϕ_2 vanish (under DBCs), or ϕ_1' and ϕ_2' vanish (NBCs). Hence $\phi_2 = C\phi_1$. ■

The same argument applies to any one-dimensional equation, whether the independent variable is a Cartesian or a radial variable.

Notice that under periodic instead of homogeneous BCs degeneracy is perfectly possible even in 1D, as illustrated by the eigenfunctions $\exp(im\phi)$ of $-\partial^2/\partial\phi^2$ discussed above. On the other hand, if L though Hermitean is complex (e.g. $L = -i\,\mathrm{d}/\mathrm{d}x$), then it is in general impossible to find a real complete orthonormal set of eigenfunctions even if L is non-degenerate.

Reasoning like the above is familiar in quantum mechanics, where L would be the Hamiltonian. This is real if it is invariant under time-reversal, which transforms into $\phi(\boldsymbol{r})$ into $\phi^*(\boldsymbol{r})$.

F The Green's functions G_0 as Fourier integrals

The unbounded-space Green's functions G_0 for our various equations were found in the text by various methods. Mathematically the most powerful method determines them through their Fourier integral representations, treating all the independent variables on an equal footing at the outset; this approach always works, and calls only for the evaluation of integrals rather than for the solutions of equations. It has been relegated to an appendix because, except for Poisson's equation, the integrals (in Sections F.3–5) require the calculus of residues, i.e. Cauchy's theorem from the theory of functions of a complex variable; and because contour integration is not always the best source of physical insight.

F.1 Isotropic Fourier integrals

Integrals of the type

$$F(\boldsymbol{R}) = \int \mathrm{d}^3k\, f(k) \exp(i\boldsymbol{k}\cdot\boldsymbol{R}) \tag{F.1.1}$$

are needed frequently in 3D. Here, $f(k)$ depends only on the magnitude k of $\boldsymbol{k}$, but not on the direction of $\boldsymbol{k}$. It is clear from the start, and calculation will confirm, that consequently F depends only on the magnitude but not on the direction of $\boldsymbol{R}$. To see this, note that the RHS is a scalar, but depends on no vector other than $\boldsymbol{R}$: the quantity $\boldsymbol{k}$ is merely an integration variable (a so-called dummy variable). Such a scalar function of a single vector $\boldsymbol{R}$ is necessarily a function only of $R \equiv (\boldsymbol{R}\cdot\boldsymbol{R})^{\frac{1}{2}}$.

The first step is to adopt spherical polar coordinates $\boldsymbol{k} \equiv (k, \Omega) \equiv (k, \theta, \phi)$:

$$F(\boldsymbol{R}) = \int_0^\infty \mathrm{d}k\, k^2 f(k) \int \mathrm{d}\Omega \exp(i\boldsymbol{k}\cdot\boldsymbol{R}). \tag{F.1.2}$$

The essential trick for performing the solid-angle integration $\int \mathrm{d}\Omega$ is to choose the polar axis in $\boldsymbol{k}$-space in the direction of $\boldsymbol{R}$, so that the exponent becomes $ikR\cos\theta$:

$$\int \mathrm{d}\Omega \exp(i\boldsymbol{k}\cdot\boldsymbol{R}) = \int_0^{2\pi} \mathrm{d}\phi \int_{-1}^{1} \mathrm{d}\cos\theta \exp(ikR\cos\theta)$$

$$= 2\pi \cdot \frac{1}{ikR}[\exp(ikR) - \exp(-ikR)],$$

$$\int d\Omega \exp(i\boldsymbol{k}\cdot\boldsymbol{R}) = 4\pi \sin(kR)/kR. \tag{F.1.3}$$

Substitution into (F.1.2) yields

$$F(\boldsymbol{R}) = \frac{4\pi}{R}\int_0^\infty dk\, kf(k)\sin(kR). \tag{F.1.4}$$

Notice that it makes no difference to the integral (F.1.3) whether $\Omega \equiv (\theta, \phi)$ is regarded as the polar angle $\Omega(\hat{\boldsymbol{k}})$ of $\boldsymbol{k}$ with fixed $\boldsymbol{R}$ (as above), or as the polar angle $\Omega(\hat{\boldsymbol{R}})$ of $\boldsymbol{R}$ for fixed $\boldsymbol{k}$; all that matters is that we integrate over the *relative* direction of $\boldsymbol{k}$ and $\boldsymbol{R}$.

In 2D, we adopt plane polar coordinates, choose the first axis of $\boldsymbol{k}$-space in the direction of $\boldsymbol{R}$, and find

$$F^{(2)}(\boldsymbol{R}) \equiv \int d^2k\, f(k)\exp(i\boldsymbol{k}\cdot\boldsymbol{R})$$

$$= \int_0^\infty dk\, kf(k)\int_0^{2\pi} d\phi \exp(ikR\cos\phi). \tag{F.1.5}$$

But $\int_0^{2\pi} d\phi \exp(i\alpha\cos\phi) = 2\pi J_0(\alpha)$, where J_0 is the standard Bessel function. Thus

$$F^{(2)}(\boldsymbol{R}) = 2\pi\int_0^\infty dk\, kf(k)J_0(kR). \tag{F.1.6}$$

F.2 Poisson's equation in 3D

The Fourier integral representation of the Poisson Green's function is just its expansion in terms of the eigenfunctions of $-\nabla^2$ (see Section 4.5.2). It is easy to evaluate in 3D, but awkward in 2D and 1D. [The reader could try these as an exercise, but should be prepared to find it quite searching.] In 3D we write

$$G_0(\boldsymbol{r}\,|\,\boldsymbol{r}') = \int d^3k\, g(\boldsymbol{k})\exp(i\boldsymbol{k}\cdot\boldsymbol{r}), \tag{F.2.1}$$

$$\delta(\boldsymbol{r}-\boldsymbol{r}') = \int d^3k\,(2\pi)^{-3}\exp(i\boldsymbol{k}\cdot(\boldsymbol{r}-\boldsymbol{r}')), \tag{F.2.2}$$

where g depends, tacitly, on $\boldsymbol{r}'$ as well as on $\boldsymbol{k}$. Substituting into the defining equation $-\nabla^2 G_0 = \delta(\boldsymbol{r}-\boldsymbol{r}')$, and then equating the coefficients of $\exp(i\boldsymbol{k}\cdot\boldsymbol{r})$, one finds

$$-\nabla^2 G_0 = \int \mathrm{d}^3k\, g(\boldsymbol{k})(-\nabla^2 \exp(i\boldsymbol{k}\cdot\boldsymbol{r}))$$

$$= \int \mathrm{d}^3k\, g(\boldsymbol{k})k^2 \exp(i\boldsymbol{k}\cdot\boldsymbol{r})$$

$$= (2\pi)^{-3} \int \mathrm{d}^3k \exp(i\boldsymbol{k}\cdot(\boldsymbol{r}-\boldsymbol{r}')), \tag{F.2.3}$$

$$k^2 g(\boldsymbol{k}) = \exp(-i\boldsymbol{k}\cdot\boldsymbol{r}')/(2\pi)^3. \tag{F.2.4}$$

This determines g only up to an addend $A(\boldsymbol{k})\,\delta(k^2)$, where A is an arbitrary polynomial in the Cartesian components of $\boldsymbol{k}$: one can see this from the fact that multiplication by k^2 (as on the left of (F.2.4)) destroys such an addend, because $k^2\,\delta(k^2) = 0$. Hence, on dividing (F.2.4) by k^2 in order to obtain g, the addend must be reinstated explicitly:

$$g(\boldsymbol{k}) = \exp(-i\boldsymbol{k}\cdot\boldsymbol{r}')/(2\pi)^3k^2 + A(\boldsymbol{k})\,\delta(k^2). \tag{F.2.5}$$

However, in our particular case of Poisson's equation in 3D, the addend can be dropped, because its contribution to G_0 vanishes automatically:

$$\int \mathrm{d}^3k\, A(\boldsymbol{k})\,\delta(k^2)\exp(i\boldsymbol{k}\cdot\boldsymbol{r}) = \int_0^\infty \mathrm{d}k\, k^2\,\delta(k^2)\int \mathrm{d}\Omega\, A(\boldsymbol{k})\exp(i\boldsymbol{k}\cdot\boldsymbol{r})$$

$$= \int_0^\infty \tfrac{1}{2}\,\mathrm{d}k^2\,(k^2)^{\frac{1}{2}}\,\delta(k^2)\int \mathrm{d}\Omega\, A(\boldsymbol{k})\exp(i\boldsymbol{k}\cdot\boldsymbol{r}) = 0.$$

The last expression vanishes because, under integration with respect to k^2, $(k^2)^{\frac{1}{2}}\,\delta(k^2) = 0$.

With $g = \exp(-i\boldsymbol{k}\cdot\boldsymbol{r}')/(2\pi)^3k^2$, (F.2.1) yields

$$G_0(\boldsymbol{r}\mid\boldsymbol{r}') = \int \mathrm{d}^3k \frac{1}{(2\pi)^3k^2}\exp(i\boldsymbol{k}\cdot\boldsymbol{R}), \qquad \boldsymbol{R} \equiv \boldsymbol{r}-\boldsymbol{r}'. \tag{F.2.6}$$

The integral is precisely of the form (F.1.1), with $f(k) = 1/(2\pi)^3k^2$. Substituting into (F.1.4) we obtain

$$G_0(\boldsymbol{r}\mid\boldsymbol{r}') \equiv G_0(R) = \frac{4\pi}{(2\pi)^3R}\int_0^\infty \mathrm{d}k\frac{\sin(kR)}{k}.$$

The integral is just Dirichlet's (Appendix B); since R is positive, it equals

$\frac{1}{2}\pi$, leading to the end-result

$$G_0(R) = \frac{4\pi}{(2\pi)^3 R}\frac{\pi}{2} = \frac{1}{4\pi R}, \tag{F.2.7}$$

as in (4.4.4).

F.3 The Helmholtz equation

To avoid confusion with the parameter k^2 of the Helmholtz equation $-(\nabla^2 + k^2)G_k = \delta(\boldsymbol{r} - \boldsymbol{r}')$, we now write the Fourier variable as $\boldsymbol{l}$ (instead of $\boldsymbol{k}$ as in Sections F.1, 2), and indicate dependence on k through a suffix. (Thus G_k here is the function called G_0 or $G_{0\omega}$ in the text.) The Poisson Green's function follows as the limit $k^2 \to 0$. The new feature is precisely the appearance of k^2, and its connection with the outgoing-wave BC.

We write, in nD,

$$G_k(\boldsymbol{r} \mid \boldsymbol{r}') = \int \mathrm{d}^n l\, g(\boldsymbol{l}) \exp(i\boldsymbol{l} \cdot \boldsymbol{r}), \tag{F.3.1}$$

$$\delta(\boldsymbol{r} - \boldsymbol{r}') = \int \frac{\mathrm{d}^n l}{(2\pi)^n} \exp(i\boldsymbol{l} \cdot (\boldsymbol{r} - \boldsymbol{r}')) \tag{F.3.2}$$

$$-(\nabla^2 + k^2)G_k = \int \mathrm{d}^n l\, (l^2 - k^2) g \exp(i\boldsymbol{l} \cdot \boldsymbol{r})$$

$$= \int \frac{\mathrm{d}^n l}{(2\pi)^n} \exp(i\boldsymbol{l} \cdot (\boldsymbol{r} - \boldsymbol{r}')),$$

$$(l^2 - k^2) g(\boldsymbol{l}) = \exp(-i\boldsymbol{l} \cdot \boldsymbol{r}')/(2\pi)^n, \tag{F.3.3}$$

$$g(\boldsymbol{l}) = \exp(-i\boldsymbol{l} \cdot \boldsymbol{r}')/(2\pi)^n (l^2 - k^2) + A(\boldsymbol{l})\, \delta(l^2 - k^2). \tag{F.3.4}$$

The second term enters for the same reason as it did in (F.2.5), but now its contribution to G_k need not vanish. Because, again, A is an arbitrary polynomial in the Cartesian components of $\boldsymbol{l}$, its Fourier transform can represent any well-behaved solution of the homogeneous equation:

$$-(\nabla^2 + k^2) \int \mathrm{d}^n l\, A(\boldsymbol{l})\, \delta(l^2 - k^2) \exp(i\boldsymbol{l} \cdot \boldsymbol{r})$$

$$= \int \mathrm{d}^n l\, A(\boldsymbol{l})(l^2 - k^2)\, \delta(l^2 - k^2) \exp(i\boldsymbol{l} \cdot \boldsymbol{r}) = 0. \tag{F.3.5}$$

In other words, this component is a complementary function. It must be chosen so that its combination with the particular integral (which stems from the first term in (F.3.4)) satisfies the outgoing-wave BC.

Exercise: Show that $\int d^3l\, A(\boldsymbol{l})\, \delta(l^2 - k^2) \exp(i\boldsymbol{l} \cdot \boldsymbol{r}) = A(-i\nabla)\{2\pi \times \sin(kr)/r\}$, and verify that this vanishes when $k^2 = 0$, as already known from Section F.2.

On the other hand, the integral $\int d^nl\, (l^2 - k^2)^{-1} \cdots$ (over the first term in (F.3.4)) is itself underdefined as yet, because in the complex l-plane the integration contour C must avoid the poles at $l = \pm k$ by infinitesimal detours that have not yet been specified. Different specifications assign to the integral different multiples of the residues at the poles, so that this underdefinition merely replicates the arbitrariness still left in the choice of A: all the terms in question stem wholly from $l^2 = k^2$. Therefore we can and now do drop $A\, \delta(l^2 - k^2)$ from $g(\boldsymbol{l})$, and write

$$G_k = \int_C \frac{d^nl}{(2\pi)^n} \cdot \frac{\exp(i\boldsymbol{l} \cdot \boldsymbol{R})}{(l^2 - k^2)}, \qquad \boldsymbol{R} \equiv \boldsymbol{r} - \boldsymbol{r}'. \tag{F.3.6}$$

It remains only to prescribe the contour C for the integration over the magnitude l of $\boldsymbol{l}$, in such a way that G_k obeys the outgoing-wave BCs.

We deal with 3D, 2D and 1D in turn.

(i) *Three dimensions.* The integral (F.3.6) is of the isotropic type (F.1.1), with $f(l) = (2\pi)^{-3}(l^2 - k^2)^{-1}$. Therefore (F.1.4) yields

$$G_k^{(3)} = \frac{4\pi}{(2\pi)^3} \frac{1}{R} \int_0^\infty dl\, l \frac{1}{l^2 - k^2} \sin(lR). \tag{F.3.7}$$

The integrand is even in l; hence we write $\int_0^\infty dl \cdots = \frac{1}{2}\int_{-\infty}^\infty dl \cdots = \frac{1}{2}\int_C dl$. Taking partial fractions, one obtains

$$G_k^{(3)} = \frac{4\pi}{(2\pi)^3} \frac{1}{2R} \int_C dl\, l \frac{1}{2l} \left[\frac{1}{l - k} + \frac{1}{l + k}\right] \frac{1}{2i} [\exp(ilR) - \exp(-ilR)], \tag{F.3.8}$$

where k is defined as positive.

The integrals with $\exp(\pm ilR)$ are closed along semicircles at infinity in the upper (lower) half complex plane respectively, where the exponentials vanish. As we shall see, in order to respect the BCs, the contour C must be chosen as in Fig. F.1a; the poles, indicated by crosses, are sidestepped along small semicircles with radii $\varepsilon \to 0$. Equivalently, one can integrate along the real axis, but displace the poles infinitesimally, by modifying the integrand of (F.3.8) according to

$$\frac{1}{l \mp k} \to \frac{1}{l \mp (k + i\varepsilon)}, \qquad \varepsilon \to 0+. \tag{F.3.9}$$

As shown in Fig. F.1.b, this shifts the poles off the contour, and to the same side of it as in Fig. F.1a. Cauchy's theorem then yields the integrals

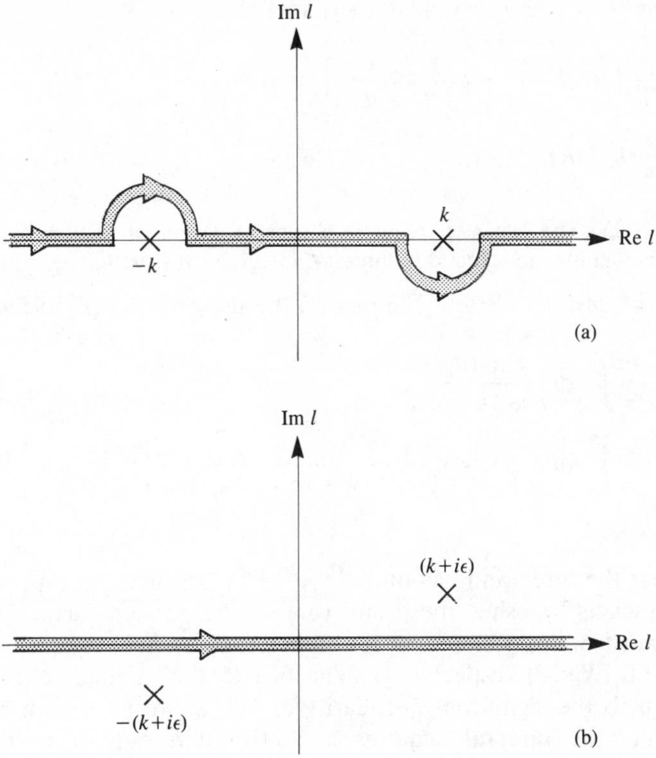

Fig. F.1

in terms of the residues at the poles:

$$G_k^{(3)} = \frac{4\pi}{(2\pi)^3}\frac{1}{2R}\frac{1}{4i}\{(2\pi i)\exp(ikR) - (-2\pi i)\exp(ikR)\}$$
$$= \exp(ikR)/4\pi R, \qquad \text{(F.3.10)}$$

as in (13.2.11). The extra minus sign enters the second term because the contour when closed in the lower halfplane runs clockwise, i.e. in the negative direction.

Exercise: Verify that the outgoing-wave BCs are contravened if either or both of the poles are shifted to the other side of the contour. For instance, if in (F.3.9) one replaces $(k + i\varepsilon)$ by $(k - i\varepsilon)$, then one obtains the Green's function obeying incoming-wave BCs. What ensues if in Fig. F.1b both poles are above (or below) the axis?

(ii) *Two dimensions.* The same choice of contour as above, or the same substitution $k \to k + i\varepsilon$, now yield, through (F.1.6),

$$G_k^{(1)} = 2\pi \int_0^\infty \mathrm{d}l\, l \left\{ \frac{1}{(2\pi)^2} \cdot \frac{1}{(l^2 - (k + i\varepsilon)^2)} \right\} J_0(lR)$$

$$= \frac{i}{4} H_0^{(1)}(kR), \tag{F.3.11}$$

as in (13.2.12). The last step requires a representation of $H_0^{(1)}$ that is rather obscure, though it can be found in tables of integrals, at least for negative k^2.

(iii) *One dimension.* From (F.3.6) with the prescription (F.3.9) we have

$$G_k^{(1)} = \frac{1}{2\pi} \int_{-\infty}^{\infty} \mathrm{d}l \frac{\exp(ilX)}{l^2 - (k + i\varepsilon)^2}$$

$$= \frac{1}{2\pi} \int_{-\infty}^{\infty} \mathrm{d}l \exp(ilX) \frac{1}{2l} \left[\frac{1}{l - k - i\varepsilon} + \frac{1}{l + k + i\varepsilon} \right], \tag{F.3.12a}$$

$$X \equiv x - x'. \tag{F.3.12b}$$

Notice that the integrand has no pole at $l = 0$, because the expression in square brackets vanishes there and cancels the explicit factor $1/l$. The contour must now be closed in the upper (lower) half complex l-plane when $X > 0$ ($X < 0$) respectively. (The fact that X, unlike R, can have either sign, is the significant peculiarity of 1D, as we have seen before.) Accordingly the integral acquires a contribution only from the first (second) term:

$$G_k^{(1)} = \frac{1}{\pi} \left\{ \theta(X) \frac{2\pi i}{2k} \exp(ikX) + \theta(-X) \frac{(-2\pi i)}{(-2k)} \exp(-ikX) \right\}$$

$$= \frac{i}{2k} \exp(ik\,|X|), \tag{F.3.13}$$

as in (13.2.14).

F.4 The wave equation

Equation (11.2.1) wrote the nD propagator $K_0^{(n)}$ as an n-fold Fourier integral with respect to the space coordinates, the time-dependent factor $c \sin(kc\tau)/k$ being determined by a separate argument. The Green's function then emerged as $G_0^{(n)} = \theta(\tau) K_0^{(n)}$. By contrast, we now determine $G_0^{(n)}$ directly as an $(n+1)$-fold Fourier integral:

$$G_0(\mathbf{r}, t \,|\, \mathbf{r}', t') = \int \mathrm{d}^n k \int \mathrm{d}\omega\, g(\mathbf{k}, \omega) \exp(-i\omega t + i\mathbf{k} \cdot \mathbf{r}). \tag{F.4.1}$$

All the integrations run from $-\infty$ to $+\infty$. Tacitly, g is a function also of t' and $\boldsymbol{r}'$. In the defining equation $\Box^2 G_0 = \delta(t-t')\,\delta(\boldsymbol{r}-\boldsymbol{r}')$ we represent the delta-functions too by their Fourier integrals, and find

$$\left(\frac{1}{c^2}\frac{\partial^2}{\partial t^2} - \nabla^2\right)G_0 = \int \mathrm{d}^n k \int \mathrm{d}\omega \left(-\frac{\omega^2}{c^2} + k^2\right) g \exp\left(-i\omega t + i\boldsymbol{k}\cdot\boldsymbol{r}\right)$$
$$= \frac{1}{(2\pi)^{n+1}} \int \mathrm{d}^n k \int \mathrm{d}\omega \exp\left(-i\omega(t-t') + i\boldsymbol{k}\cdot(\boldsymbol{r}-\boldsymbol{r}')\right). \quad \text{(F.4.2)}$$

Equating coefficients one obtains $g = -c^2/(2\pi)^{n+1}(\omega^2 - c^2k^2)$, with the integration contour C in the complex ω-plane (for fixed $\boldsymbol{k}$) still to be determined. Thus,

$$G_0 = -\frac{c^2}{(2\pi)^{n+1}} \int \mathrm{d}^n k \int_C \mathrm{d}\omega \frac{\exp\left(-i\omega\tau + i\boldsymbol{k}\cdot\boldsymbol{R}\right)}{\omega^2 - c^2k^2}$$
$$= -\frac{c^2}{(2\pi)^{n+1}} \int \mathrm{d}^n k \exp\left(i\boldsymbol{k}\cdot\boldsymbol{R}\right)$$
$$\times \int_C \mathrm{d}\omega \exp\left(-i\omega\tau\right) \frac{1}{2ck}\left\{\frac{1}{\omega - ck} - \frac{1}{\omega + ck}\right\}, \quad \text{(F.4.3a)}$$

$$\tau \equiv t - t', \qquad \boldsymbol{R} \equiv \boldsymbol{r} - \boldsymbol{r}'. \quad \text{(F.4.3b)}$$

Notice that the ω integral does not involve the dimensionality n.

We claim that the causal initial condition, namely $G_0 = 0$ for $\tau < 0$, is satisfied if C is chosen as in Fig. F.2a; this is equivalent to replacing ω by $\omega + i\varepsilon$ in the denominator (with $\varepsilon \to 0+$), and integrating along the real axis, because either prescription locates both poles on the same side of the contour, namely below it. (At first sight this prescription differs from that adopted for the Helmholtz Green's function: here it is the integration variable ω that is assigned a positive imaginary part, while in Section F.3 such an imaginary part was assigned to the parameter k. But physically the two prescriptions correspond: the frequency ω in the wave equation becomes precisely the parameter ck in the Helmholtz equation, and acquires a positive imaginary part in both cases.)

The initial conditions are verified as follows. When $\tau < 0$, the contour is closed along the semicircle at infinity in the upper halfplane, where $\exp(-i\omega\tau)$ vanishes. But then there are no poles within the contour, and by Cauchy's theorem the integral, and therefore G_0, vanish. When $\tau > 0$, the contour is closed in the lower halfplane, including both poles, and introducing an extra minus sign because the contour now runs clockwise.

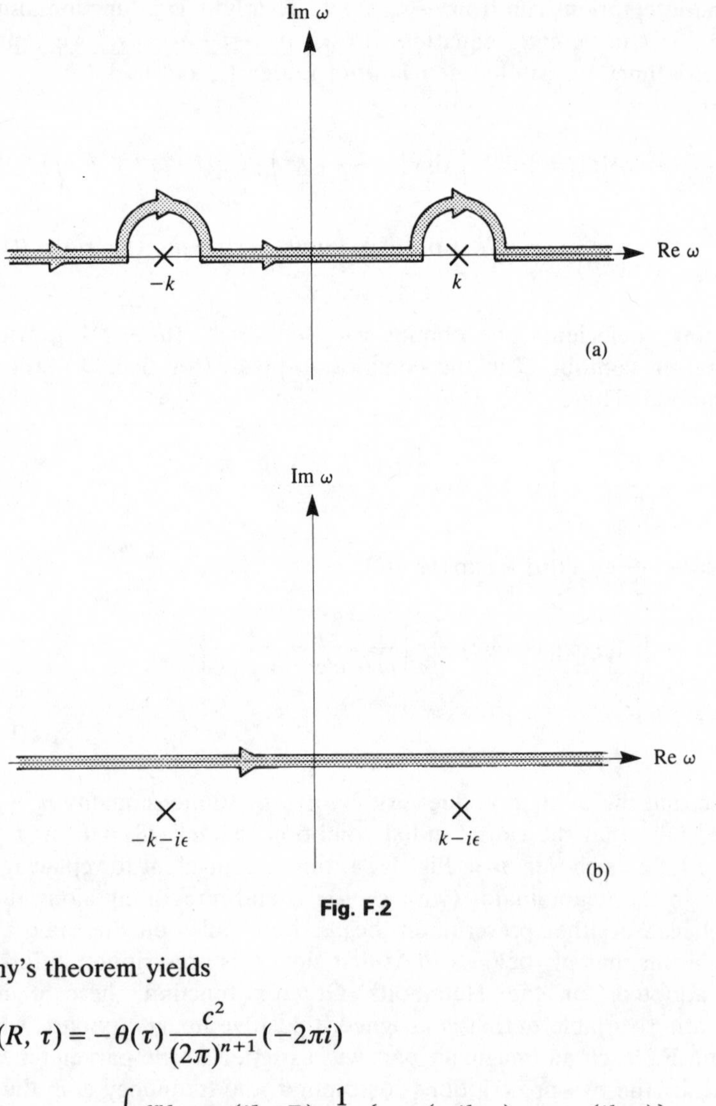

Fig. F.2

Cauchy's theorem yields

$$G_0(R, \tau) = -\theta(\tau)\frac{c^2}{(2\pi)^{n+1}}(-2\pi i)$$
$$\times \int d^n k \exp(i\boldsymbol{k}\cdot\boldsymbol{R})\frac{1}{2ck}\{\exp(-ikc\tau) - \exp(ikc\tau)\},$$

$$G_0(R, \tau) = \theta(\tau)\frac{c}{(2\pi)^n}\int d^n k \frac{1}{k}\exp(i\boldsymbol{k}\cdot\boldsymbol{R})\sin(kc\tau). \tag{F.4.4}$$

This is just the basic expression $G_0 = \theta(\tau)K_0$ with K_0 as in (11.2.1).

Exercise: Determine what ensues from the alternative prescriptions $(\omega^2 - c^2k^2) \to (\omega^2 - c^2(k \pm i\varepsilon)^2)$.

F.5 The diffusion equation

By contrast with Section 8.3.1, we now determine the 1D Green's function G_0 directly as a double Fourier integral satisfying $LG_0 \equiv (\partial/\partial t - D\nabla^2)G_0 = \delta(t-t')\,\delta(x-x')$, and $G_0 = 0$ for $t < t'$.

By the same steps as in Sections F.4 and F.5, one finds

$$G_0(x, t \mid x', t') = \int_{-\infty}^{\infty} \mathrm{d}k \int_{-\infty}^{\infty} \mathrm{d}\omega\, g(k, \omega) \exp(-i\omega t + ikx),$$

$$LG_0 = \int_{-\infty}^{\infty} \mathrm{d}k \int_{-\infty}^{\infty} \mathrm{d}\omega\, (-i\omega + Dk^2) g \exp(-i\omega t + ikx)$$

$$= \frac{1}{(2\pi)^2} \int_{-\infty}^{\infty} \mathrm{d}k \int_{-\infty}^{\infty} \mathrm{d}\omega \exp(-i\omega(t-t') + ik(x-x')), \tag{F.5.1}$$

$$g(k, \omega) = \frac{1}{(2\pi)^2} \frac{i}{(\omega + iDk^2)}. \tag{F.5.2}$$

In this case, division by $(\omega + iDk^2)$ does not lead to any arbitrariness, because the divisor cannot vanish for real k and real ω. Thus, with $\tau \equiv t - t'$ and $\xi \equiv x - x'$,

$$G_0(\xi, \tau) = \frac{i}{(2\pi)^2} \int_{-\infty}^{\infty} \mathrm{d}k \exp(ik\xi) \int_{-\infty}^{\infty} \mathrm{d}\omega \frac{1}{\omega + iDk^2} \exp(-i\omega\tau). \tag{F.5.3}$$

The ω-integrand has a pole at $\omega = -iDk^2$ on the negative imaginary axis. If $\tau < 0$, the contour is closed in the upper-half ω-plane, where there is no pole, so that the integral vanishes. If $\tau > 0$, the contour is closed in the lower halfplane, and Cauchy's theorem yields $\int \mathrm{d}\omega \cdots = (-2\pi i) \exp(-Dk^2\tau)$. Therefore

$$G_0(\xi, \tau) = \frac{i}{(2\pi)^2}(-2\pi i) \int_{-\infty}^{\infty} \mathrm{d}k \exp(ik\xi) \exp(-Dk^2\tau)$$

$$= \frac{1}{2\pi} \int_{-\infty}^{\infty} \mathrm{d}k \exp(-Dk^2\tau + ik\xi), \tag{F.5.4}$$

precisely our former result (8.3.11).

G Dilemmas with notations for boundary- and initial-value problems

In non-homogeneous BVPs there are always conflicting pressures on the notation, which no choice can reconcile with complete success. The same conflicts beset IVPs. They concern the choice of symbols (i) for boundary points, and (ii) for the data, i.e. for the prescribed values of ψ or of its normal derivative on the boundary, or initially. Both dilemmas can be exhibited say through the Dirichlet magic rule for Poisson's equation, namely through

$$\begin{aligned}\psi(\boldsymbol{r}) &= \int_V \mathrm{d}V'\ \rho(\boldsymbol{r}')G_\mathrm{D}(\boldsymbol{r}' \mid \boldsymbol{r}) - \int_S \mathrm{d}S'\ \psi_S(\boldsymbol{r}')\ \partial_n' G_\mathrm{D}(\boldsymbol{r}' \mid \boldsymbol{r}) \\ &\equiv f_\mathrm{D}(\boldsymbol{r}) + g_\mathrm{D}(\boldsymbol{r}). \qquad \text{(G.1)}\end{aligned}$$

Ultimately, the only safe course is to understand the questions involved, and to abandon the expectation that any manageable notation will illuminate all of them simultaneously.

(i) *The labelling of boundary points.* In the volume integral f_D, $\boldsymbol{r}'$ represents a point ranging throughout V. By contrast, in the surface integral g_D, $\boldsymbol{r}'$ ranges only over the boundary S. Moreover, there is a prime on $\mathrm{d}S'$, which indicates that it is a surface element at $\boldsymbol{r}'$: when $\mathrm{d}S'$ shifts, $\boldsymbol{r}'$ shifts. But the suffix S on the integration symbol carries no prime, because it merely identifies the boundary generically. This may seem a large amount of information to have to bear in mind while looking at such expressions on the page; yet without the information the expressions cease to make sense. In fact it is possible to devise notations indicating all this explicitly; but these have the disadvantage of cluttering up what is actually written, and thus making it difficult to read in a different way. The notation we have adopted is quite common in theoretical physics, and should cause no difficulties once it is understood, and after some practice in reading and writing it.

(ii) *Distinguishing input from output.* Under the surface integral g_D, $\psi_S(\boldsymbol{r}')$ stands for the prescribed boundary-value: it is data, i.e. known in advance, and not an outcome of the calculation. In this respect it differs very significantly from $\psi(\boldsymbol{r})$ on the left of (G.1), which is the solution to a problem, rather than part of the data that specifies the problem at the outset. Hence it would be tempting to reserve an altogether different symbol for the data $\psi_S(\boldsymbol{r}') \equiv (\psi(\boldsymbol{r})$ with $\boldsymbol{r}'$ on $S)$. Similarly, in an IVP, we might wish to specify, in advance, that $\psi(\boldsymbol{r}, t)$ when $t = t_0$ is given by an initial-value function $a(\boldsymbol{r})$, say.

The drawbacks of special symbols of this kind are twofold. First, and equally important for boundary values and for initial values, the notation ceases to indicate the physical nature of the input data: for instance, '$a(\boldsymbol{r})$' above has no visible association with 'ψ'. (Nor have the symbols 'f_D' and 'g_D' in (G.1).) Second, and troublesome especially for boundary values, in order to specify data defined purely on S, we should have to parametrize (coordinatize) the surface; this cannot always be done naturally and directly by means of the coordinate $\boldsymbol{r}'$ itself, because (in 3D) $\boldsymbol{r}'$ has three components, whereas S is a two-dimensional manifold, and one would expect a point on it to be specified by a surface-adapted two-component coordinate system.

In most places we try to alleviate such conflicts by writing $\psi_S(\boldsymbol{r})$ when we mean 'ψ at a point $\boldsymbol{r}$ which is on the surface', and $\partial_n\psi_S(\boldsymbol{r})$ when we mean 'the normal derivative of ψ, calculated in a direction normal to the surface at the point $\boldsymbol{r}$ on the surface'. But even this notation is already potentially misleading, because it suggests, falsely, that there exists a function $\psi_S(\boldsymbol{r})$ of which $\partial_n\psi_S(\boldsymbol{r})$ is some kind of derivative.

Occasionally we do switch to special symbols, especially for the separate integrals like f_D and g_D in the magic rules and in the Kirchhoff representations. In Section 7.1 for instance this makes it easy to formulate the crucial question about the self-consistency of the magic rule, namely 'Is it true that $(f_D(\boldsymbol{r}) + g_D(\boldsymbol{r})) \to \psi(\boldsymbol{r})$ (or equivalently$\to$ $\psi_S(\boldsymbol{r})$) as $\boldsymbol{r}$ tends to the surface?' If we had persisted in denoting $(f_D(\boldsymbol{r}) + g_D(\boldsymbol{r}))$ by $\psi(\boldsymbol{r})$, then the wording of the question, whether $\psi(\boldsymbol{r}) \to \psi_S(\boldsymbol{r})$, would (falsely) mimic a tautology.

It is probably worth spending some time and effort to make these points and potential pitfalls clear to oneself, especially because they are equally relevant to the diffusion and the wave equations.

H Poisson's equation: Green's functions for circle and sphere

H.1 Introduction

Our main purpose is to collect all these Green's functions and their polar expansions, for both interior and exterior problems, and under both DBCs and NBCs. One reason for doing this is that the results do not all appear in any single readily accessible reference; another reason is that derivations of the Neumann Green's functions for the sphere are generally very hard to follow.

As a by-product, Section H.4 generalizes to a different problem the ideas of the single-sum expansion method, introduced for rectangular boundaries in Sections 5.3.2 and 6.5. Further, also in Section H.4, we observe explicitly, and in an altogether non-trivial context, the peculiarities of the Neumann problem discussed in Sections 6.2 and 6.3.

Unfortunately, it would be far too repetitive to present a self-contained argument for each case. The equations from the present section are needed throughout. For circles, Section H.2.3 on NBCs presupposes Section H.2.2 on DBCs. For spheres, Section H.4 on NBCs similarly presupposes Section H.3, which covers the basic expansion in Legendre polynomials, and then deals with DBCs.

For brevity, in 2D and for the Dirichlet problem in 3D we start from the Green's functions in closed form as already constructed by the image method in Sections 5.3.4 and 6.4, and simply expand them in polars. By contrast, the 3D Neumann Green's functions will be constructed from their polar expansions, which can, with some effort, be summed in closed form. Indeed this appears to be the only straightforward method for G_N for a sphere. Once it has been understood, the other Green's functions should be reconstructed by the same method as an exercise.

For convenience, we start by reviewing some crucial formulae.

The origin is at the centre; the radius is a. We introduce the image point $\tilde{\boldsymbol{r}}'$, and the vector $\tilde{\boldsymbol{R}}$, shown in Figs 5.3 and 5.4. As in Section 5.3.4,

$$\tilde{\boldsymbol{r}}' = \hat{\boldsymbol{r}}'a^2/r' = \boldsymbol{r}'a^2/r'^2, \qquad r'\tilde{r}' = a^2, \tag{H.1.1a,b}$$

$$\hat{\boldsymbol{r}}\cdot\hat{\boldsymbol{r}}' \equiv \cos\gamma \equiv \mu, \tag{H.1.2}$$

$$R = |\boldsymbol{r}-\boldsymbol{r}'| = [r^2 + r'^2 - 2rr'\mu]^{\frac{1}{2}}, \tag{H.1.3}$$

$$r'\tilde{R} = r'\,|\boldsymbol{r}-\tilde{\boldsymbol{r}}'| = [r^2r'^2 + a^4 - 2a^2rr'\mu]^{\frac{1}{2}}. \tag{H.1.4}$$

Notice that $r' = a$ entails $R = \tilde{R}$.

With our standard notation $r_<(r_>) \equiv \min(\max)(r, r')$, we define

$$\zeta \equiv r_</r_>. \tag{H.1.5}$$

Then (H.1.3) can be written

$$R = r_>[1 + \zeta^2 - 2\zeta\mu]^{\frac{1}{2}}. \tag{H.1.6}$$

Notice that $r_<$, $r_>$, and their ratio ζ, when expressed in terms of r and r', involve step-functions (e.g. $r_< = \{\theta(r' - r)r + \theta(r - r')r'\}$, whence their $\partial/\partial r$ and $\partial/\partial r'$ derivatives involve delta-functions; by contrast, their product is just an ordinary (not a generalized) function:

$$r_< r_> = rr'. \tag{H.1.7}$$

Later we shall need the combinations

$$\xi \equiv rr'/a^2 = r/\bar{r}', \qquad \eta \equiv 1/\xi = a^2/rr' = \bar{r}'/r. \tag{H.1.8a,b}$$

All these formulae apply equally in 2D and in 3D.

H.2 Circles

H.2.1 The basic expansion

The polar expansion basic to all Green's functions for the circle is

$$\log\frac{a}{R} = \log\frac{a}{r_>} + \sum_{m=1}^{\infty} \frac{\cos(m\gamma)}{m}\,\zeta^m. \tag{H.2.1}$$

(In the rest of this section, $\sum_{m=1}^{\infty}$ is written as $\sum$.)

Proof: Denoting the sum in (H.2.1) by S_1, we must show that $S_1 = \{\log(a/R) - \log(a/r_>)\} = \log(r_>/R)$. One has

$$S_1 = \operatorname{Re}\sum \frac{\exp(im\gamma)}{m}\,\zeta^m = \operatorname{Re}\sum (\zeta\exp(i\gamma))^m/m$$

$$= \operatorname{Re}\{-\log(1 - \zeta\exp(i\gamma))\} = -\operatorname{Re}\log(1 - \zeta\exp(i\gamma)), \tag{H.2.2}$$

having recognized the Taylor series $-\log(1-x) = (x + x^2/2 + x^3/3 + \cdots)$. Writing $Z \equiv (1 - \zeta\exp(i\gamma)) \equiv |Z|\exp(i\delta)$, we notice that

$$|Z| = \{(1 - \zeta\cos\gamma)^2 + \zeta^2\sin^2\gamma\}^{\frac{1}{2}} = \{1 + \zeta^2 - 2\zeta\cos\gamma\}^{\frac{1}{2}}$$
$$= \{1 + \zeta^2 - 2\zeta\mu\}^{\frac{1}{2}} = R/r_>,$$

where the last step relies on (H.1.6). Thus

$$S_1 = -\operatorname{Re}\log(|Z|\exp(i\,\delta)) = -\operatorname{Re}(\log|Z| + i\,\delta)$$
$$= -\log|Z| = -\log(R/r_>) = \log(r_>/R). \qquad \blacksquare$$

H.2.2 The Dirichlet Green's functions

From (5.3.28) the interior and exterior Green's functions can be written as

$$G_{\mathrm{D}}^{(2)}(\boldsymbol{r}\mid\boldsymbol{r}') = \frac{1}{2\pi}\left\{\log\frac{a}{R} - \log\frac{a}{\tilde{R}} + \log\frac{r'}{a}\right\}. \tag{H.2.3}$$

The polar expansion of $\log(a/R)$ is given by (H.2.1) in either case, but the expansions of $\log(a/\tilde{R})$ differ.

(i) *For the interior problem*, one has $r, r' < a$ and $\tilde{r}' = a^2/r' > a$; consequently, $\tilde{r}' > r$, and one can obtain $\tilde{R} = |\boldsymbol{r} - \tilde{\boldsymbol{r}}'|$ from $R = |\boldsymbol{r} - \boldsymbol{r}'|$, or $\log(a/\tilde{R})$ from $\log(a/R)$, by making the replacements

$$r_> \to \tilde{r}' = a^2/r', \qquad r_< \to r, \qquad \zeta = r_</r_> \to rr'/a^2 = \xi, \quad \text{(interior)}. \tag{H.2.4}$$

Under these replacements, (H.2.1) yields

$$\log\frac{a}{\tilde{R}} = \log\frac{r'}{a} + \sum \frac{\cos(m\gamma)}{m}\left(\frac{rr'}{a^2}\right)^m, \quad \text{(interior)}. \tag{H.2.5}$$

Substituting from (H.2.1, 5) into (H.2.3), we find

$$\begin{aligned} G_{\mathrm{D}}^{(2)} &= \frac{1}{2\pi}\left\{\left[\log\frac{a}{r_>} + \sum\frac{\cos(m\gamma)}{m}\left(\frac{r_<}{r_>}\right)^m\right]\right. \\ &\qquad \left. - \left[\log\frac{r'}{a} + \sum\frac{\cos m\gamma}{m}\left(\frac{rr'}{a^2}\right)^m\right] + \log\frac{r'}{a}\right\}, \end{aligned}$$

$$G_{\mathrm{D,int}}^{(2)}(\boldsymbol{r}\mid\boldsymbol{r}') = \frac{1}{2\pi}\left\{\log\frac{a}{r_>} + \sum\frac{\cos(m\gamma)}{m}\left[\left(\frac{r_<}{r_>}\right)^m - \left(\frac{rr'}{a^2}\right)^m\right]\right\}. \tag{H.2.6}$$

Comparison with (H.2.1) shows that

$$\begin{aligned} G_{\mathrm{D,int}}^{(2)} &= \frac{1}{2\pi}\left\{\log\frac{a}{R} - \sum\frac{\cos(m\gamma)}{m}\left(\frac{rr'}{a^2}\right)^m\right\} \\ &\equiv G_0^{(2)}(R) + \chi(\boldsymbol{r}\mid\boldsymbol{r}'), \end{aligned} \tag{H.2.7}$$

where χ is harmonic (see (5.1.3)).

Conversely, if χ were not already known from (H.2.3), one could easily determine it through its polar expansion

$$\chi = \{\chi_0(r\mid r') + \sum\cos(m\gamma)\chi_m(r\mid r')\},$$

as follows. Since χ is harmonic for $r \leqslant a$, χ_m is proportional to r^m for all m. Further, since χ is symmetric in r and r', χ_m must have the form

$$(\alpha_m/2\pi)(rr'/a^2)^m = (\alpha_m/2\pi)(2r_>/a^2)^m,$$

where the α_m are constants. Substituting this expansion on the right of (H.2.7) together with the standard expansion of $G_0^{(2)} = (2\pi)^{-1}\log(a/R)$, one obtains

$$G^{(2)}_{\mathrm{D,int}}(\mathbf{r}\mid\mathbf{r}') = \frac{1}{2\pi}\left\{\left[\log\frac{a}{r_>} + \alpha_0\right]\right.$$

$$\left. + \sum \cos(m\gamma)\left[\frac{1}{m}\left(\frac{r_<}{r_>}\right)^m + \alpha_m\left(\frac{r_< r_>}{a^2}\right)^m\right]\right\}.$$

Finally, by virtue of the BCs, each polar component vanishes when $r = a$ (which implies $r_> = a$). Hence $\alpha_0 = 0$, $\alpha_m = -1/m$, reproducing (H.2.6). In this approach, the only un-obvious step is the next, namely the final recognition that the polar expansion of χ in fact sums to the closed form $(2\pi)^{-1}\log(r'/\tilde{R})$ visible in (H.2.3).

In the magic rule, one needs $\partial G_{\mathrm{D}}/\partial r'$ at $r' = a$; then $r_> = r' = a$, $r_< = r$, whence (H.2.6) yields

$$\left.\frac{\partial G^{(2)}_{\mathrm{D,int}}}{\partial r'}\right|_{r'=a} = \frac{1}{2\pi}\left\{-\frac{1}{a} + \sum \frac{\cos(m\gamma)}{m}\left[-m\frac{r^m}{a^{m+1}} - m\frac{r^m a^{m-1}}{a^{2m}}\right]\right\}$$

$$= -\left\{\frac{1}{2\pi a} + \frac{1}{\pi a}\sum \cos(m\gamma)(r/a)^m\right\}$$

$$= -(a^2 - r^2)/2\pi a R^2|_{r'=a}, \tag{H.2.8a}$$

as in (5.3.29). The sum in the penultimate expression is just the geometric series from (5.4.6). Finally,

$$R^2|_{r'=a} = (\mathbf{r} - \mathbf{r}')^2|_{r'=a} = r^2 + a^2 - 2ar\cos\gamma. \tag{H.2.8b}$$

(ii) *For the exterior problem,* one has $r, r' > a$ and $\tilde{r}' = a^2/r' < a$; consequently, to obtain $\tilde{R} \equiv |\mathbf{r} - \tilde{\mathbf{r}}'|$ from R we now replace

$$r_> \to r, \qquad r_< \to \tilde{r}' = a^2/r', \qquad \zeta = r_</r_> \to a^2/rr' = \eta, \quad \text{(exterior)}. \tag{H.2.9}$$

Under these replacements, (H.2.1) yields

$$\log\frac{a}{\tilde{R}} = \log\frac{a}{r} + \sum \frac{\cos(m\gamma)}{m}\left(\frac{a^2}{rr'}\right)^m, \quad \text{(exterior)}, \tag{H.2.10}$$

which may be compared with (H.2.5). Substituting from (H.2.1, 10) into (H.2.3), and using

$$\log(rr'/ar_>) = \log(r_< r_>/ar_>) = \log(r_</a),$$

we find

$$G^{(2)}_{\mathrm{D}} = \frac{1}{2\pi}\left\{\left[\log\frac{a}{r_>} + \sum \frac{\cos(m\gamma)}{m}\left(\frac{r_<}{r_>}\right)^m\right]\right.$$

$$\left. - \left[\log\frac{a}{r} + \sum \frac{\cos(m\gamma)}{m}\left(\frac{a^2}{rr'}\right)^m\right] + \log\frac{r'}{a}\right\},$$

$$G^{(2)}_{\mathrm{D,ext}}(\mathbf{r}\mid\mathbf{r}') = \frac{1}{2\pi}\left\{\log\frac{r_<}{a} + \sum \frac{\cos(m\gamma)}{m}\left[\left(\frac{r_<}{r_>}\right)^m - \left(\frac{a^2}{rr'}\right)^m\right]\right\}. \tag{H.2.11}$$

This entails

$$\left.\frac{\partial G^{(2)}_{\mathrm{D,ext}}}{\partial r'}\right|_{r'=a} = \frac{1}{2\pi a} + \frac{1}{\pi a}\sum \cos(m\gamma)\left(\frac{a}{r}\right)^m = \frac{1}{2\pi}\frac{(r^2-a^2)}{aR^2}. \tag{H.2.12}$$

Equations (H.2.11, 12) should be compared with (H.2.6, 8).

Exercises: (i) Determine $G^{(2)}_{\mathrm{D}}$ and $\partial G^{(2)}_{\mathrm{D}}/\partial r'$ as functions of γ when $r = a = r'$. This will verify that the BC is obeyed and that the magic rule is self-consistent (cf. Section 7.1. (ii) For given $\psi_S(a, \phi)$, the surface integrals g_{D} entering the magic rule (5.2.4) are related by

$$g^{(2)}_{\mathrm{D,ext}}(r, \phi) = g^{(2)}_{\mathrm{D,int}}(a^2/r, \phi). \tag{H.2.13}$$

Prove this, recalling that ∂'_n always differentiates along the *outward* normal.

H.2.3 The Neumann Green's functions

From (6.4.9), the *interior* Neumann Green's function is

$$G^{(2)}_{\mathrm{N,int}}(\boldsymbol{r}\,|\,\boldsymbol{r}') = \frac{1}{2\pi}\left\{\log\frac{a}{R} + \log\frac{a}{\bar{R}} + \log\frac{a}{r'}\right\}. \tag{H.2.14}$$

The polar expansion follows from (H.2.1, 5):

$$\begin{aligned} G^{(2)}_{\mathrm{N,int}}(\boldsymbol{r}\,|\,\boldsymbol{r}') &= \frac{1}{2\pi}\left\{\left[\log\frac{a}{r_>} + \sum\frac{\cos(m\gamma)}{m}\left(\frac{r_<}{r_>}\right)^m\right]\right. \\ &\quad \left. + \left[\log\frac{r'}{a} + \sum\frac{\cos(m\gamma)}{m}\left(\frac{rr'}{a^2}\right)^m\right] + \log\frac{a}{r'}\right\} \\ &= \frac{1}{2\pi}\left\{\log\frac{a}{r_>} + \sum\frac{\cos(m\gamma)}{m}\left[\left(\frac{r_<}{r_>}\right)^m + \left(\frac{rr'}{a^2}\right)^m\right]\right\}. \end{aligned} \tag{H.2.15}$$

This should be compared with (H.2.6).

Exercise: Evaluate $G^{(2)}_{\mathrm{N,int}}$ at $r' = a$, and verify (6.4.10).

From (6.4.11), the *exterior* Green's function is

$$G^{(2)}_{\mathrm{N,ext}}(\boldsymbol{r}\,|\,\boldsymbol{r}') = \frac{1}{2\pi}\left\{\log\frac{a}{R} + \log\frac{a}{\bar{R}} + \log\frac{r}{a}\right\}. \tag{H.2.16}$$

By virtue of (H.2.1, 9) this eventually yields

$$G^{(2)}_{\mathrm{N,ext}}(\boldsymbol{r}\,|\,\boldsymbol{r}') = \frac{1}{2\pi}\left\{\log\frac{a}{r_>} + \sum\frac{\cos(m\gamma)}{m}\left[\left(\frac{r_<}{r_>}\right)^m + \left(\frac{a^2}{rr'}\right)^m\right]\right\}. \tag{H.2.17}$$

For the magic rule (6.3.5) one needs G_{N} at $r' = a$. In the interior (exterior) problem this entails $r_> = r' = a$, $r_< = r$ ($r_> = r$, $r_< = r' = a$). Accordingly, (H.2.15, 17) yield

$$G^{(2)}_{\mathrm{N,int}}(\boldsymbol{r} \mid \boldsymbol{r}')\big|_{r'=a} = \frac{1}{\pi}\sum \frac{\cos(m\gamma)}{m}\left(\frac{r}{a}\right)^m = \frac{1}{2\pi}\log\frac{a^2}{R^2}\bigg|_{r'=a}, \tag{H.2.18}$$

$$\begin{aligned} G^{(2)}_{\mathrm{N,ext}}(\boldsymbol{r} \mid \boldsymbol{r}')\big|_{r'=a} &= \frac{1}{2\pi}\log\frac{a}{r} + \frac{1}{\pi}\sum \frac{\cos(m\gamma)}{m}\left(\frac{a}{r}\right)^m \\ &= \frac{1}{2\pi}\log\frac{ar}{R^2}\bigg|_{r'=a}. \end{aligned} \tag{H.2.19}$$

Exercises: (i) Show that for given $\partial\psi_S/\partial r$ satisfying $\int d\phi'\, \partial\psi_S/\partial r' = 0$, the surface integrals g_{N} in the magic rule (6.3.5) are related by

$$g^{(2)}_{\mathrm{N,ext}}(r, \phi) = -g^{(2)}_{\mathrm{N,int}}(a^2/r, \phi), \tag{H.2.20}$$

which should be compared with (H.2.13). (ii) Using (H.2.13, 20), determine the differences $[g_{\mathrm{ext}} - g_{\mathrm{int}}]$ and $[\partial g_{\mathrm{ext}}/\partial r - \partial g_{\mathrm{int}}/\partial r]$ at $\boldsymbol{r} = (a, \phi)$ for both the Dirichlet and the Neumann problems, and interpret them in the light of Sections 4.4.5 and 7.5.

H.3 Spheres: basic formulae and the Dirichlet Green's function

Given $\boldsymbol{r} = (r, \Omega) = (r, \theta, \phi)$ and $\boldsymbol{r}' = (r', \Omega') = (r', \theta', \phi')$, the formulae in Section H.1 continue to apply, except that (H.1.2b) is in 3D replaced by

$$\begin{aligned} \mu = \cos\gamma &= \{\cos\theta\cos\theta' + \sin\theta\sin\theta'\cos(\phi - \phi')\} \\ &= \{\cos\theta\cos\theta' + \sin\theta\sin\theta'[\cos\phi\cos\phi' + \sin\phi\sin\phi']\}. \end{aligned} \tag{H.3.1}$$

We also need the addition theorem for spherical harmonics:

$$\sum_{m=-l}^{l} Y^*_{lm}(\Omega')Y_{lm}(\Omega) = \frac{2l+1}{4\pi}P_l(\mu). \tag{H.3.2}$$

The basic polar expansion (i.e. the 3D analogue of (H.2.1)) stems from the generating function of the Legendre polynomials:

$$\frac{1}{[1 + \zeta^2 - 2\zeta\mu]^{\frac{1}{2}}} = \sum_{l=0}^{\infty} P_l(\mu)\zeta^l. \tag{H.3.3a}$$

Applied to $1/R$ in view of (H.1.6), this gives

$$\frac{1}{R} = \sum_{l=0}^{\infty} P_l(\mu)\frac{r_<^l}{r_>^{l+1}}. \tag{H.3.3b}$$

In the rest of this appendix, $\sum_{l=0}^{\infty}$ is written as $\sum$; other limits will be indicated explicitly.

On the left of (H.3.3a), $\zeta \to -\zeta$ is equivalent to $\mu \to -\mu$; the same equivalence on the right yields $P_l(-\mu) = (-1)^l P_l(\mu)$. Note in particular $P_l(1) = 1$, $P_l(-1) = (-1)^l$.

For the interior and exterior problems, we obtain $1/\tilde{R}$ from $1/R$ by the replacements (H.2.4) and (H.2.9) respectively:

$$\frac{1}{\tilde{R}} = \sum P_l(\mu) \frac{r'}{a^2}\left(\frac{rr'}{a^2}\right)^l = \frac{r'}{a^2} \sum P_l(\mu)\left(\frac{rr'}{a^2}\right)^l, \quad \text{(interior)}, \tag{H.3.4a}$$

$$\frac{1}{\tilde{R}} = \frac{1}{r} \sum P_l(\mu)\left(\frac{a^2}{rr'}\right)^l, \quad \text{(exterior)}. \tag{H.3.4b}$$

From (5.3.23, 24) the Dirichlet Green's function for either case is

$$G_D^{(3)}(\boldsymbol{r} \mid \boldsymbol{r}') = \frac{1}{4\pi}\left\{\frac{1}{R} - \frac{a}{r'\tilde{R}}\right\}. \tag{H.3.5}$$

By virtue of (H.3.3–5), the polar expansions are

$$G_{D,\text{int}}^{(3)}(\boldsymbol{r} \mid \boldsymbol{r}') = \frac{1}{4\pi} \sum P_l(\mu)\left\{\frac{r_<^l}{r_>^{l+1}} - \frac{1}{a}\left(\frac{rr'}{a^2}\right)^l\right\}, \tag{H.3.6}$$

$$G_{D,\text{ext}}^{(3)}(\boldsymbol{r} \mid \boldsymbol{r}') = \frac{1}{4\pi} \sum P_l(\mu)\left\{\frac{r_<^l}{r_>^{l+1}} - \frac{1}{a}\left(\frac{a^2}{rr'}\right)^{l+1}\right\}. \tag{H.3.7}$$

In the magic rule, one needs

$$\left.\frac{\partial G_{D,\text{int}}^{(3)}}{\partial r'}\right|_{r'=a} = -\frac{1}{4\pi a^2} \sum (2l+1) P_l(\mu)\left(\frac{r}{a}\right)^l$$
$$= -(a^2 - r^2)/4\pi a R^3|_{r'=a}, \tag{H.3.8}$$

$$\left.\frac{\partial G_{D,\text{ext}}^{(3)}}{\partial r'}\right|_{r'=a} = \frac{1}{4\pi a^2} \sum (2l+1) P_l(\mu)\left(\frac{a}{r}\right)^l$$
$$= (r^2 - a^2)/4\pi a R^3|_{r'=a}, \tag{H.3.9}$$

where the closed forms are already known from (5.3.25). R here is given by (H.2.8b).

Exercise: For given $\psi_S(a, \Omega)$, the surface integrals g_D in the magic rule (5.2.4) are related by

$$g_{D,\text{ext}}^{(3)}(r, \Omega) = \frac{a}{r} g_{D,\text{int}}^{(3)}(a^2/r, \Omega), \tag{H.3.10}$$

which may be compared with (H.2.13). Prove (H.3.10).

H.4 Spheres: the Neumann Green's functions

H.4.1 The polar expansion

Unlike the $G^{(2)}_{\mathrm{D,N}}$ and the $G^{(3)}_{\mathrm{D}}$, the $G^{(3)}_{\mathrm{N}}$ at this stage are totally unknown. As explained in Section H.1, we shall determine them through their polar expansions. This method is analogous to the 'single-sum' method applied in Section 5.3 to rectangular regions in 2D; now, it is the dependence of $G(r, \Omega \mid r', \Omega')$ on Ω and Ω' that is accommodated by a series expansion in spherical harmonics, while the dependence on r and r' is accommodated by solving a differential equation. Accordingly, we look for G_{N} in the form

$$G_{\mathrm{N}}(\boldsymbol{r} \mid \boldsymbol{r}') = \sum_{l=0}^{\infty} \sum_{m=-l}^{l} g_l(r \mid r') Y^*_{lm}(\Omega') Y_{lm}(\Omega)$$

$$= \frac{1}{4\pi} \sum g_l(r \mid r')(2l+1) P_l(\mu). \tag{H.4.1}$$

(In the present section, we omit the superscript (3) except in the end-results.) Substituting into $-\nabla^2 G_{\mathrm{N}} = \delta(\boldsymbol{r} - \boldsymbol{r}')$ from (H.4.1) and from

$$\delta(\boldsymbol{r} - \boldsymbol{r}') = r^{-2}\, \delta(r - r') \sum_l \sum_m Y^*_{lm}(\Omega') Y_{lm}(\Omega),$$

and equating the coefficients of corresponding spherical harmonics, we find for the radial Green's function g_l the differential equation

$$-\frac{\partial^2 g_l}{\partial r^2} - \frac{2}{r}\frac{\partial g_l}{\partial r} + \frac{l(l+1)}{r^2} g_l = \frac{1}{r'^2}\, \delta(r - r'). \tag{H.4.2}$$

Notice the factor $1/r'^2$ on the right.

H.4.2 Interior problem: Green's function and pseudo Green's function

The pseudo Green's function obeys the equation (6.2.3) and the BC (6.2.4); here these read

$$-\nabla^2 H = \delta(\boldsymbol{r} - \boldsymbol{r}') - \frac{3}{4\pi a^3}, \qquad \left.\frac{\partial H}{\partial r}\right|_{r=a} = 0. \tag{H.4.3a,b}$$

By contrast, $G_{\mathrm{N,int}}$ obeys (6.3.1) and (6.3.3), which read

$$-\nabla^2 G_{\mathrm{N,int}} = \delta(\boldsymbol{r} - \boldsymbol{r}'), \qquad \left.\frac{\partial G_{\mathrm{N,int}}}{\partial r}\right|_{r=a} = -\frac{1}{4\pi a^2}. \tag{H.4.4a,b}$$

Recall that H and G_{N} are both symmetric in $\boldsymbol{r}$ and $\boldsymbol{r}'$. Their difference

$$F \equiv G_{\mathrm{N,int}} - H \tag{H.4.5}$$

is likewise symmetric, and obeys

$$-\nabla^2 F = \frac{3}{4\pi a^3}, \qquad \left.\frac{\partial F}{\partial r}\right|_{r=a} = -\frac{1}{4\pi a^2}. \tag{H.4.6a,b}$$

By inspection,

$$F(\mathbf{r} \mid \mathbf{r}') = -(r^2 + r'^2)/8\pi a^3. \tag{H.1.4.7}$$

Notice that F is not harmonic. Since F is independent of $\cos\gamma = \mu$, the polar expansions of $G_{\mathrm{N,int}}$ and of H are the same except for the isotropic term with $l = 0$. From now on we consider only G_{N}.

H.4.3 The interior Green's function

Equation (H.4.2) shows that g_l as a function of r obeys the homogeneous equation except at $r = r'$. Hence, for $r < r'$ and $r > r'$, g_l must be of the form $A_l r^l$ and $(B_l r^l + C_l/r^{l+1})$ respectively. The BC (H.4.4b), being isotropic, affects the components with $l = 0$ and $l \neq 0$ differently. For $l = 0$ it imposes

$$-1/4\pi a^2 = \frac{\partial}{\partial r}\{(B_0 + C_0/r)Y_{00}^2\}|_{r=a} = -C_0/4\pi a^2,$$

whence $C_0 = 1$; thus

$$g_0(r \mid r') = \theta(r' - r)A_0 + \theta(r - r')(B_0 + 1/r). \tag{H.4.8}$$

As discussed in Section 2.1.2, g_0 is continuous at $r = r'$:

$$g_0(r'+ \mid r') = g_0(r'- \mid r') \Rightarrow A_0 = B_0 + 1/r'. \tag{H.4.9}$$

The derivative $\partial g_0/\partial r$ obeys the jump condition obtained by integrating (H.4.2) with respect to r from $r'-$ to $r'+$. This condition is the same for all l:

$$g_l'(r'+ \mid r') - g_l'(r'- \mid r') = -1/r'^2. \tag{H.4.10}$$

In view of (H.4.9), g_0 obeys the jump condition automatically. Thus,

$$\begin{aligned} g_0(r \mid r') &= \theta(r' - r)(B_0 + 1/r') + \theta(r - r')(B_0 + 1/r) \\ &= B_0[\theta(r' - r) + \theta(r - r')] + \theta(r' - r)/r' + \theta(r - r')/r, \end{aligned}$$

$$g_0(r \mid r') = B_0 + 1/r_>. \tag{H.4.11}$$

The constant B_0 is not determined by the equation and the BC: it is in fact the arbitrary additive constant which we know to expect in the solution of any Neumann problem.

For $l \geqslant 1$, the BC imposes $g_l'(a \mid r') = 0$, whence $C_l = B_l a^{2l+1} l/(l+1)$, and

$$g_l(r \mid r') = \theta(r' - r)A_l r^l + \theta(r - r')B_l\left(r^l + \frac{l}{l+1}\frac{a^{2l+1}}{r^{l+1}}\right). \tag{H.4.12}$$

Continuity yields

$$A_l = B_l\left\{1 + \frac{l}{l+1}\left(\frac{a}{r'}\right)^{2l+1}\right\}.$$

Substituting this into (H.4.12) and then imposing the jump condition (H.4.10), one finds after some rearrangement that

$$B_l = \frac{(l+1)}{l(2l+1)}\frac{r'^l}{a^{2l+1}},$$

whence, eventually,

$$g_l(r \mid r') = \frac{1}{(2l+1)}\left\{\frac{r_<^l}{r_>^{l+1}} + \left(1 + \frac{1}{l}\right)\frac{1}{a}\left(\frac{rr'}{a^2}\right)^l\right\}, \qquad (l \geqslant 1). \qquad \text{(H.4.13)}$$

$$\begin{aligned} G^{(3)}_{\mathrm{N,int}} &= \frac{1}{4\pi}\left\{\left(B_0 + \frac{1}{r_>}\right) + \sum_{l=1}^{\infty} P_l(\mu)\left[\frac{r_<^l}{r_>^{l+1}} + \left(1 + \frac{1}{l}\right)\frac{1}{a}\left(\frac{rr'}{a^2}\right)^l\right]\right\} \\ &= \frac{1}{4\pi}\left\{\frac{1}{R} + B_0 + \frac{1}{a}\sum_{l=1}^{\infty} P_l(\mu)\left(1 + \frac{1}{l}\right)\left(\frac{rr'}{a^2}\right)^l\right\} \\ &\equiv G_0^{(3)} + \chi. \end{aligned} \qquad \text{(H.4.14)}$$

It proves convenient to rewrite the harmonic component χ as follows. From (H.3.5, 6) we recognize that

$$\frac{a}{4\pi r'\tilde{R}} = \frac{1}{4\pi a}\sum_{l=0}^{\infty} P_l(\mu)\left(\frac{rr'}{a^2}\right)^l = \frac{1}{4\pi a}\left\{1 + \sum_{l=1}^{\infty} P_l(\mu)\left(\frac{rr'}{a^2}\right)^l\right\}. \qquad \text{(H.4.15)}$$

Eliminating $\sum_{l=1}^{\infty} P_l(\mu)(rr'/a^2)^l$ between (H.4.14, 15), we find

$$G_{\mathrm{N,int}} = \frac{1}{4\pi}\left\{\frac{1}{R} + B_0 + \frac{a}{r'\tilde{R}} - \frac{1}{a} + \frac{1}{a}S_2\right\}, \qquad \text{(H.4.16)}$$

$$S_2 \equiv \sum_{l=1}^{\infty} P_l(\mu)\frac{1}{l}\left(\frac{rr'}{a^2}\right)^l. \qquad \text{(H.4.17)}$$

The sum is evaluated in Section H.5 below. Substituting it from (H.5.4) into (H.4.16), we obtain the end-result

$$G^{(3)}_{\mathrm{N,int}}(\boldsymbol{r} \mid \boldsymbol{r}') = \frac{1}{4\pi}\left\{\frac{1}{R} + B_0 + \frac{a}{r'\tilde{R}} - \frac{1}{a} + \frac{1}{a}\log\left[\frac{2}{r'\tilde{R}/a^2 + 1 - rr'\mu/a^2}\right]\right\}. \qquad \text{(H.4.18)}$$

It is the logarithm that makes G_{N} so much more awkward than G_{D}.

Since B_0 is arbitrary, no physical significance attaches to the constant term $(-1 + \log 2)/4\pi a$ in (H.4.18): it has been kept merely to maintain agreement with the polar form (H.4.14). From now on we set $B_0 = 0$ to save writing.

In the magic rule, one needs

$$G^{(3)}_{\mathrm{N,int}}(\boldsymbol{r}\,|\,\boldsymbol{r}')\big|_{r'=a} = \frac{1}{4\pi}\left\{\frac{1}{a} + \frac{1}{a}\sum_{l=1}^{\infty}\frac{2l+1}{l}P_l(\mu)\left(\frac{r}{a}\right)^l\right\} \tag{H.4.19a}$$

$$= \frac{1}{4\pi}\left\{\frac{2}{R} - \frac{1}{a} + \frac{1}{a}\log\left[\frac{2}{R/a + 1 - r\mu/a}\right]\right\}\Bigg|_{r'=a}, \tag{H.4.19b}$$

where the polar and the closed forms follow from (H.4.14, 18) respectively. A somewhat abstruse interpretation of the logarithmic term is mentioned at the end of Section H.4.4 below.

Example: Point sources of strength $\pm\alpha$ are placed just inside the north and south poles of a rigid spherical enclosure of radius a. We compare the flow velocity at the centre with its value in absence of the enclosure.

By symmetry, the flow at points on the axis is along the axis. Hence all we need is ψ as a function of z on the axis, to order z, say for $z \geqslant 0$; terms of higher order make no contribution to $v_z = -\partial\psi/\partial z$ as $z \to 0$. The two sources are at $\boldsymbol{r}'_{1,2} = \pm\hat{\boldsymbol{z}}a$. By the magic rule,

$$\psi(\boldsymbol{r}) = \int_V \mathrm{d}V'\, G_{\mathrm{N}}(\boldsymbol{r}\,|\,\boldsymbol{r}')\alpha\{\delta(\boldsymbol{r}-\boldsymbol{r}_1) - \delta(\boldsymbol{r}-\boldsymbol{r}_2)\}$$

$$= \alpha\{G_{\mathrm{N}}(\boldsymbol{r}\,|\,\boldsymbol{r}_1) - G_{\mathrm{N}}(\boldsymbol{r}\,|\,\boldsymbol{r}_2)\}.$$

We need only the term with $l = 1$ in the polar form (H.4.19a) for G_{N}. For $\boldsymbol{r} = \hat{\boldsymbol{z}}z$ with $z \geqslant 0$, note that $R_1 \equiv |\boldsymbol{r} - \boldsymbol{r}_1| = a - z$, $R_2 \equiv |\boldsymbol{r} - \boldsymbol{r}_2| = a + z$, $P_1(\mu_1) = \mu_1 = \hat{\boldsymbol{r}}\cdot\hat{\boldsymbol{r}}_1 = 1$, $P_1(\mu_2) = \mu_2 = \hat{\boldsymbol{r}}\cdot\hat{\boldsymbol{r}}_2 = -1$, whence

$$\psi(0, 0, z \to 0) \sim \frac{1}{4\pi}\left\{\alpha\frac{1}{a}[3\cdot 1\cdot z/a] - \alpha\frac{1}{a}[3\cdot(-1)\cdot z/a]\right\}$$

$$= \alpha\cdot 6z/4\pi a^2.$$

Hence

$$v_z(z \to 0) = -\frac{\partial\psi}{\partial z}\bigg|_{z\to 0} = -6\alpha/4\pi a^2.$$

Without the enclosure, the inverse-square law gives the flow speed as $2\cdot\alpha/4\pi a^2$. Hence the enclosure enhances the flow at the centre threefold.

Exercise: Explain why this problem would not make sense with only one of the sources.

H.4.4 The exterior Green's function

For the exterior problem, H and G_{N} coincide. The radial Green's function still obeys (H.4.2), but now the BCs are $\partial G_{\mathrm{N}}/\partial r = 0$ at $r = a$ and

$r \to \infty$. The BCs are satisfied by writing

$$g_l = \theta(r' - r)A_l\left[r^l + \frac{l}{l+1}\frac{a^{2l+1}}{r^{l+1}}\right] + \theta(r - r')B_l\frac{1}{r^{l+1}}, \tag{H.4.20}$$

where an arbitrary additive constant has again been dropped from g_0. The continuity and jump conditions eventually lead to

$$g_l = \frac{1}{2l+1}\left[\frac{r_<^l}{r_>^{l+1}} + \left(1 - \frac{1}{l+1}\right)\frac{a^{2l+1}}{(rr')^{l+1}}\right], \tag{H.4.21}$$

$$G^{(3)}_{\mathrm{N,ext}}(\boldsymbol{r} \mid \boldsymbol{r}') = \frac{1}{4\pi}\sum_{l=0}^{\infty} P_l(\mu)\left\{\frac{r_<^l}{r_>^{l+1}} + \left[1 - \frac{1}{l+1}\right]\left(\frac{a^2}{rr'}\right)^{l+1}\right]\right\} \tag{H.4.22a}$$

$$= \frac{1}{4\pi}\left\{\frac{1}{R} + \frac{1}{a}\sum_{l=0}^{\infty} P_l(\mu)\left[1 - \frac{1}{l+1}\right]\left(\frac{a^2}{rr'}\right)^{l+1}\right\}. \tag{H.4.22b}$$

Comparison with (H.3.5, 4) yields

$$G^{(3)}_{\mathrm{N,ext}} = \frac{1}{4\pi}\left\{\frac{1}{R} + \frac{a}{r'\tilde{R}} + \frac{1}{a}S_3\right\}, \tag{H.4.23}$$

$$S_3 \equiv -\sum_{l=0}^{\infty} P_l(\mu)\frac{1}{l+1}\left(\frac{a^2}{rr'}\right)^{l+1}. \tag{H.4.24}$$

The sum is evaluated in Section H.5 below. Substitution from (H.5.7) gives the end-result

$$G^{(3)}_{\mathrm{N,ext}}(\boldsymbol{r} \mid \boldsymbol{r}') = \frac{1}{4\pi}\left\{\frac{1}{R} + \frac{a}{r'\tilde{R}} + \frac{1}{a}\log\left[\frac{rr'(1-\mu)/a^2}{r'\tilde{R}/a^2 + 1 - rr'\mu/a^2}\right]\right\}, \tag{H.4.25}$$

which may be compared with (H.4.18). Restoring the arbitrary constant omitted from (H.4.20) would add $A_0/4\pi$ on the right.

In the magic rule, one needs

$$G^{(3)}_{\mathrm{N,ext}}(\boldsymbol{r} \mid \boldsymbol{r}')|_{r'=a} = \frac{1}{4\pi}\left\{\frac{1}{a}\sum_{l=0}^{\infty} P_l(\mu)\frac{2l+1}{l+1}\left(\frac{a}{r}\right)^{l+1}\right\} \tag{H.4.26a}$$

$$= \frac{1}{4\pi}\cdot\left\{\frac{2}{R} + \frac{1}{a}\log\left[\frac{r(1-\mu)/a}{R/a + 1 - r\mu/a}\right]\right\}\Bigg|_{r'=a}, \tag{H.4.26b}$$

which may be compared with (H.4.19).

Example: A point source of strength α is placed just outside the north pole of a rigid sphere of radius a. We determine the tangential flow velocity at points just outside the sphere.

In terms of spherical polar coordinates $\boldsymbol{r} = (r, \theta, \phi)$, the source is at $\boldsymbol{r}' = (a, 0, 0)$. Nothing depends on ϕ. For $\psi = \alpha G_{\mathrm{N}}(r, \theta, \phi \mid a, 0, 0)$ we use the closed form (H.4.26b). We can immediately set $r = a$, whence

$$R = [r^2 + a^2 - 2ar\cos\theta]^{\frac{1}{2}} = [2a^2(1-\cos\theta)]^{\frac{1}{2}} = 2a\sin(\theta/2),$$

while $\mu = \hat{\boldsymbol{r}} \cdot \hat{\boldsymbol{r}}' = \cos\theta$. The argument of the logarithm reduces to

$$\frac{r(1-\mu)/a}{R/a + 1 - r\mu/a} = \frac{1-\cos\theta}{2\sin(\theta/2) + 1 - \cos\theta}$$

$$= \frac{2\sin^2(\theta/2)}{2\sin(\theta/2) + 2\sin^2(\theta/2)} = \frac{1}{1 + 1/\sin(\theta/2)}.$$

Accordingly,

$$\psi = \frac{\alpha}{4\pi a}\left\{\frac{2}{2\sin\theta/2} + \log\left[\frac{1}{1+1/\sin(\theta/2)}\right]\right\},$$

$$v_\theta = -\frac{\partial\psi}{a\,\partial\theta} = -\frac{\alpha}{4\pi a^2}\left\{\frac{1}{2}\frac{\cos(\theta/2)}{\sin^2(\theta/2)} - \frac{1}{2}\frac{\cos(\theta/2)}{\sin(\theta/2)(1+1/\sin(\theta/2))}\right\},$$

$$v_\theta(a, \theta, \phi) = \frac{\alpha}{8\pi a^2} \cdot \frac{\cos(\theta/2)}{\sin^2(\theta/2)(1+\sin(\theta/2))}.$$

As a check, consider the behaviour of V_θ as $\theta \to 0$. Then $V_\theta \sim \alpha/2\pi(a\theta)^2$, where $a\theta \sim R$. Since locally the sphere act as a flat reflector, this tangential velocity should agree with the corresponding result for a reflecting plane. For that case, G_{N} is given by (6.4.7); setting $z = z' = 0$ and $R = [(x-x')^2 + (y-y')^2]$, this yields $\psi = 2\alpha/4\pi R$, whence $-\partial\psi/\partial R = \alpha/2\pi R^2$, as expected.

An interpretation of the logarithmic term $S_3/4\pi a$ in (H.4.25) can be devised as follows. The calculation of S_3 in Section H.5 below shows that, in terms of $\eta = a^2/rr' = \tilde{r}'/r$, one can write

$$S_3/4\pi a = -(1/4\pi a)\int_0^\eta d\eta/[1+\eta^2 - 2\eta\mu]^{\frac{1}{2}}.$$

After changing the integration variable from η to $\tilde{r}'$, some manipulation yields

$$S_3/4\pi a = \int_0^{\tilde{r}'} d\tilde{r}'\,[-1/4\pi a\tilde{R}]. \tag{H.4.27}$$

The integral can be regarded as the potential at $\boldsymbol{r}$ due to a line source stretching from the centre to the image point $\tilde{\boldsymbol{r}}'$, of uniform strength $-1/a$ per unit length. Thus, according to (H.4.23), the harmonic component χ of $G^{(3)}_{\mathrm{N,ext}}$ is the sum of the potentials due to this negative line source, and to the positive point source of strength a/r' situated at $\tilde{\boldsymbol{r}}'$.

In the interior problem, the logarithmic term of (H.4.19b) may be written

$$S_2/4\pi a = (1/4\pi a)\int_{\tilde{r}'}^\infty d\tilde{r}'\,[-1/\tilde{r}' + 1/\tilde{R}]$$

$$= -(1/4\pi a)\int_{\tilde{r}'}^\infty d\tilde{r}'\,\frac{1}{\tilde{R}}\left[\frac{r'\tilde{R}}{a^2} - 1\right], \tag{H.4.28}$$

representing the potential due to a negative line source stretching from the image point $\tilde{r}'$ to infinity, but now of variable strength $-[r'\tilde{R}/a^2 - 1]/a$ per unit length.

H.5 The sums S_2 and S_3

For use in $G^{(3)}_{\text{N,int}}$, eqn (H.4.17) defines

$$S_2(\xi) = \sum_{l=1}^{\infty} P_l(\mu)\frac{1}{l}\xi^l. \tag{H.5.1}$$

We observe that

$$\begin{aligned} \xi\,\partial S_2/\partial\xi &= \sum_{l=1}^{\infty} P_l(\mu)\xi^l = \sum_{l=0}^{\infty} P_l(\mu)\xi^l - 1 \\ &= \frac{1}{[1+\xi^2-2\xi\mu]^{\frac{1}{2}}} - 1. \end{aligned} \tag{H.5.2}$$

Divide by ξ and integrate, noting that $S_2(0) = 0$; this yields

$$\begin{aligned} S_2(\xi) &= \int_0^{\xi} d\xi \left\{\frac{1}{\xi[1+\xi^2-2\xi\mu]^{\frac{1}{2}}} - \frac{1}{\xi}\right\} \\ &= \left\{-\log\left[\frac{2(1+\xi^2-2\xi\mu)}{\xi} + \frac{2}{\xi} - 2\mu\right] - \log\xi\right\}\Bigg|_{\xi=0}^{\xi} \\ &= \log\left[\frac{2}{(1+\xi^2-2\xi\mu)^{\frac{1}{2}} + 1 - \xi\mu}\right] \end{aligned} \tag{H.5.3}$$

$$S_2 = \log\left[\frac{2}{r'\tilde{R}/a^2 + 1 - rr'\mu/a^2}\right]. \tag{H.5.4}$$

For use in $G^{(3)}_{\text{N,ext}}$, eqn (H.4.24) defines

$$S_3(\eta) \equiv -\sum_{l=0}^{\infty} P_l(\mu)\frac{1}{l+1}\eta^{l+1}. \tag{H.5.5}$$

We observe that

$$\frac{\partial S_3}{\partial\eta} = -\sum_{l=0}^{\infty} P_l(\mu)\eta^l = \frac{1}{[1+\eta^2-2\eta\mu]^{\frac{1}{2}}}. \tag{H.5.6}$$

In view of $S_3(0) = 0$, integration yields

$$\begin{aligned} S_3 &= -\int_0^{\eta}\frac{d\eta}{[1+\eta^2-2\eta\mu]^{\frac{1}{2}}} = -\log\left[2(1+\eta^2-2\eta\mu)^{\frac{1}{2}} + 2\eta - 2\mu\right]\Bigg|_{\eta=0}^{\eta} \\ &= \log\left[\frac{1-\mu}{(1+\eta^2-\eta\mu)^{\frac{1}{2}} + \eta - \mu}\right], \end{aligned}$$

$$S_3 = \log\left[\frac{rr'(1-\mu)/a^2}{r'\tilde{R}/a^2 + 1 - rr'\mu/a^2}\right]. \tag{H.5.7}$$

I Laplace equation: the variational method and its paradoxes

The variational approximation to the eigenvalues of Hermitean operators, familiar in quantum mechanics, applies equally to the eigenvalues of $-\nabla^2$ under homogeneous DBCs or NBCs (cf. Section 4.5). Less familiar is the variational approach to the Laplace equation under *inhomogeneous* DBCs or NBCs. This will be sketched presently, not with a view to numerical work, but because it sets the stage for a paradox regarding the same equation under Cauchy BCs, an overspecified and therefore insoluble problem already discussed in Section 7.4.1. Even though it is overspecified, we shall see that from a naive variational point of view one would expect the Cauchy problem to be soluble in the same way as the Dirichlet and Neumann problems. The fact that an expectation that is so plausible at first sight nevertheless proves false shows that, occasionally, there are significant practical benefits from attending to existence proofs or to their absence (contrary to the attitude taken elsewhere in this book).

Under inhomogeneous DBCs, the variational principle for the Laplace equation is formulated as follows. First, the class of admissible trial functions is defined to contain all functions ϕ that are continuous and twice differentiable in V, and assume the prescribed values ψ_S on S (i.e. $\phi_S = \psi_S$). When such a function ϕ is varied within the class, $\phi(\mathbf{r}) \to \phi(\mathbf{r}) + \delta\phi(\mathbf{r})$, the change $\delta\phi(\mathbf{r})$ must obey $\delta\phi_S = 0$, since both ϕ_S and $(\phi_S + \delta\phi_S)$ must equal ψ_S.

The variational principle asserts that the energy integral $J = \frac{1}{2}\int_V dV(\nabla\phi)^2$ is minimized by the trial function that satisfies the Laplace equation. To confirm this, consider the first-order change δJ induced in J by the variation $\delta\phi$:

$$\begin{aligned}\delta J &= \int_V dV\,(\nabla\,\delta\phi)\cdot\nabla\phi = \int_V dV\,\{\mathrm{div}\,(\delta\phi\nabla\phi) - \delta\phi\nabla^2\phi\}\\ &= \int_S dS\,\delta\phi\,\partial_n\phi - \int_V dV\,\delta\phi\nabla^2\phi\\ &= -\int_V dV\,\delta\phi(\mathbf{r})\nabla^2\phi(\mathbf{r}).\end{aligned} \tag{I.1}$$

If J is stationary, then δJ must vanish for $\delta\phi(\mathbf{r})$ that are arbitrary (within the class of admissible variations); but δJ can vanish only if the coefficient of $\delta\phi(\mathbf{r})$ under the integral vanishes everywhere in V, i.e. only if $\nabla^2\phi = 0$. That the lowest stationary value if it exists must be a minimum follows automatically from the fact that J is non-negative and therefore possesses a lower bound. (In fact we know from Section 4.2 that the solution of the Dirichlet BVP is unique, whence J can have at most this one stationary value.)

At this stage it seems plausible that J, being bounded below, must be minimized by some ϕ. If correct, this argument (sometimes called 'Dirichlet's principle') would amount to an existence proof for the Dirichlet problem. The gap in the argument is the assumption that amongst the trial functions there is one for which the greatest lower bound on J is actually attained. Even if this were not so, we could still construct a sequence of trial functions whose J's have the greatest lower bound as their limit, but the limit of the sequence of functions would not itself be admissible as a trial function.

At first sight, such reservations might appear as a mere splitting of hairs, and for the Dirichlet problem it turns out that an optimum trial function does exist (though nó direct proof is given in this book).

However, the same argument suggests with apparently equal force the (false) conclusion that the Laplace equation with Cauchy BCs likewise has a solution. To appreciate the plausibility of this suggestion, restrict the class of trial functions by requiring further that $\partial_n\phi$ assume arbitrarily prescribed values $\partial_n\psi_S$ on S (i.e. $\partial_n\phi_S = \partial_n\psi_S$ as well as $\phi_S = \psi_S$). Varying ϕ we arrive at (I.1) exactly as before, and the same argument suggests that J is minimized by a solution of $\nabla^2\phi = 0$. Can we then, after all, solve the Cauchy problem by finding an optimal trial function? The answer must be no, from our previous non-variational argument in Section 7.4.1: in fact, the limit function minimizing J (i.e. attaining the greatest lower bound) is not itself an admissible trial function. We illustrate how this happens with an artificially simple example in 1D, which is nevertheless typical of the way in which a variational optimum manages to avoid solving the insoluble Cauchy problem. Incidentally, the example also serves as a remainder of some subtleties of Fourier series.

Consider the Cauchy problem defined by

$$d^2\psi/dx^2 = 0, \qquad 0 \leqslant x \leqslant \pi, \tag{I.2}$$

under the BCs

$$\psi(0) = 0, \qquad \psi(\pi) = \pi; \tag{I.3a,b}$$

$$\psi'(0) = 0, \qquad \psi'(\pi) = 0. \tag{I.4a,b}$$

Of course, inspection soon reveals that it is impossible even for the most general solution of (I.2), namely for $\psi = A + Bx$, to satisfy all the conditions (I.3, 4). But we are supposing that this obvious fact has been overlooked, and try a variational approach, aiming to minimize

$$J = \frac{1}{2}\int_0^\pi dx\,(\phi')^2 \tag{I.5}$$

with ϕ subject to (I.3, 4).

Note for reference that the corresponding Dirichlet problem, defined by (I.2) and (I.3) (but ignoring (I.4)) is solved by

$$\psi_D = x, \qquad J_D = \frac{1}{2}\int_0^\pi dx\,(\psi_D')^2 = \frac{1}{2}\int_0^\pi dx = \frac{\pi}{2}. \tag{I.6a,b}$$

Since the Cauchy problem admits only a subset of the trial functions admitted in the Dirichlet problem (they have to obey (I.4) as well as (I.3)), it is clear that $J \geqslant J_D$.

For the Cauchy problem we choose the trial function

$$\phi = \sum_{n=0}^{N} a_n \cos(nx), \qquad \phi' = -\sum_{n=1}^{N} n a_n \sin(nx). \tag{I.7a,b}$$

The variational parameters are the coefficients a_n and eventually the integer N. Thus $\delta\phi = \sum \delta a_n \cos(nx)$. The BCs (I.4) are satisfied irrespective of the values of the a_n; the BCs (I.3a, b) are constraints on the a_n, which will be accommodated by Lagrange multipliers α and β respectively. Since

$$\int_0^{\pi} dx \sin(nx)\sin(mx) = \tfrac{1}{2}\pi\,\delta_{nm},$$

we have

$$J = \tfrac{1}{2}\sum_n\sum_m nma_n a_m \int_0^{\pi} dx \sin(nx)\sin(mx),$$

$$J = \tfrac{1}{4}\pi \sum_{m=1}^{N} m^2 a_m^2, \tag{I.8}$$

$$\phi(0) = \sum_{m=0}^{N} a_m = 0, \tag{I.9}$$

$$\phi(\pi) = \sum_{m=0}^{N} (-1)^m a_m = \pi. \tag{I.10}$$

Varying a_n one finds

$$\frac{\partial}{\partial a_n}\{J - \alpha[\textstyle\sum a_m] - \beta[-\pi + \sum(-1)^m a_m)\} = 0,$$

$$\tfrac{1}{2}\pi n^2 a_n - \alpha - (-1)^n\beta = 0. \tag{I.11}$$

For $n = 0$ this yields $\beta = -\alpha$, but no information directly about a_0; then for $n \geq 1$, it yields

$$a_2 = a_4 = \cdots = 0, \qquad a_{2k+1} = 4\alpha/\pi(2k+1)^2. \tag{I.12}$$

Substituting from (I.12) into (I.9, 10) we obtain two equations for the two unknowns a_0 and $(4\alpha/\pi)$. They are solved by $a_0 = \tfrac{1}{2}\pi$, $(4\alpha/\pi) = -\pi/2S(K)$, where

$$S(K) \equiv \sum_{k=0}^{K} 1/(2k+1)^2, \qquad S(\infty) = \pi^2/8, \tag{I.13a,b}$$

K being the largest integer such that $(2K+1) \leq N$. Thus (I.7) is optimal for $a_0 = \tfrac{1}{2}\pi$, $a_{2k>0} = 0$, $a_{2k+1} = -\pi/2S(K)(2k+1)^2$, whence

$$J(K) = \frac{\pi}{4}\sum_{k=0}^{K}(2k+1)^2\frac{\pi^2}{4S^2(K)}\frac{1}{(2k+1)^4} = \frac{\pi^3}{16}\frac{1}{S(K)}, \tag{I.14}$$

$$\phi(x;K) = \frac{\pi}{2} - \frac{\pi}{2}\frac{1}{S(K)}\sum_{k=0}^{K}\frac{\cos[(2k+1)x]}{(2k+1)^2}. \tag{I.15}$$

Exercise: Verify by direct substitution that (I.15) satisfies all the BCs (I.3, 4).

The original trial function (I.7) clearly becomes more flexible, i.e. better, by admitting more terms, i.e. by increasing N and thereby K. This is indeed an improvement, because $S(K)$ increases steadily with K, whence $J(K)$ steadily falls. Therefore we look at the limit $K \to \infty$, where, by (I.14, 13b),

$$J(\infty) = \tfrac{1}{2}\pi. \tag{I.16}$$

Since $J(\infty) = \frac{1}{2}\pi = J_D$, the comment following (I.6) shows that $K \to \infty$ in fact realizes the absolute lower bound on J.

However, $\phi(x; K)$, which for any finite K satisfies all the BCs (I.3, 4), ceases to satisfy (I.4) in the limit $K \to \infty$. This is observable on comparing (I.15) (where $\pi/2S(K) \to \pi/2S(\infty) = 4/\pi$) with the standard Fourier series for $|x|$ and for its derivative $\varepsilon(x)$ in the range $-\pi \leqslant x \leqslant \pi$:

$$|x| = \frac{\pi}{2} - \frac{4}{\pi} \sum_{k=0}^{\infty} \frac{\cos[(2k+1)x]}{(2k+1)^2}, \tag{I.17}$$

$$\varepsilon(x) = \frac{4}{\pi} \sum_{k=0}^{\infty} \frac{\sin[(2k+1)x]}{(2k+1)}. \tag{I.18}$$

It is a familiar fact that the *finite* cosine series approximating the even function $|x|$ over $-\pi \leqslant x \leqslant \pi$ all have zero slope at $x = 0$ and $x = \pm\pi$. Indeed this is immediately obvious from term-by-term differentiation of any finite number of terms in (I.17), yielding the corresponding terms in (I.18), which all vanish at $x = 0$ and $\pm\pi$. However, the series (I.18), unlike (I.17), does not converge uniformly: for any fixed x, however close to 0 or π, the sum of sufficiently many terms approximates arbitrarily closely to 1, rather than to the zero value prescribed by the BCs (I.4).

Thus we have indeed illustrated how the sequence of trial functions $\phi(x; K)$ converges to the limit $\phi(x; \infty) = \psi_D(x)$, which realizes the true lower bound J_D of J, but violates half the Cauchy BCs, namely those on $\partial_n\phi$: in other words, $\phi(x; \infty)$ itself does not belong to the class of admissible trial functions. Looked at purely from within the variational approach this is a surprise; one concludes that variational arguments are safe only if one is armed with a prior proof that a solution to the problem exists.

Finally, for completeness we record the variational principle for the Laplace–Neumann problem, with prescribed $\partial_n\psi_S$. It asserts that the solution corresponds to a stationary value not of J, but of the combination

$$J - L \equiv \frac{1}{2}\int_V \mathrm{d}V\,(\nabla\phi)^2 - \int_S \mathrm{d}S\,\phi\,\partial_n\psi_S. \tag{I.19}$$

The trial functions are now subject to the BC $\partial_n\phi_S = \partial_n\psi_S$, whence $\partial_n\,\delta\phi_S = 0$, while $\delta\phi_S$ is unconstrained.

The variation $\phi \to \phi + \delta\phi$ induces a first-order change in J given by the penultimate expression in (I.1); combining this with the change in L, we obtain

$$\delta(J - L) = \left[\int_S \mathrm{d}S\,\delta\phi\,\partial_n\psi_S - \int_V \mathrm{d}V\,\delta\phi\nabla^2\phi\right] - \int_S \mathrm{d}S\,\delta\phi\,\partial_n\psi_S; \tag{I.20}$$

the two surface integrals cancel, so that demanding $\delta(J - L) = 0$ with arbitrary $\delta\phi$ again implies $\nabla^2\phi = 0$.

J Diffusion equation: simple-harmonic solutions from the magic rule

In Section 8.2, simple-harmonic solutions were found straightforwardly, but by a method not applicable to other cases. Later, in Section 9.4.4, we conjectured that the same solutions should be obtainable systematically from the magic rule, by evaluating say (9.4.14b) with $a(t) = \alpha \cos(\omega t)$ and $t_0 \to -\infty$. Here we perform this calculation. It is incomparably harder than the direct solution. The purpose of the exercise is twofold: first to show that the systematic application of the magic rule is indeed not always the easiest way; and second, as practice in such systematic work, in an example which is far from trivial when so tackled.

We must evaluate

$$\psi(x, t) = \frac{x}{(4\pi D)^{\frac{1}{2}}} \int_0^\infty \frac{d\tau}{\tau^{\frac{3}{2}}} \exp(-x^2/4D\tau)\alpha \cos[\omega(t-\tau)]$$

$$\equiv \operatorname{Re} \Psi \equiv \operatorname{Re} \frac{\alpha x}{(4\pi D)^{\frac{1}{2}}} \int_0^\infty \frac{d\tau}{\tau^{\frac{3}{2}}} \exp(-x^2/4D\tau) \exp(i\omega(t-\tau)). \tag{J.1}$$

Change variables, as in Section 8.5.1, to ξ defined by $\xi^2 = x^2/4D\tau$, $\tau = x^2/4D\xi^2$:

$$\Psi = \frac{\alpha x}{(4\pi D)^{\frac{1}{2}}} \exp(i\omega t) \int_0^\infty \frac{2x^2}{4D\xi^3} d\xi (4D)^{\frac{3}{2}} \frac{\xi^3}{x^3} \exp(-\xi^2) \exp(-i\omega x^2/4D\xi^2)$$

$$= \frac{2\alpha}{\pi^{\frac{1}{2}}} \exp(i\omega t) \int_0^\infty d\xi \exp\left\{-\left(\xi^2 + \frac{i\omega x^2}{4D}\frac{1}{\xi^2}\right)\right\}. \tag{J.2}$$

The crucial trick is to complete the square in the exponent, as in Section 8.3.1:

$$(\cdots) = \left(\left[\xi - \left(\frac{i\omega}{4D}\right)^{\frac{1}{2}} \frac{x}{\xi}\right]^2 + 2 \cdot \xi \cdot \left(\frac{i\omega}{4D}\right)^{\frac{1}{2}} \frac{x}{\xi}\right)$$

$$= \left(\left[\xi - \left(\frac{i\omega}{4D}\right)^{\frac{1}{2}} \frac{x}{\xi}\right]^2 + x\left(\frac{i\omega}{D}\right)^{\frac{1}{2}}\right). \tag{J.3}$$

Substitute (J.3) into (J.2):

$$\Psi = \frac{2\alpha}{\pi^{\frac{1}{2}}} \exp[i\omega t - x(i\omega/D)^{\frac{1}{2}}] \int_0^\infty d\xi \exp\left\{-\left(\xi - \left(\frac{i\omega}{4D}\right)^{\frac{1}{2}} \frac{x}{\xi}\right)^2\right\}. \tag{J.4}$$

Write the integral in (J.4), call it $J(\lambda)$, as

$$J(\lambda) = \int_0^\infty d\xi \exp\{-(\xi - \lambda/\xi)^2\}, \qquad \lambda \equiv (i\omega/4D)^{\frac{1}{2}}x. \tag{J.5}$$

The second crucial point is that $J(\lambda)$ is actually independent of λ. We shall prove this separately below. Accepting it for the moment, one has

$$J(\lambda) = J(0) = \int_0^\infty d\xi \exp(-\xi^2) = \tfrac{1}{2}\pi^{\frac{1}{2}}, \tag{J.6}$$

whence

$$\begin{aligned} \Psi &= \alpha \exp\{i\omega t - x(i\omega/D)^{\frac{1}{2}}\} \\ &= \alpha \exp\left\{i\omega t - x(1+i)\left(\frac{\omega}{2D}\right)^{\frac{1}{2}}\right\}, \end{aligned} \tag{J.7}$$

$$\psi = \operatorname{Re}\Psi = \alpha \exp(-x(\omega/2D)^{\frac{1}{2}}) \cos\left\{\omega t - x\left(\frac{\omega}{2D}\right)^{\frac{1}{2}}\right\}, \tag{J.8}$$

which agrees with our earlier (and much more easily found) solution (8.2.7).

Proof of (J.6): In (J.5) change the integration variable to $\eta \equiv \xi - \lambda/\xi$. Then $\xi^2 - \eta\xi - \lambda = 0$, and $\xi = \frac{1}{2}[\eta + (\eta^2 + 4\lambda)^{\frac{1}{2}})]$. Accordingly,

$$J(\lambda) = \int_{-\infty}^\infty d\eta \frac{1}{2}\left\{1 + \frac{\eta}{(\eta^2 + 4\lambda)^{\frac{1}{2}}}\right\} \exp(-\eta^2).$$

But the second part of the integrand is an odd function of η, so that it vanishes when integrated over the symmetric range $-\infty$ to $+\infty$. Hence

$$J(\lambda) = \frac{1}{2}\int_{-\infty}^\infty d\eta \exp(-\eta^2) = \tfrac{1}{2}\pi^{\frac{1}{2}}. \qquad \blacksquare$$

K Sound waves

In this book, sound is important mainly as a prototype for scalar waves. We need to establish the relations quoted in Section 10.1, and to clarify the status of the expressions for the acoustic energy density H and the acoustic energy flux $\boldsymbol{N}$. Good treatments of sound in its own right, and of its relation to fluid dynamics more generally, are given for instance by Lighthill (1978), Temkin (1981), Pierce (1981), Dowling and Ffowcs Williams (1983), and in the classic by Rayleigh (1896).

We consider only non-viscous compressible fluids, and only irrotational flow (curl $\boldsymbol{v}(\boldsymbol{r}, t) = 0$, where $\boldsymbol{v}$ is the fluid velocity), without questioning the physics of these restrictions. Then $\boldsymbol{v}$ can be written as a gradient,

$$\boldsymbol{v} = -\nabla\psi. \tag{K.1}$$

Further, we consider only small deviations from equilibrium: equilibrium values are (at first) identified by a suffix 0, and deviations from them by Δ except that (following common usage) no Δ is prefixed to $\boldsymbol{v}$, ψ, or to the source density ρ defined below. While deriving the wave equation, one keeps only first-order-small quantities, and linearizes the governing equations of the fluid accordingly; in H and $\boldsymbol{N}$, which vanish to first order, one keeps only the leading terms, which are of second order. (One must bear in mind that $\boldsymbol{v}$, ψ, and ρ are first-order small.) As to linearization, recall the penultimate paragraph of the preamble to Chapter 10.

Define mass density $\sigma = \sigma_0 + \Delta\sigma$, and pressure $p = p_0 + \Delta p$.

The equilibrium values σ_0 and p_0 are constants, so that $\sigma_t = \Delta\sigma_t$, $\nabla\sigma = \nabla\Delta\sigma$, and similarly for p. We consider only adiabatic changes (i.e. at constant entropy S, and as if the thermal conductivity were zero), so that temperature is not an independent variable. The fluid is modelled by the equation of state

$$\Delta p = (\partial p/\partial\sigma)_S \Delta\sigma = c^2 \Delta\sigma, \tag{K.2}$$

where the bulk modulus is taken to be constant, and written as c^2 by hindsight.

The mass-flux is evidently $\sigma\boldsymbol{v}$, or, to first order, $\sigma_0\boldsymbol{v}$; in other words, $\sigma_0\boldsymbol{v} \cdot \delta\boldsymbol{A}$ is the mass crossing the vector element of area $\delta\boldsymbol{A}$ per unit time. Sources are described in the first instance by specifying the mass $\tilde{\rho}(\boldsymbol{r}, t)\,\delta V$ of fluid injected per unit time into the (geometric) volume element δV. This means that the volume of fluid thus injected is $\rho\,\delta V$, where $\rho \equiv \tilde{\rho}/\sigma$. To first order we need not distinguish here between σ and σ_0, and write $\rho \approx \tilde{\rho}/\sigma_0$, which will be the source-strength distribution featured in our linearized equations.

To derive the wave equation, the exact mass conservation law $\partial\sigma/\partial t = -\text{div}\,(\sigma \boldsymbol{v}) + \tilde{\rho}$ is first linearized to

$$\frac{\partial \Delta\sigma}{\partial t} = -\sigma_0 \,\text{div}\, \boldsymbol{v} + \sigma_0 \rho = \sigma_0 \nabla^2 \psi + \sigma_0 \rho, \tag{K.3}$$

and Euler's equation (i.e. Newton's second law for a given mass-element of fluid), namely $\sigma(\partial/\partial t + \boldsymbol{v} \cdot \boldsymbol{\nabla})\boldsymbol{v} = -\boldsymbol{\nabla} p$, is linearized to

$$\sigma_0 \,\partial \boldsymbol{v}/\partial t = -\sigma_0 \frac{\partial}{\partial t} \boldsymbol{\nabla}\psi = -\sigma_0 \boldsymbol{\nabla} \frac{\partial \psi}{\partial t} = -\boldsymbol{\nabla}\, \Delta p,$$

$$\sigma_0 \boldsymbol{\nabla} \frac{\partial \psi}{\partial t} = \boldsymbol{\nabla}\, \Delta p = c^2 \boldsymbol{\nabla}\, \Delta\sigma. \tag{K.4}$$

This integrates to $\sigma_0 \,\partial\psi/\partial t = c^2 \,\Delta\sigma + F(t)$, where $F(t)$ may be a function of t but not of $\boldsymbol{r}$. However, we are free to set $F = 0$, because F is irrelevant to both the basic observables: to $\boldsymbol{v}$, because $\boldsymbol{v} = -\text{grad}\, \psi$; and to $\Delta\sigma$, because in any changes starting from equilibrium at time t_0, we can write $\Delta\sigma = \int_{t_0}^{t} \mathrm{d}t\, \partial\, \Delta\sigma/\partial t$ (at fixed $\boldsymbol{r}$). With the aid of (K.3), this determines $\Delta\sigma$ directly in terms of $\sigma_0 \nabla^2 \psi = -\sigma_0 \,\text{div}\, \boldsymbol{v}$, and of the data ρ. But all these quantities are unaffected by F. Thus we replace (K.4) by

$$\sigma_0 \frac{\partial \psi}{\partial t} = \Delta p = c^2 \,\Delta\sigma. \tag{K.5}$$

Finally, act on (K.5) with $\partial/\partial t$, substitute on the right from (K.3), and divide through by $c^2 \sigma_0$; this yields the wave equation (10.1.1), namely

$$\left(\frac{1}{c^2}\frac{\partial^2}{\partial t^2} - \nabla^2\right)\psi = \rho. \tag{K.6}$$

One can rid oneself of the unwanted integration constant $F(t)$ more elegantly as follows. Act with $\partial/\partial t$ on the relation $\sigma_0 \,\partial\psi/\partial t = \Delta p + F(t) = c^2 \,\Delta\sigma + F(t)$ that emerges from (K.4), and, as before, substitute for $\Delta\dot{\sigma}$ from (K.3). This yields $\Box^2 \psi = \rho + \dot{F}(t)/c^2 \sigma_0$. Now define a new velocity potential ψ' by

$$\psi(\boldsymbol{r}, t) \equiv \psi'(\boldsymbol{r}, t) + \int_{t_0}^{t} \mathrm{d}t'\, F(t')/\sigma_0,$$

where the lower integration limit is arbitrary. Evidently,

$$\partial\psi/\partial t = \partial\psi'/\partial t + F(t)/\sigma_0, \qquad \partial^2\psi/\partial t^2 = \partial^2\psi'/\partial t^2 + \dot{F}(t)/\sigma_0,$$

whence

$$\boldsymbol{v} = -\boldsymbol{\nabla}\psi = -\boldsymbol{\nabla}\psi', \qquad \Delta p = \sigma_0 \,\partial\psi/\partial t - F(t) = \sigma_0 \,\partial\psi'/\partial t,$$

while

$$\begin{aligned}(c^{-2}\,\partial^2/\partial t^2 - \nabla^2)\psi &= (c^{-2}\,\partial^2/\partial t^2 - \nabla^2)\psi' + \dot{F}(t)/c^2\sigma_0 \\ &= \rho + \dot{F}(t)/c^2\sigma_0,\end{aligned}$$

whence

$$\Box^2\psi' = \rho.$$

In other words, ψ' satisfies all the relations previously deduced for ψ after adopting the special (albeit perfectly legitimate) choice $F=0$. However, no special choice of $F(t)$ is necessary in order to secure these relations for ψ'. Accordingly, the answer to the question: What would we do if F were not zero? is that we would simply adopt ψ' instead of the originally envisaged ψ as the velocity potential, drop the prime, and proceed exactly as before. In the language of field theory, we have eliminated $F(t)$ from our formalism by means of *a gauge transformation* (of ψ to ψ').

Exercise (for those interested in analytical dynamics): Elucidate the physical status and the relation between the available-energy densities H and H' constructed from ψ and ψ' respectively.

It remains to identify the physics of the so-called acoustic energy density H quoted in Section 10.1. On formal grounds alone one would already expect H to prove an important function, because of the simplicity of the conservation laws (10.1.3–5), because they follow directly from the wave equation (as shown in eqns (K.17–21) below), and because of their evident similarity to the energy conservation laws appropriate to Maxwell's equation. However, for sound waves the precise status of H is not quite so obvious as most discussions imply, and we proceed rather carefully. As explained in Chapter 10, we work only to second order.

The kinetic energy per unit volume is

$$H_{\text{kin}} = \tfrac{1}{2}\sigma_0 \boldsymbol{v}^2 = \tfrac{1}{2}\sigma_0(\boldsymbol{\nabla}\psi)^2. \tag{K.7}$$

As regards energy changes due to compression, we are concerned not with the total internal energy U, but with the *available energy* A, changes in which equal the so-called *useful work* done on systems that are under given external pressure p_0 and in contact with a reservoir of fluid at given chemical potential μ_0. In our case, p_0 and μ_0 are the values appropriate to the fluid in equilibrium. The fluid reservoir is essential to the argument, because in the presence of sources ρ the total amount of fluid in the system can change. To reckon useful work, one must ignore energy supplied by the agencies maintaining the external constraints p_0 and μ_0. (For a good discussion of useful work and available energy, see Callen (1960). Adkins (1968) and Landau and Lifshitz (1958) also consider useful work, but only for a fixed amount of fluid; unfortunately, in that case one can, and they do, dispense with the final term $-\mu_0 M$ on the right of (K.8) below, which as we shall see proves crucial for sound waves with sources.)

The functions U and A are related by

$$A = U + p_0 V - \mu_0 M, \tag{K.8}$$

where V is the volume and M is the mass, so that

$$\sigma = M/V. \tag{K.9}$$

We shall need also the enthalpy $\tilde{H}$, and the chemical potential μ, defined by

$$\tilde{H} = U + p(\sigma)V = \mu(\sigma)M. \tag{K.10a,b}$$

(In our model of the fluid, defined by the equation of state (K.2), all processes are adiabatic, and temperature (or entropy) is not an independent degree of freedom. Thus all intensive variables, like p and μ, are functions of the density σ alone. For the same reason, the enthalpy plays the role that the Gibbs free energy $G = U + pV - TS$ plays in standard thermodynamics: hence the equality $\tilde{H} = \mu M$.)

Next we derive the variation of the total energy density $u \equiv (U/V)$. For fixed M one has $\mathrm{d}U = -p\,\mathrm{d}V$, and (by (K.9)) $\mathrm{d}V/V = -\mathrm{d}\sigma/\sigma$; hence

$$\mathrm{d}u = \mathrm{d}(U/V) = \mathrm{d}U/V - U\,\mathrm{d}V/V^2$$

$$= -p\,\mathrm{d}V/V - (U/V)\,\mathrm{d}V/V = -(p+u)\,\mathrm{d}V/V,$$

$$\mathrm{d}u = (p+u)\,\mathrm{d}\sigma/\sigma. \tag{K.11}$$

Rearrangement, multiplication by $1/\sigma$, and appeal to (K.2) in the form $p = p_0 + \Delta p = p_0 + c^2\,\Delta\sigma = p_0 + c^2(\sigma - \sigma_0)$ yield

$$\frac{1}{\sigma}\left(\mathrm{d}u - u\frac{\mathrm{d}\sigma}{\sigma}\right) = \frac{1}{\sigma}p\frac{\mathrm{d}\sigma}{\sigma},$$

$$\mathrm{d}(u/\sigma) = p\,\mathrm{d}\sigma/\sigma^2 = (p_0 - c^2\sigma_0)\,\mathrm{d}\sigma/\sigma^2 + c^2\,\mathrm{d}\sigma/\sigma. \tag{K.12}$$

Integrating from equilibrium and multiplying by σ we find

$$u = u_0\sigma/\sigma_0 - (p_0 - c^2\sigma_0)(1 - \sigma/\sigma_0) + c^2\sigma\log(\sigma/\sigma_0).$$

Finally we write $\sigma = \sigma_0 + \Delta\sigma$, and keep only terms up to second order in $\Delta\sigma/\sigma_0$:

$$u - u_0 \equiv \Delta u = (u_0 + p_0)\,\Delta\sigma/\sigma_0 + \frac{c^2}{2\sigma_0}(\Delta\sigma)^2 + \cdots$$

$$\equiv \Delta^{(1)}u + \Delta^{(2)}u + \cdots. \tag{K.13}$$

Exercise: Derive (K.13) from (K.12) in full detail.

It is the first-order term $\Delta^{(1)}u \equiv (u_0 + p_0)\,\Delta\sigma/\sigma_0$ that eliminates Δu as a candidate for the 'potential-energy' component of the acoustic energy density H,† which is wholly second-order, as it must be in order to be guaranteed non-negative.

Before determining the density $a \equiv A/V$, we need the chemical potential. From (K.10b),

$$\mu = (U + pV)/M = (U + pV)/\sigma V = [(U/V) + p]/\sigma,$$

$$\mu = (u + p)/\sigma. \tag{K.14a}$$

In particular, at equilibrium one has

$$\mu_0 = (u_0 + p_0)/\sigma_0. \tag{K.14b}$$

The density a now follows straightforwardly:

$$a \equiv (A/V) = (U + p_0 V - \mu_0 M)/V = u + p_0 - \mu_0 \sigma,$$

$$a = u + p_0 - (u_0 + p_0)\sigma/\sigma_0. \tag{K.15a}$$

For $\Delta a \equiv a - a_0$ this entails

$$\Delta a = \Delta u - (u_0 + p_0)\,\Delta\sigma/\sigma_0. \tag{K.15b}$$

We see that the second term precisely cancels the linear component $\Delta^{(1)}u$ of Δu as given by (K.13):

$$\Delta a = \frac{c^2}{2\sigma_0}(\Delta\sigma)^2. \tag{K.15c}$$

Finally, (K.2) and (K.5) yield

$$\Delta a = \frac{(\Delta p)^2}{2\sigma_0 c^2} = \frac{\sigma_0}{2c^2}\left(\frac{\partial\psi}{\partial t}\right)^2 \equiv H_{\text{pot}}. \tag{K.16}$$

The acoustic energy density (10.1.3)

$$H = H_{\text{pot}} + H_{\text{kin}} = \tfrac{1}{2}\sigma_0\left\{\frac{1}{c^2}\left(\frac{\partial\psi}{\partial t}\right)^2 + (\nabla\psi)^2\right\} \tag{K.17}$$

is now fully identified as the *density of available energy*: the first term stems from the (useful) work done in compressing the fluid; the second term stems from the kinetic energy of the flow (all of which is useful). Notice that for sound waves the kinetic (motional) part of H is the part that does *not* explicitly refer to the time-variation of ψ.

† In purely oscillatory disturbances the long-time-averages of such first-order (linear) terms vanish automatically. Unfortunately, this fact by itself is not sufficient reason for ignoring them always and everywhere, though impatient authors often construe it so. Another argument for ignoring the linear term is that its volume integral over the system vanishes provided the total mass is constant, since in that case $\Delta M = \int dV\,\Delta\sigma = 0$. This argument also fails. First, it does not apply in the presence of sources, when ΔM need not vanish. Second, even if $\Delta M = 0$, a global argument of this type cannot identify densities and fluxes, which depend essentially on the local distribution of the physical quantities in question.

Expressions of this form are relevant to all scalar waves, though the physical significance of the individual terms is not always the same.

Exercise: Establish the expression (10.1.7) appropriate to transverse waves on a stretched string or membrane.

The acoustic energy flux $\boldsymbol{N}$ is defined so that $\boldsymbol{N}\cdot\delta\boldsymbol{A}$ is the acoustic energy crossing the vector element of area $\delta\boldsymbol{A}$ per unit time. (It must not be confused with the mass flux $\sigma\boldsymbol{v}$.) When there are no sources ($\rho=0$), and therefore no mechanism for doing external work on the fluid, acoustic energy is conserved. Then $\delta V\,\partial H/\partial t=$ (outflow per unit time of acoustic energy from the (geometric) volume element δV) $=-\delta V\cdot\operatorname{div}\boldsymbol{N}$. $\boldsymbol{N}$ is identified by considering $\partial H/\partial t$:

$$\begin{aligned}\frac{\partial H}{\partial t}&=\sigma_0\left\{\frac{1}{c^2}\frac{\partial\psi}{\partial t}\frac{\partial^2\psi}{\partial t^2}+(\boldsymbol{\nabla}\psi)\cdot\boldsymbol{\nabla}\frac{\partial\psi}{\partial t}\right\}\\&=\sigma_0\left\{\frac{\partial\psi}{\partial t}(\nabla^2\psi+\rho)+(\boldsymbol{\nabla}\psi)\cdot\boldsymbol{\nabla}\frac{\partial\psi}{\partial t}\right\}\\&=\operatorname{div}\left(\sigma_0\frac{\partial\psi}{\partial t}\boldsymbol{\nabla}\psi\right)+\sigma_0\rho\frac{\partial\psi}{\partial t}.\end{aligned}\tag{K.18}$$

The second step eliminates $(1/c^2)\,\partial^2\psi/\partial t^2$ by appeal to the wave equation (K.6). From the special case $\rho=0$ we see that

$$\boldsymbol{N}=-\sigma_0\frac{\partial\psi}{\partial t}\boldsymbol{\nabla}\psi=\boldsymbol{v}\,\Delta p.\tag{K.19}$$

As expected, one recognizes $\boldsymbol{v}\,\Delta p\cdot\delta\boldsymbol{A}$ as the work done by the excess pressure on the fluid crossing $\delta\boldsymbol{A}$ in the unit time.

The second term on the right of (K.18) is the work done by the sources ρ against the excess pressure, i.e. their useful power output $w(\boldsymbol{r},t)$ per unit volume:

$$w(\boldsymbol{r},t)=\sigma_0\rho\frac{\partial\psi}{\partial t}=\rho\,\Delta p,\tag{K.20}$$

again as one might have foreseen. Thus (K.18) can be viewed as

$$\frac{\partial H}{\partial t}=-\operatorname{div}\boldsymbol{N}+w,\tag{K.21}$$

which establishes eqns (10.1.4–6). (Outside this apppendix we drop the suffix from σ_0 in such equations, which is correct to the order to which we are working.)

It is worth stressing that, with H, $\mathbf{N}$, w defined by (K.17, 19, 20), the conservation law (K.21) is an exact mathematical consequence of the wave equation, quite independently of any prior views one might hold about the physics of H or of its two components $(\sigma_0/2c^2)(\partial\psi/\partial t)^2$ and $(\sigma_0/2)(\nabla\psi)^2$ separately. Alternatively to the procedure just followed, for sound waves one can derive (K.21) without explicit appeal to the wave equation, by starting instead from the exact hydrodynamic equations and from the exact conservation law for the total (rather than the available) energy, implementing them to second order. Careful arguments along such lines are given by Lighthill (1978), Pierce (1981, Section 1.11), and Temkin (1981, Section 2.7), albeit only in the absence of sources. The disadvantage of this alternative approach is that it leaves unexplored the precise physical significance of H and $\mathbf{N}$, and the reason why these second-order expressions are more interesting, in their own right, than certain associated first-order terms like for instance the $\Delta^{(1)}u$ in (K.13).

We close with some remarks about sources.

(i) One way to realize volume sources is to inject or extract fluid by main force, for instance through a tube so thin that it does not appreciably affect the flow pattern, and whose end opens inside the system.

(ii) A more common kind of volume source is simply a moving boundary. For instance, a sphere centred on the fixed point $\mathbf{r}'$, with its volume Q made to vary with time, can, if it is small enough on the relevant length scale, mimic a point source $\rho(\mathbf{r}, t) = (\mathrm{d}Q/\mathrm{d}t)\,\delta(\mathbf{r} - \mathbf{r}')$. An example is considered in Section 12.4.

(iii) By contrast, a small moving sphere of constant volume mimics positive sources on the leading and negative sources (sinks) on the trailing hemisphere; to a first approximation this amounts to a point dipole source, whose strength depends on the velocity of the sphere. We shall not pursue such problems: though they are relevant to acoustics, they are less useful as analogues to electromagnetic radiation than are sources whose strength is independent of their speed.

(iv) Lastly, volume sources can be mimicked by expansion of the fluid induced by local heating. Examples are discussed by Krasil'nilov and Pavlov (1981) and by Temkin (1981, Section 5.10). One such effect is thunder.

L Wave equation: the initial conditions on $K_0^{(3)}$ verified explicitly

Section 11.2.2 derived the 3D propagator

$$K_0^{(3)}(\boldsymbol{R}, \tau) = \frac{1}{4\pi R}\{\delta(\tau - R/c) - \delta(\tau + R/c)\}$$

$$= \frac{c}{4\pi R}\{\delta(c\tau - R) - \delta(c\tau + R)\}$$

$$= \frac{c}{4\pi R}\{\delta(R - c\tau) - \delta(R + c\tau)\}, \tag{L.1a}$$

$$K_{0\tau}^{(3)}(\boldsymbol{R}, \tau) = \frac{1}{4\pi R}\{\delta'(\tau - R/c) - \delta'(\tau + R/c)\}$$

$$= \frac{c^2}{4\pi R}\{\delta'(c\tau - R) - \delta'(c\tau + R)\}$$

$$= \frac{c^2}{4\pi R}\{-\delta'(R - c\tau) - \delta'(R + c\tau)\}, \tag{L.1b}$$

and mentioned that care is needed with the limits $\tau \to 0+$ if one wishes to verify explicitly that these expressions obey the initial conditions

$$\lim_{\tau\to 0+} K_0(\boldsymbol{R}, \tau) = 0, \qquad \lim_{\tau\to 0+} K_{0\tau}(\boldsymbol{R}, \tau) = c^2\,\delta(\boldsymbol{R}). \tag{L.2a,b}$$

Very roughly, the problem is to marry the weak definition under which the delta-functions enter (L.1) to the strong definition underlying the standard 3D polar forms (see Section 1.4.2)

$$\delta(\boldsymbol{R}) = \delta(R)/4\pi R^2 = -\delta'(R)/4\pi R. \tag{L.3a,b}$$

To verify that (L.1) satisfy (L.2) we need two obvious facts and two lemmas. The obvious facts are that $\delta(c\tau + R)/4\pi R$ and $\delta'(c\tau + R)/4\pi R$ vanish if τ is strictly positive (because $R \geqslant 0$ by definition, so that $(c\tau + R)$ cannot then be zero). It follows that

$$\lim_{\tau\to 0+} \frac{c\,\delta(c\tau + R)}{4\pi R} = 0, \qquad \lim_{\tau\to 0+} \frac{c\,\delta'(c\tau + R)}{4\pi R} = 0. \tag{L.4a,b}$$

Lemma 1:

$$\lim_{\tau\to 0+} c\,\delta(c\tau - R)/4\pi R = 0. \tag{L.5}$$

Proof: We prove the lemma by showing that

$$J \equiv \lim_{\tau\to 0+} \int \mathrm{d}V f(\boldsymbol{R}) c\,\delta(c\tau - R)/4\pi R = 0, \tag{L.6}$$

where $f(\boldsymbol{R})$ is any test function well-behaved at the origin, so that, when $R \to 0$, $f(\boldsymbol{R}) = f(R, \Omega)$ tends to a unique limit $f(0, \Omega) \equiv f(0)$ independent of Ω. To prove (L.6), write $\int \mathrm{d}V \cdots = \int \mathrm{d}\Omega \int_0^\infty \mathrm{d}R\, R^2 \cdots$, whence

$$J = \int \frac{\mathrm{d}\Omega}{4\pi} \lim_{\tau \to 0+} \int_0^\infty \mathrm{d}R\, R^2 f(R, \Omega) \frac{c\,\delta(c\tau - R)}{R}$$

$$= \int \frac{\mathrm{d}\Omega}{4\pi} \lim_{\tau \to 0+} c^2 \tau f(c\tau, \Omega) = \int \frac{\mathrm{d}\Omega}{4\pi} \lim_{\tau \to 0+} c^2 \tau f(0) = 0. \qquad \blacksquare$$

Lemma 2:

$$\lim_{\tau \to 0+} c^2\, \delta'(c\tau - R)/4\pi R = c^2\, \delta(\boldsymbol{R}). \tag{L.7}$$

Proof: We show that the two sides give the same result when multiplied by a test function $f(\boldsymbol{R})$ and integrated over all space. On the right, we obtain $c^2 f(0)$. On the left, using $\delta'(c\tau - R) = -\delta'(R - c\tau)$, we obtain

$$-c^2 \int \frac{\mathrm{d}\Omega}{4\pi} \lim_{\tau \to 0+} \int_0^\infty \mathrm{d}R\, R^2 f(R, \Omega) \frac{\delta'(R - c\tau)}{R}$$

$$= c^2 \int \frac{\mathrm{d}\Omega}{4\pi} \lim_{\tau \to 0+} \frac{\partial}{\partial R} [R f(R, \Omega)] \Big|_{R = c\tau}$$

$$= c^2 \int \frac{\mathrm{d}\Omega}{4\pi} \lim_{\tau \to 0+} \left[f(c\tau, \Omega) + c\tau \frac{\partial f(R, \Omega)}{\partial R} \right] \Bigg|_{R = c\tau}$$

$$= c^2 \int \frac{\mathrm{d}\Omega}{4\pi} f(0, \Omega) = c^2 f(0). \qquad \blacksquare$$

The initial condition (L.2a) is now verified by substitution on the right from (L.4a) and (L.5). The other initial condition (L.2b) is verified by substitution from (L.4b) and (L.7). ■

It proves important in related problems that the limits of the two terms on the right of (L.1a) vanish individually and not by mutual cancellation; and that on the right of (L.1b) the first term by itself provides the expression required by (L.2b). For instance, if naively but wrongly we started by setting $\tau = 0$ on the right of (L.1b), we would obtain

$$\frac{c^2}{4\pi R} \{\delta'(-R) - \delta'(R)\} = -2 \frac{c^2}{4\pi R} \delta'(R),$$

which is double the true expression. Similar mistakes are easily made in applying d'Alembert's method (Appendix N.4), and would then yield an expression for $K_0^{(3)}$ wrong by a factor $\frac{1}{2}$.

M Wave equation: the polar expansion of $K_0^{(3)}(\boldsymbol{r}, t \mid \boldsymbol{r}', t')$

Although $K_0^{(3)}$ is an isotropic function of $\boldsymbol{R} = \boldsymbol{r} - \boldsymbol{r}'$, it is not of course isotropic as a function of $\boldsymbol{r}$ for given $\boldsymbol{r}'$, and often one needs the expansion of $K_0^{(3)}(\boldsymbol{r}, \boldsymbol{r}'; \tau)$ in the spherical harmonics $Y_{lm}(\Omega)$, i.e. the analogue of the Poisson Green's function expressed as

$$1/4\pi R = \sum_{lm} \frac{r_<^l}{r_>^{l+1}} \frac{1}{(2l+1)} Y'_{lm}(\Omega') Y_{lm}(\Omega)$$

(Appendix H.3). In this section $\sum_{lm}$ stands for $\sum_{l=0}^{\infty} \sum_{m=-l}^{l}$, and to ease the notation we often drop the suffix 0 specifying unbounded space.

The polar expansion of $K_0^{(3)}$ is useful when ψ and ρ are themselves expanded in polars:

$$\psi(\boldsymbol{r}, t) = \sum_{lm} R_{lm}(r, t) Y_{lm}(\Omega) = \frac{1}{r} \sum_{lm} \phi_{lm}(r, t) Y_{lm}(\Omega). \tag{M.1}$$

$$\rho(\boldsymbol{r}, t) = \frac{1}{r} \sum_{lm} \rho_{lm}(r, t) Y_{lm}(\Omega). \tag{M.2}$$

The prefactors $1/r$ are introduced by hindsight. The reduced radial functions $\phi_{lm} \equiv rR_{lm}$ then obey

$$\left(\frac{1}{c^2}\frac{\partial^2}{\partial t^2} - \frac{\partial^2}{\partial r^2} + \frac{l(l+1)}{r^2}\right)\phi_{lm}(r, t) = \rho_{lm}(r, t); \tag{M.3}$$

for $l = 0$, i.e. for the isotropic component, this coincides with the 1D wave equation:

$$\left(\frac{1}{c^2}\frac{\partial^2}{\partial t^2} - \frac{\partial^2}{\partial r^2}\right)\phi_{00} = \rho_{00}. \tag{M.4}$$

The difference from 1D is that the true radial functions R_{ln} must be well-behaved at the origin, whence at $r = 0$ all the ϕ_{lm} obey the homogeneous DBC

$$\phi_{lm}(0, t) = 0. \tag{M.5}$$

In view of its symmetry in $\boldsymbol{r}$ and $\boldsymbol{r}'$, we look for $K_0^{(3)}$ in the form

$$K_0^{(3)}(\boldsymbol{r}, \boldsymbol{r}'; \tau) = \frac{1}{rr'} \sum_{lm} \kappa_l(r, r'; \tau) Y^*_{lm}(\Omega') Y_{lm}(\Omega). \tag{M.6}$$

To determine the coefficients κ_l, one substitutes (M.6) into the defining relations of $K_0^{(3)}$; on the right of the initial condition on $K_{0\tau}$ one uses the polar expansion

$$\delta(\boldsymbol{r} - \boldsymbol{r}') = (1/r^2) \sum_{lm} \delta(r - r') Y^*_{lm}(\Omega') Y_{lm}(\Omega):$$

$$\Box^2 K_0^{(3)} = 0 \Rightarrow \left(\frac{1}{c^2}\frac{\partial^2}{\partial t^2} - \frac{\partial^2}{\partial r^2} + \frac{l(l+1)}{r^2}\right)\kappa_l = 0; \tag{M.7}$$

$$(K_0^{(3)} \text{ well-defined at } r = 0) \Rightarrow \kappa_l(0, r'; \tau) = 0; \tag{M.8}$$

$$K_0^{(3)}(\mathbf{r}, \mathbf{r}'; 0) = 0 \Rightarrow \kappa_l(r, r'; 0) = 0; \tag{M.9}$$

$$K_{0\tau}^{(3)}(\mathbf{r}, \mathbf{r}'; 0) = c^2\, \delta(\mathbf{r} - \mathbf{r}') \Rightarrow \kappa_{l\tau}(r, r'; 0) = c^2\, \delta(r - r'). \tag{M.10}$$

Since the defining relations (M.7–9) do not refer to m, we see that the coefficients κ_l are indeed independent of m, as anticipated tacitly in (M.6).

The integrals f, h_1, and h_2 in the magic rule now acquire the polar expansions

$$f(\mathbf{r}, t) = \frac{1}{r} \sum_{lm} \left\{ \int_{t_0}^{t} \mathrm{d}t' \int_0^\infty \mathrm{d}r'\, \kappa_l(r, r'; \tau) \rho_{lm}(r', t') \right\} Y_{lm}(\Omega), \tag{M.11}$$

$$h_1(\mathbf{r}, t) = \frac{1}{r} \sum_{lm} \left\{ \int_0^\infty \mathrm{d}r' \frac{1}{c^2} \kappa_l(r, r'; \tau) \phi_{lm,t_0}(r', t_0) \right\} Y_{lm}(\Omega), \tag{M.12}$$

$$h_2(\mathbf{r}, t) = \frac{1}{r} \frac{\partial}{\partial t} \sum_{lm} \left\{ \int_0^\infty \mathrm{d}r' \frac{1}{c^2} \kappa_l(r, r'; \tau) \phi_{lm}(r', t_0) \right\} Y_{lm}(\Omega). \tag{M.13}$$

It remains to determine the κ_l. In this appendix we consider only $l = 0$, which is readily found by the method of images. The same result will be deduced in Appendix N.3 by d'Alembert's method, which is more easily generalized to $l \geq 1$.

The isotropic component κ_0 is defined by (M.7) with $l = 0$, plus (M.8–10). But these are precisely the equations defining the 1D propagator in the halfspace $r \geq 0$, subject to the homogeneous DBC (M.8); the result is given immediately by the method of images (Section 12.8.1):

$$\begin{aligned} \kappa_0(r, r'; \tau) &= K_0^{(1)}(r, r'; \tau) - K_0^{(1)}(r, -r'; \tau) \\ &= \varepsilon(\tau) \frac{c}{2} \{ \theta(c\,|\tau| - |r - r'|) - \theta(c\,|\tau| - (r + r')) \}. \end{aligned} \tag{M.14}$$

The 1D propagator $K_0^{(1)}$ is given by (11.2.15, 16) and sketched in Figs 11.3.4.

Exercise: Write down the corresponding equations (i) for the polar expansion in 2D; (ii) for the diffusion equation in 3D.

As an example we reconsider the problem of the bursting ballon (solved in Section 12.2.2 by Poisson's method), and determine the excess pressure $\sigma\psi_t$ at exterior points $r > a$ from the initial values $\psi(\mathbf{r}, 0) = 0$, $\psi_t(\mathbf{r}, 0) = Q\theta(a - r)$, in the absence of sources. Only the integral h_1 contributes in the magic rule, and only the $l = 0$ term survives in its polar expansion (M.12). In view of (M.1) we have

$$\psi_t(r', 0) = Q\theta(a - r') = \frac{1}{r'} \phi_{00,t}(r', 0) Y_{00},$$

$$\phi_{00,t}(r', 0) Y_{00} = r' Q\theta(a - r'), \tag{M.15}$$

where the factor r' on the right should be noted. Substitute into (M.12), and act on both sides with $\sigma\, \partial/\partial t$ to obtain Δp:

$$\begin{aligned}
\Delta p(\boldsymbol{r}, t) &= \sigma \frac{\partial}{\partial t} \frac{1}{r} \int_0^\infty \mathrm{d}r' \frac{1}{c^2} \kappa_0(r, r'; t) r' \alpha \theta(a - r') \\
&= \frac{\sigma Q}{c^2} \frac{1}{r} \int_0^a \mathrm{d}r'\, r' \frac{\partial \kappa_0(r, r'; t)}{\partial t} \\
&= \frac{\sigma Q}{c^2} \frac{1}{r} \int_0^a \mathrm{d}r'\, r' \frac{c}{2} c \{\delta(ct - |r - r'|) - \delta(ct - (r + r'))\} \\
&= \frac{\sigma Q}{2r} \int_0^a \mathrm{d}r'\, r' \{\delta(ct - r + r') - \delta(ct - r - r')\}.
\end{aligned} \tag{M.16}$$

The third step follows differentiating (M.14); in the last step, $r > a$ has entailed $|r - r'| = r - r'$.

The first delta-function vanishes throughout the integration range unless $0 < (r - ct) < a$, and the second vanishes unless $0 < (ct - r) < a$. Therefore

$$\begin{aligned}
\Delta p &= \frac{\sigma Q}{2r} \theta(a - |r - ct|)(r - ct) \\
&= \sigma Q \theta(a - |r - ct|) \frac{r - ct}{2r}.
\end{aligned} \tag{M.17}$$

This agrees with (12.2.11) given by Poisson's solution, and sketched in Fig. 12.2a.

Exercise: Find Δp at the interior points $r < a$ and check the result at $r = 0$ with the solution of Problem 11.7.

N Wave equation: d'Alembert's solution

As promised in Section 11.1, we describe a classic and powerful approach to IVPs in 1D; by appropriate special pleading it covers isotropic problems in 3D as well.

N.1 d'Alembert's solution of the initial-value problem in 1D

The homogeneous wave equation in 1D can be reduced to triviality by changing the independent variables from (x, t) to

$$\xi \equiv x + ct, \qquad \eta \equiv x - ct. \tag{N.1.1}$$

Consequently, one has

$$\begin{aligned}
\frac{\partial}{\partial x} &= \frac{\partial \xi}{\partial x}\frac{\partial}{\partial \xi} + \frac{\partial \eta}{\partial x}\frac{\partial}{\partial \eta} = \left(\frac{\partial}{\partial \xi} + \frac{\partial}{\partial \eta}\right),\\
\frac{\partial}{\partial t} &= \frac{\partial \xi}{\partial t}\frac{\partial}{\partial \xi} + \frac{\partial \eta}{\partial t}\frac{\partial}{\partial \eta} = c\left(\frac{\partial}{\partial \xi} - \frac{\partial}{\partial \eta}\right),\\
\Box^2 &= \frac{1}{c^2}\frac{\partial^2}{\partial t^2} - \frac{\partial^2}{\partial x^2} = \left(\frac{\partial}{\partial \xi} - \frac{\partial}{\partial \eta}\right)^2 - \left(\frac{\partial}{\partial \xi} + \frac{\partial}{\partial \eta}\right)^2\\
&= -4\frac{\partial^2}{\partial \xi\, \partial \eta}.
\end{aligned} \tag{N.1.2}$$

Thus the wave equation $\Box^2\psi = 0$ becomes

$$\frac{\partial^2}{\partial \xi\, \partial \eta}\psi = 0. \tag{N.1.3}$$

This equation is satisfied by

$$\psi(x, t) = f(\eta) + g(\xi) = f(x - ct) + g(x + ct), \tag{N.1.4}$$

where f and g can be any functions of their respective arguments. (Do not confuse f and g here with the integrals f and g in the magic rule.) Because this expression contains two adjustable functions, it can, as we shall see, fit any prescribed initial values

$$\psi_t(x, 0) \equiv \alpha(x), \qquad \psi(x, 0) \equiv \beta(x). \tag{N.1.5a, b}$$

Therefore (N.1.4) solves the most general initial-value problem. Since the wave equation is linear, one can write such a solution as $\psi = \tilde{h}_1 + \tilde{h}_2$, where $\tilde{h}_1(x, 0) = \beta$, $\tilde{h}_{1t}(x, 0) = 0$, while $\tilde{h}_2(x, 0) = 0$, $\tilde{h}_{2t} = \alpha$. We deal with $\tilde{h}_1$ and $\tilde{h}_2$ in turn.

Writing $\tilde{h}_1(x, t) = f_1(x - ct) + g_1(x + ct)$, the initial condition (N.1.5a), with $\alpha = 0$, requires

$$\tilde{h}_{1t}(x, 0) = -cf'_1(x) + cg'_1(x) = 0.$$

Therefore $g'_1(x) = f'_1(x)$, and, by integration,

$$g_1(x) = f_1(x) + A, \tag{N.1.6}$$

where A is a constant. (A prime always signifies the derivative of a function with respect to its argument, whatever that may be. For instance, the chain rule implies $\frac{\partial}{\partial t} f_1(x - ct) = f'_1(x - ct) \cdot (-c)$.) Thus

$$\tilde{h}_1(x, t) = f_1(x - ct) + f_1(x + ct) + A, \tag{N.1.7}$$

and the other initial condition (N.1.5b) requires

$$\tilde{h}_1(x, 0) = 2f_1(x) + A = \beta(x), \qquad f_1(x) = \tfrac{1}{2}\beta(x) - \tfrac{1}{2}A.$$

On substitution into (N.1.7), A cancels, and we find the end-result

$$\tilde{h}_1(x, t) = \tfrac{1}{2}[\beta(x - ct) + \beta(x + ct)]. \tag{N.1.8}$$

Writing $\tilde{h}_2(x, t) = f_2(x - ct) + g_2(x + ct)$, we first impose (N.1.5b), with $\beta = 0$:

$$\tilde{h}_2(x, 0) = f_2(x) + g_2(x) = 0, \qquad g_2(x) = -f_2(x),$$

whence

$$\tilde{h}_2(x, t) = f_2(x - ct) - f_2(x + ct). \tag{N.1.9}$$

Then the other IC requires

$$\tilde{h}_{2t}(x, 0) = -2cf'_2(x) = \alpha(x), \qquad f'_2(x) = -\frac{1}{2c}\alpha(x),$$

$$f_2(x) = -\frac{1}{2c}\int_a^x \mathrm{d}x'\, \alpha(x'), \tag{N.1.10}$$

the lower limit a being arbitrary. Substitution into (N.1.9) yields the end-result

$$\tilde{h}_2(x, t) = -\frac{1}{2c}\int_a^{x-ct} \mathrm{d}x'\, \alpha(x') + \frac{1}{2c}\int_a^{x+ct} \mathrm{d}x'\, \alpha(x'),$$

$$\tilde{h}_2(x, t) = \frac{1}{2c}\int_{x-ct}^{x+ct} \mathrm{d}x'\, \alpha(x'). \tag{N.1.11}$$

Combining (N.1.8) and (N.1.11), we can now write the solution of the general IVP, $\psi = \tilde{h}_1 + \tilde{h}_2$, as

$$\psi(x, t) = \tfrac{1}{2}[\beta(x - ct) + \beta(x + ct)] + \frac{1}{2c}\int_{x-ct}^{x+ct} \mathrm{d}x' \, \alpha(x'), \qquad \text{(N.1.12)}$$

which is, naturally, the same as the expression delivered in Section 12.1 by the magic rule: $\tilde{h}_{1,2}$ coincide with the integrals h_{10}, h_{20}.

N.2 The propagator in 1D

Like any other solution of the homogeneous equation, the propagator $K_0^{(1)}(X, \tau)$ is obtainable from (N.1.12) by inserting the appropriate initial values, in this case

$$\alpha(X) = \delta(X), \qquad \beta(X) = 0:$$

$$K_0^{(1)}(X, \tau) = \frac{1}{2c}\int_{X-c\tau}^{X+c\tau} \mathrm{d}x' \, \delta(x'). \qquad \text{(N.2.1)}$$

For $\tau > 0$, the integral is $+1$ if $(X + c\tau) > 0$ and $(X - c\tau) < 0$, i.e. if $-c\tau < X < c\tau$; otherwise the integral vanishes. For $\tau < 0$, the integral is -1 or 0, respectively, under these conditions. Thus

$$K_0^{(1)}(X, \tau) = \varepsilon(\tau)\frac{1}{2c}\,\theta(c\,|\tau| - |X|), \qquad \text{(N.2.2)}$$

in agreement with the result of the Fourier analysis in Section 11.2.2.

N.3 The isotropic component of $K_0^{(3)}$

Appendix M found the component κ_0 of the 3D propagator $K_0^{(3)}$ by using the image method to solve eqns (M.7–10) with $l = 0$. The same equations define the propagator in a halfspace under DBCs, as discussed in Section 12.8.1. Here we solve them by d'Alembert's method, as if the image method had never been thought of. The same idea can be applied to the κ_l with $l \geqslant 1$, along the lines indicated in Problem 11.11.

The most general solution of (M.3) with $l = 0$ is, accordingly, looked for in the form

$$\kappa_0(r, r'; \tau) = f(r - c\tau) + g(r + c\tau). \qquad \text{(N.3.1)}$$

The boundary condition (M.8) requires

$$\kappa_0(0, r'; \tau) = 0 = f(-c\tau) + g(c\tau)$$

for all τ; hence (N.3.1) becomes

$$\kappa_0(r, r'; \tau) = f(r - c\tau) - f(-r - c\tau). \qquad \text{(N.3.2)}$$

The initial condition (M.9) requires

$$\kappa_0(r, r'; 0) = 0 = f(r) - f(-r)$$

for all $r > 0$; thus $f(r)$ is an even function of its argument, and (N.3.2) becomes

$$\kappa_0(r, r'; \tau) = f(r - c\tau) - f(r + c\tau), \tag{N.3.3}$$

which shows that κ_0 is odd as a function of τ, as it should be.

The other initial condition (M.10) then requires

$$\kappa_{0\tau}(r, r'; 0) = -2cf'(r) = c^2\,\delta(r - r'), \qquad (r, r' \geqslant 0).$$

Integration (still for $r \geqslant 0$) leads to

$$f(r) = -\frac{c}{2}\,\theta(r - r') + A, \qquad (r, r' \geqslant 0), \tag{N.3.4}$$

where A is an arbitrary constant which will cancel from κ_0. The fact that f is an even function allows it to be re-expressed in a form valid for all values of the argument:

$$f(r) = -\frac{c}{2}\,\theta(|r| - r') + A. \tag{N.3.5}$$

Substituting this functional form into (N.3.3) we obtain the end-result

$$\kappa_0(r, r'; \tau) = \frac{c}{2}\{-\theta(|r - c\tau| - r') + \theta(|r + c\tau| - r')\}. \tag{N.3.6}$$

By taking $\tau > 0$ and sketching the two step-functions, and their difference, as functions of τ first for $r < r'$ and then for $r > r'$, one can convince oneself that (N.3.6) is equivalent to

$$\kappa_0(r, r'; \tau) = \varepsilon(\tau)\frac{c}{2}\{\theta(c\,|\tau| - |r - r'|) - \theta(c\,|\tau| - (r + r'))\}. \tag{N.3.7}$$

This agrees with the result (12.8.1) or (M.14) of the image method. It is worth noting how very different at first sight are the two actually equivalent expressions (N.3.6) and (N.3.7).

N.4 d'Alembert's method for $K_0^{(3)}(\boldsymbol{R}, \tau)$

To determine the 3D propagator itself (rather than just is isotropic component) we write it as

$$K_0^{(3)}(\boldsymbol{R}, \tau) \equiv H(R, \tau)/4\pi R. \tag{N.4.1}$$

Then

$$\left(\frac{1}{c^2}\frac{\partial^2}{\partial \tau^2} - \nabla_R^2\right)K_0^{(3)} = 0$$

entails

$$\left(\frac{1}{c^2}\frac{\partial^2}{\partial \tau^2} - \frac{\partial^2}{\partial R^2}\right)H(R, \tau) = 0. \tag{N.4.2}$$

Since $K_0^{(3)}$ is well-behaved at $R=0$, H must satisfy the boundary condition

$$H(0, \tau) = 0 \tag{N.4.3}$$

for all τ; and it satisfies the ICs reflecting

$$\lim_{\tau\to 0+} K_0^{(3)}(\boldsymbol{R}, \tau) = 0, \qquad \lim_{\tau\to 0+} K_{0\tau}^{(3)}(\boldsymbol{R}, \tau) = c^2\,\delta(\boldsymbol{R}) = -c^2\,\delta'(R)/4\pi R,$$

namely

$$\lim_{\tau\to 0+} H(R, \tau) = 0, \qquad \lim_{\tau\to 0+} H_\tau(R, \tau) = -c^2\,\delta'(R). \tag{N.4.4a,b}$$

(Comparison with (M.7–10) for $l=0$ shows that the differential equation, the boundary condition, and the first initial condition (N.4.4a) are the same as those for the isotropic component $\kappa_0(r, r'; \tau)$, while the second initial condition (N.4.4b) is different.)
In the now familiar way (N.4.2) gives

$$H(R, \tau) = f(c\tau - R) + g(c\tau + R); \tag{N.4.5}$$

the boundary condition (N.4.3) requires $f(c\tau) + g(c\tau) = 0$, whence

$$H(R, \tau) = f(c\tau - R) - f(c\tau + R). \tag{N.4.6}$$

The initial conditions (N.4.4) then impose

$$\lim_{\tau\to 0+} \{f(c\tau - R) - f(c\tau + R)\}/4\pi R = 0, \tag{N.4.7}$$

$$\lim_{\tau\to 0+} c\{f'(c\tau - R) - f'(c\tau + R)\}/4\pi R = -c^2\,\delta'(R)/4\pi R. \tag{N.4.8}$$

At this point one must be very careful in order to avoid the pitfall anticipated at the end of Appendix L. Obviously if somewhat loosely, (N.4.8) suggests that $c\{f'(-R) - f'(R)\} = -c^2\,\delta'(R)$. If we were to infer that $f'(R) \propto \delta'(R)$, then (in view of the fact that δ' is weakly odd) we might be led on to write $-2cf'(R) = -c^2\,\delta'(R)$, whence $f(R) = \frac{1}{2}c\,\delta(R) + \text{constant}$. This is wrong. The correct solution of (N.4.7, 8) can be written down in light of eqns (L.4, 5, 7):

$$f(c\tau) = c\,\delta(c\tau), \tag{N.4.9}$$

differing by a factor of 2 from the incorrect result just mentioned. Then (N.4.7) becomes

$$\lim_{\tau\to 0+} \{\delta(c\tau - R) - \delta(c\tau + R)\}/4\pi R = 0,$$

which is satisfied by virtue of (L.4a) and (L.5); and (N.4.8) becomes

$$\lim_{\tau\to 0+} c^2\{\delta'(c\tau - R) - \delta'(c\tau + R)\}/4\pi R = -c^2\,\delta'(R)/4\pi R,$$

which is satisfied by virtue of (L.4b) and (L.7).

Substitution of the functional form (N.4.9) into (N.4.1, 6) yields the end-result, namely

$$K_0^{(3)}(\boldsymbol{R}, \tau) = \{\delta(c\tau - R) - \delta(c\tau + R)\}/4\pi R, \tag{N.4.10}$$

which agrees with the result (11.2.4) obtained by Fourier analysis.

O Wave equation: Relativistic methods for the propagator

O.1 Introduction

There is an unmistakably relativistic air about many of the expressions of Chapters 11 and 12, relating to the wave equation in unbounded space. As regards sound waves this is deceptive, because the wave equation governs sound waves only in the inertial frame where the undisturbed medium is at rest. Referred to other frames the governing equation must take different forms: for instance, the speed of a plane wave then depends on its direction, so that d'Alembert's formula $\psi(\boldsymbol{r}, t) = \{f(ct - \boldsymbol{n}\cdot\boldsymbol{r}) + g(ct + \boldsymbol{n}\cdot\boldsymbol{r})\}$ can no longer be a solution for arbitrary unit vectors $\boldsymbol{n}$. (The footnote on p. 312 points to another difference.)

Nevertheless, formally relativistic but purely mathematical arguments applied to the basic Fourier representation (11.2.1) of the wave propagator $K_0^{(n)}(\boldsymbol{R}, \tau)$ can establish the causality property from the outset ($K = 0$ if $c^2\tau^2 < R^2$), and yield the closed expressions for $K_0^{(n)}$ very simply. To present such arguments in familiar terms, we ignore, temporarily, the fact that c is the speed of sound rather than of light. Define (for $n = 3$) a four-vector $\vec{R} \equiv (R_0, \boldsymbol{R}) \equiv (c\tau, \boldsymbol{R})$; write Lorentz transformations (with the customary choice of axes) as

$$R_0' = [R_0 - (u/c)R_1]/[1 - u^2/c^2]^{\frac{1}{2}}, \quad R_1' = [R_1 - (u/c)R_0]/[1 - u^2/c^2]^{\frac{1}{2}},$$
$$R_2' = R_2, \qquad R_3' = R_3;$$

and define as a four-vector any set of four quantities that we choose to transform according to the same rules. (Only $u < c$ will be considered.) The dimensionality n is varied simply by omitting the surplus space components: the component R_3 is dropped in 2D, and the components R_2, R_3 are dropped in 1D.

Recall that the scalar product between two four-vectors, defined as $\vec{A}\cdot\vec{B} \equiv A_0B_0 - \boldsymbol{A}\cdot\boldsymbol{B}$, is invariant ('scalar') in the sense that $\vec{A}\cdot\vec{B} = \vec{A}'\cdot\vec{B}'$. In particular

$$\vec{R}\cdot\vec{R} = R_0^2 - \boldsymbol{R}^2 = c^2\tau^2 - \boldsymbol{R}^2 \equiv \lambda^2$$

is invariant; moreover, if (but only if) $\vec{R}\cdot\vec{R} \geq 0$, then the sign of R_0, i.e. of τ, is also invariant. For instance, $\delta(\lambda^2)\varepsilon(\tau)$ and $\delta(\lambda^2)\theta(\tau)$ are invariants, but $\theta(\pm\tau)$ on their own are not. The same applies to all four-vectors. We shall call $\vec{R}$ space-like if $\vec{R}\cdot\vec{R} < 0$; time-like if $\vec{R}\cdot\vec{R} > 0$; and light-like if $\vec{R}\cdot\vec{R} = 0$. The invariant regions $\vec{R}\cdot\vec{R} \geq 0$, $R_0 \gtrless 0$ are called the forward and backward light cones, respectively.

Recall further that the wave operator

$$\Box^2 \equiv \left(\frac{1}{c^2}\frac{\partial^2}{\partial \tau^2} - \nabla^2\right)$$

is invariant: one verifies that $\Box^2 = \Box'^2$ by regarding the change from $\vec{R}$ to $\vec{R}'$ as an ordinary change of independent variables, and then expressing $\Box^2$ in terms of $\partial/\partial R_0'$, $\partial/\partial R_1'$, . . . by repeated application of the chain rule (e.g.

$$\partial/\partial R_0 = (\partial R_0'/\partial R_0)\,\partial/\partial R_0' + (\partial R_1'/\partial R_0)\,\partial/\partial R_1' + \cdots).$$

Similarly, one verifies that the Jacobian of the transformation is unity: $J \equiv |\partial(R_0', \boldsymbol{R}')/\partial(R_0, \boldsymbol{R})| = 1$, whence

$$c\int d\tau \int d^n R = \int d^{n+1}R = c\int d\tau' \int d^n R' = \int d^{n+1}R'.$$

(According to Section 1.4.4, $J = 1$ also entails that $\delta(\tau)\,\delta(\boldsymbol{R}) = \delta(\tau')\,\delta(\boldsymbol{R}')$, i.e. that $\delta(\vec{R}) = \delta(\vec{R}')$, though we shall not need to use this explicitly.)

We shall exploit these ideas in a standard way, by showing first that $K_0(\vec{R})$ is a scalar, namely that $K_0(\vec{R}) = K_0(\vec{R}')$, whence it must be a function only of the scalars constructed from $\vec{R}$. (The word 'scalar' is here preferred to 'invariant', in order to avoid talking about 'invariant variables'.) Once the invariance of K_0 is guaranteed, we can evaluate K_0 in terms of any set of Lorentz-transformed (primed) variables that proves convenient (we say: 'in any convenient inertial frame'); then the result may be written in terms of scalar variables, or re-expressed in terms of the original (unprimed) variables ('transformed back to the original frame').

O.2 The propagators $K_0^{(n)}$

O.2.1 The Lorentz-invariance of $K_0^{(n)}$

The crucial step in proving that $K_0^{(n)}(\vec{R}) = K_0^{(n)}(\vec{R}')$ is to rewrite the basic n-fold Fourier representation (11.2.1) (where $k \equiv |\boldsymbol{k}|$),

$$K_0^{(n)}(\vec{R}) = \frac{c}{(2\pi)^n}\int d^n k\,\frac{\sin(kc\tau)}{k}\exp(i\boldsymbol{k}\cdot\boldsymbol{R}), \qquad \text{(O.2.1)}$$

by introducing a seemingly gratuitous integration over a new Fourier variable k_0. We notice that

$$\frac{\sin(kc\tau)}{k} = \frac{i}{2k}(\exp(-ikc\tau) - \exp(ikc\tau))$$

$$= i\int_{-\infty}^{\infty} dk_0 \left\{\frac{\delta(k_0 - k) - \delta(k_0 + k)}{2k}\right\}\exp(-ik_0 c\tau)$$

$$= i\int_{-\infty}^{\infty} dk_0 \left\{\frac{\delta(k_0 - k) + \delta(k_0 + k)}{2k}\right\}\varepsilon(k_0)\exp(-ik_0c\tau)$$

$$= i\int_{-\infty}^{\infty} dk_0\, \delta(k_0^2 - k^2)\varepsilon(k_0)\exp(-ik_0c\tau).$$

When this is substituted into (O.2.1), it proves convenient to write $\vec{K} \equiv (k_0, \boldsymbol{k})$. Then $(k_0^2 - k^2) = \vec{K}\cdot\vec{K}$; the exponentials combine into $\exp(-i\vec{K}\cdot\vec{R})$; and $\int d^n k \int dk_0 = \int d^{n+1}K$:

$$K_0^{(n)}(\vec{R}) = \frac{ic}{(2\pi)^n}\int d^{n+1}K\, \delta(\vec{K}\cdot\vec{K})\varepsilon(K_0)\exp(-i\vec{K}\cdot\vec{R}). \qquad \text{(O.2.2)}$$

Finally, we change the integration variables from $\vec{K}$ to the Lorentz-transformed $\vec{K}'$. As explained in Section O.1,

$$\int d^{n+1}K = \int d^{n+1}K', \qquad \delta(K\cdot K)\varepsilon(K_0) = \delta(K'\cdot K')\varepsilon(K_0');$$

and, in terms of the transformed coordinate $\vec{R}'$, we have $\vec{K}\cdot\vec{R} = \vec{K}'\cdot\vec{R}'$. Therefore

$$K_0^{(n)}(\vec{R}) = \frac{ic}{(2\pi)^n}\int d^{n+1}K'\, \delta(\vec{K}'\cdot\vec{K}')\varepsilon(K_0')\exp(-i\vec{K}'\cdot\vec{R}')$$
$$= K_0^{(n)}(\vec{R}'). \qquad \text{(O.2.3)}$$

The last equality follows because the symbol used for the integration variable is irrelevant to the value of the integral; hence we can simply drop the prime from $\vec{K}'$, obtaining precisely the representation (O.2.2) but with $\vec{R}'$ in place of $\vec{R}$. ■

Being invariant, $K_0^{(n)}(\vec{R})$ can be a function only of the Lorentz-scalars $\lambda^2 = \vec{R}\cdot\vec{R} = c^2\tau^2 - \boldsymbol{R}^2$, and, if $\lambda^2 > 0$, of $\varepsilon(\tau)$. What makes this conclusion so powerful is that the expression (O.2.1) for $K_0^{(n)}$ does not at all appear to be a function only of these variables, but nevertheless is.

O.2.2 Causality

'Causality' is the name given to the property that $K_0^{(n)}$ vanishes for space-like $\vec{R}$, i.e. when $\lambda^2 < 0$. We can now prove this directly from (O.2.1) for any n, and without doing any calculations.

Since $K_0(\boldsymbol{R}, \tau)$ is invariant, we can evaluate it in any Lorentz frame. When $\lambda^2 < 0$, we choose the special frame, indicated by tildes, where $\tilde{\tau} = 0$ and $\lambda^2 = -\boldsymbol{R}^2$. But (O.2.1) vanishes when $\tilde{\tau} = 0$, because $\sin(kc\tilde{\tau})$ then vanishes. ■

Speaking relativistically, one says that $K_0^{(n)}$ vanishes outside the light cone. It follows that $\theta(\tau)$ when multiplied into $K_0^{(n)}$ is invariant, i.e. that $G_0^{(n)}$ is. Thus $G_0^{(n)}$ vanishes outside the forward light cone.

O.2.3 Evaluation of $K_0^{(n)}$

When $\lambda^2 > 0$, we choose another special frame, indicated by carets, where $\hat{\boldsymbol{R}} = 0$, so that $\lambda^2 = c^2\hat{\tau}^2$. In this frame (O.2.1) reduces to

$$K_0^{(n)} = \varepsilon(\hat{\tau})\frac{c}{(2\pi)^n}\int \mathrm{d}^n k \frac{1}{k}\sin(kc\,|\hat{\tau}|). \tag{O.2.4}$$

Our programme is to evaluate this integral for $n = 1, 2, 3$; the results are then rewritten by setting $|\hat{\tau}| = \lambda/c = (\tau^2 - R^2/c^2)^{\frac{1}{2}}$, and multiplied by $\theta(\tau^2 - R^2/c^2)$ in order to yield the desired expression for $K_0^{(n)}$ in the original frame (i.e. in the rest frame of the medium).

(i) *One dimension.* With $n = 1$, the integral in (O.2.4) is just the Dirichlet integral (Appendix B):

$$\hat{K}_0^{(1)}(0, \hat{\tau}) = \varepsilon(\hat{\tau})\frac{c}{2\pi}\int_{-\infty}^{\infty} \mathrm{d}k \frac{1}{k}\sin(kc\,|\hat{\tau}|) = \varepsilon(\hat{\tau})\frac{c}{2\pi}\pi,$$

$$K_0^{(1)}(\boldsymbol{R}, \tau) = \varepsilon(\tau)\theta(\tau^2 - R^2/c^2)\frac{c}{2}, \tag{O.2.5}$$

which is precisely the previous result (11.2.16b).

(ii) *Two dimensions.* Equation (O.2.4) now yields the apparently ill-defined expression

$$\hat{K}_0^{(2)} = \varepsilon(\hat{\tau})\frac{c}{(2\pi)^2}\int_0^{\infty} 2\pi\,\mathrm{d}k\, k\frac{1}{k}\sin(kc\,|\hat{\tau}|).$$

We interpret this by inserting a convergence factor $\exp(-\eta k)$ into the integrand; at the end of the calculation we take the limit $\eta \to 0$, under which the convergence factor tends to unity. This procedure is traditional in physics, and we use it without further apology whenever convenient; a formal justification calls for the theory of generalized functions (see e.g. Lighthill 1958). Accordingly,

$$\hat{K}_0^{(2)} = \varepsilon(\hat{\tau})\frac{c}{2\pi}\lim_{\eta\to 0}\int_0^{\infty} \mathrm{d}k \sin(kc\,|\hat{\tau}|)\exp(-\eta k).$$

But

$$\lim \int \mathrm{d}k \cdots = \lim \{c\,|\hat{\tau}|/(c^2\hat{\tau}^2 + \eta^2)\} = 1/c\,|\hat{\tau}|,$$

whence

$$\hat{K}_0^{(2)}(\hat{\tau}) = \varepsilon(\hat{\tau})/2\pi\,|\hat{\tau}|,$$

$$K_0^{(2)}(\boldsymbol{R}, \tau) = \varepsilon(\tau)\theta(\tau^2 - R^2/c^2)/2\pi(\tau^2 - R^2/c^2)^{\frac{1}{2}}, \tag{O.2.6}$$

which is precisely (11.2.11), obtained here without recourse to Bessel functions or to embedding.

(iii) *Three dimensions.* Using the same convergence factor as in 2D, we find

$$\hat{K}_0^{(3)}(0,\, \hat{\tau}) = \varepsilon(\hat{\tau})\frac{c}{(2\pi)^3}\lim_{\eta\to 0}\int_0^\infty 4\pi\, \mathrm{d}k\, k^2\frac{1}{k}\sin\,(kc\,|\hat{\tau}|)\exp\,(-\eta k).$$

But

$$\lim_{\eta\to 0}\int_0^\infty \mathrm{d}k\, k\sin\,(kc\,|\hat{\tau}|)\exp\,(-\eta k) = \lim_{\eta\to 0}\frac{2c\,|\hat{\tau}|\eta}{(c^2\hat{\tau}^2+\eta^2)^2}$$
$$= \frac{\pi}{c^2}\,\delta(\hat{\tau}^2); \qquad (\text{O.2.7})$$

the last step will be justified presently. Substitution then yields

$$\hat{K}_0^{(3)}(0,\, \hat{\tau}) = \varepsilon(\hat{\tau})\,\delta(\hat{\tau}^2)/2\pi c,$$
$$K_0^{(3)}(\boldsymbol{R},\, \tau) = \varepsilon(\tau)\,\delta(\tau^2 - R^2/c^2)/2\pi c, \qquad (\text{O.2.8})$$

which is equivalent to (11.2.5).

It remains to justify the last step in (O.2.7). Since $\delta(\hat{\tau}^2)$ in the special frame may be regarded as the limit of $\delta(\hat{\tau}^2 - \hat{R}^2/c^2)$ when $\hat{R}^2 \to 0$, it should in this context be subjected to the strong definition requiring $\int_0^\infty \mathrm{d}\hat{\tau}^2\,\delta(\hat{\tau}^2)f(\hat{\tau}) = f(0)$ for any test function f well-behaved at the origin. Accordingly we shall establish (O.2.7) by proving that

$$\lim_{\eta\to 0}\int_0^\infty \mathrm{d}\hat{\tau}^2 f(\hat{\tau})\frac{2c\hat{\tau}\eta}{(c^2\hat{\tau}^2+\eta^2)^2} = \frac{\pi}{c^2}f(0).$$

Changing the integration variable to $x \equiv c\hat{\tau}/\eta$, the integral becomes

$$\lim_{\eta\to 0}\int_0^\infty \mathrm{d}x^2\frac{\eta^2}{c^2}f\left(\frac{\eta x}{c}\right)\frac{2c\eta(\eta/c)x}{\eta^4(x^2+1)^2} = \lim_{\eta\to 0}\frac{2}{c^2}\int_0^\infty \mathrm{d}x\, f\left(\frac{\eta x}{c}\right)\frac{x}{(x^2+1)^2}$$
$$= \frac{2}{c^2}f(0)\int_0^\infty 2\,\mathrm{d}x\frac{x^2}{(x^2+1)^2} = \frac{\pi}{c^2}f(0). \quad \blacksquare$$

(Experience suggests that a hasty or a less explicit attempt to evaluate the basic integral (O.2.4) with $n = 3$ is likely to give wrong numerical factors.)

References

Daggers indicate books that the writer has found particularly useful in preparing this one.

Asterisks draw attention to recommended further reading at a higher mathematical level, which the present book may help to introduce and motivate.

† Abramowitz, M. and Stegun, I. A. (1968). *Handbook of mathematical functions*. Dover, New York.

Adkins, C. J. (1968). *Equilibrium thermodynamics*. McGraw-Hill, London. (Section 10.2: useful work and available energy.)

Arfken, G. (1973). *Mathematical methods for physicists* (2nd edn). Academic Press, New York.

Baker, B. B. and Copson, E. T. (1950). *The mathematical theory of Huygens' principle* (2nd edn). Clarendon Press, Oxford.

Baldock, G. R. and Bridgman, T. (1981). *The mathematical theory of wave motion*. Ellis Horwood, Chichester.

† Bender, C. M. and Orszag, S. A. (1978). *Advanced mathematical methods for scientists and engineers*. McGraw-Hill, New York. (Sections 1.3–1.5: Green's functions for ordinary differential equations.)

Born, M. and Wolf, E. (1975). *Principles of optics* (5th edn). Pergamon, Oxford. (Sections 8.1–3: Huygens' principle and diffraction.)

† Butkov, E. (1968). *Mathematical physics*. Addison-Wesley, Reading, Massachusetts. (Chapter 6: delta-function; Sections 12.1–5: ordinary differential equations; Chapter 8: types of partial differential equations; Sections 8.4–5, 12.5–6: diffusion equation; 8.9, 12.8: Helmholtz equation.)

Callen, H. B. (1960). *Thermodynamics*. Wiley, New York. (Chapter 6: useful work and available energy.)

Carslaw, H. S. and Jaeger, J. C. (1959). *Conduction of heat in solids* (2nd edn). Clarendon Press, Oxford.

Champeney, D. C. (1987). *A handbook of Fourier theorems*. Cambridge University Press, Cambridge.

Cohen-Tannoudji, C., Diu, B. and Laloe, F. *Quantum mechanics*. Wiley, New York. (Volume 2, Appendix 2: delta-function.)

Coulson, C. A. and Jeffrey, A. (1977). *Waves* (2nd edn). Longman, London. (Nothing on Green's functions, but excellent and very accessible.)

Courant, R. (1937). *Differential and integral calculus*, Vol. 1 (2nd edn). Blackie, London. (Chapter 9: Fourier series.)

†* Courant, R. and Hilbert, D. (1953). *Methods of mathematical physics*, Vol. 1. Interscience, New York. (Chapter 5: eigenvalue problems, Sturm–Liouville theory, and Green's functions for Poisson's equation.)

Courant, R. and Hilbert, D. (1962). *Methods of mathematical physics*, Vol. 2. Interscience, New York.

† Ditchburn, R. W. (1952). *Light*. Blackie, London. (Sections 7.1–11, Appendix VI.A, eventually Sections 7.15–21: diffraction.)

Dowling, A. P. and Ffowcs Williams, J. E. (1983). *Sound and sources of sound*. Ellis Horwood, Chichester.

Dwight, H. B. (1957). *Tables of integrals and other mathematical data* (4th edn). Macmillan, New York.

Feller, W. (1970). *An introduction to probability theory and its applications*, Vol. 1 (3rd edn). Wiley, New York.

Feynman, R. P., Leighton, R. B., and Sands, M. (1964). *The Feynman lectures on physics*. Addison-Wesley, Reading, Massachusetts. (Volume 2, Chapters 4–6, 12: Poisson's equation.)

†* Garabedian, P. R. (1964). *Partial differential equations*. Wiley, New York.

Glasser, M. L. (1970). *American Journal of Physics* **30,** 415.

Gradshteyn, I. S. and Ryzhik, I. M. (1980). *Table of integrals, series and products*. Academic Press, New York.

Groebner, W. and Hofreiter, N. (1957, 1958). *Integraltafel* I, II. Springer-Verlag, Wien.

†* Hadamard, J. (1964). *La théorie des équations aux dérivées partielles*. Editions Scientifiques, Pekin. (A marvellous book by one of the very few mathematical giants who can talk intelligibly to physicists prepared to listen. The present book might serve as background and motivation for tackling Hadamard.)

Hecht, E. and Zajac, A. (1974). *Optics*. Addison-Wesley, Reading, Massachusetts. (Chapter 10, esp. Section 10.4: diffraction.)

† Jackson, J. D. (1962). *Classical electromagnetism*. Wiley, New York. (Chapters 1, 2, 3: delta-functions, especially in 3D; Poisson's equation; Section 9.5: diffraction.)

* Kellogg, O. D. (1954). *Foundations of potential theory*. Dover, New York. (Originally published 1929. Rigorous but relatively readable; almost everything that can be said about Poisson's equation without benefit of delta-functions.)

Krasil'nikov, V. A. and Pavlov, V. I. (1981). *Soviet Physics Doklady* **26,** 60. (Acoustic sources due to local heating.)

Lamb, H. (1932). *Hydrodynamics* (6th edn). Cambridge University Press.

* Lanczos, C. (1961). *Linear differential operators*. Van Nostrand, London. (Chapter 2: Fourier analysis; Chapter 5: Green's functions. Unorthodox and stimulating; rather more mathematical than this course.)

Landau, L. D. and Lifshitz, E. M. (1958). *Statistical mechanics*. Pergamon, London. (Sections 19–21: useful work.)

Landau, L. D. and Lifshitz, E. M. (1975). *The classical theory of fields* (4th edn). Pergamon, Oxford.

Levine, H. A. and Weinberger, H. F. (1986). *Archive for Rational Mechanics and Analysis* **94,** 193.

* Lighthill, J. (1958). *Introduction to Fourier analysis and generalised functions.* Cambridge University Press.

Lighthill, J. (1978). *Waves in fluids.* Cambridge University Press. (Chapter 1, sound waves).

Lighthill, J. (1987). Fourier analysis and generalised functions. In *Tributes to Paul Dirac,* (ed.) J. G. Taylor. Hilger, Bristol.

Marchand, E. W. and Wolf, E. (1966). *Journal of the Optical Society of America* **56,** 1712. (Diffraction.)

Morse, P. M. and Feshbach, H. (1953). *Methods of theoretical physics.* McGraw-Hill, New York. (Section 5.1: separable coordinates; Section 6.1: types of partial differential equations; pp. 791–802: GFs for Poisson's equation; Section 7.4: diffusion equation; 7.3, esp. pp. 834–849: wave equation; 7.2: Helmholtz equation.)

† Page, C. H. (1955). *Physical mathematics.* Van Nostrand, Princeton.

Panofsky, W. K. H. and Phillips, M. (1955). *Classical electricity and magnetism.* Addison-Wesley, Reading, Massachusetts. (Chapters 1, 3, 5: Poisson's equation.)

Pierce, A. D. (1981). *Acoustics.* McGraw-Hill, New York.

Pumplin, J. (1969). *American Journal of Physics* **37,** 734.

Rayleigh (1896). *The theory of sound,* Vols. 1 and 2 (2nd edn). Macmillan, London. (Reprinted Dover, New York, 1945.)

Rees, W. G. (1987). *European Journal of Physics* **8,** 44.

Reif, F. (1965). *Fundamentals of statistical and thermal physics.* McGraw-Hill, New York. (Sections 15.5–10: Brownian motion.)

Riley, K. F. (1974). *Mathematical methods for the physical sciences.* Cambridge University Press. (Chapter 9: types of partial differential equations.)

Roach, G. F. (1982). *Green's functions* (2nd edn). Cambridge University Press. (Section 9.2: delta-functions in more than one dimension and in curvilinear (polar) coordinates; Section 9.6: the Neumann pseudo Green's function.)

* Schwartz, L. (1966). *Mathematics for the physical sciences.* Hermann, Paris. (Rigorous introduction to delta-functions and their applications.)

Smith, M. G. (1967). *Introduction to the theory of partial differential equations.* Van Nostrand, London.

† Sneddon, I. N. (1957). *Elements of partial differential equations.* McGraw-Hill, New York. (Chapter 4, Sections 1–8: Poisson's equation; Chapter 6: diffusion equation; Chapter 5, Sections 1, 2, 5–8: wave and Helmholtz equations.)

* Sobolev, S. L. (1964). *Partial differential equations of mathematical physics.* Pergamon, Oxford. (Accessibly written but much more mathematical than this book; ideal for a rigorous underpinning of much of it. A truly exceptional text, head and shoulders above most others with similar titles.)

Stamnes, J. J. (1986). *Waves in focal regions.* Hilger, Bristol. (Diffraction.)

Stratton, J. A. (1941). *Electromagnetic theory.* McGraw-Hill, New York. (Sections 3.3–7, 3.14–17: Kirchhoff's representation for the Laplace equation; 8.13: diffraction.)

Temkin, S. (1981). *Elements of acoustics.* Wiley, New York.

Van Kampen, N. G. (1981). *Stochastic processes in physics and chemistry.* North-Holland, Amsterdam.

Weinberger, H. F. (1965). *Partial differential equations.* Blaisdell, New York.

Witham, G. B. (1974). *Linear and nonlinear waves.* Wiley, New York.

Index

For entries in CAPITALS see also the TABLE OF CONTENTS